Innate Immunity

METHODS IN MOLECULAR BIOLOGY™

John M. Walker, SERIES EDITOR

417. **Tissue Engineering, Second Edition,** edited by *Hannsjörg Hauser and Martin Fussenegger, 2007*
416. **Gene Essentiality:** *Protocols and Bioinformatics,* edited by *Andrei L. Osterman, 2008*
415. **Innate Immunity,** edited by *Jonathan Ewbank and Eric Vivier, 2008*
414. **Apoptosis in Cancer:** *Methods and Protocols,* edited by *Gil Mor and Ayesha Alvero, 2008*
413. **Protein Structure Prediction, Second Edition,** edited by *Mohammed Zaki and Chris Bystroff, 2008*
412. **Neutrophil Methods and Protocols,** edited by *Mark T. Quinn, Frank R. DeLeo, and Gary M. Bokoch, 2007*
411. **Reporter Genes for Mammalian Systems,** edited by *Don Anson, 2007*
410. **Environmental Genomics,** edited by *Cristofre C. Martin, 2007*
409. **Immunoinformatics:** *Predicting Immunogenicity In Silico,* edited by *Darren R. Flower, 2007*
408. **Gene Function Analysis,** edited by *Michael Ochs, 2007*
407. **Stem Cell Assays,** edited by *Vemuri C. Mohan, 2007*
406. **Plant Bioinformatics:** *Methods and Protocols,* edited by *David Edwards, 2007*
405. **Telomerase Inhibition:** *Strategies and Protocols,* edited by *Lucy Andrews and Trygve O. Tollefsbol, 2007*
404. **Topics in Biostatistics,** edited by *Walter T. Ambrosius, 2007*
403. **Patch-Clamp Methods and Protocols,** edited by *Peter Molnar and James J. Hickman, 2007*
402. **PCR Primer Design,** edited by *Anton Yuryev, 2007*
401. **Neuroinformatics,** edited by *Chiquito J. Crasto, 2007*
400. **Methods in Lipid Membranes,** edited by *Alex Dopico, 2007*
399. **Neuroprotection Methods and Protocols,** edited by *Tiziana Borsello, 2007*
398. **Lipid Rafts,** edited by *Thomas J. McIntosh, 2007*
397. **Hedgehog Signaling Protocols,** edited by *Jamila I. Horabin, 2007*
396. **Comparative Genomics,** *Volume 2,* edited by *Nicholas H. Bergman, 2007*
395. **Comparative Genomics,** *Volume 1,* edited by *Nicholas H. Bergman, 2007*
394. **Salmonella:** *Methods and Protocols,* edited by *Heide Schatten and Abe Eisenstark, 2007*
393. **Plant Secondary Metabolites,** edited by *Harinder P. S. Makkar, P. Siddhuraju, and Klaus Becker, 2007*
392. **Molecular Motors:** *Methods and Protocols,* edited by *Ann O. Sperry, 2007*
391. **MRSA Protocols,** edited by *Yinduo Ji, 2007*
390. **Protein Targeting Protocols, Second Edition,** edited by *Mark van der Giezen, 2007*
389. **Pichia Protocols, Second Edition,** edited by *James M. Cregg, 2007*
388. **Baculovirus and Insect Cell Expression Protocols, Second Edition,** edited by *David W. Murhammer, 2007*
387. **Serial Analysis of Gene Expression (SAGE):** *Digital Gene Expression Profiling,* edited by *Kare Lehmann Nielsen, 2007*
386. **Peptide Characterization and Application Protocols,** edited by *Gregg B. Fields, 2007*
385. **Microchip-Based Assay Systems:** *Methods and Applications,* edited by *Pierre N. Floriano, 2007*
384. **Capillary Electrophoresis:** *Methods and Protocols,* edited by *Philippe Schmitt-Kopplin, 2007*
383. **Cancer Genomics and Proteomics:** *Methods and Protocols,* edited by *Paul B. Fisher, 2007*
382. **Microarrays, Second Edition:** *Volume 2, Applications and Data Analysis,* edited by *Jang B. Rampal, 2007*
381. **Microarrays, Second Edition:** *Volume 1, Synthesis Methods,* edited by *Jang B. Rampal, 2007*
380. **Immunological Tolerance:** *Methods and Protocols,* edited by *Paul J. Fairchild, 2007*
379. **Glycovirology Protocols,** edited by *Richard J. Sugrue, 2007*
378. **Monoclonal Antibodies:** *Methods and Protocols,* edited by *Maher Albitar, 2007*
377. **Microarray Data Analysis:** *Methods and Applications,* edited by *Michael J. Korenberg, 2007*
376. **Linkage Disequilibrium and Association Mapping:** *Analysis and Application,* edited by *Andrew R. Collins, 2007*
375. **In Vitro Transcription and Translation Protocols:** *Second Edition,* edited by *Guido Grandi, 2007*
374. **Quantum Dots:** *Applications in Biology,* edited by *Marcel Bruchez and Charles Z. Hotz, 2007*
373. **Pyrosequencing® Protocols,** edited by *Sharon Marsh, 2007*
372. **Mitochondria: Practical Protocols,** edited by *Dario Leister and Johannes Herrmann, 2007*
371. **Biological Aging:** *Methods and Protocols,* edited by *Trygve O. Tollefsbol, 2007*
370. **Adhesion Protein Protocols,** *Second Edition,* edited by *Amanda S. Coutts, 2007*
369. **Electron Microscopy:** *Methods and Protocols, Second Edition,* edited by *John Kuo, 2007*
368. **Cryopreservation and Freeze-Drying Protocols,** *Second Edition,* edited by *John G. Day and Glyn Stacey, 2007*
367. **Mass Spectrometry Data Analysis in Proteomics,** edited by *Rune Matthiesen, 2007*
366. **Cardiac Gene Expression:** *Methods and Protocols,* edited by *Jun Zhang and Gregg Rokosh, 2007*
365. **Protein Phosphatase Protocols:** edited by *Greg Moorhead, 2007*
364. **Macromolecular Crystallography Protocols:** *Volume 2, Structure Determination,* edited by *Sylvie Doublié, 2007*
363. **Macromolecular Crystallography Protocols:** *Volume 1, Preparation and Crystallization of Macromolecules,* edited by *Sylvie Doublié, 2007*
362. **Circadian Rhythms:** *Methods and Protocols,* edited by *Ezio Rosato, 2007*
361. **Target Discovery and Validation Reviews and Protocols:** *Emerging Molecular Targets and Treatment Options, Volume 2,* edited by *Mouldy Sioud, 2007*
360. **Target Discovery and Validation Reviews and Protocols:** *Emerging Strategies for Targets and Biomarker Discovery, Volume 1,* edited by *Mouldy Sioud, 2007*

METHODS IN MOLECULAR BIOLOGY™

Innate Immunity

Edited by

Jonathan Ewbank

*Centre d'Immunologie de Marseille-Luminy, INSERM, CNRS,
Université de la Méditerranée, Marseille, France*

and

Eric Vivier

*Centre d'Immunologie de Marseille-Luminy, INSERM, CNRS,
Université de la Méditerranée, Marseille, France*

HUMANA PRESS ✳ TOTOWA, NEW JERSEY

© 2008 Humana Press Inc.
999 Riverview Drive, Suite 208
Totowa, New Jersey 07512

www.humanapress.com

All rights reserved. No part of this book may be reproduced, stored in a retrieval system, or transmitted in any form or by any means, electronic, mechanical, photocopying, microfilming, recording, or otherwise without written permission from the Publisher. Methods in Molecular Biology™ is a trademark of The Humana Press Inc.

All papers, comments, opinions, conclusions, or recommendations are those of the author(s), and do not necessarily reflect the views of the publisher.

This publication is printed on acid-free paper. ∞
ANSI Z39.48-1984 (American Standards Institute) Permanence of Paper for Printed Library Materials.

Cover Illustration: Tissue section of a mouse spleen showing the compartmentalization of innate immune cells. NK cells (green) are in the red pulp and macrophages (red) in the marginal zone, outside the T cell zone (blue). The B cell zone is the dark area between the T cell zone and the marginal zone. Image courtesy of Claude Grégoire and Thierry Walzer (CIML).

Cover design by Karen Schulz

Production Editor: Michele Seugling

For additional copies, pricing for bulk purchases, and/or information about other Humana titles, contact Humana at the above address or at any of the following numbers: Tel.: 973-256-1699; Fax: 973-256-8341; E-mail: humana@humanapr.com; or visit our Website: www.humanapress.com

Photocopy Authorization Policy:
Authorization to photocopy items for internal or personal use, or the internal or personal use of specific clients, is granted by Humana Press Inc., provided that the base fee of US $30.00 per copy is paid directly to the Copyright Clearance Center at 222 Rosewood Drive, Danvers, MA 01923. For those organizations that have been granted a photocopy license from the CCC, a separate system of payment has been arranged and is acceptable to Humana Press Inc. The fee code for users of the Transactional Reporting Service is: [978-1-58829-746-4/08 $30.00].

10 9 8 7 6 5 4 3 2 1

e-ISBN: 978-1-59745-570-1

Library of Congress Control Number: 2007932539

Preface

Not so long ago, immunology was a discipline almost entirely dedicated to the study of T and B cells. Commonly used antigens were proteins, such as albumin or hen egg lysozyme, or peptides, and the immune responses of non-vertebrate species were largely neglected. Immunologists today are interested in all of the diverse cell types involved in host defense and have a deeper appreciation of the importance of innate immune mechanisms as a first line of protection against pathogens. As a substantial part of the molecular bases of innate immunity appear to be conserved, invertebrate models are being used more and more frequently to answer questions about evolution, and also because of their tractability to provide leads for investigation in higher organisms.

This volume thus contains chapters dealing with the isolation and functional characterization of cells involved in innate immunity in mouse and man, including mast cells and eosinophils, and with several chapters focusing on natural killer cells. Other chapters describe methods in statistics, in vivo imaging, genome engineering, mutagenesis, and culture that are particularly adapted to the study of innate immunity in these hosts. These are complemented with a series of chapters dealing with alternative models: plants, worms, mosquitoes, flies, and fish. Together, these approaches and models are being used to dissect the complex interplay between hosts and pathogens and contribute to developing strategies to help fight infection.

Jonathan Ewbank
Eric Vivier

Contents

Preface v
Contributors xi

1 ENU Mutagenesis in Mice
   Philippe Georgel, Xin Du, Kasper Hoebe, and Bruce Beutler *1*

2 Innate Immunity Genes as Candidate Genes: *Searching for Relevant Natural Polymorphisms in Databases and Assessing Family-Based Association of Polymorphisms with Human Diseases*
   Pascal Rihet *17*

3 KIR Locus Polymorphisms: *Genotyping and Disease Association Analysis*
   Maureen P. Martin and Mary Carrington *49*

4 Experimental Model for the Study of the Human Immune System: *Production and Monitoring of "Human Immune System" $Rag2^{-/-}\gamma_c^{-/-}$ Mice*
   Nicolas Legrand, Kees Weijer, and Hergen Spits *65*

5 Bacterial Artificial Chromosome Transgenesis Through Pronuclear Injection of Fertilized Mouse Oocytes
   Kristina Vintersten, Giuseppe Testa, Ronald Naumann, Konstantinos Anastassiadis, and A. Francis Stewart *83*

6 Bioluminescence Imaging to Evaluate Infections and Host Response In Vivo
   Pamela Reilly Contag *101*

7 Intravital Two-Photon Imaging of Natural Killer Cells and Dendritic Cells in Lymph Nodes
   Susanna Celli, Béatrice Breart, and Philippe Bousso *119*

8 Dissection of the Antiviral NK Cell Response by MCMV Mutants
   Stipan Jonjic, Astrid Krmpotic, Jurica Arapovic, and Ulrich H. Koszinowski *127*

9 Analyzing Antibody–Fc-Receptor Interactions
   Falk Nimmerjahn and Jeffrey V. Ravetch *151*

10 The Isolation and Identification of Murine Dendritic Cell
    Populations from Lymphoid Tissues and Their Production
    in Culture
    David Vremec and Ken Shortman 163

11 Analysis of Individual Natural Killer
    Cell Responses
    Wayne M. Yokoyama and Sungjin Kim 179

12 Isolation and Analysis of Human Natural Killer
    Cell Subsets
    Guido Ferlazzo 197

13 Innate Immune Function of Eosinophils: *From Antiparasite to
    Antitumor Cells*
    *Fanny Legrand, Gaetane Woerly, Virginie Driss, and Monique
    Capron* 215

14 Ex Vivo and In Vitro Primary Mast Cells
    *Michel Arock, Alexandra Le Nours, Odile Malbec, and Marc
    Daëron* 241

15 Murine Macrophages: *A Technical Approach*
    Luisa Martinez-Pomares and Siamon Gordon 255

16 Clinical Analysis of Dendritic Cell Subsets: *The Dendritogram*
    Anne Hosmalin, Miriam Lichtner, and Stéphanie Louis 273

17 Clinical Analysis of Human Natural Killer Cells
    Pascale André and Nicolas Anfossi 291

18 Lentiviral Transduction of Immune Cells
    *Louise Swainson, Cedric Mongellaz, Oumeya Adjali, Rita Vicente,
    and Naomi Taylor* 301

19 Axenic Mice Model
    Antoine Giraud 321

20 In Vivo Analysis of Zebrafish Innate Immunity
    *Jean-Pierre Levraud, Emma Colucci-Guyon, Michael J. Redd,
    Georges Lutfalla, and Philippe Herbomel* 337

21 Reverse Genetics Analysis of Antiparasitic Responses
    in the Malaria Vector, *Anopheles gambiae*
    Stephanie A. Blandin and Elena A. Levashina 365

22 *Drosophila* Immunity: *Methods for Monitoring the Activity of Toll
    and Imd Signaling Pathways*
    Yves Romeo and Bruno Lemaitre 379

Contents

23 Investigating the Involvement of Host Factors Involved in Intracellular Pathogen Infection by RNAi in *Drosophila* Cells
   Hervé Agaisse *395*

24 Models of *Caenorhabditis elegans* Infection by Bacterial and Fungal Pathogens
   Jennifer R. Powell and Frederick M. Ausubel *403*

25 Genetic Analysis of *Caenorhabditis elegans* Innate Immunity
   Michael Shapira and Man-Wah Tan *429*

26 Measuring Cell-Wall-Based Defenses and Their Effect on Bacterial Growth in *Arabidopsis*
   Min Gab Kim and David Mackey *443*

Index 453

Contributors

OUMEYA ADJALI • *Institut de Génétique Moléculaire de Montpellier, Montpellier, France*
HERVÉ AGAISSE • *Section of Microbial Pathogenesis, Yale University School of Medicine, Boyer Center for Molecular Medicine, New Haven, CT*
KONSTANTINOS ANASTASSIADIS • *Genomics, BioInnovationsZentrum, Technische Universität Dresden, Dresden, Germany*
PASCALE ANDRÉ • *Innate Pharma SAS, Marseille, France*
NICOLAS ANFOSSI • *Innate Pharma SAS, Marseille, France*
JURICA ARAPOVIC • *Department of Histology and Embryology, Faculty of Medicine, University of Rijeka, Rijeka, Croatia*
MICHEL AROCK • *Ecole Normale Supérieure de Cachan, Laboratoire de Biotechnologies et Pharmacologie Génétique Appliquées, Cachan, France*
FREDERICK M. AUSUBEL • *Department of Molecular Biology, Massachusetts General Hospital, Department of Genetics, Harvard Medical School, Boston, MA*
BRUCE BEUTLER • *Department of Immunology, The Scripps Research Institute, La Jolla, CA*
STEPHANIE A. BLANDIN • *IBMC, Strasbourg, France*
PHILIPPE BOUSSO • *G5 Dynamiques des Réponses Immunes, Equipe Avenir, Département d'Immunologie, Institut Pasteur, Paris Cedex, France*
BÉATRICE BREART • *G5 Dynamiques des Réponses Immunes, Equipe Avenir, Département d'Immunologie, Institut Pasteur, Paris Cedex, France*
MONIQUE CAPRON • *Université de Lille, Institut Pasteur de Lille, France*
MARY CARRINGTON • *Laboratory of Genomic Diversity, SAIC-Frederick, Inc., NCI-Frederick, Frederick, MD*
SUSANNA CELLI • *G5 Dynamiques des Réponses Immunes, Equipe Avenir, Département d'Immunologie, Institut Pasteur, Paris Cedex, France*
EMMA COLUCCI-GUYON • *Unité Macrophages et Développement de l'Immunité, Institut Pasteur, Paris, France*
PAMELA REILLY CONTAG • *Xenogen Corporation, Alameda, CA*
MARC DAËRON • *Institut Pasteur, Département d'Immunologie, Unité d'Allergologie Moléculaire et Cellulaire, Paris, France*

VIRGINIE DRISS • *Université de Lille, Institut Pasteur de Lille, France*
XIN DU • *Department of Immunology, The Scripps Research Institute, La Jolla, CA*
JONATHAN EWBANK • *Centre d'Immunologie de Marseille-Luminy, INSERM, CNRS, Université de la Méditerranée, Marseille, France*
GUIDO FERLAZZO • *School of Medicine, University of Messina, Italy*
PHILIPPE GEORGEL • *Laboratoire d'Immunogénétique Moléculaire Humaine, Centre de Recherche en Immunologie et Hématologie, Faculté de Médecine, Strasbourg Cedex, France*
ANTOINE GIRAUD • *Department of Medical Biochemistry and Microbiology, Uppsala University, Uppsala, Sweden*
SIAMON GORDON • *Sir William Dunn School of Pathology, South Parks Road, Oxford, United Kingdom*
PHILIPPE HERBOMEL • *Unité Macrophages et Développement de l'Immunité, Institut Pasteur, Paris, France*
KASPER HOEBE • *Department of Immunology, The Scripps Research Institute, La Jolla, CA*
ANNE HOSMALIN • *Institut Cochin, Département d'Immunologie, Paris, France*
STIPAN JONJIC • *Department of Histology and Embryology, Faculty of Medicine, University of Rijeka, Rijeka, Croatia*
MIN GAB KIM • *Department of Plant Cellular and Molecular Biology, Program in Molecular Cellular and Developmental Biology, The Ohio State University, Columbus, OH*
SUNGJIN KIM • *Howard Hughes Medical Institute, Division of Rheumatology, Washington University School of Medicine, St. Louis, MO*
ULRICH H. KOSZINOWSKI • *Max von Pettenkofer Institute, Ludwig Maximilians University of Munich, Munich, Germany*
ASTRID KRMPOTIC • *Department of Histology and Embryology, Faculty of Medicine, University of Rijeka, Rijeka, Croatia*
FANNY LEGRAND • *Inserm U547-Université de Lille 2, Institut Pasteur de Lille, France*
NICOLAS LEGRAND • *Department of Cell Biology and Histology, Academic Medical Center of the University of Amsterdam (AMC-UvA), Amsterdam, The Netherlands*
BRUNO LEMAITRE • *Centre de Génétique Moléculaire du CNRS, Avenue de la terrasse, Gif-sur-Yvette, France*

Contributors

ALEXANDRA LE NOURS • Ecole Normale Supérieure de Cachan, Laboratoire de Biotechnologies et Pharmacologie Génétique Appliquées, Cachan, France
ELENA A. LEVASHINA • IBMC, Strasbourg, France
JEAN-PIERRE LEVRAUD • Unité Macrophages et Développement de l'Immunité, Institut Pasteur, Paris, France
MIRIAM LICHTNER • Department of Infectious and Tropical Diseases, "La Sapienza" University, Rome, Italy
STÉPHANIE LOUIS • Institut Cochin, Département d'Immunologie, Paris, France
GEORGES LUTFALLA • CNRS/Université Montpellier II, Montpellier, France
DAVID MACKEY • Department of Plant Cellular and Molecular Biology, Program in Molecular Cellular and Developmental Biology, The Ohio State University, Columbus, OH
ODILE MALBEC • Institut Pasteur, Département d'Immunologie, Unité d'Allergologie Moléculaire et Cellulaire, Paris, France
MAUREEN P. MARTIN • Laboratory of Genomic Diversity, SAIC-Frederick, Inc., NCI-Frederick, Frederick, MD
LUISA MARTINEZ-POMARES • Institute of Infection, Immunity and Inflammation, Queen's Medical Centre, Floor A West Block, Nottingham, United Kingdom
CEDRIC MONGELLAZ • Institut de Génétique Moléculaire de Montpellier, Montpellier, France
RONALD NAUMANN • Max-Planck-Institute for Cell Biology and Genetics, Dresden, Germany
FALK NIMMERJAHN • Laboratory of Experimental Immunology and Immunotherapy, University of Erlangen-Nuernberg, Nikolaus-Fiebiger-Center for Molecular Medicine, Erlangen, Germany
JENNIFER R. POWELL • Department of Molecular Biology, Massachusetts General Hospital, Department of Genetics, Harvard Medical School, Boston, MA
JEFFREY V. RAVETCH • Laboratory of Molecular Genetics and Immunology, The Rockefeller University, New York, NY
MICHAEL J. REDD • Huntsman Cancer Institute, Salt Lake City, UT
PASCAL RIHET • Université de la Méditerranée, Marseille Cedex, France
YVES ROMEO • Centre de Génétique Moléculaire du CNRS, Gif-sur-Yvette, France
MICHAEL SHAPIRA • Department of Genetics, Stanford University School of Medicine, Stanford, CA

KEN SHORTMAN • *The Walter and Eliza Hall Institute of Medical Research, Melbourne, Victoria, Australia*
HERGEN SPITS • *Department of Cell Biology and Histology, Academic Medical Center of the University of Amsterdam (AMC-UvA), Amsterdam, The Netherlands*
A. FRANCIS STEWART • *Genomics, BioInnovationsZentrum, Technische Universität Dresden, Dresden, Germany*
LOUISE SWAINSON • *Institut de Génétique Moléculaire de Montpellier, Montpellier, France*
MAN-WAH TAN • *Department of Genetics, Stanford University School of Medicine, Stanford, CA*
NAOMI TAYLOR • *Institut de Génétique Moléculaire de Montpellier, Montpellier, France*
GIUSEPPE TESTA • *The FIRC Institute of Molecular Oncology Foundation Via Adamello 16, Milan, Italy*
RITA VICENTE • *Institut de Génétique Moléculaire de Montpellier, Montpellier, France*
KRISTINA VINTERSTEN • *Mount Sinai Hospital, Samuel Lunenfeld Research Institute, Toronto, Ontario, Canada*
ERIC VIVIER • *Centre d'Immunologie de Marseille-Luminy, INSERM, CNRS, Université de la Méditerranée, Marseille, France*
DAVID VREMEC • *The Walter and Eliza Hall Institute of Medical Research, Melbourne, Victoria, Australia*
KEES WEIJER • *Department of Cell Biology and Histology, Academic Medical Center of the University of Amsterdam (AMC-UvA), Amsterdam, The Netherlands*
GAETANE WOERLY • *Inserm U547-Université de Lille, Institut Pasteur de Lille, France*
WAYNE M. YOKOYAMA • *Howard Hughes Medical Institute, Division of Rheumatology, Washington University School of Medicine, St. Louis, MO*

1

ENU Mutagenesis in Mice

Philippe Georgel, Xin Du, Kasper Hoebe, and Bruce Beutler

Summary

Forward genetics has led to many "breakthrough" discoveries, and with the mouse genome almost fully sequenced, the creation of phenotypes through random germline mutagenesis has become an efficient means by which to find the function of yet undescribed genes. In this chapter, we will provide a practical guideline for performing germline mutagenesis in mice. In particular, we will focus on the application of this technology to identify genes that are essential to innate immune defense.

Key Words: Mouse; ENU; mutagenesis; screening; phenotype; immunity; TLR; MCMV; VSV; mapping.

1. Introduction

To date, more than 25,000 mammalian genes have been identified, but the function of most genes has yet to be discovered. Therefore, the central challenge in all areas of biological inquiry is to uncover gene function. Both forward and reverse genetic approaches have been successfully applied to this end. Forward (classical) genetics proceeds from phenotype to the identification of a causal genetic change (mutation). Reverse genetics, on the contrary, begins with the creation of a genetic change and ends with the identification of a phenotype.

The observable characteristics of an individual organism result from the interaction between its genotype and the environment. As often genotypes differ between individuals, for many biological phenomena, two or more "phenotypes" exist, for example, white versus black coat color; calm versus agitated behavior; large versus small body size. In the past, phenotypes were primarily studied in the context of existing inbred strains that contained distinct phenotypic traits. The collection of more than 450 *Mus musculus* strains *(1)*, in addition to wild-derived mice *(2)*, encompasses considerable genetic diversity that continues to fuel the forward genetic approach. In addition, the occurrence of spontaneous mutations in inbred strains provided an important source of genetic variation leading to the identification of numerous loci involved in specific phenotypic traits. Since the 1980s, it has been possible to find the ultimate cause of phenotypes through positional cloning. As such, forward genetics has allowed major "breakthroughs" in many fields. For example, it has allowed the decipherment of complex immune mechanisms involved in the host response against microbial pathogens *(3–7)*. Although positional cloning, or "cloning by phenotype," was once considered an arcane art, two key developments have made it far simpler. The successful determination of the mouse genome sequence has eliminated the need for contig construction and trivialized the identification of informative markers for high-resolution mapping. And the steady increase in the speed and decline in the cost of DNA sequencing has made it far easier to find mutations. Where several years were once required for a well-equipped laboratory to positionally clone one mouse mutation, many mutations may now be positionally cloned each year by a laboratory with equivalent resources, reflecting at least a 50-fold increase in speed. The limiting factor in positional cloning has now become the availability of strong monogenic phenotypes and has been referred to as the "phenotype gap" *(8)*.

In *Drosophila*, the chemical mutagen ethylmethanesulfonate (EMS) has contributed to the understanding of the genetic regulation of embryonic development as well as innate immunity *(9)*. In mice, *N*-ethyl-*N*-nitrosourea (ENU) has been described as a powerful mutagen that at optimal doses can result in more than 1 mutation per locus per 700 gametes *(10–15)*. ENU introduces point mutations in spermatogonial stem cells, predominantly affecting A/T base pairs (44% A/T→T/A transversions and 38% A/T→G/C transitions), whereas at the protein level, ENU primarily results in missense mutations (64%), and 26% splicing errors and 10% of mutations resulting in nonsense mutations *(16)*. ENU is believed to induce about one base-pair change per million base pairs of genomic DNA *(17)*.

Here, we provide a practical guide to ENU mutagenesis in mice and describe how it can uncover genes that are essential to innate immunity. The latter entails an ex vivo approach to study macrophage function, including

Toll-like Receptor (TLR) signaling, and a conditional lethal screen intended to identify genes essential for survival after mouse cytomegalovirus (MCMV) infection in vivo. We also detail the procedures required to identify the causal mutations.

2. Mice and Materials

2.1. ENU Mutagenesis and Breeding

ENU is used in animal studies to elicit a mutagenic and/or carcinogenic response. ENU is classified as a potential human mutagen, carcinogen, and teratogen. Therefore, researchers working with the material must be aware of potential hazards while handling and disposing of the material and/or treated animals/tissues.

1. C57BL/6 male mice (Jackson laboratories, Bar Harbor, ME, USA) 5–6 weeks old.
2. Isopak ENU (Sigma Aldrich St Louis, MO, USA, N3385), stored at –20°C.
3. 95% EtOH.
4. Phosphate citrate buffer: 0.1 M NaH_2PO_4, 0.05M sodium citrate, pH 5.0–6.0.
5. ENU-inactivation solution: 5% $Na_2S_2O_3$ in 0.1 M NaOH.
6. 18-gauge needles, 10- and 60-ml syringes for solution preparation.
7. Container with warm water.
8. 1-ml insulin syringe for injections.
9. Double gloves, face shields for injections.
10. Anesthetics: 10 mg/ml ketamine HCl (Fort Dodge, IA, USA) and 2 mg/ml xylazine (Butler Animal Health, Dublin, OH, USA) in phosphate-buffered saline (PBS) solution.

2.2. Functional Screen I: TLR Signaling in Macrophages

1. 3% thioglycollate medium, Brewer modified (Becton Dickinson, Sparks, MD, USA). Prepared and autoclaved according to manufacturer's instructions.
2. Anesthetics: 10 mg/ml ketamine HCl (Fort Dodge) and 2 mg/ml xylazine (Butler Animal Health) in PBS solution.
3. Heating pad.
4. 5-ml syringes with 18-gauge needles containing PBS.
5. Hank's Buffered Salt Solution (HBSS) supplemented with 10% fetal bovine serum (FBS) and 2% penicillin (10,000 U/ml)/streptomycin (10,000g/ml) (P/S) solution.
6. Dubecco's Modified Eagle's Medium (DMEM), 5% FBS, and 2% P/S.
7. TLR-ligands: lipopolysaccharide from *Salmonella minnesota* R595 (Re) (TLRgrade™) (liquid; ALX-581-008-L002, Alexis Biochemicals, Lausen, Switzerland). Poly I: poly C; Amersham (Piscataway, NJ, USA). Pam_3Cys-SK_4, Pam_2Cys-SK_4, and MALP-2; EMC microcollections GmbH (Tübingen, Germany).

Phosphorothioate-stabilized CpG oligodeoxynucleotide (CpG ODN) 5′-TCC-ATG-ACG-TTC-CTG-ATG-CT-3′: Integrated DNA Technologies (Coralville, IA USA).

2.3. Functional Screen II: Susceptibility to MCMV Infection

1. BALB/c mice; 3 weeks (Jackson Laboratories).
2. MCMV Smith strain.
3. DMEM, 3% FBS, and 2% P/S.
4. Tissue tearor.
5. 24-well tissue culture plates.
6. 0.8% carboxymethylcellulose (CMC); autoclave 4 g of CMC C-50B high viscosity in an empty 500-ml bottle. Fill the bottle with 500 ml of sterile DMEM supplemented with 3% FBS, 2% P/S, 2% gentamycin. Store at 4°C with occasional shaking to allow complete dissolution of CMC.
7. Buffered formalin.
8. Staining solution: 2.8 g crystal violet in 10 ml ethanol and complete to 800 ml with distilled water.
9. NIH/3T3 (or BALB/3T3 cells, ATCC N° CCL-163).

2.4. Preparation of Genomic DNA

1. Lysis buffer; 100 mM Tris–HCl pH 7.5–8.0, 0.2% sodium dodecyl sulfate (SDS), 5mM ethylenediaminetetraacetic acid (EDTA), 200 mM NaCl.
2. 96-well polymerase chain reaction (PCR) plates.
3. Proteinase K; 20 mg/ml in H_2O.
4. SDS-out (Pierce, Rockford, IL, USA).

2.5. Genome-Wide Mapping and Establishing a Critical Region

1. Out cross-strain; C3H/HeN mice (Taconic).
2. Thermocycler (Perkins Elmer, Waltham, USA) and PCR reagents.
3. A panel of microsatellite or single-nucleotide polymorphisms (SNPs) informative for background and outcross strain, covering the entire mouse genome with regular intervals (~60 for backcross or ~120 for intercross mapping).
4. Microsatellites and/or SNPs to further define the critical region.
5. Appliances and software for SNP/microsatellite detection (i.e., ABI PRISM 3100/3700, MegaBACE 1000, ABI PRISM 7900H, and/or Pyrosequencing).

3. Methods
3.1. Generation of G0 ENU Mice

Six-week-old male C57BL/6 mice (*see* **Note 1**) are anesthetized to prevent struggling or kicking and then treated with ENU administered in three weekly doses (90 mg/kg body weight) through intraperitoneal injection. Wear protective clothing double gloves and face shield and perform this procedure in a fume hood.

3.1.1. Preparation of ENU Solution and ENU Injection

1. Before preparation, bring ENU Isopak to room temperature (RT).
2. Inject 10 ml of 95% EtOH using an 18-gauge needle and 10-ml syringe (dispose the needle and syringe immediately).
3. Swirl the Isopak in the container with warm water until ENU powder is entirely dissolved.
4. Add 90 ml phosphate-citrate buffer using 60-ml syringes and 18-gauge needles.
5. ENU is easily degraded, and the exact concentration needs to be determined by spectrophotometer using the formula: $0.72 \ OD_{398nm} = 1$ mg/ml of ENU.
6. After preparation of the ENU solution, anesthetize the mice with xylazine/ketamine injection, 100 μl i.m. While they are asleep, weigh them and inject 90 mg/kg of ENU i.p.
7. Leave the mice overnight in disposable cages in the fume hood.

3.1.2. Breeding of G1 and G3 ENU Mice

After the last dose, mice are placed in isolation (one mouse per cage) for 12 weeks to allow recovery of fertility. Fertility testing is performed at 8 weeks to assure that all mice tested are sterile. Fertility at this stage indicates that a suboptimal dose of ENU was administered. After the recovery period, each G0 male is bred to normal C57BL/6 female mice to generate a maximum of 20 G1 offspring. These G1 animals are used either for phenotypic screens or to produce G2 mice, which in turn are backcrossed to the G1 to generate G3 offspring (*see* **Fig. 1**). Usually, male G1 animals are used to propagate mutations to homozygosity, and two daughters are backcrossed to each G1 founder.

3.2. Considerations for Phenotypic Screening of ENU Mice

Phenotypic screening assays should have limited variance, should target a large genomic footprint if possible (i.e., many genes with non-redundant function), and should ideally probe a poorly understood phenomenon. Both G1 and G3 ENU mice can be used in any screen to identify phenodeviants (*see* **Note 2**). A non-lethal screen is advantageous in that it allows for secondary confirmation and breeding of phenodeviants directly, but lethal screens of G3 mice are feasible if the G1 and G2 parents are retained to propagate mutations. Here, we provide two examples of immunological screens that have successfully revealed gene function in innate immunity.

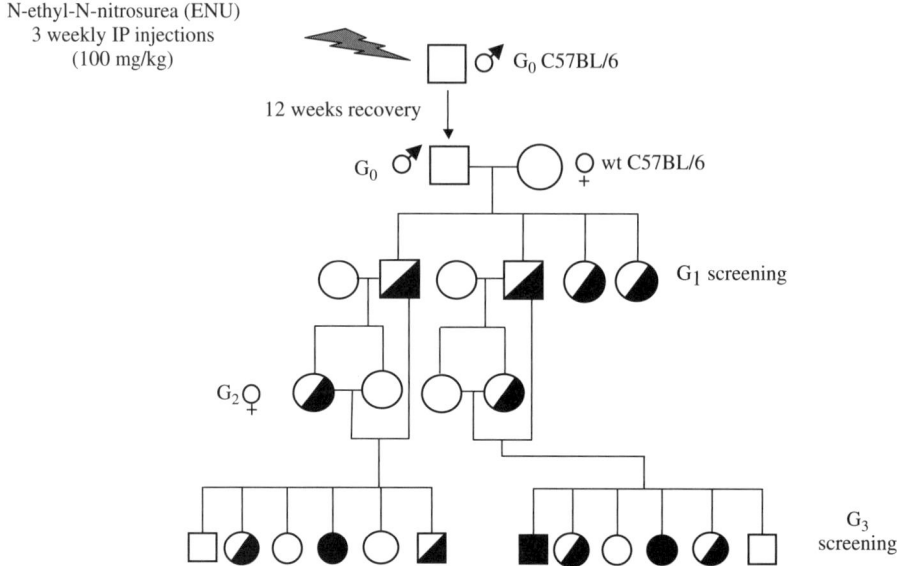

Fig. 1. ENU treatment of C57BL/6 mice and micropedigree construction for screening of dominant (G1) and recessive (G3) screening. Each G1 male is bred with two G2 females and subsequently from each G2 female three G3 offspring are tested in recessive screens.

3.3. Functional Screen I: TLR Signaling in Macrophages

3.3.1. Isolation and Plating of Peritoneal Macrophages

1. Inject 7–8-week-old mice with 1 ml of 3% thioglycollate solution i.p.
2. After 3 days, anaesthetize mice by injecting 100 µl of ketamine/xylazine solution i.m.
3. Harvest peritoneal macrophages through lavage of the peritoneum with 5 ml of sterile PBS using a syringe and 18-gauge needle and allow the mice to recover by placing their cages on heating pads when finished.
4. Collect cells in 15 ml conical polypropylene tubes containing 5 ml of HBSS medium and centrifuge at 500 g for 10 min.
5. Resuspend cells in DMEM medium and bring to a final density of 1×10^6 cells/ml.
6. Add 100 µl of cell suspension per well to 96-well flat-bottom plates and after 1 h of incubation at 37°C and 5% CO_2 expose cells to TLR ligands at a suboptimal concentration (for dose–response curves, see **ref. 18**). After an additional 4-h incubation, supernatants are collected and either stored at −80°C or used directly for measurement of secreted cytokines (*see* **Fig. 2**).

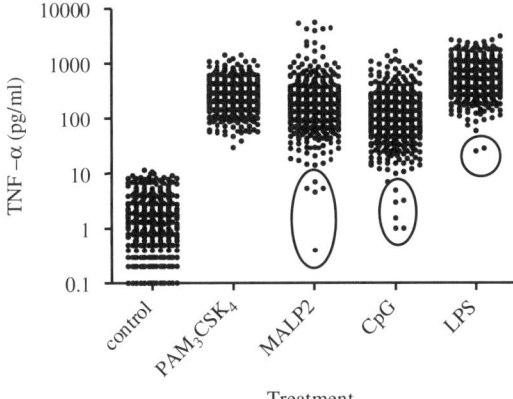

Fig. 2. Screening of ex vivo peritoneal macrophages derived from G3 ENU mice for their ability to respond to specific toll-like receptor (TLR) ligands. Macrophages are incubated with Pam3CSK4 (100 ng/ml), MALP2 (100 ng/ml), Phosphorothioate-stabilized CpG oligodeoxynucleotide (CpG ODN) 5′-TCC-ATG-ACG-TTC-CTG-ATG-CT-3′ (1 μM) or LPS (100 ng/ml). After 4 h incubation, supernatants were collected, and the TNF-α concentration was determined using a L929 bioassay. Encircled points represent phenodeviant responses of mice that have been selected for confirmation. Graph represents the distribution of TNF production by macrophages obtained from 2400 G3 mice.

3.4. Functional Screen II: Susceptibility to MCMV Infection

3.4.1. MCMV Stock Preparation and Titration

About 10^3 plaque-forming units (pfu) of MCMV in 200 μl of DMEM are injected i.p. into 3-week-old female BALB/c mice. After 12–14 days, the animals are euthanized, and the salivary glands are harvested and collected in 50-ml conical Falcon tubes containing 25 ml of DMEM medium on ice. Glands are homogenized with a tissue tearor, which was previously cleaned with bleach and rinsed in ethanol followed by PBS. The homogenate is centrifuged at 200 *g* for 3 min to pellet cellular debris, and the supernatant is divided among 15-ml tubes. After centrifugation (5 min at 1500 *g*), the upper lipidic layer is eliminated by aspiration, and the supernatant is aliquoted into eppendorf tubes (100 μl/tube) that are stored at −80°C.

To titrate the viral stock, NIH3T3 cells are plated in 24-well plates (50,000 cells/well) for 24 h in DMEM 5% FCS, 2% P/S. Culture medium is removed by aspiration and replaced by 200 μl of serial dilutions (10^{-1}–10^{-12}) made from

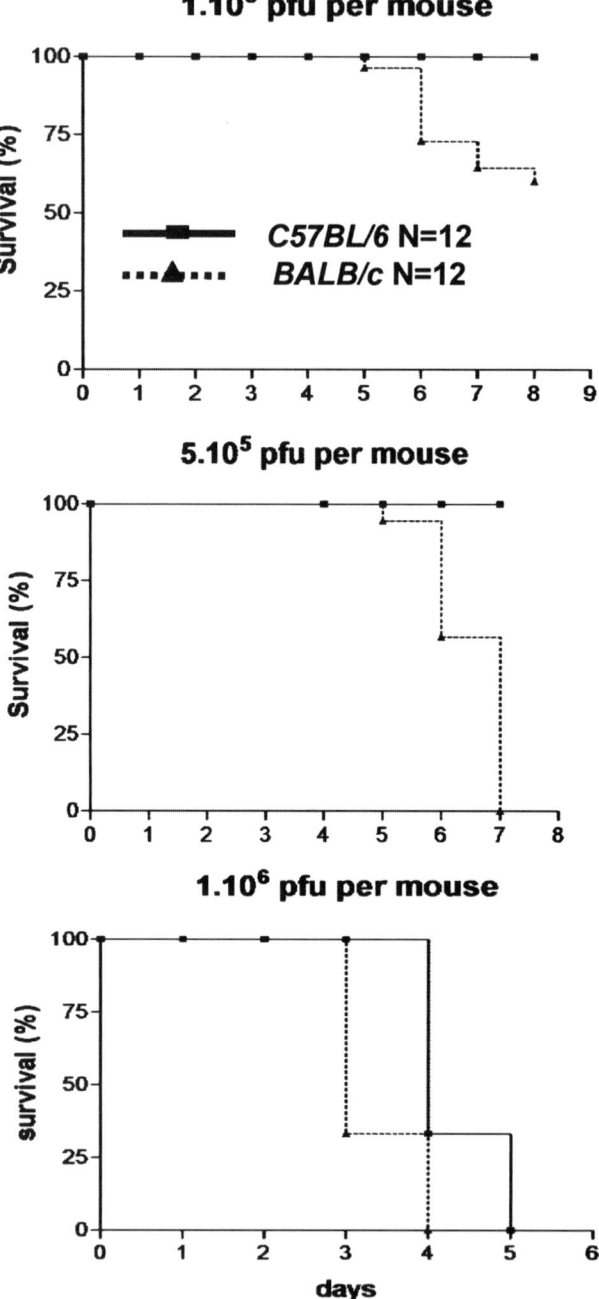

1 tube of the viral stock (chilled on ice) in DMEM. After 2 h incubation at 37°C, 5% CO_2, 1 ml of pre-heated (at 42°C) 0.8% CMC solution is added. Plates are then placed for 4–5 days in the incubator. When plaques are apparent after microscopic inspection, cells are fixed by addition of 200 µl of formalin (overnight at RT). CMC solution and formalin are removed by aspiration, and the cells are stained with 500-µl crystal violet (3 min at RT). After extensive washing with distilled water, plaques are counted for the different dilutions.

3.4.2. Measuring the Infectivity of the Viral Suspension

Because the viral inoculum is amplified in vivo, the infectivity of each preparation must be evaluated in resistant and susceptible animals to determine an LD50. C57BL/6 mice are considered resistant, whereas BALB/c animals that lack the LY49H receptor are highly sensitive to MCMV infections. Typically, groups of 12 mice for each strain are infected with different viral doses (10^5, 5×10^5 and 10^6 pfu in 200 µl of DMEM, as determined after the plaque assay). As seen one example illustrated **Fig. 3**, the lowest dose induces little mortality in BALB/c mice and permits 100% survival in control C57BL/6, whereas the highest inoculum is lethal for both strains.

3.4.3. In Vivo Screening for MCMV-Susceptible Mice

Because an intermediate infectious dose (5×10^5 pfu/mouse) permits easy discrimination between resistant and susceptible strains, this inoculum is used in screening. The read-out to identify mutant mice is death or severe signs of sickness (visible weight loss, rough fur) as endpoints. Only G3 animals are screened.

3.4.4. Confirmation of the Susceptibility and Generation of a Homozygous Stock

Once a susceptible G3 mouse has been identified, additional litters produced by its G1 and G2 parents are tested to confirm the presence of a transmissible phenotype. Next, additional siblings are randomly mated through several

---

Fig. 3. Determination of the viral doses for an in vivo MCMV susceptibility screen. C57BL/6 and Balb/C mice are injected with either 1×10^5, 5×10^5, or 1×10^6 pfu/ml MCMV, as described in the text. In each group, the survival was followed up to 8 days after infection.

generations to fix the mutation. Fixation, verified by expanding an inbred line and testing for uniform susceptibility, is achieved as a prelude for mapping the mutation.

3.5. Genome-Wide Mapping and Defining the Critical Region

Once a mutation is fixed, it has first to be mapped to a chromosomal interval and then to a small region of the chromosome with a limited number of genes (the "critical region"). The critical region is then explored to find the mutation that causes the phenotype by sequencing all the genes that it contains. In the vast majority of cases, phenotypes emanate from changes in the amino acid sequence of proteins, resulting from missense, nonsense, or splicing mutations. As a first approach, coding exons or the coding regions of cDNAs are therefore examined to find the causative mutation. As ENU creates mutations with low frequency (only one per million bp of DNA) and as coding regions comprise only about 1.5% of total genomic DNA, a small critical region (1–2 Mb in length) is very unlikely to have more than one coding change.

3.5.1. Choosing an Outcross Strain and Breeding Strategies

To map a mutation, the mutant stock (C57BL/6 background) is outcrossed to a second *M. musculus* strain (e.g., C3H/HeN), and the F1 hybrids that result are backcrossed either to the mutant stock (for a recessive mutation) or to the outcross strain (for a dominant mutation). In choosing the strain used as a mapping partner, it is important to ensure that there is strong phenotypic similarity between this partner and normal C57BL/6 mice and moreover that hybrid animals are similar to both parents. Most *M. musculus* strains have sufficient genetic variation (SNPs or micro-satellite markers) to permit the mapping of a mutation produced on the C57BL/6 background, and even C57BL/10 mice, which are very similar to C57BL/6, can now be used for mapping at least to a chromosomal interval. Recessive mutations can also be mapped using an intercross strategy: that is, the F1 hybrids are mated with each other rather than backcrossing them to the mutant stock. This will result in F2 mice produced from twice the number of meioses. Each F2 mouse is therefore more informative, but a higher density of markers throughout the genome is required to assure linkage between the phenotype and one or more markers in the panel. Once F2 offspring are produced, they are tested for the phenotype of interest, and genomic DNA is collected from each mouse.

3.5.2. Isolation of Genomic DNA

There are numerous DNA preparation/isolation kits commercially available. The following protocol is inexpensive and can be applied to high-throughput isolation of genomic DNA samples.

1. Collect a small tail or ear sample from mice using anesthetics (i.e., isoflurane). Add the sample to a 96-well PCR plate.
2. Prepare DNA lysis solution by adding proteinase K to the lysis buffer (proteinase K: lysis buffer, 1:10 v/v). Mix gently.
3. Add 100 µl DNA lysis solution to each well of the PCR plate containing sample. Seal the plate and mix gently.
4. Let the sample digest at 50°C until no piece of tail/ear is visible. Mix gently.
5. Heat/inactivate at 95°C for 15 min using a thermocycler.
6. Centrifuge the plate at 200 *g* for 5 min to pellet impurities.
7. Collect approximately 75 µl supernatant (without disturbing the pellet) and transfer to a new PCR plate.
8. Add 10 µl SDS-out, seal the plate, mix gently, and immediately place the plate in a thermocycler at 4°C for 30 min.
9. Spin the plate at 400 *g* for 5 min in a pre-cooled (4°C) centrifuge.
10. Collect 50 µl of supernatant (without disturbing the pellet) and transfer to a new plate.
11. Store samples at 4 or –20°C and dilute the sample 1:10 in PCR grade H_2O or 0.1× Tris/EDTA-(TE) buffer. Before use, determine the DNA concentration.

3.5.3. Establishing Chromosomal Linkage for a Phenotype

The ENU background (i.e., C57BL/6) and outcross strains (i.e., C3H/HeN) are both inbred strains and thus homozygous at all loci (with the exception of the ENU-induced mutations on the C57BL/6 background). The C57BL/6 mice can be distinguished from C3H/HeN mice through germline differences at more than 1 million loci scattered along the chromosomes. These differences may be considered as potential markers and, in an F2 mouse, can be used to trace the parental origin of a chromosomal interval. These markers can involve single base pair differences (SNPs) or differences in the length of simple repetitive sequences or microsatellites (poly[CA], poly[AG], or poly[GATA], for example) (*see* **Note 3**). In the case of a backcross approach, F1 hybrids (heterozygous for all markers) are crossed to the homozygous C57BL/6 mutant stock. We can therefore expect each of the offspring to be homozygous C57BL/6 at about half of the marker loci and heterozygous at the remaining loci. To localize the mutation, all animals of the F2 generation are examined phenotypically to determine whether they are homozygous or heterozygous for the mutation of interest. In addition, they are examined genotypically to determine whether they are homozygous or heterozygous for each of the genome-wide markers. Concordance between the presence of the phenotype and homozygosity for a marker and the absence of the phenotype and heterozygosity for that same marker is evaluated for each of the markers in the mapping panel. A log odds distance (LOD) score is calculated for each marker and is used as an objective index of linkage. Usually, the analysis of approximately 20 mice (including those with phenotype and those without) is sufficient to obtain significant linkage to one or more of the chromosomal markers.

The LOD score (Z) developed by Newton E. Morton in 1955 is equivalent to:

$$Z = \frac{\text{probability of a linkage}}{\text{probability of non} - \text{linkage}}$$

between the mutation and the marker. Given non-linkage, we expect 50% concordance between the marker genotype and the phenotype of interest.

The LOD score can be calculated for each genome-wide marker using the formula:

$$Z(\Phi) = \log\left[\frac{(1-\Phi)^{n-r}\Phi^r}{(1/2)^n}\right]$$

where Φ = fraction that is concordant, $1 - \Phi$ = fraction that is disconcordant, n = the total number of observations (mice), $(n - r)$ = number of discordant mice, and r = number of concordant mice. In principal, an LOD score of 3 or higher is considered as significant ($P < 0.01$) linkage, but ultimately one should aim for an LOD score of 4 or 5 to be absolutely confident. In a backcross, using a panel of 60 evenly spaced markers, 20–30 F2 mice are usually sufficient to achieve this.

3.5.4. Further Confinement of the Mutation to a "Critical Region"

Once significant linkage to one or more markers on a chromosome has been obtained, new markers within the chromosomal interval are examined. At this point, the original F2 as well as additional F2 animals should be included in the analysis of chromosomal markers, so that "critical region" can be established. This can be done by identifying one proximal and one distal marker that are each separated from the mutation by at least one crossover event. In a cross between two *M. musculus* strains, one centiMorgan (cM; 1% crossover frequency) corresponds to approximately 1.6 Mb of DNA, the average critical region that is derived from 100 G2 mice would theoretically correspond to approximately 1.6 Mb in length, given a high density of informative markers. Many thousands of SNP and microsatellite markers have been published (*see* **Subheading 3.4.6., step 3**), but new informative microsatellite markers may often be found by randomly sequencing CA or GATA repeats as needed (*see* **Note 3**).

The mutation is confined to a region small enough to permit sequencing of all coding exons within the region, and for some laboratories, this will mean tighter confinement than it will for others. Candidate genes may be examined either at the cDNA level or the genomic level. The average Mb of genomic DNA contains 12 genes and the average gene contains 8.8 exons, and hence, approximately 100 exons with coding function. However, it is often found that many more genes reside within a small critical region, because genes tend to be clustered rather than randomly dispersed in the genome.

3.5.5. Identification of the Causative Mutation within the Critical Region

In some instances, a "best" candidate gene presents itself for sequencing, and the causal mutation is readily found. In other cases (perhaps the most interesting cases), there is little basis for choice, and prioritization is difficult. Sequencing of candidate genes can be carried out at the cDNA level (preferable if a gene contains many exons and a small number of splice forms) or at the genomic level (if there are many exons and an unknown number of splice forms, or if there is uncertainty as to whether the mRNA is expressed). In addition to sequencing coding regions, the expression levels of the genes in the critical region can be assessed by Northern blot or quantitative PCR analysis. This may be helpful to identify mutations outside the coding region that affect the expression of a gene and that would otherwise be missed (fortunately a very rare circumstance).

3.5.6. Useful Tools or Websites to Efficiently Identify a Mutation in the Critical Region

To identify efficiently the mutation in a critical region, various bioinformatic tools can help prioritize candidate genes and/or are elementary to analyze successfully genomic information. Here, we will describe some of the programs that are freely accessible and their specific application to efficiently analyze any critical region.

1. Ensembl (http://www.ensembl.org/): Ensembl is a genome browser, jointly created by the EMBL-European Bioinformatics Institute (EBI) and the Sanger Institute that contains databases of selected eukaryotic genomes. The website provides

essential information with regard to genomic sequences as well as available micro-satellite markers and/or SNPs for any given region. The databases are regularly updated and as of this writing, annotation changes frequently. Other genome browsers include the Vertebrate Genome Annotation database (also known as VEGA; http://vega.sanger.ac.uk/index.html), which supplies manual annotation for specific regions of the mouse genome, and the NCBI Map Viewer (http://www.ncbi.nlm.nih.gov/mapview/map_search.cgi?taxid=10090).

2. Repeatmasker (developed by A.F.A. Smit, R. Hubley, and P. Green; http://repeatmasker.org): This program is essential for processing sequencing data derived from the public database (Ensembl) or the Celera database. It screens DNA sequences for interspersed repeats and low complexity DNA sequences. The output of the program is a detailed annotation of the repeats that are present in the query sequence as well as a modified version of the query sequence in which all the annotated repeats have been masked (masked nucleotides are depicted as Ns). This program is useful for the design of unique amplification primers for genomic DNA.

3. Websites for the identification of SNPs/microsatellite markers: http://www.ncbi.nlm.nih.gov/projects/SNP/MouseSNP.cgi, http://www.broad.mit.edu/snp/mouse/, and http://snp.gnf.org/, which also contain an extensive gene expression atlas.

4. The mouse phenome database (http: //phenome.jax.org /pub-cgi /phenome/): A broad collection of phenotypic and genotypic data for laboratory mice.

4. Notes

1. In principle, any strain can be used for mutagenesis, but one should consider the availability of genomic information as well as relevant background issues for the phenotype that will be tested and examine the sensitivity of the strain to ENU if it is not already known (considerable interstrain variation has been observed). The C57BL/6 strain has been a popular choice for many ENU laboratories.

2. With 1 bp change per million bp and a total length of approximately 2998 Mb for the mouse genome, one can calculate that each G1 male carries approximately 3000 bp changes genome wide. With the coding region being 1.3% of total genomic sequence and 76% of random bp changes creating a coding change, it follows that each G1 mouse carries about 30 coding changes genome wide. These exist in a heterozygous form and do not necessarily lead to phenotypic changes. Many ENU-induced mutations are recessive or co-dominant at best, and phenotypic changes are then observed only in homozygotes. For any ENU mutagenesis project, it is important to decide whether one will generate and screen G1 mice only (dominant mutations) and/or whether one will generate and screen G3 mice (recessive mutations). In the G3 generation, each mouse will carry approximately four mutations in its homozygous form. Although increasing the numbers of G3 mice for screening of each pedigree will ultimately reveal all mutations in its homozygous form, a point of diminishing return is reached, and many small pedigrees are more cost efficient than a few large ones for revealing phenotype. The

probability of transmission of each G1 mutation to homozygosity can be calculated as $P = 1 - [0.5 + (0.5)*(3/4)^{g3}]^{g2}$, where g2 is the number of G2 and g3 is the number of G3 offspring per G2 mother. If pedigrees are constructed to produce two G2 daughters per G1 male, and three G3 pups per G2 daughter are directed to a given screen, approximately 50% capture of G1 mutations in homozygous form is achieved.

3. In some cases, no published markers reside within the critical region, yet further meiotic mapping is desirable. New markers must then be found. First, it is essential to determine whether the region is stringently conserved between the strains that were used in the mapping cross. An SNP database (http://snp.gnf.org/) can disclose whether coinheritance has occurred within the region. If coinheritance has *not* occurred, it may be most efficient to sequence parts of the region at random, whereon one can expect to find an SNP every 1–2 kb. If coinheritance *has* occurred, one can still look for simple sequence length polymorphism by analyzing simple repeats such as CA (=TG), TA, (=AT), and CT (=GA) or GATA (=TATC). GC repeats are exceedingly rare. All repeats can be polymorphic, but the frequency polymorphism is, in our experience, GATA > CA > CT > TA. On average, about one in six CA repeats is informative when two disparate *M. musculus* strains are compared.

Acknowledgments

PG is supported by grants from the Fondation pour la Recherche Médicale (FRM), la Ligue pour la Recherche sur le Cancer and the Association pour la Recherche sur le Cancer (ARC).

Reference

1. Beck, J. A., Lloyd, S., Hafezparast, M., Lennon-Pierce, M., Eppig, J. T., Festing, M. F., & Fisher, E. M. (2000). Genealogies of mouse inbred strains. *Nat. Genet.* **24,** 23–25.
2. Guenet, J. L. & Bonhomme, F. (2003). Wild mice: an ever-increasing contribution to a popular mammalian model. *Trends Genet.* **19,** 24–31.
3. Brunkow, M. E., Jeffery, E. W., Hjerrild, K. A., Paeper, B., Clark, L. B., Yasayko, S. A., Wilkinson, J. E., Galas, D., Ziegler, S. F., & Ramsdell, F. (2001). Disruption of a new forkhead/winged-helix protein, scurfin, results in the fatal lymphoproliferative disorder of the scurfy mouse. *Nat. Genet.* **27,** 68–73.
4. Lee, S. H., Webb, J. R., & Vidal, S. M. (2002). Innate immunity to cytomegalovirus: the Cmv1 locus and its role in natural killer cell function. *Microbes. Infect.* **4,** 1491–1503.
5. Poltorak, A., He, X., Smirnova, I., Liu, M. Y., Van Huffel, C., Du, X., Birdwell, D., Alejos, E., Silva, M., Galanos, C., Freudenberg, M., Ricciardi-Castagnoli, P., Layton, B., & Beutler, B. (1998). Defective LPS signaling in C3H/HeJ and C57BL/10ScCr mice: mutations in Tlr4 gene. *Science* **282,** 2085–2088.

6. Scalzo, A. A., Wheat, R., Dubbelde, C., Stone, L., Clark, P., Du, Y., Dong, N., Stoll, J., Yokoyama, W. M., & Brown, M. G. (2003). Molecular genetic characterization of the distal NKC recombination hotspot and putative murine CMV resistance control locus. *Immunogenetics* **55**, 370–378.
7. Webb, J. R., Lee, S. H., & Vidal, S. M. (2002). Genetic control of innate immune responses against cytomegalovirus: MCMV meets its match. *Genes Immun.* **3**, 250–262.
8. Balling, R. (2001). ENU mutagenesis: analyzing gene function in mice. *Annu. Rev. Genomics Hum. Genet.* **2**, 463–492.
9. Nusslein-Volhard, C. & Wieschaus, E. (1980). Mutations affecting segment number and polarity in Drosophila. *Nature* **287**, 795–801.
10. Favor, J., Neuhauser-Klaus, A., & Ehling, U. H. (1988). The effect of dose fractionation on the frequency of ethylnitrosourea-induced dominant cataract and recessive specific locus mutations in germ cells of the mouse. *Mutat. Res.* **198**, 269–275.
11. Favor, J., Neuhauser-Klaus, A., Ehling, U. H., Wulff, A., & van Zeeland, A. A. (1997). The effect of the interval between dose applications on the observed specific-locus mutation rate in the mouse following fractionated treatments of spermatogonia with ethylnitrosourea. *Mutat. Res.* **374**, 193–199.
12. Rinchik, E. M., Carpenter, D. A., & Selby, P. B. (1990). A strategy for fine-structure functional analysis of a 6- to 11-centimorgan region of mouse chromosome 7 by high-efficiency mutagenesis. *Proc Natl. Acad. Sci. U. S. A* **87**, 896–900.
13. Rinchik, E. M. & Carpenter, D. A. (1999). N-ethyl-N-nitrosourea mutagenesis of a 6- to 11-cM subregion of the Fah-Hbb interval of mouse chromosome 7: completed testing of 4557 gametes and deletion mapping and complementation analysis of 31 mutations. *Genetics* **152**, 373–383.
14. Russell, W. L., Hunsicker, P. R., Raymer, G. D., Steele, M. H., Stelzner, K. F., & Thompson, H. M. (1982). Dose–response curve for ethylnitrosourea-induced specific-locus mutations in mouse spermatogonia. *Proc Natl. Acad. Sci. U. S. A* **79**, 3589–3591.
15. Russell, W. L., Hunsicker, P. R., Carpenter, D. A., Cornett, C. V., & Guinn, G. M. (1982). Effect of dose fractionation on the ethylnitrosourea induction of specific-locus mutations in mouse spermatogonia. *Proc Natl. Acad. Sci. U. S. A* **79**, 3592–3593.
16. Justice, M. J., Noveroske, J. K., Weber, J. S., Zheng, B., & Bradley, A. (1999). Mouse ENU mutagenesis. *Hum. Mol. Genet.* **8**, 1955–1963.
17. Concepcion, D., Seburn, K. L., Wen, G., Frankel, W. N., & Hamilton, B. A. (2004). Mutation rate and predicted phenotypic target sizes in ethylnitrosourea-treated mice. *Genetics* **168**, 953–959.
18. Hoebe, K., Du, X., Goode, J., Mann, N., & Beutler, B. (2003). Lps2: a new locus required for responses to lipopolysaccharide, revealed by germline mutagenesis and phenotypic screening. *J. Endotoxin. Res.* **9**, 250–255.

2

Innate Immunity Genes as Candidate Genes
Searching for Relevant Natural Polymorphisms in Databases and Assessing Family-Based Association of Polymorphisms with Human Diseases

Pascal Rihet

Summary

The identification of genes underlying complex traits is a challenging task, and there are a limited number of confirmed genes that influence human complex diseases. In particular, few genes involved in complex diseases related to immune response, such as infectious diseases and inflammatory diseases, have been identified. Recent advances in genotyping technology lead to the depository of millions of single-nucleotide polymorphisms (SNPs) into public databases, and SNPs are considered powerful tools in the search for genes involved in complex diseases. A number of SNP-genotyping methods are available, and two critical points are to select the SNPs required for a comprehensive analysis and to perform association analyses that avoid statistical biases because of population substructure. This chapter describes a way to take advantage of the mass of known SNPs and to evaluate family-based association between polymorphisms and phenotypes related to diseases, with special emphasis on innate immunity genes. After summarizing relevant aspects of genetic epidemiology, I describe how to

- obtain SNP data from ENSEMBL
- visualize an annotated sequence containing SNPs with SNPper
- select SNPs on the basis of population frequency and functional information
- explore SNP data in the IIGA database focused on innate immunity genes
- evaluate the association of SNPs with quantitative phenotypes by using Quantitative trait Transmission/Disequilibrium Tests (QTDT)
- evaluate the association of SNPs with binary and quantitative phenotypes by using Family-Based Association Tests (FBAT).

From: *Methods in Molecular Biology, vol. 415: Innate Immunity*
Edited by: J. Ewbank and E. Vivier © Humana Press Inc., Totowa, NJ

All the procedures use publicly available servers and free statistical programs for academic users.

Key Words: SNP; linkage disequilibrium; transmission/disequilibrium tests; family-based association tests; binary traits; quantitative traits; innate immunity.

1. Introduction
1.1. Genetic Epidemiology and Innate Immunity Genes

Genetic epidemiology aims to detect the inheritance pattern of a particular human disease, localize the gene, find a marker associated with disease susceptibility, and finally identify the genes and their variants involved in the disease. The approach has been successful for studying genetic predisposition to monogenic diseases. There are, nevertheless, a limited number of confirmed alleles affecting human multi-factorial diseases, and linkage analyses have confined genes involved to chromosomal regions. One of the major objectives of human genetics is now to identify genes and their alleles involved in complex human diseases. These include common diseases related to immune responses, such as infectious diseases, inflammatory diseases, allergy, or hypersensitivity.

The genes involved in innate immunity can be considered as candidate genes in such diseases, especially when such genes are located within chromosomal regions linked to a disease. The pivotal role of innate immunity molecules in the host response to the environment makes relevant the hypothesis that natural polymorphisms in innate immunity genes influence human susceptibility to several common diseases. Thus, a number of innate immunity genes show a wide range of genetic variation within human populations, even though they are highly conserved across a large number of species. For example, *Toll* genes are involved in host immune response against pathogens in Drosophila, mice, and humans *(1)*. Initial studies suggest an association of *TLR2*, *TLR4*, and *TLR5* polymorphisms with resistance or susceptibility to bacterial infections *(2)*. Besides, *TLR2*, *TLR3*, *TLR4*, *TLR6*, and *TLR9* polymorphisms have been found to be associated with asthma or atopy, although the results remain to be replicated *(1,3)*. Other genes involved in the Toll pathway are also conserved between species, and it is expected that genetic variation in these genes influence susceptibility to human diseases. Another gene family of interest is the family of *KIR* genes, which encode activating and inhibiting receptors expressed by natural killer (NK) cells and which are grouped on chromosome 19q13. The *KIR* locus displays genetic variation both in gene content and in DNA sequence, and several associations have been reported between *KIR* genes and human diseases *(4)*. Ongoing research in animal models will provide new insights into the knowledge of genes and pathways involved in innate immunity and may make relevant candidate genes of human orthologs. It is likely that new associations between innate immunity genes and human common diseases will be evidenced in the near future. If gene targeting provides direct evidence of the role of a gene, using natural polymorphisms may provide more subtle information on how genes influence physiological pathways and diseases. Also, genetic epidemiology may contribute to a better understanding of the biology of innate immunity molecules.

1.2. Single-Nucleotide Polymorphisms and Other Natural Polymorphisms, as Genetic Markers or Causal Mutations

The most widely used genetic markers are microsatellites and single-nucleotide polymorphisms (SNPs), which denote any polymorphic variation at a single nucleotide. The microsatellites are di-, tri-, and tetra-nucleotide repeats. Different numbers of repeats give different alleles, and alleles are identified by sizing. The microsatellites are highly polymorphic. However, the density of microsatellites is lower than that of SNPs. SNPs may occur approximately every 500 bp in humans, and more than 5 million SNPs have been described. Genotyping technology has been recently improved, and SNPs can be typed through large-scale automated methods (5). Besides, linked SNPs can be analyzed together after reconstructing haplotypes, and this provides a genetic information content that is similar to the one provided by microsatellites.

For a genetic marker (M), one can estimate the degree of heterozygosity (H) or the polymorphism information content (PIC)

$H = 1 - \sum_i p_i^2$, where p_i is the frequency of the ith allele.

$PIC = 1 - \sum_{ij} p_i^2 - \sum_{ij} 2p_i^2 p_j^2$, which accounts for the loss of information because of homozygous parents ($M_i M_i$) and heterozygous parents both with the same genotype ($M_i M_j$).

SNPs are usually considered as genetic markers that are genotyped to locate susceptibility genes on the human genome. To this aim, linkage or linkage disequilibrium mapping is performed to uncover either a marker genetically linked to the disease or a marker associated with it. This leads to the location of the mutation causing the disease. The causal mutations can be SNPs altering the sequence of the protein or the expression of the gene. This allows to conduct two strategies: first, one can select SNPs with a high genetic information content, and second, one can search for SNPs that likely alter the properties of the protein, the binding of a transcription factor, the splicing, and/or that are located within conserved regions across species. The second strategy is based on the hypothesis that the candidate polymorphism causes the disease and/or influences the phenotype. A huge amount of SNP data are available in various sources, such as NCBI dbSNP (http://www.ncbi.nlm.nih.gov/projects/SNP/) and ENSEMBL *(6)*. In addition, web tools help to retrieve SNP annotations and to select SNP on the basis of various parameters, such as allele frequency or functional information *(7–10)*. These tools will provide a very useful help in selecting SNPs for genetic association studies. SNPs can be genotyped by using several methods, such as denaturating high-performance liquid chromatography (DHPLC), single-strand conformation polymorphism (SSCP), oligonucleotide microarrays, or direct sequencing. These procedures are described elsewhere *(11)*.

1.3. Linkage, Association, and Linkage Disequilibrium Analyses

The underlying principle in genetic epidemiology is that genetic variation partly explains phenotypic variability. The phenotypes are generally binary (affected or unaffected) and/or quantitative. Familial clustering of a disease suggests that the disease (or a disease-related trait) is under genetic influence. For a binary trait, it can be expressed by the risk to relative R of an affected proband compared with the population risk. For quantitative traits, it can be tested by using analysis of variance. Furthermore, one can determine the degree to which a disease or trait is heritable. The heritability is the proportion of phenotypic variance because of genetic effects. "He" is the broad heritability, and "h^2" is the narrow heritability.

$$He = \frac{V_g}{V_p}$$

with $V_p = V_e + V_g$
or

$$h^2 = \frac{V_a}{V_p}$$

With $V_a = V_g - V_d$ and $V_p = V_e + V_d + V_a$
and V_g genetic variance, V_p phenotypic variance, V_e environmental variance, V_a additive genetic variance, and V_d dominant genetic variance.

Linkage analysis tests whether a genetic marker is co-inherited with disease (or disease-related trait) within families more often than expected by chance. Linkage analyses do not focus on a particular allele, and alleles co-inherited with disease may vary from one family to another. Multi-locus linkage analyses lead to the identification of chromosomal regions that likely contain a causal genetic variant. It should be stressed that the analysis includes neither families with homozygous parents nor families with heterozygous parents having the same genotype.

An association study at the population level is the next step for fine mapping. Population-based association studies test whether case and control groups differ in marker allelic frequencies. Family-based association studies test whether a particular allele of the marker is transmitted from parents to affected offspring in families more often than expected by chance. Case–control association studies are thought to be more powerful than family-based ones. Population-based association can, however, arise from biases because of population stratification or population admixture. Family-based association studies avoid such biases.

The power of association tests depends on the size of the sample, on the frequency of the causal mutation, and on the magnitude effect of the causal mutation. So far, it is not known whether alleles influencing common human diseases and complex traits are rare, and whether alleles have a high-magnitude effect *(12)*. It has been argued that susceptibility alleles for monogenic or multi-factorial diseases have been subjected to purifying selection. In this way, monogenic diseases are largely caused by rare alleles that disrupt a protein-coding sequence and that have a high-magnitude effect. There is, nevertheless, an exception to this pattern, when heterozygous carriers have a selective advantage. For example, homozygosity for Hemoglobin S that causes sickle-cell anemia is highly deleterious, and heterozygosity for Hemoglobin S protects against malaria *(13)*; as a result, Hemoglobin S is frequent in Africa. Besides, common diseases may be partly due to mutations having more subtle effects and influencing homeostasis. In this way, Mitchison divided genes into extrovert genes, concerned with the outside world, and introvert genes, concerned with cellular homeostasis *(14)*. He suggested that variation in extrovert genes occurs more frequently in coding regions and that variation in introvert genes occurs in regulatory regions. He further suggested that genetic variation in introvert genes has mild effects and that such alleles may be common. In addition, alleles that cause common diseases occurring after the age of reproduction are unlikely to be subject to negative selection. Although our knowledge of the variations explaining common diseases is very limited, it can be assumed that some alleles causing such diseases may be common and that such alleles may have a low-magnitude effect.

The power of association tests depends also on the linkage disequilibrium (LD) between the causal mutation and the genetic marker tested. If the polymorphism tested is the causal genetic variant, there is no loss of power. If the polymorphism tested is in linkage disequilibrium with the causal genetic variant, the genotyped polymorphism can be significantly associated with the disease. Nevertheless, there is a loss of power related to the strength of the LD. Association studies are actually based on the association between allelic variants at neighboring loci, that is, LD between them. The strength of LD depends both on the history of haplotypes and on the physical distance.

Lewontin's LD coefficient D is: $D_{ij} = p_{ij} - p_i p_j$
where p_i, p_j and p_{ij} are the frequency of the allele at locus i, the frequency of the allele at locus j, and the frequency of variants associated, respectively.

A standardized value of D is $D' = D/D_{max}$
where D_{max} depends on p_i and p_j. D' ranges from 0 to 1.

The r^2 coefficient is another popular LD coefficient for selecting informative SNPs. r^2 is the standard χ^2 statistic divided by the number of chromosomes in the sample. It ranges from 0 to 1. When r^2 is 1, SNPs are in complete LD.

$$r^2 = \frac{(p11p22 - p12p21)^2}{p1p2q1q2}$$

where p11, p22, p12, and p21 are two-locus haplotype frequencies, and p1, p2, q1, and q2 are allele frequencies.

D' and r^2 are useful in genetic epidemiology. D' is used in the HapMap project *(15)* to describe haplotype blocks. Simulation studies have suggested that D' better localizes the true locus than r^2 *(16)*. Nevertheless, D' can be inflated in small samples, and values of $r^2 < 1$ are more interpretable than values of $D' < 1$. In addition, the amount of information provided by one locus about another is well reflected by r^2. In other words, r^2 is related to the power to detect a disease allele through studying neighboring SNPs and is therefore a guide in association studies *(17,18)*.

1.4. Combined Linkage and Association Analyses: Family-Based Association Tests

1.4.1. The Transmission/Disequilibrium Test

The transmission/disequilibrium test (TDT) has been introduced by Spielman et al. *(19)*. The test that has been proposed for binary traits compares the allele transmitted by a heterozygous parent to the affected child with the non-transmitted allele.

Considering the family trio containing a homozygous father ($M_1 M_1$), a heterozygous mother ($M_1 M_2$), and a heterozygous affected child ($M_1 M_2$), the mother yields one observation in the 2 × 2 contingency table below in the cell (M_2 transmitted, M_1 non-transmitted). **Table 1** summarizes the number of

Table 1
Transmitted and Non-Transmitted Alleles M1 and M2 Among 2n Parents of n Affected Children

Transmitted allele	Non-transmitted allele		
	M1	M2	Total
M_1	a	b	a + b
M_2	c	d	c + d
Total	a + c	b + d	2n

times an allele is transmitted from $2n$ parents to n affected children. The null hypothesis is H_0: no linkage and/or no association. Thus, the rejection of H_0 provides evidence of linkage and association. If the null hypothesis is true, the probability of transmitting M_2 will be 0.5. Only data from heterozygous M_1M_2 parents should be used in the test. The statistic is given by

$$\chi^2 = \frac{(b-c)^2}{(b+c)}$$

which follows a χ^2 distribution with 1 degree of freedom. The test is a Mac Nemar test.

Nuclear families with multiple sibs can also be analyzed. However, the rejection of H_0 provides evidence of linkage. It is not a TDT.

A number of refinements have been proposed for LD-based mapping of binary and quantitative traits, and general approaches that accommodate multi-allelic markers and families of any size with or without parental information are available. Family-Based Association Tests (FBAT) and Quantitative trait Transmission/Disequilibrium Tests (QTDT) are two of the most popular packages used.

1.4.2. QTDT

Abecasis et al *(20)* proposed the following orthogonal model that is a generalization of the Fulker model *(21)* and that is used in a variance components framework.

$$\mu_{ij} = \mu + \beta_b b_i + \beta_w w_{ij}$$

where b_i and w_{ij} are orthogonal between and within family components of the genotype score g_{ij} for the jth offspring ($j = 1...n_i$) in the ith family

($i = 1\ldots K$). μ_{ij} is the expected value of the phenotype for individual j. β_b and β_w are the regression coefficients.

And for the offspring in each family i, the $n_i \times n_i$ variance-covariance matrix, Ω_i, which has the elements Ω_{ijk} fo each sib pair observation:

$$V_e + V_g + V_a \text{ if } j = k$$
$$\pi_{ijk} a + Vg \text{ if } j \neq k$$

π_{ijk} is the proportion of alleles shared identical by descent between sibs j and k.

$g_{ij} = m_{ij} - 1$, where m_{ij} is the number of "1" alleles at locus M.

$b_i = \Sigma_j g_{ij}/n_i$ if parental genotypes are unknown. b_i is thus the average of sibling genotype scores.

$b_i = (g_{iF} + g_{iM})/2$ if parental genotypes are known, where g_{iF} and g_{iM} are the genotype scores for the female and male parent, respectively. b_i then is the average of parental genotype scores.

$$w_{ij} = g_{ij} - b_i$$

Thus, b_i is the expectation of each g_{ij} conditional on family data, and w_{ij} is the deviation from this expectation for offspring j. Positive values of w_{ij} mean that a child inherits more copies of the allele than expected by chance given the family.

Covariates (x_{i1} and x_{i2}) that influence the phenotype, such as age or sex, can be added to the model:

$$\mu_{ij} = \mu + \beta_1 x_{ij1} + \beta_2 x_{ij2} + \beta_b b_i + \beta_w w_{ij}$$

The model should be used in a variance components framework, when the sibship size is higher than 1 (e.g., in the case of family of various sizes). In this context, the β_w regression coefficient is a direct estimate of the genetic effect at the marker M, whereas bias because of population substructure is accounted for by β_b. Additive and dominant effects can be evaluated. In addition, the model can estimate the association of multi-allelic markers with the phenotype.

The likelihood of the data for the complete set of parameters is

$$L = \Pi_i (2\pi)^{-ni/2} |\Omega_i|^{-1/2} e^{-1/2[(y_i - \mu_i)/\Omega_i^{-1}(y_i - \mu_i)]}$$

where y_i and μ_i are the observed phenotype and the expected phenotype, respectively.

Evidence of association can be evaluated by likelihood-ratio test with the constraint $\beta_w = 0$ (null hypothesis likelihood L_0) and without constraints on β_w (alternative hypothesis likelihood L_1). Asymptotically, the quantity $2(L_n L_1 - L_n L_0)$ is distributed as χ^2 with degrees of freedom equal to the difference in number of parameters estimated.

1.4.3. FBAT

The general FBAT statistic has been proposed *(22,23)*.

$$S = \Sigma_{ij} T_{ij} X_{ij}$$
$$E(S) = \Sigma_{ij} T_{ij} E(X_{ij}/P_{iF}, P_{iM})$$
$$V = \text{Var}(S) = \Sigma_{ij} T_{ij}^2 \text{Var}(X_{ij}/P_{iF}, P_{iM})$$

where X_{ij} and T_{ij} denote some function of the genotype of the jth offspring in the ith family and some function of a trait or phenotype, respectively. P_{iF} and P_{iM} denote parental genotypes, and $E(X_{ij}/P_{iF}, P_{iM})$ is the expected marker score under the null hypothesis.

The default null hypothesis tested by FBAT is H01: no linkage and no association. To avoid biases because of population stratification, mis-specification of the trait distribution, or selection based on the phenotype, the distribution of S is calculated using the distribution of offspring genotypes, conditional on the trait T, and on the parental genotype. $E(S)$ and V are calculated under the null hypothesis conditioned on the parental genotypes (treating X_{ij} as random and T_{ij} as fixed). If X_{ij} is a scalar summary of an individual genotype, then

$$Z = \frac{[S - E(S)]}{V^{1/2}} \approx N(0,1)$$

FBAT gives the value of Z and a two-sided p-value.
When X_{ij} is a vector, then

$$G = [S - E(S)]^T V^{-} [S - E(S)]$$

has an approximate χ^2 distribution with degree of freedom equal to the rank of V. V^{-} denotes the inverse of V. FBAT gives the value of χ^2 and a one-sided p-value.

Alternatively, the distribution of S under the null hypothesis can be computed by using a Monte Carlo permutation procedure. The procedure is recommended when the number of informative families is small.

An alternative null hypothesis H02 can be specified: linkage but no association. In this case, the genotypes of sibs or half-sibs will be correlated. FBAT will compute an empirical variance without making assumptions about the recombination parameter or degree of correlation between sibs or half-sibs in a family.

The main issues to be addressed in using FBAT are how to specify X_{ij} and how to specify T_{ij}. The specification of X_{ij} is determined by the choice of the genetic model (additive, dominant, or recessive) and the choice of the association analysis mode (single or multiple testing). Choosing the wrong model does not invalidate the test, but it may reduce its power. If the marker has more than two alleles, each allele can be tested separately against all others, or all alleles can be simultaneously compared. In the former case, X_{ij} is a scalar summary, and in the later case, X_{ij} is a vector of n alleles ($X_{ij1}, \ldots X_{ijk}, \ldots X_{ijn}$).

FBAT can handle any kinds of trait. When the trait is dichotomous, one can use

$$T = Y - \mu$$

where Y is a dichotomous indicator (affected or non-affected), and μ is the prevalence. Choosing $0 < \mu < 1$ allows both affected and non-affected individuals to contribute to the statistics. It should be noted that a classical TDT *(19)* can be achieved by setting T= 1 for affected subjects and T= 0 for unaffected subjects.

When the trait is quantitative, T should be centered:

$$T = Y - \mu$$

where Y is a quantitative phenotype, and μ is the mean phenotype in the population.

If covariates influencing Y are known, Y can be adjusted for the covariates by using an appropriate regression *(24)*. Multiple linear regression and multiple logistic regression can be used for quantitative and dichotomous traits, respectively. In these cases, T will be the residual of the regression model. This procedure minimizes the variance and increases the power of the test.

In addition, FBAT can handle multiple correlated traits available on each sib *(25)*. In this case, T_{ij} is a vector of traits ($T_{ij1}, \ldots T_{ijk}, \ldots T_{ijn}$). If traits show highly skewed distributions, the use of rank or transforming data to normalize scores is recommended.

FBAT can also handle multiple markers, with a vector X_{ij} of K markers $(X_{ij1}, \ldots X_{ijK})$. Besides, FBAT constructs haplotypes from multiple tightly linked markers and computes an haplotype version of FBAT statistic *(26)*.

2. Materials

2.1. Searching for Known Polymorphisms: Databases and Bioinformatics Tools

Searching information in databases can be done on a computer equipped with Sun Solaris, windows XP, or Mac OS connected to an Internet service provider and running an internet browser.

2.2. Programs for Family-Based Association Tests

FBAT, QTDT package (pedstats, prelude, finale, qtdt), GENEHUNTER gh2, and MERLIN can run on a Sun workstation ULTRA10 equipped with the Sun OS. FBAT, QTDT, and MERLIN can also be run on a computer equipped with the windows XP operating system or Mac OS X system.

Programs and documentation are available at
http://www.fhcrc.org/science/labs/kruglyak/Downloads/index.html
http://www.sph.umich.edu/csg/abecasis/QTDT/
http://www.sph.umich.edu/csg/abecasis/Merlin/
http://www.biostat.harvard.edu/~fbat/fbat.htm

3. Methods

3.1. Searching for Known Polymorphisms

3.1.1. Visualizing and Exporting SNP Data from ENSEMBL

1. Enter the ENSEMBL website by pointing at http://www.ensembl.org. Select the genome of interest.
2. Enter the gene name. Select the gene of interest by clicking on the gene Id.
3. Click on the "gene variation info" button. This allows to visualize the SNPs and to obtain a list of known SNP (*see* **Fig. 1**) (*see* **Note 1**).
4. Further information is obtained by clicking on the SNP Id. If available, genotype frequencies are given. To explore LD patterns in populations, click on "Links to LDview per population." This returns a graphical display of D' and r^2. The size of the window, which is initially 100 kb, can be modified by clicking on the "zoom" button.
5. Gene variation information can also be exported by using the Biomart tool. Click on the "export SNPinfo in the region." You are presented with a few options: you can select region information, gene and SNP attributes, and the output file format (html, text, or excel) (*see* **Note 1**).

ID	Type	Chr: bp	Alleles	Ambiguity	AA change	AA co-ordinate	class	Source	status
rs760361	5PRIME_UTR	9: 119506503	G/A	R	-	-	snp	dbSNP	-
rs2737192	5PRIME_UTR	9: 119506515	G/A	R	-	-	snp	dbSNP	freq
rs760363	5PRIME_UTR	9: 119506517	G/A	R	-	-	snp	dbSNP	-
rs2737193	5PRIME_UTR	9: 119506549	G/A	R	-	-	snp	dbSNP	-
rs2737194	5PRIME_UTR	9: 119506552	G/C	S	-	-	snp	dbSNP	-
rs5900307	INTRONIC	9: 119506753	A/-	-	-	-	deletion	dbSNP	-
rs11536871	INTRONIC	9: 119510319	A/C	M	-	-	snp	dbSNP	freq
rs11536872	INTRONIC	9: 119510636	C/G	S	-	-	snp	dbSNP	-
rs2770148	INTRONIC	9: 119510879	C/T	Y	-	-	snp	dbSNP	freq
rs5030643	INTRONIC	9: 119514477	T/-	-	-	-	deletion	dbSNP	-
rs5030710	5PRIME_UTR	9: 119514542	T/C	Y	-	-	snp	dbSNP	cluster, freq, doublehit
rs16906079	5PRIME_UTR	9: 119514750	A/G	R	-	-	snp	dbSNP	freq
rs4986790	NON_SYNONYMOUS_CODING	9: 119515123	A/G	R	D/G	99 (2)	snp	dbSNP	cluster, freq
rs2770145	NON_SYNONYMOUS_CODING	9: 119515145	A/C	M	C/W	106 (3)	snp	dbSNP	cluster
rs2770144	NON_SYNONYMOUS_CODING	9: 119515156	A/C	M	V/G	110 (2)	snp	dbSNP	cluster
rs5031050	NON_SYNONYMOUS_CODING	9: 119515252	T/A	W	F/Y	142 (2)	snp	dbSNP	freq, hapmap
rs11536884	NON_SYNONYMOUS_CODING	9: 119515382	G/T	K	L/F	185 (3)	snp	dbSNP	-
rs4986791	NON_SYNONYMOUS_CODING	9: 119515423	C/T	Y	T/I	199 (2)	snp	dbSNP	cluster, freq
rs4987233	NON_SYNONYMOUS_CODING	9: 119515426	G/A	R	S/N	200 (2)	snp	dbSNP	freq, hapmap
rs5030716	NON_SYNONYMOUS_CODING	9: 119515556	C/A/T	H	F/L	243 (3)	snp	dbSNP	cluster
rs5030718	NON_SYNONYMOUS_CODING	9: 119515647	G/A	R	E/K	274 (1)	snp	dbSNP	cluster, freq
rs5030719	NON_SYNONYMOUS_CODING	9: 119515757	G/T	K	Q/H	310 (3)	snp	dbSNP	cluster, freq, hapmap
rs5030721	SYNONYMOUS_CODING	9: 119516186	G/A	R	K	453 (3)	snp	dbSNP	cluster
rs11536885	3PRIME_UTR	9: 119516903	A/C	M	-	-	snp	dbSNP	-
rs11536886	DOWNSTREAM	9: 119517150	C/T	Y	-	-	snp	dbSNP	-

Fig. 1. ENSEMBL result screen. ENSEMBL is a non-specialized tool that provides a list of SNPs in an intuitive way. For each SNP, identifier (rs number), type, location, nucleotide change, and consequences are given. A graphical representation of genes and SNPs is also displayed.

3.1.2. Visualizing an Annotated Sequence Containing SNPs with SNPper

1. Enter the SNP per website by pointing at http://snpper.chip.org/. Although registration is suggested, free guest access is available.
2. At the SNP per homepage, click on the link "GeneFinder." Then, enter the gene symbol.
3. To display the annotated DNA sequence, click on the "annotated" button (*see* **Fig. 2**). **Figure 3A** shows the result obtained (*see* **Note 2**).
4. To display the annotated protein sequence, click on the "protein" button (*see* **Fig. 2**). **Figure 3B** shows the result obtained (*see* **Note 2**).
5. To display a graphical view of a gene with SNPs, click on the "SNP graph" button (*see* **Note 2**).

3.1.3. Selecting SNPs with PupasSNP and Pupasview

1. Enter the PupaSNP website by pointing at http://pupasnp.bioinfo.ochoa.fib.es/
2. Enter the gene name as en external Id. You obtain SNPs in the genomic region, SNPs up to 4000 bp upstream, SNPs located at intron boundaries, SNPs located at exonic splicing silencers, and coding SNPs located at interpro domains (*see* **Note 3**).
3. Click on the "Pupasview" button to send the data to the Pupasview tool. This allows one to select SNPs on the following criteria: (1) the validation status, (2) the

Genetic Epidemiology and Natural Polymorphisms

Gene: TLR4				
Name:	toll-like receptor 4 isoform D		XmlXport	
Sequence:	Fasta - Annotated - Genbank - Protein		Strand:	+
Transcript Position:	chr9:117,546,138-117,556,781 (9q33.1)		Length:	10644 bp
Coding Sequence Position:	chr9:117,554,561-117,556,480		Length:	640 aa
Look up this gene in:				
Genbank (refSeq):	(unknown)	Genbank (mRNA): NM_138557	Genbank (prot):	NP_612567
Entrez:	TLR4	LocusLink: 7099	Unigene:	Hs.174312
PubMed:	TLR4	OMIM: 603030	Homologene:	41317
Ensembl:	TLR4	SwissProt: O00206	GeneCards:	TLR4
GeneOntology:	-- Biological Process -- (Select)		ASAP:	Hs.174312

Gene layout:		
	Position	Length
Exon #1	117546138 - 117546397	260
Intron #1	117546398 - 117554220	7823
Exon #2	117554221 - 117556781	2561
XmlXport	Exons total:	2821

Known SNPs:	
SNPset: SS793	
Source:	Gene TLR4
Created on:	05/30/2006 05:41:00
SNPs:	157 (avg dist: 182)
Export:	SNPset data Flanking sequences XmlXport
Commands:	Refine this SNPset SNP graph View all 157 SNPs...
(157 SNPs)	

Fig. 2. SNPper result screen. A query for the gene symbol "TLR4" returns the result screen above. To continue, one may click on the links "Annotated" and "Protein" in sequence row to obtain annotated coding and non-coding sequences. One may also refine the SNPset and export it by clicking on the links "SNPsetdata" and "Refine this SNPset", respectively. A graphical representation of genes and SNPs can be obtained by clicking on the link "SNP graph".

location, (3) the frequency in various populations, and (4) the functional properties (*see* **Fig. 4**).

4. To visualize the selected SNPs, click on the "run" button." As the viewer has been constructed using ENSEMBL APIs, Pupasview and ENSEMBL views are very similar. **Figure 5** shows the result obtained.
5. To obtain a table containing selected SNPs and available information, click on the "get list of SNPs" button.

3.1.4. Exploring SNP Data in a Database Focused on Genes Involved in Innate Immunity

1. Enter the IIPGA website by pointing at http://innateimmunity.net/
2. Enter or select the gene name. You obtain a gene card containing general information (*see* **Fig. 6**) and links to pages with raw data obtained by the IIPGA consortium, allele and genotype frequencies, haplotype, LD coefficients (D' and r^2), and case–control association study data (*see* **Note 4**).
3. Click on the "SNPper analysis" button to visualize annotated sequence (*see* **Figs 2** and **3**).

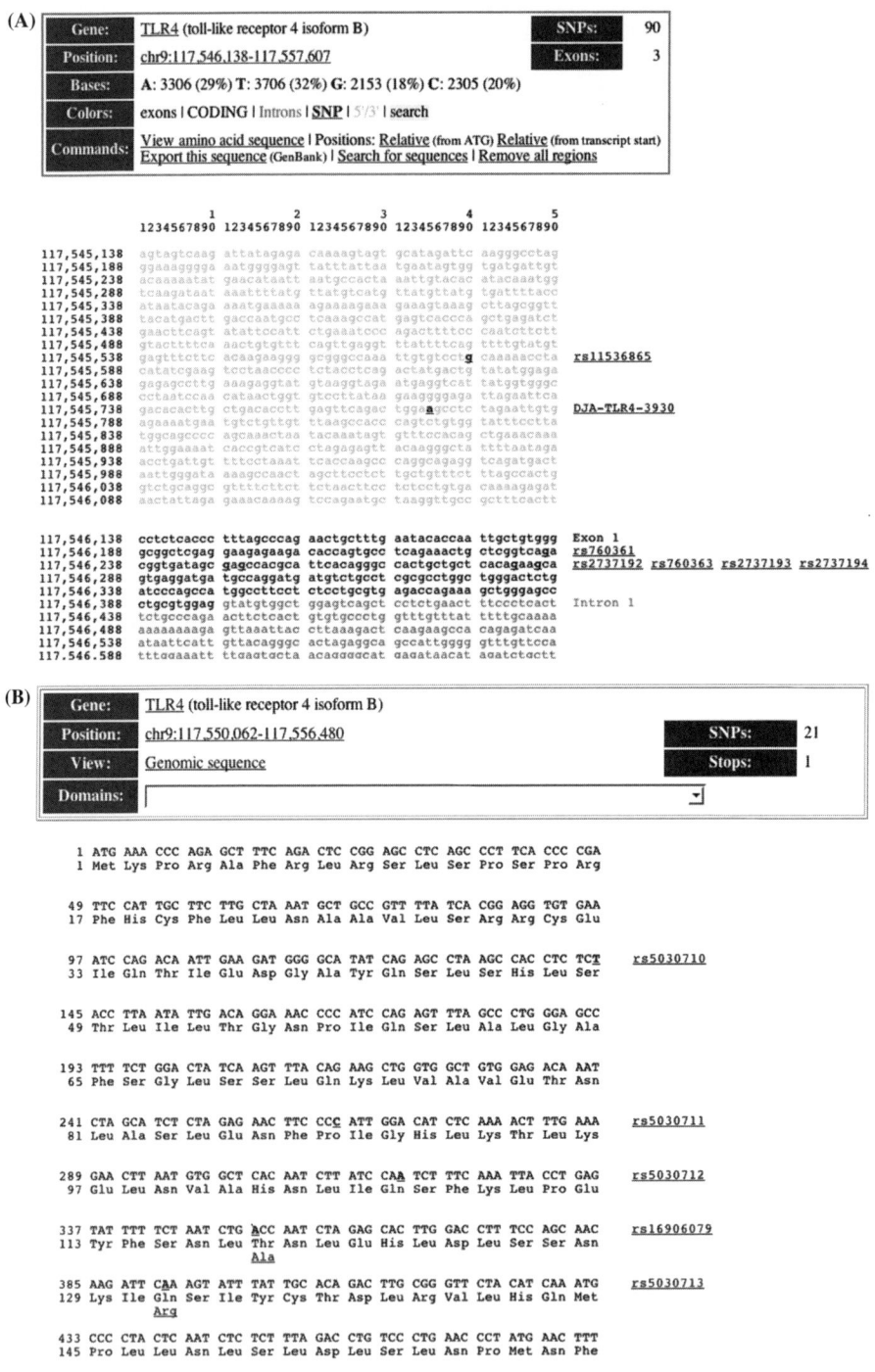

Genetic Epidemiology and Natural Polymorphisms

-- Validation Status (dbSNP) --	-- Type --
☑ by-frequency ☑ by-2hit-2allele ☑ by-cluster ☑ by-other-pop ☑ no-info	☑ coding ☑ intron ☑ utr ☑ local

-- Frequency & Population --

Filter Minor Allele Frequency from `0` to `0.5`

☑ Europe   ☑ Central/South America   ☑ Central Asia   ☑ Unknown
☑ Europe, Multi-National   ☑ North/East Africa & Middle East   ☑ East Asia   ☑ HAPMAP
☑ Europe, North America   ☑ Central/South Africa   ☑ Pacific
☑ North America   ☑ West Africa   ☑ Multi-National

-- Functional Properties --

- Non-synonymous SNPs
  - ● All non-syn mutations
  - ○ Only Pathological non-syn mutations

- Triplex
  - Minimum length of triplex sequences `10` bp
  - ● All regions
  - ○ Only Rattus Norvegicus conserved regions
  - ○ Only Rattus Norvegicus high conserved regions

- TFBS (10000 bp upstream)
  - ● All regions
  - ○ Only Rattus Norvegicus conserved regions
  - ○ Only Rattus Norvegicus high conserved regions

- Intron Boundaries

- Exonic Splicing Enhancer
  - ● All regions
  - ○ Only Rattus Norvegicus conserved regions
  - ○ Only Rattus Norvegicus high conserved regions

- Rattus Norvegicus conserved regions
  - BLASTZ NET (cons)

-- Block finding algorithms --
- ● Confidence Intervals (Gabriel et al.)
- ○ Four Gamete Rule (Wang et al.)
- ○ Solid Spine of LD

Fig. 4. Pupasview search screen. One may enter identifiers (gene symbol or ENSEMBL identifier) of any gene into the text field. SNPs can be filtered on the basis of their validation status, their frequencies, and their functional properties.

4. Click on the "Phase Output" button to obtain haplotypes reconstructed from experimental data.
5. Click on the "African American" or the "European American" button to visualize LD in the population of interest.

Fig. 3. **(A)** SNPper-annotated genomic sequence screen. SNPs are located in the sequence and their identifiers are stated on the right. **(B)** SNPper-annotated coding sequence screen. DNA and protein sequences are displayed. Amino acid changes are specified. Non-synonymous and synonymous SNPs are located on the sequence, and their identifiers are stated on the right. One may obtain additional information by clicking on the SNP identifier.

3.2. Testing Family-Based Association

3.2.1. QTDT

3.2.1.1. INPUT FILES

QTDT requires matched data, pedigrees, and IBD information files for variance component analyses (*see* **Notes 5** and **6**).

1. The first five columns of the pedigree file describe family relationships. The other columns include marker genotypes, covariate, and phenotype values (additional data). All data are separated by a tab. Each line in the pedigree file has the following format:

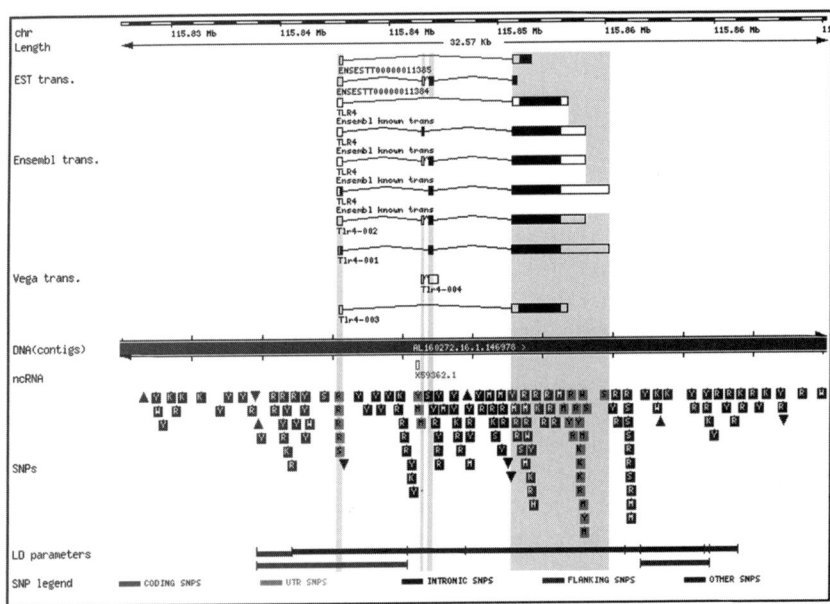

Fig. 5. Pupasview result screen. The viewer has been constructed using ENSEMBL API. A query for the gene symbol "TLR4" returns the result screens above without filtering SNPs (**A**) and after selecting non-synonymous SNPs with minor frequency higher than 0.05 (**B**). A list of SNPs can be obtained by clicking on the button "get list of SNPs".

Genetic Epidemiology and Natural Polymorphisms

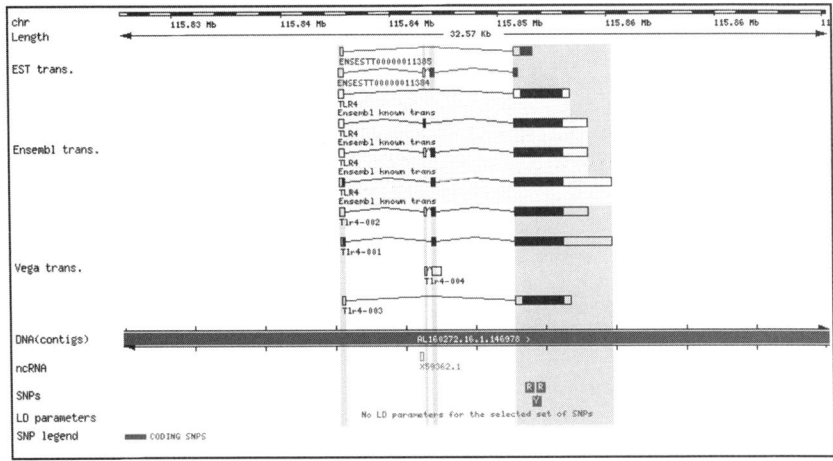

Fig. 5. *(Continued)*

```
<family identifier> <personnal id> <father id> <mother id>
<sex> <additional data>
```

The datafile describes the additional data of the pedigree file. For example, a data file contains

M Marker_1
M Marker_2
C Covariate
T Trait
A Affection status

In the pedigree file, marker genotypes (M) are encoded as two integers, separated by a tab, space, or a "/" (forward slash); missing values can be encoded as 0. Quantitative phenotypes (T) and covariates (C) are encoded as numbers; missing values should be encoded with an appropriate text label such as –99.999. Affection status (A) denotes a binary phenotype that will be ignored by QTDT.

2. To prepare IBD files, you may use simwalk2 or genehunter2. Prelude and finale provide a convenient interface between QTDT and these programs.

To set up a genehunter2 run for the sibs data set, run

```
> prelude -d file.dat -p file.ped -aa -t 0.001
```

You can specify that you want use all the individuals (`-aa`) or only founders (`-af`) to calculate allelic frequencies. The `-t` option specificies the default recombination fraction between markers. A map file (`-m`) in MERLIN format can also be used. Map positions are assumed to be in Haldane centiMorgans.

A map file has the following format

```
<chromosome> <marker> <map position>
```

To create the IBD file using genehunter2, run

gh<genehunter.in

and

>finale genehunter.in

(A)

TLR4

The following information is based on the unmasked version of the consensus sequence. We have also generated data for the **masked** version of the assembly. There is also an **Introduction** available if you are looking for a place to get started.

	Information
Name	toll-like receptor 4 isoform B
Source	InnateImmunity
Chromosome	chr9 (+) (chr9:117546138-117557607)
Accession	NM_138556
SNPs	44
Indels	0
Populations	2
Subjects	0
Links	[SNPper] [GoldenPath] [Gene Image] [LocusLink] [Omim] [PubMed]
Biological Significance	TLR4 is a critical component of the heteromeric receptor complex that transduces signals delivered by lipopolysaccharide (LPS) of Gram-negative bacteria. CD14, a molecule selectively expressed by monocytes and granulocytes, an d MD-2 are also involved in LPS-mediated signaling. C3H/HeJ mice which bear mutations in the **Tlr4** gene exhibit defective LPS signaling, and **TLR4** knockout mice are selectively impaired in their ability to recognize Gram-negative bacteria. In humans, **TLR4** mutations have been found to be associated with endotoxin hyporesponsiveness. Recognition of viral products has also been proposed to be mediated by **TLR4**, and studies in animal models point to a major role of **TLR4** in the response to respiratory syncytial virus infection. (See Omim for more ...)

Fig. 6. IIGA result screen. Gene information is viewed (**A**). The links provide access to raw data, allele information, genotypes, haplotypes, and linkage disequilibrium (**B**). One may display an annotated sequence by clicking on the links "SNPper Analysis" (*see* **Fig. 3**).

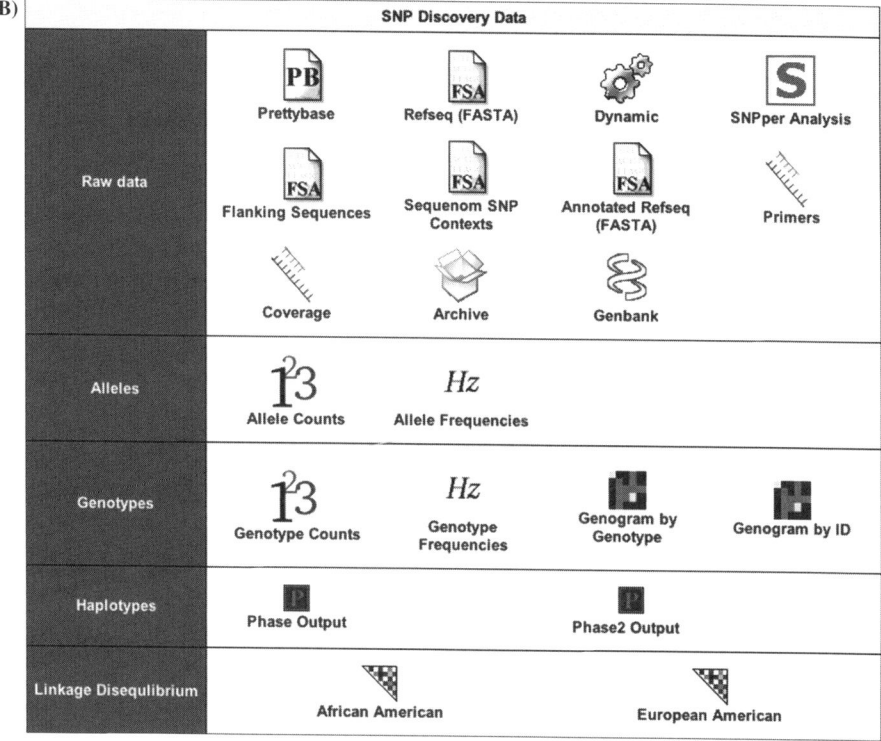

Fig. 6. *(Continued)*

finale collects ibd estimates and writes the IBD file to qtdt.ibd.

Alternatively, the IBD file can be obtained by using MERLIN by typing

```
merlin -d file.dat -p file.ped -m file.map -markernames-ibd
```

The format for the IBD file is

```
<family> <person1> <person2> <marker> <z0> <z1> <z2>
```

where <z0>, <z1>, and <z2> denote the probability that <person1> and <person2> in <family> share 0, 1, or 2 alleles IBD at the marker locus <marker>.

3. One can check data, pedigree, and IBD files by running

```
> pedstats -d file.dat -p file.ped -i file.ibd
```

The output describes the family composition and specifies the number of individuals and founders genotyped and heterozygosity. It also gives the mean and variance of each trait, the number of individuals and founders phenotyped, and the number of individuals for whom IBD probabilities have been calculated.

3.2.1.2. VARIANCE COMPONENT MODELS OF ASSOCIATION BY RUNNING QTDT

1. One can estimate the significance of individual components of variance, such as Vg (genetic variance). For example, the heritability of the trait can be estimated by running

```
> qtdt -d file.dat -p file.ped -a- -we -veg
```

To disable the association model, you must use −a− option. To specify the alternative variance models of interest, you must use −we −veg options. Variance components are specified by −w option. −d and −p describe the input files (data file and ped file).

Summary outputs appear on screen

```
The following models will be evaluated... _
NULL MODEL
Means = Mu
Variances = Ve
FULL MODEL
Means = Mu
Variances = Ve + Vg
Testing trait: trait1
Allele df(0) -LnLk(0) df(V) -LnLk(V) ChiSq p
N/A 220 114.21 219 108.60 11.22 0.0008 (222 probands)
```

The results provide evidence of a significant Vg component. Additional information that includes variance component estimates is placed in the regress.tbl file. Estimates for the means and variances appear in the same order as in the model description of the QTDT output. For instance, the estimates for Ve (environmental variance) and for Vg (genetic variance) were 0.882 and 0.411, respectively. The heritability was, therefore, $0.411/(0.411 + 0.882) = 0.317$. In other words, the results suggest that 31.7% of the phenotypic variance was explained by genetic variation.

Linkage can be estimated by running

```
> qtdt -d file.dat -p file.ped -a- -weg -vega
```

Environmental and polygenic variances are included in the default model (-weg). Environmental, polygenic, and additive major locus variances are included in the tested model (-vega).

Summary outputs appear on screen

```
The following models will be evaluated... _
NULL MODEL
Means = Mu
Variances = Ve + Vg
FULL MODEL
Means = Mu
Variances = Ve + Vg + Va
Testing trait: trait1
Allele df(0) -LnLk(0) df(V) -LnLk(V) ChiSq p
All 219 108.60 218 104.71 7.78 0.005 (222 probands)
```

The results provide evidence of linkage.

2. To evaluate association in the presence of linkage, run the following command:

```
> qtdt -d file.dat -p file.ped -ao -wega
```

This specifies that environmental, polygenic, and additive major locus variances are included in the model (-wega) and that the orthogonal association is used (-ao) (*see* **Notes 7** and **8**).

Summary outputs appear on screen.

```
The following models will be evaluated...
NULL MODEL
Means = Mu + B
Variances = Ve + Vg + Va
FULL MODEL
Means = Mu + B + W
Variances = Ve + Vg + Va
Testing trait: trait1
Allele df(0) -LnLk(0) df(V) -LnLk(V) ChiSq p
All 182 93.18 181 86.58 13.21 0.0002 (120/189 probands)
```

There is evidence of association. The estimates of Ve, Vg, and Va, which were given in regress.tbl, were 0.733, 0.104, and 0.105, respectively. This means that the heritability because of the locus was $0.105/(0.733 + 0.104 + 0.105) = 0.111$.

3. Violation of phenotypic multivariate normality assumption warrants a Monte Carlo permutation test (*see* **Note 6**). Empirical *p*-values can be calculated using a Monte Carlo permutation framework by running

> qtdt -d file.dat -p file.ped -ao -m9999 -wega

This specifies that 9999 permutations will be done.

4. To evaluate whether a candidate marker could be the disease mutation (*see* **Note 8**), you can model linkage and association simultaneously by running

> qtdt -d file.dat -p file.ped -ao -weg -vega

Summary outputs appear on screen

```
The following models will be evaluated... _
NULL MODEL
Means = Mu + B + W
Variances = Ve + Vg
FULL MODEL
Means = Mu + B + W
Variances = Ve + Vg + Va
Testing trait: trait1
Allele df(0) -LnLk(0) df(V) -LnLk(V) ChiSq p
All 182 86.76 181 86.58 0.37 (120/189 probands)
```

In the example, there is no evidence of linkage after modeling association. Therefore, the possible linkage at the locus was accounted for by association. This suggests that the polymorphism is the causal mutation or that the polymorphism is in strong linkage disequilibrium with it.

5. Population-based association can also be evaluated by running

> qtdt -d file.dat -p file.ped -at -wega

Summary outputs appear on screen

```
The following models will be evaluated... _
NULL MODEL
Means = Mu
Variances = Ve + Vg + Va
FULL MODEL
Means = Mu + X
Variances = Ve + Vg + Va
Testing trait: ACE
Allele df(0) -LnLk(0) df(V) -LnLk(V) ChiSq p
All 185 97.58 184 91.66 11.84 0.0005 (189 probands)
```

This further provides evidence of association between the locus and the trait. It should be stressed that the test is no longer a TDT and that it is sensitive to population stratification.

6. To evaluate whether a candidate marker could be the disease mutation, linkage after modeling association can be tested by running

```
> qtdt -d file.dat -p file.ped -at -weg -vega
```

Summary outputs appear on screen

```
The following models will be evaluated... _
NULL MODEL
Means = Mu + X
Variances = Ve + Vg
FULL MODEL
Means = Mu + X
Variances = Ve + Vg + Va
Testing trait: ACE
Allele df(0) -LnLk(0) df(V) -LnLk(V) ChiSq p
All 185 91.95 184 91.66 0.58     (189 probands)
```

After modeling association, there was no evidence of linkage, suggesting that the polymorphism is in strong linkage disequilibrium with the causal mutation.

3.2.2. FBAT

3.2.2.1. INPUT FILES

Matched pedigree and phenotype files are needed (*see* **Notes 5** and **6**). The pedigree file describes family relationships and affection status. Lines in the pedigree file have the following format:

```
First line:
Name of the markers
Other lines:
<family identifier> <personnal id> <father id> <mother id>
<sex> <status> <marker genotype >
```

Missing genotype data are recoded "0 0". Affected and unaffected individuals are recoded 2 and 1, respectively. Individuals with unknown affection status are recoded 0.

The phenotype file contains phenotype values.

```
First line:
Name of the phenotypes
Other lines:
<family identifier> <personnal id> <phenotype 1> <phenotype
```

```
2>…<phenotype N>
Missing phenotype data are recoded ''-''.
```

All data are separated by space.

3.2.2.2. GENETIC ASSOCIATION BY RUNNING FBAT

1. The program is interactively run. After opening FBAT, one needs to load the pedigree file by writing

    ```
    >> load file.ped
    ```

    In the presence of Mendelian inconsistencies, the output returns

    ```
    read in: 5 markers from 35 pedigrees (35 nuclear
    families,223 persons)
    mendelian error: SNP1, pedigree 1 [1,2]
    A total of 1 mendelian error has been found
    genotypes of families with mendelian error have been
    reset to 0
    ```

2. To set trait values for affected, unaffected, and subjects with unknown status, write

    ```
    >> setafftrait aff_t unaff_t unknown_t
    ```

    When the default values (1,0,0) are used, only affected sibs are included in the analysis.

    Alternatively, you may choose an offset, such as the disease prevalence. For example, with $\mu = 0.5$, affected subjects will have $T_{ij} = 0.5$ and unaffected subjects will have $T_{ij} = -0.5$. You may also choose $\mu = n_{aff}/(n_{aff} + n_{unaff})$, where n_{aff} is the number of transmissions to affected subjects and n_{unaff} the number of transmissions to unaffected subjects. If the ascertainment does not depend on the phenotype, this is approximately the sample prevalence, and this offset minimizes Var(S).

    To select an offset, type

    ```
    >>offset [offset_value]
    ```

3. To select the genetic model (a: additive, d: dominant, r: recessive), type

    ```
    >>model a (or d or r)
    ```

4. To select the biallelic (b), multiallelic (m), or both testing procedure (a), type

    ```
    >>mode b (or m or a)
    ```

5. To evaluate linkage and association between the binary trait and all the markers, run the following command

   `>> fbat`

   Summary outputs appear on screen

   ```
   trait affection; offset 0.300; model additive; test bi-
   allelic; minsize 10; p 1.000
   Marker Allele afreq fam# S E(S) Var(S) Z P
   SNP1 1 0.565 25 70.000 63.750 16.188 1.553 0.120322
   SNP1   2 0.435 25 36.000 42.250 16.188 -1.553 0.120322
   SNP2 1 0.860 10 29.000 28.000 5.000 0.447 0.654721
   SNP2 2 0.140 10 9.000 10.000 5.000 -0.447 0.654721
   SNP3 1 0.761 18 56.000 51.000 12.500 1.414 0.157299
   SNP3 2 0.239 18 20.000 25.000 12.500 -1.414 0.157299
   SNP4 1 0.754 18 55.000 49.500 12.750 1.540 0.123485
   SNP4 2 0.246 18 21.000 26.500 12.750 -1.540 0.123485
   SNP5 1 0.571 22 49.000 47.667 15.222 0.342 0.732544
   SNP5 2 0.429 22 37.000 38.333 15.222 -0.342 0.732544
   ```

   For bi-allelic codings, genotype information is a scalar, and the FBAT statistic is based on a Z score. For multi-allelic codings, genotype information X_{ij} is a vector, and FBAT is based on a χ^2 statistic. On the basis of the results, we could not reject the null hypothesis "no linkage and/or no association."

6. To evaluate association in the presence of known linkage between the binary trait and all the markers, you should use an empirical variance (*see* **Notes 8** and **9**). To this aim, run the following command

   `>> fbat -e`

   Summary outputs appear on screen

   ```
   trait affection; offset 0.300; model additive; test bi-
   allelic; minsize 10; p 1.000
   Marker Allele afreq fam# S E(S) Var(S) Z P
   SNP1    1 0.565 19 56.000 49.750 14.062 1.667 0.095581
   SNP1    2 0.435 19 26.000 32.250 14.062 -1.667 0.095581
   SNP2    1 0.860 8 *****
   SNP2    2 0.140 8 *****
   SNP3 1 0.761 14 48.000 43.000 8.500 1.715 0.086348
   SNP3 2 0.239 14 14.000 19.000 8.500 -1.715 0.086348
   SNP4    1 0.754 15 49.000 43.500 8.750 1.859 0.062979
   SNP4    2 0.246 15 15.000 20.500 8.750 -1.859 0.062979
   ```

```
SNP5    1 0.571 13 30.000 28.667 9.444 0.434 0.664389
SNP5    2 0.429 13 20.000 21.333 9.444 -0.434 0.664389
```

No significant association in the presence of linkage was found.

7. To evaluate the association of haplotypes with the trait (*see* **Notes 8** and **9**), type

```
>>hbat
```

In this case, all markers are used to compute haplotype. To select markers, you must provide the name of the markers by typing

```
>>hbat SNP2 SNP3 SNP4 SNP5
```

Summary outputs appear on screen

```
trait affection; offset 0.300; model additive; test multi-
allelic; minsize 10; p 1.000
haplotype analysis for the following markers:
SNP1 SNP3 SNP4 SNP5
haplotypes and EM estimates of frequency:
h1 : 1 1 1 1 0.385
h2 : 1 1 1 2 0.204
h3 : 2 2 2 1 0.164
h4 : 2 1 1 1 0.076
h5 : 2 2 2 2 0.074
h6 : 2 1 1 2 0.073
h7 : 2 1 2 2 0.012
h8 : 1 1 2 1 0.008
h9 : 2 1 2 1 0.004
Allele afreq fam# S E(S) Var(S) Z P Offset
-----------------------------------------------------------
h1 0.385 22.0 16.914 14.727 8.143 0.766 0.443400 0.300000
h2 0.204 18.0 11.486 9.503 6.227 0.795 0.426853 0.300000
h3 0.164 15.1 2.252 3.776 4.621 -0.709 0.478404 0.300000
h4 0.076 8.0 *****
h5 0.074 9.5 *****
h6 0.073 7.3 *****
h7 0.012 1.0 *****
h8 0.008 1.0 *****
h9 0.004 0.0 *****
Allele# DF CHISQ P Offset
-----------------------------------------------------
9 3 1.841 0.606058 0.300000
```

When each haplotype is separately analyzed, genotype information is a scalar, and the FBAT statistic is based on a Z score. When all haplotypes are simultaneously analyzed by using the multi-allelic mode, genotype information is a vector, and FBAT performs a χ^2 statistic. Neither the former procedure nor the later one yielded significant results. Both procedures test linkage and association. You may also test linkage in the presence of association by typing "≫hbat −e".

8. To compute an empirical *p*-value using Monte Carlo samples from the null distribution, type

```
>>hbat -p SNP1 SNP3 SNP4 SNP5
```

Summary outputs appear on screen

```
trait affection; offset 0.300; model additive; test bi-
allelic; minsize 10; p 1.000
haplotype permutation test for the following markers:
SNP1 SNP3 SNP4 SNP5
Permutation cycles = 3044
haplotype afreq fam# rank(S_obs) P_2side
h1 : 1 1 1 1 0.385 24.0 676.0 0.444152
h2 : 1 1 1 2 0.204 19.0 670.0 0.440210
h3 : 2 2 2 1 0.164 16.1 2308.0 0.483574
h4 : 2 1 1 1 0.076 8.0 1853.0 0.782523
h5 : 2 2 2 2 0.074 9.5 2525.0 0.340999
h6 : 2 1 1 2 0.073 8.3 1502.5 0.987188
h7 : 2 1 2 2 0.012 1.0 2944.0 0.065703
h8 : 1 1 2 1 0.008 1.0 936.0 0.614980
h9 : 2 1 2 1 0.004 *** too few informative obs ***
whole marker 1763.0 0.579172
```

Empirical *p*-value for each haplotype and empirical *p*-value for the whole dataset are given. The results showed a trend for haplotype h8, and the statistic for the whole dataset was not significant. Therefore, no significant association was found.

9. Empirical *p*-values are provided for each haplotype analyzed and for the whole marker permutation test (*see* **Notes 6** and **8**). Interestingly, empirical *p*-values can also be obtained for a single marker by typing

```
>>hbat -p [marker]
```

10. To load the quantitative phenotype file, type

```
>>load file.phe
```

and the output returns

```
7 quantitative traits have been successfully read
222 persons have been phenotyped
```

11. Then, you must select the trait you want to analyze by typing

    ```
    >>trait phenotype3
    ```

    and the output returns

    ```
    affection phenotype1 phenotype2 phenotype3** phenotype4
    phenotype5 phenotype6 phenotype7
    ```

12. Type "fbat", "fbat –e", or "hbat" to analyze the association of the quantitative trait with all the markers. The procedures used for binary and quantitative traits are the same. The model of genetic models, the mode of testing procedure, and the offset value can be selected. If the quantitative trait is centered or adjusted for covariates or normalized and standardized, setting the trait offset is, nevertheless, useless.
13. To select several traits, type

    ```
    >>trait phenotype1 phenotype2 phenotype3 phenotype4
    phenotype5
    ```

    and the output returns

    ```
    affection phenotype1** phenotype2** phenotype3**
    phenotype4** phenotype5** phenotype6 phenotype7
    ```

    In this case, trait information T_{ij} is a vector, and FBAT is based on a χ^2 statistic. You can carry out association analyses by typing "fbat". If necessary, an offset vector can be set to mean-center the phenotype vector using the trait sample means $\mu = (\mu_1, \ldots \mu_K)$, with μ_i the mean of the ith trait.

    After typing "fbat", the output returns

    ```
    5 traits selected: phenotype1 phenotype2 phenotype3
    phenotype4 phenotype5
    model dominant; test multi-allelic; minsize 5; p 1.000000
    Marker Allele# DF CHISQ P
    SNP1 2   10 5.428321 0.860793
    SNP2 2    5 2.820312 0.727665
    SNP3 2   10 6.860566 0.738540
    SNP4 2   10 5.494947 0.855763
    SNP5 2   10 10.016937 0.439009
    ```

    No significant association was found.

4. Notes

1. ENSEMBL provides a huge amount of genomic information, which includes SNPs (*see* **Subheading 3.1.1.**). Interestingly, it provides SNPs annotation from the HGVBASE database (http://hgvbase.cgb.ki.se/), which is not provided by SNPper or Pupasview. SNP functional information is, nevertheless, limited.
2. The main data source used by SNPper is dbSNP (http://www.ncbi.nlm.nih.gov/projects/SNP/). SNPper is the most convenient tool to visualize SNP in genomic and coding sequence (*see* **Subheading 3.1.2.**). In addition, it provides tools for extracting and exporting SNP data and links to PCR primer design programs. Except for SNP in coding sequences, SNPper does not evaluate the potential functional effect of SNPs, such as on gene expression; this tool might, nevertheless, be incorporated in the future.
3. The main data source used by Pupasview is also dbSNP. Pupasview evaluates the potential functional effect of SNPs in coding and non-coding sequences and allows to select SNPs on this basis (*see* **Subheading 3.1.3.**). In addition, it provides LD coefficients between SNPs, when available. When using warehouse databases, some SNPs available at the dbSNP website may be lacking. Checking this point is recommended, when the study focuses on a gene of major interest.
4. The IIGA database provides an accurate source of information on innate immunity genes. In particular, one can obtain haplotypes and LD between SNPs within the genes (*see* **Subheading 3.1.4.**). This allows one to select a combination of SNPs with high genetic information content. The number of genes included is, however, low. Hence, the use of other databases will be required. These include general databases, such as SNPper or Pupasnp, or specific databases, such as KIR databases (http://www.allelefrequencies.net/ and http://www.ebi.ac.uk/ipd/kir/index.html).
5. It is recommended to check for Mendelian errors at a single locus before evaluating genetic association (*see* **Subheadings 3.2.1.1. and 3.2.2.1.**). Checking Mendelian transmission is required before running QTDT. This can be done by using FBAT. In addition, checking for Mendelian errors may be performed by using multi-locus procedure. The aim is to look for tight double recombinants: when very close SNPs are typed, it is expected that recombination events are rare, and it is likely that a double recombinant represents a genotyping error. The procedure can be carried out by using MERLIN.
6. Exploring phenotypic distribution is recommended (*see* **Subheadings 3.2.1.1. and 3.2.2.1.**). Tests of normality, such as Kolmorov–Smirnov test, can be carried out by using classical statistical programs. Transformation data, such as log or square root, may be necessary to make the distribution normal. The orthogonal model used by QTDT is sensitive to trait distribution; violation of multivariate normality assumption warrants a Monte Carlo permutation test. Phenotypic distribution affects the power of FBAT statistics. The PBAT program may be used to rank or to compute normal scores (http://www.biostat.harvard.edu/~fbat/pbat.htm). FBAT can also compute a Monte Carlo permutation test.

7. All the options are not described here, and it is recommended to read carefully the user manual for the program being used (*see* **Subheading 3.2.1.2.**). A dominant option (by typing "−dominance"), and other variance components representing common environment shared by related individuals (by typing "−wegan" or "−wegac") are available by using the QTDT package. Population stratification can be evaluated by using the "−ap" option; it is recommended when population-based association tests ("−at" option) that are not TDT are being used. Some options are set as defaults: pedigree, data, and ibd files are called qtdt.ped, qtdt.dat, and qtdt.ibd, respectively. The association model used is the orthogonal model if it is not specified.
8. Comparison of QTDT with FBAT. The methods are based on different statistics, and using both methods is informative. Nevertheless, QTDT well manages quantitative traits, whereas FBAT analyzes binary, quantitative, and also censored traits. Association in the presence of linkage is tested by both packages (*see* **Subheadings 3.2.1.2. and 3.2.2.2.**). Permutation procedure and scoring allelic transmission in extended pedigrees are implemented in both packages. QTDT analyzes linkage after modeling association and, therefore, tests whether a candidate marker could be the disease mutation. Unlike QTDT, FBAT directly computes association test between haplotypes and phenotypes. Unlike FBAT, QTDT manages covariates.
9. It is also recommended to read carefully user manual for the program being used. New options have been introduced in the last version, FBAT 1.7.1 (*see* **Subheading 3.2.2.2.**). Multimarker tests are performed by typing "fbat −m" or "fbat −m −e" without resolving phase. These procedures test the null hypotheses H01 (no linkage or association between any marker and phenotype) and H02 (no linkage in the presence of association between any marker and phenotype), respectively. The linear combination test, which is another multimarker test, is performed by typing "fbat −l" and forms a linear combination of the individual Z statistics. Typing "hapfreq −d" estimates haplotype frequencies and Lewontin's D and D' LD coefficients.

References

1. Lazarus, R., Vercelli, D., Palmer, L. J., et al. (2002) Single nucleotide polymorphisms in innate immunity genes: abundant variation and potential role in complex human disease. *Immunol Rev* **190,** 9–25.
2. Turvey, S. E., and Hawn, T. R. (2006) Towards subtlety: understanding the role of Toll-like receptor signaling in susceptibility to human infections. *Clin Immunol* **120,** 1–9.
3. Yang, I. A., Fong, K. M., Holgate, S. T., and Holloway, J. W. (2006) The role of Toll-like receptors and related receptors of the innate immune system in asthma. *Curr Opin Allergy Clin Immunol* **6,** 23–28.
4. Parham, P. (2005) MHC class I molecules and KIRs in human history, health and survival. *Nat Rev Immunol* **5,** 201–214.

5. Sobrino, B., Brion, M., and Carracedo, A. (2005) SNPs in forensic genetics: a review on SNP typing methodologies. *Forensic Sci Int* **154**, 181–194.
6. Birney, E., Andrews, D., Caccamo, M., et al. (2006) Ensembl 2006. *Nucleic Acids Res* **34**, D556–561.
7. Riva, A., and Kohane, I. S. (2002) SNPper: retrieval and analysis of human SNPs. *Bioinformatics* **18**, 1681–1685.
8. Riva, A., and Kohane, I. S. (2004) A SNP-centric database for the investigation of the human genome. *BMC Bioinformatics* **5**, 33.
9. Conde, L., Vaquerizas, J. M., Santoyo, J., et al. (2004) PupaSNP Finder: a web tool for finding SNPs with putative effect at transcriptional level. *Nucleic Acids Res* **32**, W242–248.
10. Conde, L., Vaquerizas, J. M., Ferrer-Costa, C., de la Cruz, X., Orozco, M., and Dopazo, J. (2005) PupasView: a visual tool for selecting suitable SNPs, with putative pathological effect in genes, for genotyping purposes. *Nucleic Acids Res* **33**, W501–505.
11. Kwok, P. (2003) *Single Nucleotide Polymorphisms: Methods and Protocols.* Humana Press, Totowa, NJ.
12. Smith, D. J., and Lusis, A. J. (2002) The allelic structure of common disease. *Hum Mol Genet* **11**, 2455–2461.
13. Kwiatkowski, D. (2000) Genetic susceptibility to malaria getting complex. *Curr Opin Genet Dev* **10**, 320–324.
14. Mitchison, A. (1997) Partitioning of genetic variation between regulatory and coding gene segments: the predominance of software variation in genes encoding introvert proteins. *Immunogenetics* **46**, 46–52.
15. (2003) The International HapMap Project. *Nature* **426**, 789–796.
16. Devlin, B., and Risch, N. (1995) A comparison of linkage disequilibrium measures for fine-scale mapping. *Genomics* **29**, 311–322.
17. Pritchard, J. K., and Przeworski, M. (2001) Linkage disequilibrium in humans: models and data. *Am J Hum Genet* **69**, 1–14.
18. Ardlie, K. G., Kruglyak, L., and Seielstad, M. (2002) Patterns of linkage disequilibrium in the human genome. *Nat Rev Genet* **3**, 299–309.
19. Spielman, R. S., McGinnis, R. E., and Ewens, W. J. (1993) Transmission test for linkage disequilibrium: the insulin gene region and insulin-dependent diabetes mellitus (IDDM). *Am J Hum Genet* **52**, 506–516.
20. Abecasis, G. R., Cardon, L. R., and Cookson, W. O. (2000) A general test of association for quantitative traits in nuclear families. *Am J Hum Genet* **66**, 279–292.
21. Fulker, D. W., Cherny, S. S., Sham, P. C., and Hewitt, J. K. (1999) Combined linkage and association sib-pair analysis for quantitative traits. *Am J Hum Genet* **64**, 259–267.
22. Laird, N. M., Horvath, S., and Xu, X. (2000) Implementing a unified approach to family-based tests of association. *Genet Epidemiol* **19 (Suppl 1)**, S36–42.

23. Lake, S. L., Blacker, D., and Laird, N. M. (2000) Family-based tests of association in the presence of linkage. *Am J Hum Genet* **67,** 1515–1525.
24. Lunetta, K. L., Faraone, S. V., Biederman, J., and Laird, N. M. (2000) Family-based tests of association and linkage that use unaffected sibs, covariates, and interactions. *Am J Hum Genet* **66,** 605–614.
25. Lange, C., Silverman, E. K., Xu, X., Weiss, S. T., and Laird, N. M. (2003) A multivariate family-based association test using generalized estimating equations: FBAT-GEE. *Biostatistics* **4,** 195–206.
26. Horvath, S., Xu, X., Lake, S. L., Silverman, E. K., Weiss, S. T., and Laird, N. M. (2004) Family-based tests for associating haplotypes with general phenotype data: application to asthma genetics. *Genet Epidemiol* **26,** 61–69.

3

KIR Locus Polymorphisms
Genotyping and Disease Association Analysis

Maureen P. Martin and Mary Carrington

Summary

The genes encoding the killer immunoglobulin-like receptors (*KIR*) are situated within a segment of DNA that has undergone expansion and contraction over time due in large part to unequal crossing over. Consequently, individuals exhibit considerable haplotypic variation in terms of gene content. The highly polymorphic human leukocyte antigen (HLA) class I loci encode ligands for the KIR; thus, it is not surprising that *KIR* genes also show significant allelic polymorphism. As a result of the receptor–ligand relationship between KIR and HLA, functionally relevant *KIR–HLA* combinations need to be considered in the analysis of these genes as they relate to disease outcomes. This chapter will describe a genotyping method for identifying the presence/absence of the *KIR* genes and general approaches to data analysis in disease association studies.

Key Words: *KIR* genotyping; sequence-specific priming (SSP); natural killer (NK) cells; HLA class I ligands; *KIR* haplotypes.

1. Introduction

Natural killer (NK) cells comprise about 15% of peripheral blood lymphocytes and are an important component of the innate immune system. They also play a critical role in bridging the innate and adaptive immune response to infection by the production of cytokines and chemokines that mediate activation of effector cells of the adaptive immune system *(1,2)*. NK cell activity is controlled by the balance of inhibitory and stimulatory signals generated upon ligand binding to a plethora of receptors located on their cell surfaces. The killer immunoglobulin-like receptor (*KIR*) locus comprises a family of related genes that encode a group of these activating and inhibitory receptors expressed on NK cells. Recognition of self-human leukocyte antigen (HLA) class I ligands by inhibitory KIR allows NK cells to identify and inhibit the response to normal cells *(3–5)*. On the contrary, activation of NK cells through stimulatory receptors, including some KIR molecules, is directed toward cells with little or no expression of major histocompatibility complex (MHC) class I, consistent with the so-called "missing self" hypothesis, which was originally proposed by Ljunggren and Karre *(6)*. A contemporary version of the "missing self" hypothesis proposes that, rather than completely terminating NK cell effector function, inhibitory receptors might instead dampen it depending on the amount of MHC class I expression. Thus, when multiple activating receptors are engaged, the signal might be sufficiently potent to overcome inhibition. In other words, NK cell cytotoxicity is essentially controlled by a balance between activating and inhibitory signals mediated by corresponding NK cell receptors *(7)*.

The *KIR* locus maps to chromosome 19q13.4 within the leukocyte receptor complex (*LRC*) and comprises a family of polymorphic and highly homologous genes that are tandemly arrayed over about 150 kb. On the basis of gene content, the haplotypes have been divided into two types termed A and B *(8)* (*see* **Fig. 1**). Haplotype A is invariant in terms of gene content, whereas haplotype B is quite variable, due in large part to the presence or absence of activating *KIR*, because most of the inhibitory *KIR* are present on nearly all haplotypes *(8)*.

Given the role of *KIR* in the immune response and their extensive genomic diversity, it is conceivable that *KIR* gene variation affects resistance and susceptibility to the pathogenesis of a number of diseases, and as such, they have become an attractive target for disease association studies. This chapter will describe a method for *KIR* genotyping, which detects the presence or absence of each gene in a given individual, thus providing a *KIR* profile. The method involves gene-specific polymerase chain reaction (PCR) amplification using sequence-specific primers (PCR-SSP). A general approach to the analysis of *KIR* data in disease association studies is also provided.

2. Materials
2.1. PCR
1. High-quality DNA isolated from tissue or cells using standard protocols, at a concentration of 50–100 ng/µl (*see* **Notes 1–2**).

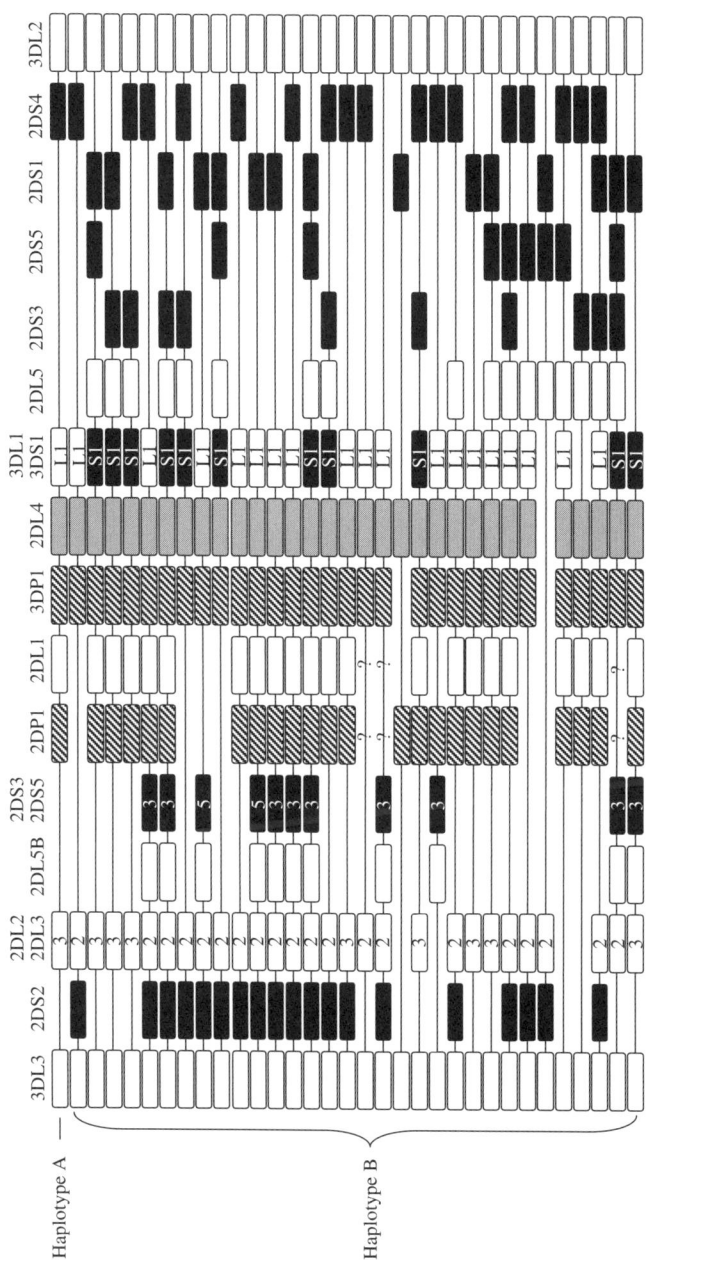

Fig. 1. *KIR* haplotypes identified by segregation analysis. Haplotypes are derived from **refs. 26–30** and unpublished observations (M.C.). Haplotypes A and B are labeled as such. White boxes represent inhibitory receptors, black boxes represent activating receptors, and hatched boxes represent pseudogenes. *KIR2DL4* is shown as grey boxes. This receptor has features of both inhibitory and activating receptors, although functional data suggest that it is primarily activating *(31,32)*.

2. Oligonucleotide primer mixes containing forward and reverse primers (5 µM) specific for each *KIR* gene and internal control (*see* **Table 1** for primer sequences) (*see* **Note 3**).
3. 50 mM $MgCl_2$.
4. 25 mM dNTP mix (mix equal quantities of 100 mM each dATP, dTTP, dGTP, and dCTP).
5. 10× PCR buffer: 200 mM Tris–HCl, pH 8.4, 500 mM KCl.
6. *Taq* DNA polymerase (*see* **Note 4**).
7. 384-well PCR plates. Each sample will use 30 wells, thus allowing typing of 12 samples per plate.
8. Multichannel pipettors (8 and 16 channel).

2.2. Gel Electrophoresis

1. Orange G loading buffer: 0.5% Orange G, 20% Ficoll, 100 mM ethylenediaminetetraacetic acid (EDTA).
2. 10× TAE electrophoresis buffer: 400 mM Tris, 200 mM acetic acid, 10 mM EDTA.
3. 100 bp DNA ladder.
4. Electrophoresis grade agarose.
5. Horizontal electrophoresis chamber.
6. Gel-casting tray and 50-well combs with teeth appropriately separated for use with multichannel pipettors. Note that the Owl Centipede™ horizontal model D3-14 wide gel system (Woburn, MA) will accommodate four combs per gel, allowing electrophoresis of 200 wells per gel.
7. Ethidium bromide solution (10 mg/ml).
8. High-voltage power supply.
9. Photographic means of gel documentation [ultraviolet (UV) light source, digital or conventional camera].

3. Methods
3.1. PCR

1. Using an 8-multichannel pipettor, dispense 1 µl of each primer mix into separate wells in 384-well plates. Each sample will require two vertical rows of wells, thus allowing genotyping of 12 samples per plate.
2. Prepare PCR cocktails for each sample in a total volume of 132 µl (enough for 33 reactions) as follows: 200 ng DNA, 16.5 µl 10× PCR buffer (final concentration 1×), 4.95 µl $MgCl_2$ (final concentration 1.5 mM), 1.32 µl dNTP (final concentration 200 mM), 0.825 µl *Taq* polymerase.
3. Add 4 µl PCR cocktail to each primer mix (total PCR volume = 5 µl).
4. Cover plates with acetate film and centrifuge briefly.
5. Amplify in a programmable thermal cycler with a heated lid using the following parameters: 3 min at 94°C, 5 cycles of 15 s at 94°C, 15 s at 65°C, 30 s at 72°C;

Table 1
KIR Genotyping Primers

Gene	Primers	Sequences	Exon	Position*	Size (bp)	Comments
2DL1	F1	GTTGGTCAGATGTCATGTTTGAA	4	437–459	146	
	R1	GGTCCCTGCCAGGTCTTGCG	4	563–582		
	F2	TGGACCAAGAGTCTGCAGGA	8	1125–1144	~330	Misses *005, Sees 2DL2*004
	R2	TGTTGTCTCCTAGAAGACG	3'UTR	1369–1388		
2DL2	F1	CTGGCCCACCCAGGTCG	4	385–401	173	Misses *004
	R1	GGACCGATGGAGAAGTTGGCT	4	537–557		
	F2	GAGGGGAGGCCCATGAAT	5	778–796	151	
	R2	TCGAGTTTGACCACTCGTAT	5	909–928		
2DL3	F1	CTTCATCGCTGGTGCTG	7	1084–1100	~550	Misses *007
	R1	AGGCTCTTGGTCCATTACAA	8	1119–1138		
	F2	TCCTTCATCGCTGGTGCTG	7	1082–1100	~800	
	R2	GGCAGGAGACAACTTTGGATCA	9	1316–1337		
2DL4	F1	CAGGACAAGCCCTTCTGC	3	82–99	254	
	R1	CTGGGTGCCGACCACT	3	320–335		
	F2	ACCTTCGCTTACAGCCCG	5	675–692	288	
	R2	CCTCACCTGTGACAGAAACAG	5	941-intron		
2DS2	F1	TTCTGCACAGAGAGGGGAAGTA	4	467–488	175	
	R1	GGGTCACTGGGAGCTGACAA	4	622–641		
	F2	CGGGCCCACGTTT	5	695–709	240	
	R2	GGTCACTCGAGTTTGACCACTCA	5	912–934		
2DS3	F1	TGGCCCACCCAGGTCG	4	386–401	242	
	R1	TGAAAACTGATAGGGGGAGTGAGG	4	604–627		

(*Continued*)

Table 1
(Continued)

Gene	Primers	Sequences	Exon	Position*	Size (bp)	Comments
	F2	CTATGACATGTACCATCTATCCAC	5	753–776	190	
	R2	AAGCAGTGGGTCACTTGAC	5	924–942		
2DS4	F1	CTGGCCCTCCCAGGTCA	4	385–401	204	
	R1	TCTGTAGGTTCCTGCAAGGACAG	4	566–588		
	F2	GTTCAGGCAGGAGAGAAT	5	706–723	197/219	
	R2	GTTTGACCACTCGTAGGGAGC	5	904–924		
2DS5	F1	TGATGGGGTCTCCAAGGG	4	522–539	126	Misses *003
	R1	TCCAGAGGGTCACTGGGC	4	630–647		
	F2	ACAGAGAGGGACGTTTAACC	4	473–493	178	
	R2	ATGTCCAGAGGGTCACTGGG	4	631–650		
3DL1	F1	CGCTGTGTGGTGCCTCGA	3	114–129	191	Misses *009
	R1	GGTGTGAACCCGACATG	3	287–304		
	F2	CCCTGGTGAAATCAGGAGAGAG	4	401–422	186	
	R2	TGTAGGTCCCTGCAAGGGCAA	4	566–586		
3DL2	F1	CAAACCCTTCCTGTCTGCCC	3	121–140	211	Misses *013, *014
	R1	GTGCCGACCACCCAGTGA	3	314–331		
	F2	CCCATGAACGTAGGCTCCG	5	788–806	130	
	R2	CACACGCAGGGCAGGG	5	902–917		
3DS1	F1	AGCCTGCAGGGAACAGAAG	3	1134–1154	~300	
	R1	GCCTGACTGTGGTGCTCG	3'UTR	1340–1357		
	F2	CCTGGTGAAATCAGGAGAGAG	4	402–422	180	
	R2	GTCCCTGCAAGGCAC	4	566–581		
3DL3	F1	GTCAGGACAAGCCCTTCCTC	3	441–461	232	
	R1	GAGTGTGGGTGTGAACTGCA	3	530–552		

	F2	TTCTGCACAGAGAGGGATCA	4	737–753	165
	R2	GAGCCGACAACTCATAGGGTA	4	910–926	
2DL5	F1	GCGCTGTGGTGCCTCG	3	113–128	214
	R1	GACCACTCAATGGGGAGC	3	308–326	
	F2	TGCAGCTCCAGGAGCTCA	5	736–753	191
	R2	GGGTCTGACCACTCATAGGGT	5	906–926	
2DP1	F1	GTCTGCCTGGCCCAGCT	3	99–115	205
	R1	GTGTGAACCCGACATCTGTAC	3	282–303	
	F2	CCATCGGTCCCATGATGG	4	548–565	89
	R2	CACTGGGAGCTGACAACTGATG	4	615–637	
2DS1	F1	CTTCTCCATCAGTCGCATGAA	4	543–563	102
	F2	CTTCTCCATCAGTCGCATGAG	4	543–563	
	R1	AGAGGGTCACTGGGAGCTGAC	4	624–644	
DRB1 (control) (21)	F1	TGCCAAGTGGAGCACCCAA	Intron 3		796
	R1	GCATCTTGCTCTGTGCAGAT	Intron 3		

*Numbering based on KIR alignment from http://www.ebi.ac.uk/ipd/kir/

21 cycles of 15 s at 94°C, 15 s at 60°C, 30 s at 72°C; 4 cycles of 15 s at 94°C, 1 min at 55°C, 2 min at 72°C, with a final 7-min extension step at 72°C (*see* **Notes 5 and 6**).

3.2. Gel Electrophoresis

1. Prepare 150 ml of 3% agarose in 1× TAE per gel and heat until the agarose has completely gone into solution.
2. Cool gel mixture to 65°C, add ~5 µl ethidium bromide, and gently mix to avoid bubble formation.
3. Pour gel mixture into the gel-casting tray, insert four 50-well combs, and allow the gel to solidify for 20 min.
4. Fill the electrophoresis chamber with 800 ml of 1× TAE buffer and submerge the gel into the chamber. Remove the combs.
5. Add 5 µl Orange G loading buffer to each PCR and centrifuge briefly.
6. Load 2 µl of the 100 bp DNA ladder to the first and last well of each row of wells.
7. Using a 16-channel pipettor, load 5 µl of each PCR product into the gel. Save the remaining PCR product from the wells containing primers for 2DS4 if subtyping of this gene is planned (see **Subheading 3.3**).
8. Electrophorese for 30 min at 100 V or until the Orange G has migrated 3 cm.
9. Visualize the gel using a UV light source and photograph the gel for a permanent record (*see* **Fig. 2A and B**).

3.3. KIR2DS4 Analysis

A variant of the *KIR2DS4* gene that has a 22-bp deletion in exon 5 can be detected by the following protocol. The primer pair in exon 5 amplifies a 219-bp product if the full-length gene is present and a 197-bp product if the deletion variant is present (*see* **Fig. 2C** and **Note 7**).

1. Prepare the gel as described in **Subheading 3.2.**, but use a single 50-well comb per gel.
2. Load 2 µl of the 100 bp DNA ladder to the first and last well in the row.
3. Load the remainder of the *KIR2DS4* PCR products (~5 µl) into the gel.
4. Electrophorese for 1 h at 100 V or until the Orange G has migrated ~7 cm. It is necessary to run the gel for a sufficient period of time to allow separation of the 197-bp and 219-bp products.
5. Visualize the gel using a UV light source and photograph the gel for a permanent record.

3.4. Interpretation of Results

Interpretation of the PCR-SSP results is relatively straightforward. Positive gene-specific amplifications are identified by PCR products of the correct sizes,

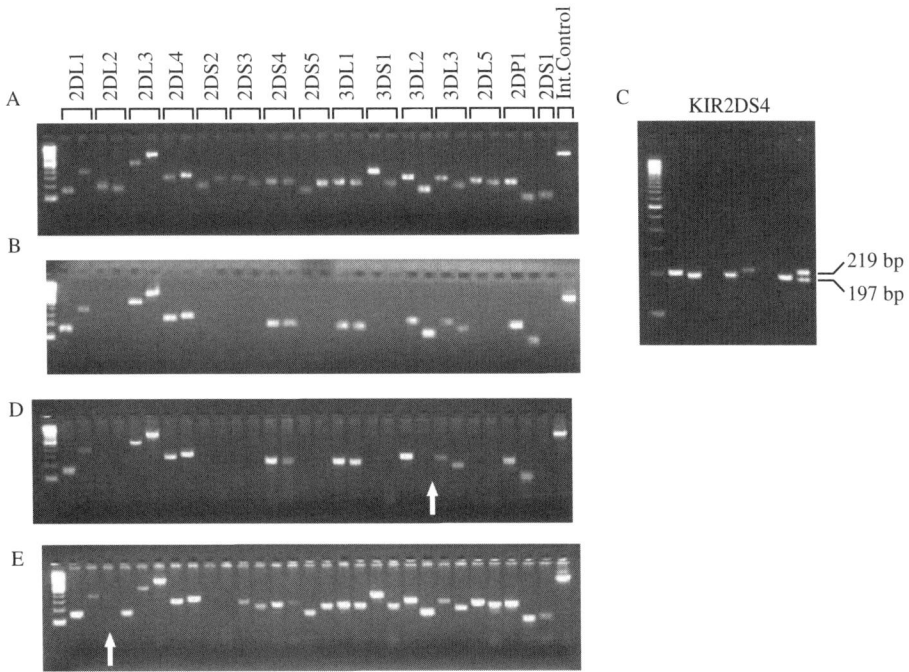

Fig. 2. *KIR* genotyping results by PCR-SSP. (**A**) Gel electrophoresis showing the presence of all genes tested. (**B**) Gel electrophoresis results from an individual homozygous for haplotype A. (**C**) Gel electrophoresis of *KIR2DS4* showing the 197 and 219 bp alleles. (**D**) A gel showing the absence of amplicon (arrow) for one of the two primer pairs for *KIR3DL2*. The PCR for *KIR3DL2* was repeated, indicating the presence of amplicon for both pairs of primers (i.e., one pair of primers did not amplify in the initial PCR due to technical error). (**E**) This gel shows the absence of amplicon for one of the primer pairs for *KIR2DL2* (arrow), which was confirmed after repeating the PCR. The sample was sequenced for *KIR2DL2* and was found to be allele *004, which is not amplified by this primer pair (*see* **Table 1**). The order of the genes for **B**, **D**, and **E** is as shown in **A**.

whereas absence of a product implies absence of the gene specific for the corresponding primer mix (*see* **Fig. 2** and **Notes 8–10**). There are two specific pairs of primers for each gene (except for *KIR2DS1*, *see* **Table 1**), and both primers should have a visible product if the gene is present or no product if the gene is absent (*see* **Notes 11 and 12**). Samples exhibiting discrepant results for primer pairs that recognize the same gene are repeated, and if discrepant results are confirmed, the gene is sequenced (*see* **Fig. 2D** and **E**). The internal control amplicon should be visible in all samples. The results are recorded in an appropriate spreadsheet, such as Microsoft Excel, indicating which genes are present and which genes are absent for each sample.

3.5. Data Analysis

Some general guidelines regarding analysis of *KIR* data in disease association studies are given in this section (*see* **ref. 9** for a recent review). There are no specific rules; rather, the type of analysis performed will depend on the questions being asked. HLA class I ligands for several of the inhibitory KIR molecules have been identified; so, an account of *KIR* genotypes in combination with *HLA* class I data is often necessary. One problem with studying polymorphic loci such as *KIR* and *HLA* is the inherent lack of statistical power, particularly when it is necessary to consider functionally relevant combinations of variants at the two unlinked gene complexes. In some cases, it is possible to take a specific hypothesis-driven approach in which only specific *KIR* genes are considered, as exemplified by our previous study of *KIR3DS1* effects on HIV disease progression *(10)*. This approach is reasonable when specific *KIR/HLA* associations can be addressed based on previously defined functional or genetic data, but it is often necessary to test all genes for any potential associations. Thus, it is recommended that testing for *KIR/HLA* effects be performed in study groups that are large enough to provide sufficient statistical power. An approach to disease association analysis is given below.

3.5.1. Disease Association Analysis

1. An obvious starting point is to test for disease associations with presence/absence of individual *KIR* genes. For example, this approach was used in the analysis of *KIR* in scleroderma *(11)*, where the combination of the presence of *2DS2* and the absence of *2DL2* was associated with susceptibility.
2. The effect of combinations of *KIR* genes with *HLA* alleles based on known receptor–ligand relationships should be tested. *KIR/HLA* associations have been described for a number of diseases such as psoriasis *(12)*, type 1 diabetes *(13)*, HIV-1 disease progression *(10)*, and rheumatoid vasculitis *(14)*.
3. Grouping of *KIR* genes based on functional characteristics, while also accounting for *HLA*, may provide additional information regarding the nature of the disease association. This process is exemplified in our study of psoriatic arthritis (PsA) *(15)*. This study also illustrates the challenges involved in the analysis of multiple linked and unlinked loci. Initially, we concluded that individuals with certain activating KIR were susceptible to developing PsA when HLA ligands for the corresponding homologous inhibitory receptors were missing. With our increased knowledge of KIR function in the regulation of NK cell activity, however, we were able to propose a more logical model to explain our results (*see* **Fig. 3**). In the old model, the assumption was made

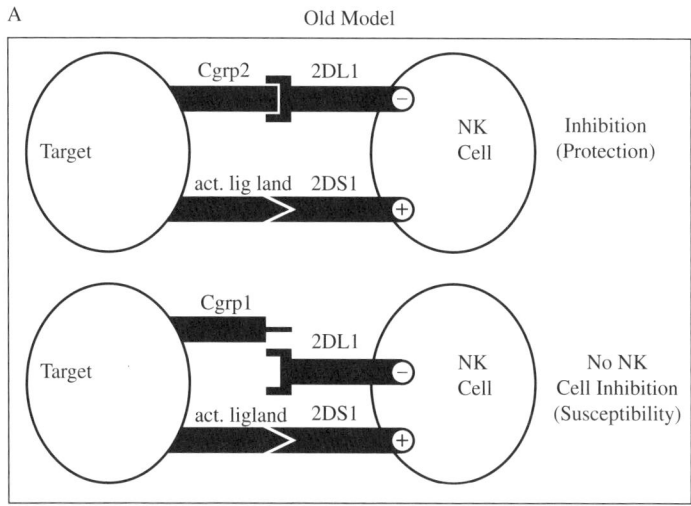
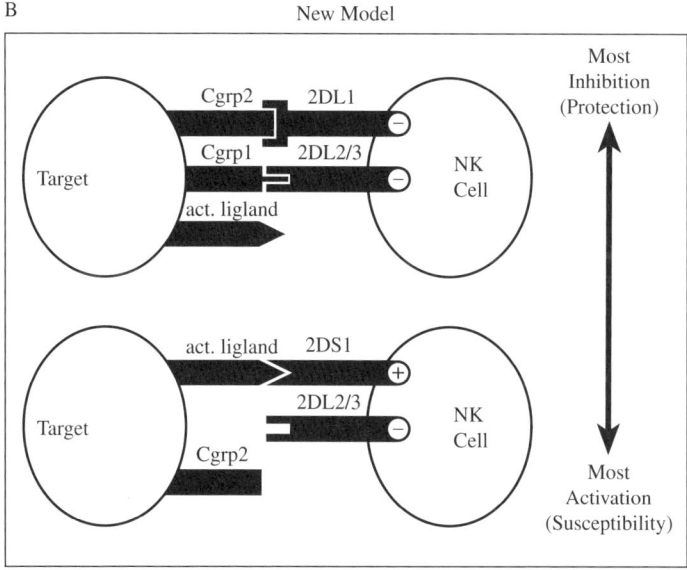

Fig. 3. Susceptibility to psoriatic arthritis. (**A**) Old model. NK cell activation (by KIR2DS1 in this case) is quenched when HLA ligand (HLA-C group 2) for the corresponding homologous inhibitory receptor (KIR2DL1) is present (protection), but not when ligand is absent (susceptibility). (**B**) New model. There is a trend in susceptibility to PsA such that genotypes conferring the most inhibition are protective, whereas those conferring the most activation are susceptible (adapted from **ref. *16***).

that only the corresponding inhibitory receptor could counteract the activity of the homologous activating receptor *(15)*. The new model on the contrary proposes that genotypes conferring the most inhibition are protective against PsA, whereas those conferring the most activation are susceptible *(16)* (*see* **Note 13** and **Fig. 3**).
4. *KIR* profile or haplotypic analyses might also be appropriate in some cases. One example is the analysis of *KIR* profiles with increasing numbers of activating *KIR*. We recently showed that there was a trend toward increasing susceptibility to nasopharyngeal carcinoma with ≥5 activating *KIR* *(17)*. Analysis of the effect of haplotypes A and B on disease outcomes might also be useful. Hiby et al. *(18)* reported that maternal homozygosity for haplotype A (which lacks most or all activating *KIR*) combined with the presence of *HLA-C* group 2 in the fetus was associated with increased risk of pre-eclampsia, a disorder attributed to poor placental perfusion. The authors suggested that the combination of fetal HLA-C group 2 and maternal Haplotype A results in strong NK cell inhibition, which impairs the remodeling of maternal blood vessels, a process which is in part due to NK cell activation.
5. Allelic analysis of specific genes can also be performed if allelic subtyping is available. This is particularly useful for loci that are highly polymorphic, such as the *KIR3DL1* and *KIR3DL2* loci. An allele of *KIR3DL2* was recently shown to be associated with an increased interaction of NK cells with red blood cells infected with the malaria parasite in vitro, and this interaction resulted in induction of interferon-γ synthesis *(19)*.

4. Notes

1. Good quality DNA is essential for reliable PCR-SSP results. Heparinized blood should be avoided because heparin is a PCR inhibitor *(20)*.
2. The protocol requires approximately 5 ng of DNA per specific primer reaction, hence, the requirement for a total of 150–200 ng DNA. If this quantity of DNA is not available, it is possible (although not desirable) to use less DNA. In such a case, the total number of cycles should be increased to 35 by increasing the second-cycle sequence from 21 to 25 cycles. One drawback to this protocol alteration is that increasing the number of cycles is also likely to increase the likelihood of false-positive reactions.
3. All primer mixes should be tested with positive and negative controls when new working dilution aliquots are made. In addition, careful attention should be given to the concentration of primers to ensure maximum amplification efficiency.
4. The use of Platinum® *Taq* DNA polymerase (Invitrogen, Carlsbad, CA) is recommended. This enzyme is complexed with a proprietary antibody that inhibits polymerase activity at ambient temperatures. Polymerase activity is regained after the initial denaturation step, thus providing an automatic "hot start" for the PCR and resulting in improved PCR efficiency and specificity.

5. A modified "stepdown" program is utilized to increase the specificity of the PCR *(21)*. It might be necessary to alter the PCR conditions slightly depending on the PCR machine being used. A fast ramping thermocycler with a heated cover is recommended for best results.
6. PCR machines should be checked regularly to ensure that all wells are amplifying uniformly. It is also important to ensure that the PCR plates fit snugly into the machine.
7. *KIR2DS4* is the only short-tailed receptor gene present on haplotype A, which is also the most common *KIR* haplotype (in terms of gene content) across most populations. The deletion allele is also quite common with a frequency of ~80% in Caucasians. Detection of the *KIR2DS4* deletion allele might be useful when performing data analysis, because individuals who are homozygous for haplotype A and the deletion allele have no expressed activating *KIR*, apart from *KIR2DL4*, a gene that may have tissue-specific function *(22)*. If activating *KIR* genes are beneficial against certain types of diseases, then individuals homozygous for the null *KIR2DS4* allele on an AA genotype background may be at increased risk for the development of those diseases.
8. To ensure the highest degree of accuracy for PCR-SSP typing, which in many cases involves using locus-specific primers that recognize a single base difference relative to the other loci, it is important to routinely check the *KIR* database (http://www.ebi.ac.uk/ipd/kir/) for new alleles. This helps to ensure that the primers being used for a gene will amplify all known alleles of that gene and will not amplify alleles of another gene. As new alleles are discovered, it may be necessary to repeat genotyping performed before the knowledge of the new alleles.
9. The *KIR* genes are highly (80–90%) similar and appear to have arisen by gene duplication from a common ancestor. Furthermore, the gene content and sequence of some *KIR* alleles strongly suggest that unequal crossing over in the region may account for a substantial amount of the diversity at this locus *(23–25)*. Therefore, the presence of a chimeric gene for example could result in an unusual PCR profile.
10. As a result of linkage disequilibrium, some genes are found together at much higher frequencies than expected, and this information can be used to validate the results obtained. **Figure 1** shows a list of haplotypes based on segregation analysis, and it is clear that there are certain combinations of genes that are almost always found together across different haplotypes, for example, *KIR2DL2* and *KIR2DS2*, *KIR2DP1* and *KIR2DL1*, and *KIR3DL1* and *KIR2DS4*.
11. As previously noted above, the *KIR* genes are highly similar, and this poses a challenge in designing primers that will amplify all alleles of a single locus specifically. As shown in **Table 1**, some primers miss one or two alleles of a gene and others cross-react with a non-target gene.
12. Most of the *KIR* gene sequences available have been obtained largely from Caucasians, and it is likely that as more populations are studied, the number of

known alleles for each gene will increase, some of which may be population-specific. Consequently, it is also likely that some of the primers currently being used for genotyping might not be appropriate for all populations.
13. It is essential to continually consider appropriate models when analyzing data pertaining to complex genetic loci such as *KIR* in human diseases. As our knowledge and understanding of KIR biology advances, modification of models proposed previously may be necessary.

Acknowledgments

The authors thank Arman Bashirova for helpful comments. This project has been funded in whole or in part with federal funds from the National Cancer Institute, National Institutes of Health, under contract N01-CO-12400. The content of this publication does not necessarily reflect the views or policies of the Department of Health and Human Services nor does mention of trade names, commercial products, or organizations imply endorsement by the US Government. This Research was supported by the Intramural Research Program of the NIH, National Cancer Institute, Center for Cancer Research.

References

1. Cooper, M. A., Fehniger, T. A., and Caligiuri, M. A. (2001) The biology of human natural killer-cell subsets. *Trends Immunol* **22,** 633–40.
2. Nguyen, K. B., Salazar-Mather, T. P., Dalod, M. Y., Van Deusen, J. B., Wei, X. Q., Liew, F. Y., Caligiuri, M. A., Durbin, J. E., and Biron, C. A. (2002) Coordinated and distinct roles for IFN-alpha beta, IL-12, and IL-15 regulation of NK cell responses to viral infection. *J Immunol* **169,** 4279–87.
3. Storkus, W. J., Alexander, J., Payne, J. A., Dawson, J. R., and Cresswell, P. (1989) Reversal of natural killing susceptibility in target cells expressing transfected class I HLA genes. *Proc Natl Acad Sci USA* **86,** 2361–4.
4. Karre, K., Ljunggren, H. G., Piontek, G., and Kiessling, R. (1986) Selective rejection of H-2-deficient lymphoma variants suggests alternative immune defence strategy. *Nature* **319,** 675–8.
5. Shimizu, Y., and DeMars, R. (1989) Demonstration by class I gene transfer that reduced susceptibility of human cells to natural killer cell-mediated lysis is inversely correlated with HLA class I antigen expression. *Eur J Immunol* **19,** 447–51.
6. Ljunggren, H. G., and Karre, K. (1990) In search of the 'missing self': MHC molecules and NK cell recognition. *Immunol Today* **11,** 237–44.
7. Lanier, L. L. (2005) NK cell recognition. *Annu Rev Immunol* **23,** 225–74.
8. Uhrberg, M., Valiante, N. M., Shum, B. P., Shilling, H. G., Lienert-Weidenbach, K., Corliss, B., Tyan, D., Lanier, L. L., and Parham, P. (1997) Human diversity in killer cell inhibitory receptor genes. *Immunity* **7,** 753–63.

9. Carrington, M., and Martin, M. P. (2006) The impact of variation at the KIR gene cluster on human disease. *Curr Top Microbiol Immunol* **298,** 225–57.
10. Martin, M. P., Gao, X., Lee, J. H., Nelson, G. W., Detels, R., Goedert, J. J., Buchbinder, S., Hoots, K., Vlahov, D., Trowsdale, J., Wilson, M., O'Brien, S. J., and Carrington, M. (2002) Epistatic interaction between KIR3DS1 and HLA-B delays the progression to AIDS. *Nat Genet* **31,** 429–34.
11. Momot, T., Koch, S., Hunzelmann, N., Krieg, T., Ulbricht, K., Schmidt, R. E., and Witte, T. (2004) Association of killer cell immunoglobulin-like receptors with scleroderma. *Arthritis Rheum* **50,** 1561–5.
12. Luszczek, W., Manczak, M., Cislo, M., Nockowski, P., Wisniewski, A., Jasek, M., and Kusnierczyk, P. (2004) Gene for the activating natural killer cell receptor, KIR2DS1, is associated with susceptibility to psoriasis vulgaris. *Hum Immunol* **65,** 758–66.
13. van der Slik, A. R., Koeleman, B. P., Verduijn, W., Bruining, G. J., Roep, B. O., and Giphart, M. J. (2003) KIR in type 1 diabetes: disparate distribution of activating and inhibitory natural killer cell receptors in patients versus HLA-matched control subjects. *Diabetes* **52,** 2639–42.
14. Yen, J. H., Moore, B. E., Nakajima, T., Scholl, D., Schaid, D. J., Weyand, C. M., and Goronzy, J. J. (2001) Major histocompatibility complex class I-recognizing receptors are disease risk genes in rheumatoid arthritis. *J Exp Med* **193,** 1159–67.
15. Martin, M. P., Nelson, G., Lee, J. H., Pellett, F., Gao, X., Wade, J., Wilson, M. J., Trowsdale, J., Gladman, D., and Carrington, M. (2002) Cutting edge: susceptibility to psoriatic arthritis: influence of activating killer Ig-like receptor genes in the absence of specific HLA-C alleles. *J Immunol* **169,** 2818–22.
16. Nelson, G. W., Martin, M. P., Gladman, D., Wade, J., Trowsdale, J., and Carrington, M. (2004) Cutting edge: heterozygote advantage in autoimmune disease: hierarchy of protection/susceptibility conferred by HLA and killer Ig-like receptor combinations in psoriatic arthritis. *J Immunol* **173,** 4273–6.
17. Butsch Kovacic, M., Martin, M., Gao, X., Fuksenko, T., Chen, C. J., Cheng, Y. J., Chen, J. Y., Apple, R., Hildesheim, A., and Carrington, M. (2005) Variation of the killer cell immunoglobulin-like receptors and HLA-C genes in nasopharyngeal carcinoma. *Cancer Epidemiol Biomarkers Prev* **14,** 2673–7.
18. Hiby, S. E., Walker, J. J., O'Shaughnessy K, M., Redman, C. W., Carrington, M., Trowsdale, J., and Moffett, A. (2004) Combinations of maternal KIR and fetal HLA-C genes influence the risk of preeclampsia and reproductive success. *J Exp Med* **200,** 957–65.
19. Artavanis-Tsakonas, K., Eleme, K., McQueen, K. L., Cheng, N. W., Parham, P., Davis, D. M., and Riley, E. M. (2003) Activation of a subset of human NK cells upon contact with Plasmodium falciparum-infected erythrocytes. *J Immunol* **171,** 5396–405.
20. Satsangi, J., Jewell, D. P., Welsh, K., Bunce, M., and Bell, J. I. (1994) Effect of heparin on polymerase chain reaction. *Lancet* **343,** 1509–10.

21. Bunce, M., O'Neill, C. M., Barnardo, M. C., Krausa, P., Browning, M. J., Morris, P. J., and Welsh, K. I. (1995) Phototyping: comprehensive DNA typing for HLA-A, B, C, DRB1, DRB3, DRB4, DRB5 & DQB1 by PCR with 144 primer mixes utilizing sequence-specific primers (PCR-SSP). *Tissue Antigens* **46**, 355–67.
22. Rajagopalan, S., Bryceson, Y. T., Kuppusamy, S. P., Geraghty, D. E., van der Meer, A., Joosten, I., and Long, E. O. (2006) Activation of NK cells by an endocytosed receptor for soluble HLA-G. *PLoS Biol* **4**, e9.
23. Gomez-Lozano, N., Estefania, E., Williams, F., Halfpenny, I., Middleton, D., Solis, R., and Vilches, C. (2005) The silent KIR3DP1 gene (CD158c) is transcribed and might encode a secreted receptor in a minority of humans, in whom the KIR3DP1, KIR2DL4 and KIR3DL1/KIR3DS1 genes are duplicated. *Eur J Immunol* **35**, 16–24.
24. Martin, M. P., Bashirova, A., Traherne, J., Trowsdale, J., and Carrington, M. (2003) Cutting edge: expansion of the KIR locus by unequal crossing over. *J Immunol* **171**, 2192–5.
25. Shilling, H. G., Lienert-Weidenbach, K., Valiante, N. M., Uhrberg, M., and Parham, P. (1998) Evidence for recombination as a mechanism for KIR diversification. Immunogenetics 48, 413–6.
26. Gomez-Lozano, N., Gardiner, C. M., Parham, P., and Vilches, C. (2002) Some human KIR haplotypes contain two KIR2DL5 genes: KIR2DL5A and KIR2DL5B. *Immunogenetics* **54**, 314–9.
27. Hsu, K. C., Chida, S., Dupont, B., and Geraghty, D. E. (2002) The killer cell immunoglobulin-like receptor (KIR) genomic region: gene-order, haplotypes and allelic polymorphism. *Immunol Rev* **190**, 40–52.
28. Hsu, K. C., Liu, X. R., Selvakumar, A., Mickelson, E., O'Reilly, R. J., and Dupont, B. (2002) Killer Ig-like receptor haplotype analysis by gene content: evidence for genomic diversity with a minimum of six basic framework haplotypes, each with multiple subsets. *J Immunol* **169**, 5118–29.
29. Shilling, H. G., Guethlein, L. A., Cheng, N. W., Gardiner, C. M., Rodriguez, R., Tyan, D., and Parham, P. (2002) Allelic polymorphism synergizes with variable gene content to individualize human KIR genotype. *J Immunol* **168**, 2307–15.
30. Uhrberg, M., Parham, P., and Wernet, P. (2002) Definition of gene content for nine common group B haplotypes of the Caucasoid population: KIR haplotypes contain between seven and eleven KIR genes. *Immunogenetics* **54**, 221–9.
31. Kikuchi-Maki, A., Yusa, S., Catina, T. L., and Campbell, K. S. (2003) KIR2DL4 is an IL-2-regulated NK cell receptor that exhibits limited expression in humans but triggers strong IFN-gamma production. *J Immunol* **171**, 3415–25.
32. Rajagopalan, S., Fu, J., and Long, E. O. (2001) Cutting edge: induction of IFN-gamma production but not cytotoxicity by the killer cell Ig-like receptor KIR2DL4 (CD158d) in resting NK cells. *J Immunol* **167**, 1877–81.

4

Experimental Model for the Study of the Human Immune System

Production and Monitoring of "Human Immune System" $Rag2^{-/-}\gamma_c^{-/-}$ Mice

Nicolas Legrand, Kees Weijer, and Hergen Spits

Summary

Since the late 1980s, the study of the function and development of the human immune system has made intensive use of humanized animal models, among which mouse models have been proven extremely efficient and handy. Recent advances have lead to the establishment of new models with improved characteristics, both in terms of engraftment efficiency and in situ multilineage human hematopoietic development. In particular, the use of newborn BALB/c $Rag2^{-/-}\gamma_c^{-/-}$ mice as recipients for human hematopoietic stem cells has proven particularly efficient. We describe here how to produce and monitor such "human immune system" (HIS) (BALB-Rag/γ) mice, which offer large prospects for experimental study of the human immune system and as a preclinical screening tool.

Key Words: Human immune system; BALB/c $Rag2^{-/-}\gamma_c^{-/-}$ mouse; HIS (Rag/γ); $CD34^+$ hematopoietic stem cell; stem cell transplantation; SCID mouse model; fetal liver; cord blood.

1. Introduction

Experimental study of the human immune system (HIS) in vivo necessitates the establishment of efficient and reliable animal models. To address this need, mouse models have been privileged over the last 25 years, as mice are easy to handle and breed, and less expensive than primates. Confronted with xenograft transplantation barriers, several pioneering groups have screened

From: *Methods in Molecular Biology, vol. 415: Innate Immunity*
Edited by: J. Ewbank and E. Vivier © Humana Press Inc., Totowa, NJ

various immunodeficient mouse strains, and multiple engraftment approaches have been used over the past decades *(1–3)*. Among these, the severe combined immunodeficiency (SCID)-hu (Thy/Liv), which combines both human fetal liver and thymus engraftment into SCID mice, has been shown to be valuable for the study of HIV pathogenesis in vivo *(4)*, but intrinsically leads to a strong bias toward T-cell development *(5, 6)*. Alternatively, immunodeficient mice in the NOD genetic background (e.g., NOD/SCID, NOD/SCID/$\beta 2m^{-/-}$ and NOD/SCID/$\gamma_c^{-/-}$ mice) have been efficiently reconstituted with hematopoietic stem cells (HSC) isolated from human umbilical cord blood (UCB), but human T-cell development and accumulation in situ was described as being limited, especially in the peripheral lymphoid organs of the recipient mice *(7–11)*.

Recently, we and others have developed a new experimental strategy, namely the inoculation of human HSC into newborn immunodeficient mice *(12–15)*. In particular, the use of newborn BALB/c Rag2$^{-/-}\gamma_c^{-/-}$ immunodeficient mice for injection of human HSC-enriched (CD34$^+$) cell populations gives rise to robust human reconstitution, with de novo multilineage hematopoietic reconstitution and marked human thymopoiesis *(13, 14)*. The resulting animals are referred to as "HIS (BALB-Rag/γ) mice" *(3)*, or "human adaptative immune system Rag2$^{-/-}\gamma_c^{-/-}$ mice" (huAIS-RG) *(16)*. Such chimeric animals are able to mount both cellular and humoral adaptative immune responses, contain functional human antigen-presenting cells *(13)* and were recently used as a model for the study of human plasmacytoid dendritic cell development *(17)* and analysis of in vivo treatment with a superagonist anti-human CD28 antibody *(18)*.

Efficient engraftment of BALB/c Rag2$^{-/-}\gamma_c^{-/-}$ mice is age dependent *(14)* and necessitates sublethal total body irradiation before intrahepatic inoculation of CD34$^+$ HSC-enriched cell suspensions. Such suspensions can be prepared from various origins, among which fetal liver and UCB are the most commonly used. Once HIS (BALB-Rag/γ) mice are produced, their effective engraftment level is monitored using flow cytometry on blood samples, not before 6 weeks post-inoculation, to identify the animals with satisfying reconstitution, for subsequent experimental use.

2. Materials
2.1. Preparation of Recipient Mice
1. Newborn BALB/c Rag-2$^{-/-}\gamma_c^{-/-}$ mice (*see* **Note 1**).
2. Sterile laminar flow cabinet, and autoclaved cages, water bottles and food pellets.
3. Sterile irradiation device in which newborn mice fit.

2.2. Isolation of Nucleated Cells from Human HSC Source

1. A source of human HSC. UCB is isolated from pregnant patients at delivery, and fetal liver is obtained from elective abortions with gestational age ranging from 12 to 20 weeks (*see* **Note 2**). The raw material can be maintained at 0–8°C overnight in a cold room or in a fridge.
2. Dulbecco's phosphate-buffered saline (PBS) buffer without calcium/magnesium (20× stock solution): dissolve 8 g KCl, 8 g KH_2PO_4, 57.6 g $Na_2HPO_4.2H_2O$ and 320 g NaCl in 2 l of distilled water and adjust the pH to 7.2 with HCl. Dilute the 20× stock solution with distilled water. The 20× and 1× solutions are kept at room temperature.
3. Roswell Park Memorial Institute-1640 (RPMI-1640) medium supplemented with 25 mM 2-[4-(2-hydroxyethyl)-1-piperazinyl]ethanesulfonic acid (HEPES), L-glutamine and 0.2% (volume : volume) antibiotics. Antibiotics are prepared from powdered penicillin and streptomycin dissolved in Dulbecco's PBS buffer to reach a concentration of 50×10^3 U/ml penicillin and 50 mg/ml streptomycin. RPMI medium and powder antibiotics are kept at 4–8°C. Store aliquots of dissolved antibiotics at –20°C.
4. Ethanol 70% and scissors.
5. Stomacher®-80 Biomaster lab system (Seward Ltd., Worthing, United Kingdom), Stomacher® bags, and a bag-sealing device (e.g., Sealboy 236 Audion Elektro) (*see* **Note 3**).
6. Plastic disposables: 10 and 25 ml pipettes, 100 × 20 mm polystyrene Petri dishes, 50 ml polypropylene conical tubes.
7. Ficoll-Paque™ Plus (GE-Health Care, Chalfont St. Giles, United Kingdom), endotoxin tested. Store away from light at 4–8°C.
8. Complete RPMI-1640 as previously described in **Subheading 2.2, item 3**, supplemented with 2% fetal calf serum (FCS). Store at 4–8°C.

2.3. Enrichment for $CD34^+$ Hematopoietic Stem Cells

1. Complete RPMI-1640 as previously described in **Subheading 2.2, item 3**, supplemented with 2% FCS. Store at 4–8°C.
2. CD34 progenitor cell isolation kit, human (Miltenyi Biotec, Bergisch Gladbach, Germany).
3. MACS buffer is prepared as follows: Dulbecco's PBS pH 7.2 (**Subheading 2.2., item 2**), 0.5% bovine serum albumin, 2 mM ethylenediaminetetraacetic acid (EDTA). Keep at 4–8°C.
4. MACS Separator, Large Scale (LS) MACS separation columns, MACS pre-separation filters and recovery tubes (Miltenyi Biotec).

2.4. Cytometry Cell Sorting for $CD34^+CD38^-$ Hematopoietic Stem Cells

1. Complete RPMI-1640 as described in **Subheading 2.2, item 3**, supplemented with 2% FCS. Store at 4–8°C.

2. Fluorescence-activated cell sorter, for isolation of the HSC-enriched population (*see* **Note 4**).
3. Fluorochrome-coupled monoclonal antibodies (*see* **Note 5**): anti-huCD38 (clone HB-7, BD Biosciences), anti-huCD34 (clone 581, BD Biosciences, Franklin Lakes, NJ, USA). Store the antibodies in the dark at 4–8°C.
4. 5-ml polystyrene round-bottom 12 × 75 mm tubes.
5. Complete RPMI-1640 as described in **Subheading 2.2, item 3**, supplemented with 8% FCS. Store at 4–8°C.
6. 70-μm nylon mesh. Cut into 1 × 1 cm square pieces and autoclave before use.

2.5. Inoculation of Hematopoietic Stem Cells into Recipient Mice

1. Complete RPMI-1640 as described in **Subheading 2.2, item 3**, supplemented with 2% FCS. Store at 4–8°C.
2. BD Micro-Fine+ U-100 Insulin 0.5 ml 0.33(29G) × 12.7 mm syringes (BD Biosciences).

2.6. Monitoring of "HIS" (BALB-Rag/γ) Mice

1. Microvette® CB300 lithium heparin coated for capillary blood collection (Sarstedt, Nümbrecht, Germany), supplemented with one drop of heparin (5×10^3 IU/ml). Each Microvette® is composed of an outer tube and an inner tube embedded in it. The inner tube has a top cap and a plug at the bottom part of the tube. Blood is collected by capillarity from the bottom, and the plug is put back in position afterwards.
2. 5-ml polystyrene round-bottom 12 × 75 mm tubes.
3. Dulbecco's PBS buffer as described in **Subheading 2.2, item 2**.
4. Ficoll-Paque™ Plus (GE-Health Care BioScience AB), endotoxin tested. Store at 4–8°C.
5. Complete RPMI-1640 medium as described in **Subheading 2.2, item 3**, supplemented with 2% FCS. Store at 4–8°C.
6. Fluorochrome-coupled monoclonal antibodies (*see* **Note 6**): anti-huCD45 (clone 2D1, BD Biosciences) for the detection of human hematopoiesis-derived cells and other antibodies according to the desired populations to be detected. Store the antibodies in the dark at 4–8°C.
7. 4´,6-diamidino-2-phenylindole (DAPI) for dead cell exclusion (Sigma-Aldrich, Spruce St., St. Louis, MO, USA): aliquot 100× (20 mM) stock tubes at –20°C and keep the Dulbecco's PBS diluted 1× (0.2 mM) stock in the dark at 4–8°C.
8. FACS buffer: Dulbecco's PBS buffer (*see* **Subheading 2.2., item 2**), 2% FCS, 0.02% sodium azide (NaN_3). Store at 4–8°C.
9. Round bottom 96-well plates.
10. 1.4-ml U-shaped FACS tubes.
11. Fluorescence-activated cell sorter, for analysis of the cell populations (*see* **Note 4**).

3. Methods

Production of animals humanized for the immune system can be achieved by inoculation of cell populations enriched for HSC. Owing to xenograft transplantation barriers, this approach was proven successful to sufficient extents only in case of immunodeficiency, e.g., because of age of the recipients or genetic disorder *(3)*. We and others have shown that inoculation of human HSC into newborn BALB/c Rag-$2^{-/-}\gamma_c^{-/-}$ immunodeficient mice improves the engraftment efficiency *(13–15)* compared with what was previously obtained with adult immunodeficient recipient mice.

Production of HIS (BALB-Rag/γ) mice requires injection of human HSC-enriched cell population into newborn BALB/c Rag-$2^{-/-}\gamma_c^{-/-}$ mice (*see* **Fig. 1**). In this method, we describe isolation of human HSC from fetal liver, but UCB can be used as well. All isolation steps have to be performed sterily under laminar flow. First, nucleated cells have to be purified from cell suspension and

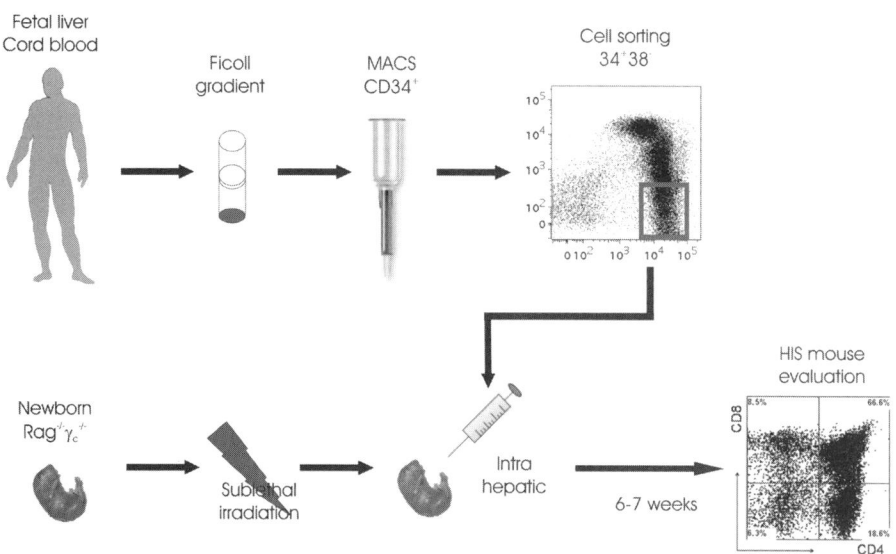

Fig. 1. Schematic protocol for the production of HIS (BALB-Rag/γ) mice. Human hematopoietic progenitors are prepared from fetal or peri-natal sources. Live nucleated cells are isolated on a Ficoll gradient, CD34$^+$ are then magnetically isolated and further purified for the CD34$^+$CD38$^-$ fraction. Sublethally irradiated newborn Balb/c Rag-$2^{-/-}\gamma_c^{-/-}$ mice are injected intrahepatically with this cell preparation. The resulting HIS (BALB-Rag/γ) mice are assessed after 6–7 weeks for the presence of human cells in the blood.

magnetically enriched for the CD34-expressing fraction. Next, these CD34$^+$ cells can be further purified by fluorescence-activated cell sorting for the CD34$^+$CD38$^-$ fraction. Although CD34$^+$ cells have proven to be sufficient for good engraftment efficiency, the use of CD34$^+$CD38$^-$ cells is more reliable and more efficient. Therefore, CD34$^+$ (step 3.3) or CD34$^+$CD38$^-$ cells (step 3.4) can be injected in the recipient mice, depending on the available amount and the desired degree of purity. As CD34$^+$CD38$^-$ cell yield from UCB is usually limited, we advise to inject CD34$^+$ cells from UCB, whereas CD34$^+$CD38$^-$ cells can routinely be isolated from fetal liver.

It usually takes 6–7 weeks before significant amounts of human lymphocytes can be detected in the peripheral blood of the HIS (BALB-Rag/γ) mice. Blood samples are analyzed using flow cytometry to determine which mice are efficiently reconstituted by the human HSC, by measuring the frequency of hematopoiesis-derived (CD45$^+$) human cells and various immune cell populations.

3.1. Preparation of Recipient Mice

1. BALB/c Rag-2$^{-/-}$γ$_c^{-/-}$ mice (*see* **Note 1**) are strongly immunodeficient and the colony has to be maintained in strict healthy conditions, for example, isolator. To avoid any contamination problems, always work sterily, for example, under laminar flow, wearing lab coat, disinfected gloves, and clean material.
2. Production of newborn mice has to be planned depending on the frequency and amount of available human HSC. As engraftment efficiency is age-dependent *(14)*, we recommend the use of newborn mice as early as possible (*see* **Note 7**). Once pregnant, female mice are removed from the breeding cage and male mice are kept in the breeding colony.
3. After birth of the pups, the female and her nest is taken out of the breeding colony (e.g., out of the isolator) and kept under laminar flow during all subsequent manipulations. After inoculation of human cells, the mice can be kept in filtered-top cages; therefore, space and sterile material has to be prepared accordingly.
4. Isolate newborn mice from the breeding cage in the isolator and transfer to sterile irradiation device. Apply sublethal irradiation to the newborn mice (∼3Gy/300rad). Bring the newborns back to the cage containing the mother.

3.2. Isolation of Nucleated Cells from Human HSC Source (Fetal Liver)

1. Prepare material under laminar flow as follows (per fetal liver): one Stomacher® bag; one 100 × 20 mm dish; six 50-ml tubes; one bottle (500 ml) of RPMI-1640 medium (*see* **Note 8** for alternative procedure using UCB as a source of HSC).
2. Pieces of fetal liver are collected in 15 ml plastic tubes containing 2–4 ml of RPMI-1640 medium and have to be processed into cell suspension. Transfer the contents

of one tube into one Stomacher® bag, add 10–15 ml of RPMI-1640 medium and seal the bag twice to avoid any leakage. Install the bag inside the Stomacher®-80 Biomaster lab system and let it mechanically process the liver pieces for one minute at high speed. The bag now contains a cell suspension mixed with non-dissociated liver stroma.
3. Wash the bag with ethanol 70% and open it with sterile scissors. Transfer the contents of the bag into a 100 × 20 mm Petri dish: no need to lose time washing the bag with medium, just get the maximum you can from it.
4. Pour RPMI-1640 medium directly from the bottle to the dish to dilute the cell suspension. Using a 25-ml pipette, recover the supernatant from the dish and distribute equally among the six 50-ml tubes. Always avoid the pieces of stroma at the bottom of the dish. When it becomes difficult to avoid stroma, pour new RPMI-1640 medium into the plate and repeat this procedure until the recovered medium is clear. Usually, pouring fresh medium three times is enough.
5. Add RPMI-1640 medium to the tubes until you reach approximately 20 ml. Wash the cell suspensions by centrifugation (450 g, 5 min). Aspirate the fat-containing supernatant with a pipette or a vacuum system.
6. Resuspend the cell pellet with a 10-ml pipette by adding 13 ml RPMI-1640 medium into each tube. Once a homogeneous cell suspension is obtained by repeated pipetting, bring 10-ml Ficoll-Paque™ underneath the cell suspension with another 10-ml pipette, carefully avoiding air bubbles that could disturb the formation of a clearly distinguishable interface (*see* **Note 9**).
7. Carry the tubes to the centrifuge, carefully avoiding disturbing the interface. Centrifuge the tubes at 1100 g for 15 min with low acceleration and no break (the centrifugation lasts for ∼25 min). After centrifugation, the pellet mostly contains erythrocytes and dead cells, whereas live nucleated cells are concentrated as a cell ring at the interface. Recover the supernatant and the cell ring from each tubes and pool into new tubes (three tubes into one).
8. Wash the cell suspensions by centrifugation (450 g, 5 min) and pool pellets in a total of 10 ml of RPMI-1640 2% FCS medium. Count the amount of nucleated cells in the suspension (*see* **Note 10**).

3.3. Enrichment for CD34$^+$ Hematopoietic Stem Cells

1. Pellet the nucleated cells by centrifugation (450 g, 5 min).
2. Enrichment for the CD34-expressing fraction is done by magnetic sorting, using a two-step strategy: (1) anti-CD34 hapten-coupled antibody; (2) anti-hapten antibody conjugated with colloidal paramagnetic beads (*see* **Note 11**). Perform the first step of indirect MACS labeling: for 10^8 nucleated cells in the pellet, use 75 μl of RPMI-1640 2% FCS medium to resuspend the pellet and add 25 μl of each "A" reagent (FcR blocking human immunoglobulins; monoclonal hapten-conjugated anti-huCD34 antibody). Keep the tube at 6–12°C for 15 min (in the fridge, as recommended by the kit's manufacturer).

3. Add 10 ml of RPMI-1640 2% FCS medium in the tube and wash the cell suspensions by centrifugation (450 g, 5 min).
4. Perform the second step of indirect MACS labeling: for 10^8 nucleated cells in the pellet, use 100 µl of RPMI-1640 2% FCS medium to resuspend the pellet and add 25 µl of the "B" reagent (colloidal super-magnetic MACS MicroBeads conjugated to an anti-hapten antibody). Keep the tube at 6–12°C for 15 min (in the fridge, as recommended by the kit's manufacturer).
5. During the incubation, install the magnetic separation column and a pre-separation filter on top of it. Prepare the collection tube for the column flow-through. Proceed to filter/column washing with MACS buffer, following the manufacturer's instructions (*see* **Note 12**). For an LS separation column, apply 3 ml of MACS buffer and let it run through.
6. Add 10 ml of RPMI-1640 2% FCS medium in the tube and wash the cell suspensions by centrifugation (450 g, 5 min).
7. Resuspend the cell pellet in MACS buffer, using 500 µl of buffer per 10^8 cells.
8. Apply cell suspension through pre-separation column to remove clumps and let the cells pass through the column. Perform column washings with MACS buffer, as indicated in the manufacturer's hand-guide. For an LS column, apply three times 3 ml of MACS buffer on the filter/column assemblage, adding buffer only when the column reservoir is completely empty.
9. Harvest the magnetically labeled cells from the column. Place the column on top of a new 15-ml tube. For an LS column, pipette 5 ml of MACS buffer onto the column and apply the plunger on the column. Immediately flush out the CD34-enriched fraction (*see* **Note 13**).
10. Add 5 ml of RPMI-1640 2% FCS medium to the tube and wash the cell suspension by centrifugation (450 g, 5 min). Resuspend the pellet in a volume of 1 ml of RPMI-1640 2% FCS medium and count the amount of nucleated cells in the suspension (*see* **Note 14**).

3.4. Cytometry Cell Sorting for $CD34^+CD38^-$ Hematopoietic Stem Cells

1. Pellet the nucleated cells by centrifugation (450 g, 5 min).
2. Meanwhile, prepare the monoclonal antibody mixture for cell sorting: mix 1 µl of the anti-CD34 antibody and 1 µl of the anti-CD38 antibody per 10^6 cells.
3. Remove as much supernatant as possible after centrifugation of the cells. Apply the antibody mixture on the dry pellet and resuspend the cells by repeated pipetting. Incubate 10 min on ice (to avoid antibody capping) and in the dark (to avoid fluorochrome bleaching).
4. Add 5 ml of RPMI-1640 2% FCS medium in the tube and wash the cell suspensions by centrifugation (450 g, 5 min).

5. During centrifugation, prepare a set of 5-ml polypropylene tubes: one "sorting tube" that will contain the sample to be sorted, with a square piece of 70 µ mesh on top of it; one recovery tube with 1 ml RPMI-1640 8% FCS medium.
6. After centrifugation, discard the supernatant and resuspend the pellet in RPMI-1640 2% FCS medium to obtain a cell concentration of $5-10 \times 10^6$ cells/ml (minimum 0.5 ml) (*see* **Note 15**).
7. Using a 1-ml pipette and tip, aspirate the cell suspension and filter it through the piece mesh disposed on top of the sorting tube.
8. Sort the human HSC-enriched $CD34^+CD38^-$ population.
9. After the sorting, transfer the content of the "recovery tube" into a new 15-ml tube. Add 5 ml of RPMI-1640 2% FCS medium in the tube and wash the cell suspensions by centrifugation (450 g, 5 min). Resuspend the pellet in a volume of 1 ml of RPMI-1640 2% FCS medium and count the amount of nucleated cells in the suspension (*see* **Note 16**).

3.5. Inoculation of Hematopoietic Stem Cells into Recipient Mice

1. For optimal conditions, plan to inoculate 10^6 $CD34^+$ or 10^5 $CD34^+CD38^-$ cells per newborn. Pellet the desired amount of cells by centrifugation (450 g, 5 min). Resuspend the pellet in the desired volume of RPMI-1640 2% FCS medium (35 µl per newborn mouse to be injected) (*see* **Note 17**).
2. Transfer the cells into an Eppendorf tube, to ensure full access of the needle to the cell suspension.
3. Working sterily in a laminar airflow cabinet, isolate the newborn mice on a piece of absorbent paper on the bench. Hold newborn mouse between thumb and index fingers, with the head down. Inject 30–35 µl of cell suspension by intra-hepatic route, that is, between the thoracic cage and the milk-filled stomach, which appears white through the skin (*see* **Fig. 2**).
4. Once injected, replace the newborn mice with the mother, food, and water (no antibiotics required). If necessary, toe mark the newborns with scissors.

3.6. Monitoring of "HIS" (BALB-Rag/γ) Mice

1. Peripheral reconstitution of the inoculated mice by human HSC is checked by flow cytometry 6–7 weeks after injection. The fraction of human hematopoiesis-derived is determined by anti-CD45 staining. For each HIS (Rag/γ) mouse, shave one of the hind legs between knee and ankle (lateral side) with a scalpel and make blood arise from the saphenous vein using a needle for limited puncture (*see* **Fig. 3**). Collect the blood drops (~50 µl) by capillarity with the Microvette® inner tubes (*see* **Note 18**). Mark the mice for numbering and label the Microvette® accordingly.
2. The blood samples have to be enriched for nucleated cells, using "small scale Ficoll" purification. Prepare two series of numbered 5-ml round-bottomed tubes. Distribute 1.5 ml of Dulbecco's PBS in the group of "dilution tubes", and 1.5 ml of Ficoll-Paque™ in the group of "gradient tubes."

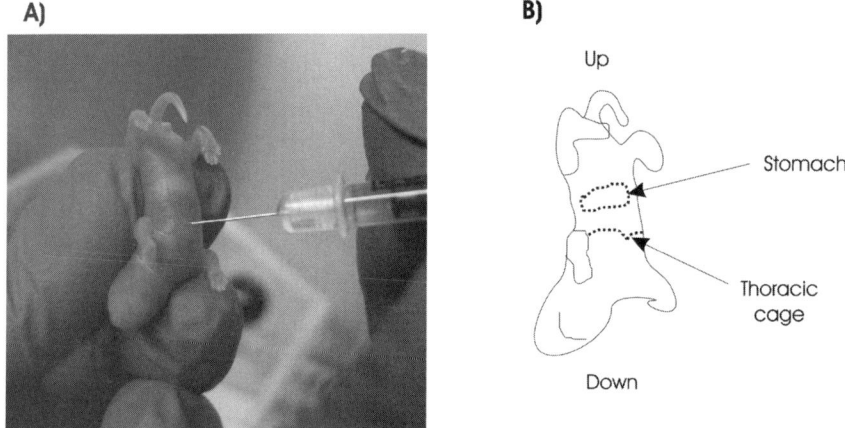

Fig. 2. Intrahepatic inoculation of human progenitor cells in newborn mice. (**A**) The newborn animals are maintained upside down between two fingers. Using an insulin syringe, the cell mixture is injected in the liver (dark red), between the stomach (white), and the thoracic cage. (**B**) Schematic positions of the stomach and thoracic cage are indicated.

3. Pipette 1 ml of Dulbecco's PBS from the first 5-ml round-bottomed tube. Get the inner tube from the Microvette®, open the top cap, horizontally maintain the tube, and remove the bottom plug. Put the bottom end of the open Microvette® inner tube against the wall of the corresponding "dilution tube." Pipette out the 1 ml of Dulbecco's PBS through the inner Microvette® tube: all the blood that was contained will be washed away. Check inside the plug for remaining blood. Repeat this step for every sample.
4. Transfer approximately 1.5 ml of PBS-diluted blood to the "gradient tube," on top of the Ficoll-Paque™ layer, by careful pipetting. For instance, use a plastic 2-ml pipette and flush drop wise. Repeat this step for every sample.
5. Carry the tubes to centrifuge, carefully avoiding disturbing the interface. Centrifuge the tubes at 1100 g for 15 min with low acceleration and no brake (the centrifugation lasts for ~25 min). After centrifugation, the pellet mostly contains erythrocytes and dead cells, whereas live nucleated cells are concentrated as a cell ring at the interface. Recover the supernatant and the cell ring into series of new numbered 15-ml tubes.
6. Add 5 ml of RPMI-1640 2% FCS medium in the tube and wash the cell suspensions by centrifugation (450 g, 5 min).
7. Meanwhile, prepare the monoclonal antibody mixture for cell staining: per sample, mix 4 µl of the FITC-coupled and PerCP-Cy5.5 antibodies, and 2 µl of the R-PE-coupled, PE-Cy7, APC, and APC-Cy7 antibodies (that is to say 16 µl total per sample in this example).

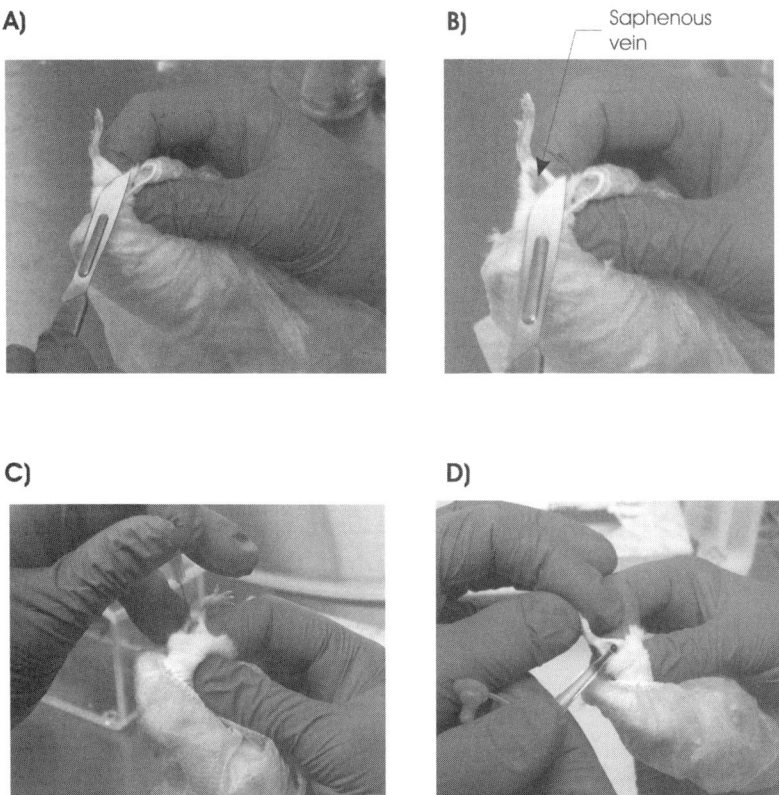

Fig. 3. Puncture of blood at the saphenous vein. (**A**) The mice are maintained with the head and part of the body inside a cap. The hand should firmly hold the hind leg outside the cap, so that it can be shaved laterally, between the knee and the ankle. (**B**) The saphenous vein is exposed, and (**C**) a needle is used to make a small puncture. (**D**) Blood drops are collected in the tubes by capillarity.

8. Remove the supernatant after centrifugation of the cells. Resuspend each pellet with 200 μl of FACS buffer and immediately transfer the suspension to individual contiguous wells in a round-bottomed 96-well plate.
9. Pellet the cell suspensions by centrifugation (450 g, 2 min) and remove the supernatant by quick inversion over the sink. Maintain the plate in this position and dry it against a piece of absorbent paper on the bench. Place the plate on ice.
10. Apply the antibody mixture on the dry pellet by distributing 15.5 μl per well. Resuspend the cells by short vortex of the plate. Incubate 10 min on ice (to avoid antibody capping) and in the dark (to avoid fluorochrome bleaching).

11. During incubation, prepare one FACS tube per sample on the appropriate rack. Prepare a solution composed by 50 µl of FACS buffer and 2.5 µl of 1× DAPI solution (1/20 dilution) per sample. Keep in dark at 4–8°C.
12. At the end of the incubation, distribute 100 µl of FACS buffer per well and wash the plate by centrifugation (450 g, 2 min). Remove the supernatant by quick inversion over the sink (*see* **step 9**). Bring the plate back on ice and distribute 50 µl of the DAPI-containing FACS buffer per well. Transfer the content of the wells to their respective FACS tube (e.g., with multi-channel 200-µl pipette) and perform cytometry analysis (*see* **Fig. 4**).

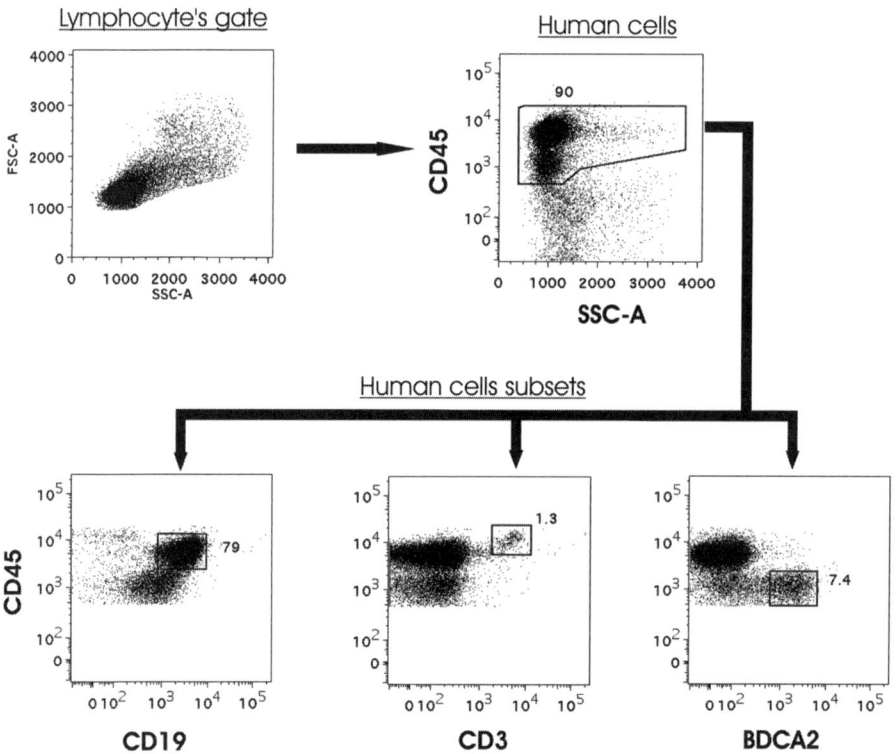

Fig. 4. Monitoring of HIS (BALB-Rag/γ) mice. The blood collected at 6–7 weeks after HSC inoculation is analyzed by flow cytometry. Human hematopoiesis-derived cells are detected with a CD45-specific antibody. Subsets of interest are subsequently analyzed by the use of antibodies directed to specific markers (e.g., CD19 for B lymphocytes, CD3 for T lymphocytes, BDCA2 for plasmacytoid dendritic cells). Most of human cells in the blood are small-sized B lymphocytes.

4. Notes

1. At least two Rag-deficient *(19, 20)* and four γ_c-deficient mouse strains *(21–24)* have been independently generated so far. Two inbred strains of Rag2$^{-/-}\gamma_c^{-/-}$ mice are currently available, respectively in the C57Bl/6 (black) and BALB/c (white) genetic background. For reasons still under investigation, newborn C57Bl/6 Rag2$^{-/-}\gamma_c^{-/-}$ mice are not efficiently reconstituted by human HSC. Therefore, the use of BALB/c Rag2$^{-/-}\gamma_c^{-/-}$ mice is strongly recommended, although we have been successfully using Rag2$^{-/-}\gamma_c^{-/-}$ mice in a mixed BALB/c × 129/OLA background with identical results.
2. Fetal liver is the richest source of human HSC but, depending on local legislation and necessity of informed consent, it may not be available for research purposes. UCB is easier to obtain, but the yield of HSC is much lower. These aspects raise the question of the size of the groups of producible HIS (Rag/γ) mice, in relation to the intrinsic donor-dependant variability. There are several alternatives: (1) produce HIS (Rag/γ) mice each time that a source of human HSC is available; (2) freeze each individual source of human HSC and make series of age synchronized HIS (Rag/γ) mice with the individual sources; (3) pool several source donors and make series of age-synchronized HIS (Rag/γ) mice with the same "normalized" pool. The last possibility saves a lot of time and reagents, especially in the situation of recurrent availability of HSC source from numerous (≥ 3) donors. In **Subheading 3**, we describe the protocol used for processing of one individual fetal liver.
3. We list here the material required for automated mechanical preparation of cell suspension. The preparation can also be made manually, simply using a metallic mesh on which the pieces of fetal liver are mechanically processed.
4. We use a FACSAria™ machine (BD Biosciences) for cell populations sorting and a BD LSR II™ machine (BD Biosciences) for cell population analysis, both with 9-parameters (size, scatter, and seven fluorescence colors including DAPI). It has to be noted that any five parameters (size, scatter, and three fluorescence colors) cell sorter is sufficient to isolate HSC-enriched populations.
5. The choice of fluorochromes has to be adapted to the available light filters in the cell sorter. We routinely use anti-huCD38 antibody coupled to R-phycoerythrine (R-PE) and anti-huCD34 antibody coupled to phycoerythrine-cyanine-7 (PE-Cy7).
6. Similar to **Note 5**, the fluorochromes have to be chosen in accordance with the available light filters in the FACS analyzer. As an example, we routinely use fluorescein isothiocyanate (FITC) anti-huCD19, R-PE anti-huCD8, peridinin chlorophyll protein-cyanin-5.5 (PerCP-Cy5.5) anti-huCD4, PE-Cy7 anti-huCD3, allophycocyanin (APC) anti-BDCA2, allophycocyanin-cyanin-7 (APC-Cy7) anti-huCD45. The proposed staining mixture for reconstitution analysis is adapted to a nine-parameter cytometer, for example, the BD LSR II™ machine (BD Biosciences) (*See* **Note 4**), and is designed for the determination of frequency of B lymphocytes (CD19), plasmacytoid dendritic cells (BDCA2) and T lymphocyte

subpopulations (CD3, CD4, CD8) within the human hematopoiesis-derived cells (CD45). In the case of a five-parameter cytometer, we advise to use only anti-CD45, anti-CD3, and anti-CD19 antibodies with the proper fluorochrome combination. The choice of the antibodies has to be adapted to the cell populations of interest but should always include a CD45-specific antibody to measure to what extent human reconstitution took place.

7. There is no strict need for "time-pregnant" female mice, especially in the case of a recurrent human HSC source. Reconstitution is optimal between 1 and 4 days of age, and efficiency usually drops severely after 5 days of age. We therefore recommend to use BALB/c Rag2$^{-/-}\gamma_c^{-/-}$ newborn mice not older than 5 days of age. Still, we have already—although rarely—observed good reconstitution using 7-day-old newborns.

8. Alternatively, UCB can be used instead of fetal liver. In brief, transfer approximately 60 ml of UCB into a 250-ml flask. Dilute the UCB with approximately 120 ml of Dulbecco's PBS, reaching a total volume of approximately 180 ml. Prepare twelve 50-ml tubes and distribute approximately 15 ml of diluted UCB to each tube. Similar to fetal liver, carefully bring 10-ml Ficoll-Paque™ underneath the cell suspension to get a clear interface. Centrifuge the tubes at 1100 g for 15 min with low acceleration and no break. Recover the supernatant and the cell ring from each tube, pool into new tubes. Wash the cell suspensions by centrifugation (450 g, 5 min) and pool pellets in a total of 15 ml of RPMI-1640 2% FCS medium. After Ficoll-Paque™ gradient, the yield from 60 ml UCB is usually around 50–100 × 10^6 nucleated cells. In contrast to the suspension obtained from fetal liver, contamination by erythrocytes is usually still clearly noticeable by eye (cell suspension is red) but is not a problem for the next steps.

9. To avoid waste of plastic disposables, prepare two 50-ml tubes, which contain RPMI-1640 medium and Ficoll-Paque™, respectively. Resuspend the cell pellet with the same pipette in the whole six-tube series, change your 15-ml pipette and next bring Ficoll-Paque™ with the same pipette in the whole series of tubes. Simply pour new medium or Ficoll-Paque™ directly from the bottle into the 50-ml tube when needed.

10. After Ficoll-Paque™ gradient, the yield from one medium-sized fetal liver is usually around 100–400 × 10^6 nucleated cells. Important variations are observed, depending on the size of the material, age of the donor, and care during tube manipulation.

11. Be careful that the "Indirect CD34 MicroBead kit" is used at this step and not the direct kit (anti-CD34 bead-conjugated antibody). We have observed that the efficiency of the direct kit is low on fetal liver and UCB cell suspensions, whereas high on post-natal thymocytes suspensions. Furthermore, manufacturer's recommendations are to use 1 ml of each reagent per 10^9 cells in the suspension, but we are routinely using 25% of the recommended volumes with no yield loss. These adapted volumes are indicated in **Subheading 3**.

12. The protocol is described for manual MACS separation. We advise to use LS separation columns, which can hold 10^8 magnetically labeled cells from up to 2×10^9 total cells. In theory, smaller columns (e.g., MS separation columns) should fit for UCB samples, but we have observed that such samples easily clog them. If you are using an automatic device such as AutoMACS™ (Miltenyi Biotec), adapt the protocol accordingly.
13. According to the manufacturer, one can expect a degree of purification of 85–98% CD34$^+$ cells. We routinely reach purification of >95% CD34$^+$ cells after the manual MACS separation.
14. After MACS separation, the yield of CD34$^+$ cells is around 0.5–2% of the initial amount of nucleated cells. Therefore, you can expect $1–10 \times 10^6$ CD34$^+$ cells per liver and $0.1–1 \times 10^6$ CD34$^+$ cells per UCB.
15. Volumes and cell concentrations for the sorting are convenient for the FACSAria™ sorter (BD Biosciences) and should be adapted to other machines according to manufacturer's instructions.
16. Be aware that the sorter cell counts are rarely fully accurate. It is therefore reasonable to expect after sorting of CD34$^+$CD38$^-$ cells, a yield around 10% of the initial CD34$^+$ cell counts, although CD34$^+$CD38$^-$ cells usually represent 20–40% of CD34$^+$ fetal liver cells.
17. Variability in reconstitution efficiency increases when lower numbers of progenitors are injected. Still, we have routinely obtained similar levels of reconstitution by inoculating 5×10^5 CD34$^+$ or 5×10^4 CD34$^+$CD38$^-$ cells per mouse (i.e., half of the "optimal conditions").
18. One single manipulator is enough to proceed through the whole procedure, but a second manipulator is very helpful to hold the tube and recover the blood.

Acknowledgments

We thank Joran Volmer for proofreading of the manuscript and daily technical assistance, as well as Jenny Meerding. We also thank Berend Hooibrink for expertise in cell sorting and maintenance of the FACS facility and Angelique Epping, Anneke Ditewig, and Lydia Wolterman for excellent animal care at the Academic Medical Center, Amsterdam. We also acknowledge our laboratory colleagues who helped in the improvement of the HIS (BALB-Rag/γ) mouse model. Last, we are grateful to the Bloemenhove Clinic (Heemstede, the Netherlands) and the Department of Internal Medicine of the AMC for providing fetal tissues and UCB, respectively. This work is supported by the Grand Challenges in Global Health Initiative (Bill & Melinda Gates Foundation, Wellcome Trust, Foundation for the National Institute of Health, Canadian Institutes of Health Research) as part of the "Human Vaccine Consortium" (http://www.hv-consortium.org/).

Reference

1. Greiner, D. L., Hesselton, R. A., and Shultz, L. D. SCID mouse models of human stem cell engraftment (1998) *Stem Cells* **16**, 166–77.
2. Macchiarini, F., Manz, M. G., Palucka, A. K., and Shultz, L. D. Humanized mice: are we there yet? (2005) *J Exp Med* **202**, 1307–11.
3. Legrand, N., Weijer, K., and Spits, H. Experimental models to study development and function of the human immune system in vivo (2006) *J Immunol* **176**, 2053–8.
4. McCune, J. M. Animal models of HIV-1 disease (1997) *Science* **278**, 2141–2.
5. McCune, J. M., Namikawa, R., Kaneshima, H., Shultz, L. D., Lieberman, M., and Weissman, I. L. The SCID-hu mouse: murine model for the analysis of human hematolymphoid differentiation and function (1988) *Science* **241**, 1632–9.
6. Krowka, J. F., Sarin, S., Namikawa, R., McCune, J. M., and Kaneshima, H. Human T cells in the SCID-hu mouse are phenotypically normal and functionally competent (1991) *J Immunol* **146**, 3751–6.
7. Hesselton, R. M., Greiner, D. L., Mordes, J. P., Rajan, T. V., Sullivan, J. L., and Shultz, L. D. High levels of human peripheral blood mononuclear cell engraftment and enhanced susceptibility to human immunodeficiency virus type 1 infection in NOD/LtSz-scid/scid mice (1995) *J Infect Dis* **172**, 974–82.
8. 2Larochelle,A.,Vormoor,J.,Hanenberg,H.,Wang,J.C.,Bhatia,M.,Lapidot,T.,Moritz, T., Murdoch, B., Xiao, X. L., Kato, I., Williams, D. A., and Dick, J. E. Identification of primitive human hematopoietic cells capable of repopulating NOD/SCID mouse bone marrow:implicationsforgenetherapy(1996)*NatMed***2**,1329–37.
9. Kollet, O., Peled, A., Byk, T., Ben-Hur, H., Greiner, D., Shultz, L., and Lapidot, T. Beta2 microglobulin-deficient (B2m(null)) NOD/SCID mice are excellent recipients for studying human stem cell function (2000) *Blood* **95**, 3102–5.
10. Kerre, T. C., De Smet, G., De Smedt, M., Zippelius, A., Pittet, M. J., Langerak, A. W., De Bosscher, J., Offner, F., Vandekerckhove, B., and Plum, J. Adapted NOD/SCID model supports development of phenotypically and functionally mature T cells from human umbilical cord blood CD34(+) cells (2002) *Blood* **99**, 1620–6.
11. Hiramatsu, H., Nishikomori, R., Heike, T., Ito, M., Kobayashi, K., Katamura, K., and Nakahata, T. Complete reconstitution of human lymphocytes from cord blood CD34+ cells using the NOD/SCID/gammacnull mice model (2003) *Blood* **102**, 873–80.
12. Ishikawa, F., Livingston, A. G., Minamiguchi, H., Wingard, J. R., and Ogawa, M. Human cord blood long-term engrafting cells are CD34+ CD38 (2003) *Leukemia* **17**, 960–4.
13. Traggiai, E., Chicha, L., Mazzucchelli, L., Bronz, L., Piffaretti, J. C., Lanzavecchia, A., and Manz, M. G. Development of a human adaptive immune system in cord blood cell-transplanted mice (2004) *Science* **304**, 104–7.
14. Gimeno, R., Weijer, K., Voordouw, A., Uittenbogaart, C. H., Legrand, N., Alves, N. L., Wijnands, E., Blom, B., and Spits, H. Monitoring the effect of

gene silencing by RNA interference in human CD34+ cells injected into newborn RAG2-/- gammac-/- mice: functional inactivation of p53 in developing T cells (2004) *Blood* **104,** 3886–93.

15. Ishikawa, F., Yasukawa, M., Lyons, B., Yoshida, S., Miyamoto, T., Yoshimoto, G., Watanabe, T., Akashi, K., Shultz, L. D., and Harada, M. Development of functional human blood and immune systems in NOD/SCID/IL2 receptor {gamma} chainnull mice (2005) *Blood* **106,** 1565–73.

16. Chicha, L., Tussiwand, R., Traggiai, E., Mazzucchelli, L., Bronz, L., Piffaretti, J. C., Lanzavecchia, A., and Manz, M. G. Human adaptive immune system Rag2-/- {gamma}c-/- mice (2005) *Ann N Y Acad Sci* **1044,** 236–43.

17. Schotte, R., Nagasawa, M., Weijer, K., Spits, H., and Blom, B. The ETS transcription factor Spi-B is required for human plasmacytoid dendritic cell development (2004) *J Exp Med* **200,** 1503–9.

18. Legrand, N., Cupedo, T., van Lent, A. U., Ebeli, M. J., Weijer, K., Hanke, T., and Spits, H. Transient accumulation of human mature thymocytes and regulatory T cells with CD28 superagonist in "human immune system" Rag2-/-{gamma}c-/- mice (2006) *Blood* **108,** 238–45.

19. Mombaerts, P., Iacomini, J., Johnson, R. S., Herrup, K., Tonegawa, S., and Papaioannou, V. E. RAG-1-deficient mice have no mature B and T lymphocytes (1992) *Cell* **68,** 869–77.

20. Shinkai, Y., Rathbun, G., Lam, K. P., Oltz, E. M., Stewart, V., Mendelsohn, M., Charron, J., Datta, M., Young, F., Stall, A. M., et al. RAG-2-deficient mice lack mature lymphocytes owing to inability to initiate V(D)J rearrangement (1992) *Cell* **68,** 855–67.

21. Cao, X., Shores, E. W., Hu-Li, J., Anver, M. R., Kelsall, B. L., Russell, S. M., Drago, J., Noguchi, M., Grinberg, A., Bloom, E. T., Paul, W. E., Katz, S. I., Love, P. E., and Leonard, W. J. Defective lymphoid development in mice lacking expression of the common cytokine receptor gamma chain (1995) *Immunity* **2,** 223–38.

22. DiSanto, J. P., Muller, W., Guy-Grand, D., Fischer, A., and Rajewsky, K. Lymphoid development in mice with a targeted deletion of the interleukin 2 receptor gamma chain (1995) *Proc Natl Acad Sci USA* **92,** 377–81.

23. Ohbo, K., Suda, T., Hashiyama, M., Mantani, A., Ikebe, M., Miyakawa, K., Moriyama, M., Nakamura, M., Katsuki, M., Takahashi, K., Yamamura, K., and Sugamura, K. Modulation of hematopoiesis in mice with a truncated mutant of the interleukin-2 receptor gamma chain (1996) *Blood* **87,** 956–67.

24. Blom, B., Spits, H., and Krimpenfort, P. The role of the common gamma chain of the IL-2, IL-4, IL-7 and IL15 receptors in the developmental of lymphocytes: constitutive expression of Bcl-2 does not rescue the development defects in gamma common-deficient mice (1996) in *Cytokines and Growth Factors in Blood Transfusion*, C.Th. Smit Sibinga, P.C. Das, B. Lowenberg (Eds.) Developments in hematology and immunology, Vol. 32, pp. 3–11, Kluwer Academic Publishers, Dordrecht, The Netherlands.

5

Bacterial Artificial Chromosome Transgenesis Through Pronuclear Injection of Fertilized Mouse Oocytes

Kristina Vintersten, Giuseppe Testa, Ronald Naumann, Konstantinos Anastassiadis, and A. Francis Stewart

Summary

In the mouse, conventional transgenes often produced unpredictable results mainly because they were too small to recapitulate a natural gene context. Bacterial artificial chromosomes (BACs) are large enough to encompass the natural context of most mammalian genes and consequently deliver more reliable recapitulations of their endogenous counterparts. Furthermore, recombineering methods now make it easy to engineer precise changes in a BAC transgene. Consequently, BACs have become the preferred vehicle for mouse transgenesis. Here, we detail methods for BAC transgenesis through pronuclear injection of fertilized oocytes.

Key Words: BACs; pronuclear injection; microinjection; superovulation; pseudopregnant host.

1. Introduction

The mouse has become the premier model for medical studies in part because of the development of advanced ways to engineer its genome (*1*). Here, we discuss the use of large transgenes, as provided by bacterial artificial chromosomal (BAC) resources. In the past, transgenes were limited in size by the constraints of DNA-engineering methods. Mostly, artificial minitransgenes were made because larger constructs were impractical. However, the average size of a mammalian gene, including its cis regulatory elements, is usually greater than that of a minitransgene. Consequently, most early transgenes suffered from exquisite sensitivity to position effects exerted by chromatin surrounding their genomic integration site. Transgenes based on large sections of genomic DNA generally do not suffer from position effects because they encompass entire genes with the cis regulatory elements in the natural position and configuration.

An increasing number of completely sequenced genomes are accompanied online by annotated libraries of BAC genomic clones. Because these BACs are precisely defined and can be readily obtained, a large number of starting options are available. Furthermore, the development of recombineering methodologies permits the precise mutagenesis or alteration of BACs in almost any way *(2)*. Hence, BAC transgenics are now amongst the most useful reagents for applications in mouse experimental biology. Transgenesis through microinjection of DNA into the pronuclei of fertilized oocytes is one of two commonly used methods for gene transfer into the mouse genome *(3,4)*. The first successful attempt to perform this technique was carried out by Lin in 1966 *(5)*, who could show that the early fertilized embryo could survive the mechanical damage of inserting a glass needle into the pronucleus. It was not until 1981, however, that small DNA fragments were integrated into the genome *(6)*. This technique is well described and has now become a standard procedure *(7,8)*, which has permitted its application to larger DNA fragments. Yeast artificial chromosome (YAC) *(9,10)*, P1 artificial chromosome (PAC) *(11,12)*, and BAC DNA *(13,14)* can all be used for the generation of transgenic mice. See Giraldo and Montoliu *(15)* and Ristevski *(16)* for reviews. Although the basic technique for microinjection of large constructs are similar to those used for shorter DNA segments, there are some special requirements *(17)*. In this chapter, we describe relevant steps for microinjection using large DNA constructs.

It is important to keep in mind that, as with any kind of pronuclear DNA injection, the integration takes place in a random manner; hence, the integration site and even the copy number of the integrated transgene are not possible to predict. All the resulting founder animals will carry a unique integration pattern. In most cases, the integration will take place at only one site; occasionally, however, separate integration sites on two different chromosomes can occur *(18)*. When these founder animals are mated for transmission of the construct through the germline, segregating integration patterns may be observed among the offspring. Transgene integration will generally take place shortly after microinjection, before DNA replication in the one-cell stage embryo, also termed the zygote. In some cases though, the integration is delayed, resulting in a mosaic transgenic embryo/animal *(19)*. Occasionally, mosaicism can interfere with the germline transmission of the transgene.

The BAC construct can be injected in one of three forms: supercoiled, linearized, or as a purified insert released from the BAC vector *(15)*. Injection of the supercoiled form is the fastest method and presents the advantage that the DNA, being more compacted, may be less prone to shearing and fragmentation during the injection procedure. Furthermore, suitable restriction sites for cleavage within the BAC vector are not required. Also, solutions of linearized BAC DNA tend to be more viscous, and hence more difficult to inject, than when the BAC is in its supercoiled form.

The main drawback of using intact (supercoiled) BACs is that integration of the BAC transgene will occur upon random linearization, thus possibly interrupting the transgene in a deleterious manner. On the contrary, injection of a linearized transgene is more predictable for reliable integration of the intact transgene. A variation of this approach has been the use of linearized BAC inserts released from the vector and purified by pulsed field gel electrophoresis (PFGE). This is admittedly the cleanest experimental avenue; it is, however, fairly laborious and apparently unnecessary, because, as it has been shown with YACs, the presence of vector sequences integrated to one side of large genomic transgenes does not seem to affect expression *(20)*.

If one wishes to linearize the BAC in the absence of suitable restriction sites, recombineering *(2)* represents a fast and convenient way to precisely introduce a unique site at any desired place in the BAC.

The handling and general conditions surrounding both embryo donor and recipient mice are of major importance for the successful outcome of microinjection experiments. For further reading about optimal husbandry conditions, the reader is referred to Foster et al.*(21)*.

Handling and culture of embryos, the microinjection process, and the final transfer of the embryos back in vivo, all require experience and optimal culture conditions *(8,22)*. The microinjection process requires expensive, specialized equipment. Therefore, these experiments should not be undertaken without careful consideration (*see* **Note 1**).

1.1. Production of Zygotes for Microinjection

The number of animals needed per experiment will depend on several factors: age and quality of the embryo donors and stud males, physical parameters in the animal facility and the microinjection lab, embryo culture conditions, DNA purity and concentration, and the manual skills of the operator.

Generally, 25–30% of the injected zygotes will immediately lyse because of the mechanical damage. On average, about 30% of the surviving embryos will develop to term, and only 15–20% of those will be transgenics. Therefore, if all conditions are met, it is usually sufficient to inject 150 fertilized zygotes to produce 3–6 founder animals.

The choice of genetic background of the host embryo should be considered carefully. In most cases, a cross between two F1 animals, such as (C57BL/6J × CBA) F1, (C57BL/6J × SJL) F1, or (C57BL/6J × DBA/2) F1, is used, because these combinations have proven to provide large numbers of good quality embryos. In these cases, however, the genetic background of the founder will be mixed (*see* **Note 2**). It has been shown that transgene expression can be modulated or suppressed by the genetic background of the donor embryo *(23,24)*.

Embryo donor females should be superovulated to increase the number of embryos obtained per mouse (*see* **Subheadings 3.1. and 3.2.**). The exact timing of the developmental stage of the zygotes is of major importance and can only be achieved by careful adjustment of the hormone treatment timing, light cycle in the animal facility, and time of zygote collection (*see* **Note 3**).

1.2. Recovery, Handling, and In Vitro Culture of Pre-implantation Stage Embryos

Great care should be taken to assure the best possible in vitro culture conditions for the embryos. A humidified 5% CO_2 incubator, 37°C is used for the embryo culture. This incubator should be reserved for embryos only and not shared with other tissue culture. The most commonly used culture medium in 5% CO_2 is M16, and the HEPES-buffered equivalent for use at the bench is M2 *(25)*. These media can be purchased from commercial suppliers (Sigma, Specialty Media) or prepared according to published protocols *(5)* (*see* **Note 4**).

The embryos are easiest handled (collected, moved, washed, sorted, and manipulated) by careful mouthpipeting, using a special device *(26)* (*see* **Subheading 3.3.**) (*see* **Note 5**).

Mice tend to ovulate and mate at the midpoint of the dark period (night). The zygotes will soon thereafter reach the swollen ampulla region of the oviduct, where fertilization takes place. The following morning (12 h after fertilization, at embryonic stage E0.5), the ampulla is greatly enlarged, and can easily be located amongst the oviduct coils, which makes the recovery of the zygotes fairly easy (*see* **Subheading 3.4. and Note 6**).

The zygotes are tightly packed together and surrounded by a large amount of cumulus cells at the time of recovery. Cumulus cell masses must be removed from the zygotes before injection can take place (**Subheading 3.5. and 3.6.**) (*see* **Note 7**).

1.3. Microinjection Equipment

An inverted microscope with either Differential Interference Contrast (DIC) or Hoffman optics is necessary to locate the pronuclei. A low-magnification lens (×2.5 or ×5) is used to get an overview of the injection, and a high-magnification lens (×32 or ×40) is used for the actual injection process. Two micromanipulators are needed for the movement of the holding and injection capillaries. Commercially available micromanipulators such as Leitz, Eppendorf, or Narishige are all suitable, and the choice of brand is a matter of personal preference. We prefer the Eppendorf TransferMan NK as it is an electronic system with which set positions can be pre-programmed. This feature greatly enhances the efficiency and speed of injection. The control of the holding capillary (capture and release of the embryos) can be achieved by using a micrometer screw-controlled, oil-filled glass syringe, or a commercially available control unit (Eppendorf CellTram air), or by simple mouthpipeting. The flow of DNA in the injection needle is best controlled by an injector (Eppendorf FemtoJet) but can also be done (by experienced experimentators) by using a glass syringe. In all the above cases, an absolutely airtight connection has to be established between the control-device and capillary. Air-filled thin polythene tubing should be used between electronic injectors and the injection needle. If injection is performed by hand, a thick hard silicon tubing (Tygon R3603) is the best alternative. For the connection of the holding capillary, hard silicon (Tygon R3603) or thin polythene tubing can be used, either oil- or air-filled. Oil-filled tubings provide a finer (slower) control but have to be free of any air bubbles to work well. Air-filled tubings provide faster movements, which can be compensated by filling oil only in the holding capillary itself.

The injection takes place in an injection chamber: a small drop of M2 medium covered by embryo-tested light paraffin oil. We recommend to use a simple aluminum frame attached to a clean glass slide with high vacuum grease (Dow Corning) to hold the media drop and oil in place. It is also possible to use a glass slide with compression well or even plastic tissue culture dishes. When using supercoiled BAC DNA preparations, microinjection needles can be obtained from commercial suppliers (Eppendorf FemtoTipII) or prepared on a needle puller such as the model P97 from Sutter Instruments (*see* **Note 8**). For linearized BAC DNAs, we only use pulled needles pulled by our glass equipment because the Eppendorf needles appear to shear linearized BAC DNA. To do this, the needle tip is pulled out further to make it longer than the Eppendorf needle at the smallest bore size. This appears to help laminar flow of the DNA solution through the smallest bore and thereby reduce shearing.

A comprehensive reference about glass capillaries and needle pullers has been published by Flaming and Brown *(27)*.

1.4. Dilution of BAC DNA Solution

After BAC DNA preparation and optional restriction digestion, the DNA should be diluted in BAC injection buffer (*see* **Subheading 3.8.**), which aids to stabilize BAC DNA *(28)*. It is very important to prepare an ultra-pure buffer, completely free of any particles, solvents, detergents, and so on. Even the smallest debris will inevitably clog the microinjection needle, and any other substances then DNA or buffer may be toxic to the embryos. The final concentration should be kept as close as possible to 1.5 ng/μl. It has been shown that integration rate increases, but embryo viability decreases when the DNA concentration is raised *(8)*. It is to some extent possible to increase the rate of single-copy integrations by decreasing the DNA concentration *(7)*. Overall efficiency will, however, be reduced.

1.5. Microinjection Process

Groups of zygotes are moved to the microinjection chamber, not more than can easily be injected within 20 min. The zygotes are picked up one by one with the holding needle and orientated, so that the two pronuclei are readily visible.

The injection needle is inserted through the cell membrane and cytoplasm into the target pronucleus. DNA is injected until a clear swelling of the pronucleus can be observed (*see* **Notes 9–15**).

1.6. Transfer of Injected Embryos to Pseudopregnant Female Mice

Most female mice have an ovulation cycle of 3–5 days, but to get into the hormonal status of pregnancy, there is a need for physical mating to take place. Female mice, which are to be used as recipients for transferred embryos, are therefore mated to sterile (vasectomized) male mice. By detection of a copulation plug in the following morning, pseudopregnant females can be selected for transfer surgery. The vasectomy of male mice *(7)* is a simple surgical procedure, where an incision is made in the scrotal sac or abdominal wall, and an approximately 10-mm long piece of the vas deferens is removed by cauterization. Males can be used for mating 2 weeks after surgery and then for as long as xx months.

Injected embryos are placed in the reproductive tract of pseudopregnant recipient females *(29)*. Anesthesia is induced by an intraperitoneal (i.p.) injection of a suitable anesthetic solution (*see* **Subheading 3.10.**). The oviduct is exposed by a surgical procedure (*see* **Subheading 3.11.**) and the embryos placed in the ampulla region, through the infundibulum (*see* **Notes 15 and 16**).

2. Materials

2.1. Preparation of Hormones for Superovulation

1. Pregnant Mares Serum Gonadotropin (PMSG) (Intervet, Intergonan, Toenisvorst, Germany).
2. Human Corionic Gonadotropin (HCG) (Sigma CG-10).
3. Sterile distilled H_2O, chilled to 4°C.

2.2. Superovulation of Young Female Mice

1. PMSG, readymade solution 0.05 IU/ml.
2. HCG, readymade solution 0.05 IU/ml.
3. Disposable 1-ml syringe and 27 or 30 G needle.

2.3. Preparation of Transfer Capillaries

1. 50-ml glass capillaries (Brand).
2. Bunsen burner.
3. Capilette (Selzer, Germany, Capilette pipetierhilfe).
4. Flat mouthpieces (HPI Hospital products, Altamonte Springs, Orlando, FL, USA 1501).
5. Flexible silicon tubing with an inner diameter of 3 mm.

2.4. Dissection of Oviducts

1. Anatomical forceps.
2. Fine straight scissors.
3. Fine straight forceps.
4. 3-cm tissue-culture dish.
5. M2 culture medium.

2.5. Preparation of Hyaluronidase Solution

1. Hyaluronidase type IV from bovine testis (Sigma H-4272)
2. M2 embryo culture medium

2.6. Recovery and In Vitro Culture of Fertilized Oocytes

1. 3.5-cm Falcon tissue culture dishes (35-3001).
2. M2 culture medium.
3. M16 culture medium.
4. Embryo-tested light paraffin oil (Sigma)
5. Hyaluronidase solution (*see* **Subheading 2.5.**)
6. Watchmakers forceps.
7. Disposable 1-ml syringe with 30-G needle.

2.7. Preparation of Holding Capillary and Injection Needle

1. Borocilicate glass capillary (Leica 520119) for holding capillary.
2. Bunsen burner.
3. Diamond point pen.
4. Microforge (Bachofer, Alcatel, Narishige), alternatively Bunsen burner.
5. Borocilicate glass capillary with inner filament (Clark Electronic Instruments, GC100TF) for injection needle.
6. Horizontal glass capillary puller (Sutter Instruments, P97 Novato, CA, USA).
7. Alternatively, purchased needles/capillaries can be used (Eppendorf VacuTips and FemtotipsII)

2.8. Preparation of Injection Buffer and Dilution of BAC DNA

1. Spermine (Sigma, tetrahydrochloride S-1141).
2. Spermidine (Sigma, trihydrochloride S-2501).
3. 1 M Tris–HCl pH 7.5, autoclaved.
4. 0.5 M ethylenediaminetetraacetic acid (EDTA) pH 8.0, autoclaved.
5. 5 M NaCl, autoclaved.
6. Sterile ddH$_2$O.

2.9. Microinjection

1. Inverted microscope as described.
2. Two micromanipulators.
3. Holding capillary (*see* **Subheading 2.7.**).
4. Injection needle (*see* **Subheading 2.7.**).
5. Microloader (Eppendorf Microloader).
6. M2 culture medium.
7. Fertilized oocytes.
8. DNA solution, 1–2 ng per µl.
9. Injector (Eppendorf FemtoJet), alternatively, 10-ml glass syringe (Becton Dickinson, 10CC, 2590,6458)

2.10. Preparation of Ketamine/Xylazine Anesthetics

1. Ketamine hydrochloride (Ketanest, Park Davis).
2. Xylazine 2% (Rompune, Roche).
3. Sterile ddH$_2$O.

2.11. Embryo Transfer to the Oviduct of Pseudopregnant Female Mice

1. Ketamine/Xylazine (*see* **Subheading 2.10.**).
2. Sterile phosphate-buffered saline (PBS).
3. Transfer capillary.
4. M2 medium.

5. Embryos.
6. 70% EtOH.
7. Paper tissues.
8. Sterile blunt forceps.
9. Two sterile watchmakers forceps.
10. Sterile fine scissors.
11. Sterile seraffin clip.
12. Wound clip with applicator (Clay Adams) or suture needles with ligature.
13. Pseudopregnant female mouse 8–12 weeks old, E0.5.
14. Stereo microscope with cold light lamp.

3. Methods

3.1. Preparation of Hormones for Superovulation

1. Thaw the hormones by adding pre-chilled sterile H_2O directly to the frozen vial (*see* **Note 17**).
2. Mix carefully and dilute the hormones to a final concentration of 50 IU/ml.
3. Aliquot and freeze immediately at –20°C until use.

3.2. Superovulation of Young Female Mice

Day 1, 11 a.m. (*see* **Note 17**):

1. Thaw an appropriate amount of PMSG immediately before use—do not keep the solution at room temperature for longer than 30 min.
2. Inject each female mouse intraperitonealy with 100 µl (5 IU) PMSG solution.

Day 3, 10 a.m. (*see* **Note 17**):

1. Thaw an appropriate amount of HCG as described above.
2. Inject each female mouse intraperitonealy with 100 µl (5 IU) HCG solution.
3. Mate the superovulated females with stud males and check for copulation plugs the following morning.

3.3. Preparation of Transfer Capillaries

1. Hold a glass capillary in the flame of a Bunsen burner until the glass starts to melt.
2. Withdraw the capillary from the flame and immediately apply a sharp pull to the ends. The faster the pull, the thinner the diameter of the capillary will be.
3. Break the thin part of the capillary approximately 5 cm from the thicker shaft.
4. Using a diamond pen, make a mark on the capillary straight across the thin end, 2–3 cm from the thickening shaft.
5. Break the thin end at the mark.
6. Hold the end of the capillary in the flame of the Bunsen burner until it is no longer sharp (*see* **Note 5**).

7. Capillaries that will be used for embryo transfer should be sterilized by baking in an oven at 180°C for 3 h.

3.4. Dissection of Oviducts

1. Kill the plug positive female mice by cervical dislocation or CO_2 gas (*see* **Note 6**).
2. Wet the abdominal wall with 70% EtOH.
3. Wipe off excess EtOH with a paper tissue.
4. Make a cut through the skin across the midline at the lower part of the abdomen with scissors.
5. Tear the skin off the abdomen by grasping the two skin flaps above and below the cut, and simply pulling them apart.
6. Using the fine scissors and forceps cut open the abdominal wall until all the organs are visible.
7. Move the intestines by scooping them aside with the forceps.
8. Grasp the top part of the uterus with the forceps.
9. Carefully tear the connective tissue on the ovary and oviduct with the scissors.
10. Make a cut between the ovary and the oviduct coils, and a second cut across the top part of the uterus horn (*see* **Note 6**).
11. Place the oviducts in a tissue culture dish with M2 medium.

3.5. Preparation of Hyaluronidase Solution

1. Dissolve the hyaluronidase in M2 culture medium obtaining a stock solution of 10 mg/ml.
2. Filter sterilize, aliquot, and store at –20°C (the stock solution is stable for several months).

3.6. Recovery of Fertilized Oocytes from the Oviducts

1. Place the oviducts in a tissue culture dish with 2 ml of fresh M2 medium.
2. Grasp one oviduct at a time with watchmakers forceps and locate the swollen ampulla (*see* **Note 7**).
3. Rip up the ampulla with the tip of the needle and make sure that the zygotes surrounded by the cumulus masses float out in the medium.
4. Thaw the hyaluronidase solution, and add to the embryo containing M2 medium to a final concentration of 300 µg/ml.
5. Incubate at room temperature for 2–3 min.
6. Collect the zygotes using a transferpipet (*see* **Note 7**).
7. Move the zygotes through three drops of M2 medium.
8. Repeat **step 7**, but using M16 medium, and place the zygotes in the CO_2 incubator.

3.7. Preparation of Holding Capillary

1. Follow **steps 1–4 in Subheading 3.3.** but use borosilicate glass capillaries. Pull the glass relatively thin and absolutely straight.

BAC Transgenesis Through Pronuclear Injection 93

2. Measure the dimensions of the capillary at the cut end; the outer diameter should be 100 μm.
3. Melt the tip of the capillary on the microforge (or alternatively in the flame of the Bunsen burner) until only a small 15-μm-wide opening remains.
4. Pull injection needles on a microcapillary puller. The shaft should be long and very thin, with an inner diameter (ID) of 0.5 μm at the tip (*see* **Note 8**).
5. Alternatively, use purchased needles and capillaries (Eppendorf VacuTips and FemtotipsII)

3.8. Preparation of Injection Buffer and Dilution of BAC DNA

1. 100-mm polyamine mix (1000× stock solution) (*see* **Note 18**): Dissolve the Spermine and Spermidine together in sterile ddH$_2$O, so that the end concentration is 30 mm Spermine and 70 mm Spermidine. Filter sterilize (0.22 μm) aliquot and store at –20°C.
2. Basic injection buffer: Mix the following in a plastic disposable 50-ml Falcon tube: 0.5 ml of 1m Tris–HCl, pH 7.5, 10 μl of 0.5 m EDTA, pH 8.0, 1 ml of 5 m NaCl. Add sterile H$_2$O up to 50 ml. Aliquot, filter sterilize (0.22 μm), and store at 4°C.
3. Ready to use injection buffer: 50 ml basic injection buffer, add 50 μl polyamine mix. Use directly, do not store.

3.9. Microinjection

1. Fill approximately 5 μl of the BAC DNA solution into the injection capillary using a microloader.
2. Attach the injection needle to the right-hand side micromanipulator handle, and position it correctly in the media drop in the microinjection chamber.
3. Attach the holding capillary to the other micromanipulator handle and position it in the media drop.
4. Move a group of 10–15 embryos from the incubator into the media drop.
5. Open up the tip of the injection needle by carefully tapping it on the holding needle (*see* **Note 9**). It is important to break up the tip to a size where the DNA readily flows without having to apply too high pressure. The size should not be so large, however, that it damages the embryos.
6. Clear the needle tip by flushing a small amount of DNA at high pressure. Set a constant flow of DNA to a low level (*see* **Note 10**).
7. Pick up one embryo with the holding capillary and position it, so that both pronuclei can be seen at the midplane of the embryo.
8. Set the focus on the nearest pronucleus, making sure that the border can be seen clearly (*see* **Note 11**).

9. Move the injection needle to the same y-axis position as the targeted pronucleus (either 6 O'clock or 12 O'clock of the embryo) and adjust the height of the needle, so that the tip of the needle appears completely sharp (without changing the focus!). This procedure will allow for the needle to exactly target the pronucleus.
10. Move the injection needle to a 3 O'clock position.
11. Insert the injection needle straight into the embryo, pushing through the zona pellucida, cell wall, cytoplasm, and into the targeted pronucleus (*see* **Notes 12 and 13**).
12. Apply a higher pressure to the DNA flow and keep injecting until a clear increase in the size of the pronucleus can be seen.
13. Withdraw the injection needle and release the embryo.
14. Repeat with the remaining embryos and move them back into the incubator, placed in M16 media (*see* **Notes 14 and 15**).

3.10. Preparation of Ketamine/Xylazine Anesthetics

1. Dissolve 100 mg Ketamin hydrochloride and 0.8 ml Xylazine 2% in 11 ml sterile ddH$_2$O.
2. Mix and store at 4°C for up to 2 weeks.

3.11. Embryo Transfer to the Oviduct of Pseudopregnant Female Mice

1. Anesthetize the mouse by injecting 10 µl Ketamine/Xylazine solution per gram bodyweight (*see* **Note 19**).
2. Load the transfer capillary with M2 medium, adding two large air bubbles for better control of the movement. Add one small air bubble, pick up the embryos, and add one more small air bubble. The embryos should now be enclosed in a small amount of medium, close to the tip of the capillary, and surrounded by two small air bubbles (*see* **Note 16**).
3. Wait until the mouse has reached surgical anesthesia. This can be checked by pinching the tail or a hind leg. The mouse should not react to these stimuli.
4. Moisten the eyes of the mouse with a drop of PBS (*see* **Note 20**).
5. Place the mouse on its right side, with the legs toward the operator.
6. Remove the fur around the area just behind the last rib.
7. Wipe with 70% EtOH.
8. Make a 12-mm incision in the skin right under the muscles surrounding the vertebra and immediately behind the last rib. A blood vessel and a nerve should be seen running in the abdominal wall, vertically across this area.
9. Make a 10-mm incision in the abdominal wall. Locate the fat pad between the ovary and the kidney.
10. Lift out the fat pad and secure it with a seraffin clip outside the body wall. Take care not to touch the ovary or oviduct coils!
11. Place the mouse under a stereomicroscope and focus on the oviduct area.

12. Tear the bursa surrounding the ovary and oviduct coils with the two pairs of watchmaker forceps.
13. Locate the infundibulum. If bleeding is obscuring the view, place a very small piece of paper tissue in the cleft between ovary and oviduct coils.
14. Insert the transfer capillary into the infundibulum and expel the embryos. Check that the two small air bubbles have entered into and behind the first turn of the coils (the air bubbles can easily be seen through the wall of the oviduct).
15. Remove the seraffine clip and place the organs back into the abdominal cavity.
16. Seal the skin incision with a wound clip.
17. Repeat the procedure on the other side.
18. Place the mouse in a quiet, dark and warm area until it recovers (*see* **Notes 21** and **22**).

4. Notes

1. The reader is strongly recommended to gather further information and experience with mouse embryo micromanipulation before attempting the actual microinjection. Useful protocols, advice, and consideration are available at the following Websites: Thom Saunders, http://www.med.umich.edu/tamc/BACDNA.html, and Dr. Lluis Montoliu, http://www.cnb.uam.es/~montoliu/prot.html. in **ref. 7**
2. We have used embryos obtained by crossing CD1 outbred females with C57 inbred males, and this combination has worked very efficiently in our hands. The number of embryos that can be obtained per donor female in these crosses is usually high, the embryo viability is good, the embryos are easy to inject and survive the microinjection process well. If an inbred background is desirable for the transgene, it is possible to use for example C57BL/6J embryos, although the microinjection process will be more difficult, and the viability of the embryos is lower.
3. The highest numbers are usually obtained from very young female mice, just at weaning age. The embryo quality is, however, often variable and low, and very small females may suffer significantly during the stressful experience of being mated by a large and often less careful male. For this reason, we recommend using 5- to 6-week-old females. Both the time-point at which each of the hormone injections are given and the light cycle in the room where the animals are kept will have influence on the ovulation. The optimal time during which microinjection can take place is as short as 2–4 h. If the embryo development is not advanced enough, the pronuclei will be very small and difficult to target. In this case, the hormones should be given earlier in the morning, and/or the light cycle should be set back. On the contrary, if the embryos are at too late a stage of development, the pronuclei will already be fused, and the first division will have taken place. The solution here is to give the hormones later, and/or set the light cycle forward. We use a light cycle of 14 h light/10 h dark, with the light switching on at 5 a.m.

4. Each batch of oil and all ingredients in homemade embryo culture media should be tested for supporting optimal embryo development before use. M16 should be pre-equilibrated in 5% CO_2 incubator for a minimum of 2 h before use.
5. Care should be taken to use glass capillaries with a smooth end, as sharp edges can harm the zona pellucida. It is also important to minimize the time which the embryos spend outside the incubator as much as possible. One-cell stage embryos should not spend more then 30 min in room air and temperature.
6. Care should be taken not to damage the ampulla during dissection, as this would inevitably result in losing the embryos. The choice of method to kill the donor mice (cervical dislocation or CO_2 gas) does not have any influence on embryo viability and is therefore a matter of personal preference and local legislation.
7. The cumulus masses can easily be seen through the wall of the swollen ampulla. By grasping the oviduct with watchmaker's forceps and tearing the wall of the ampulla with a sharp 27-G needle, the embryos are allowed to float out in the M2 media-filled tissue-culture dish. It is essential to remove the embryos from the hyaluronidase solution as soon as the cumulus masses are dissolved.
8. Settings for P97:

    a. Borosilicate glass capillaries with inner filament; outer diameter 1.0 mm, inner diameter 0.78 mm (Clark Electronical Instruments GC100TF-15).
    b. Filament = B032TF.
    c. Pressure = 100.
    d. Heat = 305 (~10 values below ramp test result).
    e. Pull = 100.
    f. Velocity = 150.
    g. Time = 100.

9. Before attempting to inject the embryos, the tip of the injection needle should be broken up to a larger size. Because BAC DNA is more viscous than solutions with smaller DNA fragments, it is more likely to clog the needle during injection. There is also a risk for shredding the DNA in case it is pushed through a small opening with a high pressure. The final size of the needle tip should be around 3–4 µm.
10. If the Eppendorf FemtoJet is used, the Pc should be set to 5–10 and the Pi to 40–50. The injection should take place in "Manual" mode, where the injection time is adjusted to each pronucleus, and kept until a clear swelling is achieved.
11. It is easier to target the larger of the two pronuclei and/or the one that is closest to the injection needle. If the embryos are well timed, the pronuclei should both appear in the center of the embryo, both be large and clearly visible. At the optimal time for injection, it is often difficult to tell which pronucleus is the male and which is the female. There is no advantage in choosing one or the other. The only criteria for the choice should be the ease of injection: the pronucleus that is largest and/or nearest to the injection needle should be targeted. If the pronuclei

are small, the embryo has not yet reached the optimal time-point for injection. It may help to bring the embryos back to the incubator for an hour and start the injection process later. If ever the embryos are left for too long at 37°C before the injection is started, the two pronuclei will fuse, and injection will be impossible. The timing of the developmental stage can be influenced by the superovulation protocol (*see* **Subheading 3.2.**).

12. Care should be taken to avoid touching any of the nucleoli within the pronucleus. These structures are extremely sticky and will immediately attach to the needle. If this should happen, the injected embryo must be discarded, and the injection needle changed.
13. The cell membrane is sometimes very difficult to penetrate. Because it appears very elastic, it will simply follow the tip of the injection needle into the embryo and sometimes all the way into the pronucleus. If this happens, a small "bubble" will form at the tip of the needle, and the pronucleus will not expand. The needle can in these cases be pushed further on, through the pronucleus, and then slowly withdrawn into the pronucleus again, which usually solves the problem.
14. Embryos of lower quality should be discarded. Under the high magnification during injection, it is easy to detect those embryos with only one or more than two pronuclei, embryos with sperm under the zona, and so on. It is important to sort away these embryos and those that could not be successfully injected. Later, it will be impossible to distinguish between well-injected healthy embryos and those of poor quality.
15. Embryos can be transferred immediately after injection. It is also possible to culture the embryos overnight in vitro and transfer only the well-developing two-cell stage embryos the next day. Great care should be taken to make sure all embryos are well in the oviduct and to minimize the damage during the surgery. The anesthesia should be carefully optimized to give a short but deep sleep, and the mouse should be kept warm until recovery.
16. Generally not more than 25–30% of the transferred one-cell stage embryos (slightly more if two-cell stage embryos are transferred) will develop to term. To keep the litter size at an optimal level (6–8 pups), the number of embryos to transfer should be kept in the range of 25–30. We highly recommend transferring half of the embryos to each oviduct, although it is possible to put all into one side.
17. It is essential to prepare the hormones as soon as possible after thawing, to work fast, and to freeze the solution as soon as it is prepared. Repeated freeze-thawing should be avoided. Hormones are stable in −20°C for at least 3 months.
18. Polyamine mix is stabile at −20°C for several months. The microinjection buffer without polyamines added is stabile at +4 for several months. The ready-to-use buffer (polyamines added) should be prepared fresh for each experiment and not stored.
19. We recommend using a mixture of Ketamine and Xylazine, as this combination has proven to be nontoxic, has a high safety marginal in terms of overdoses and

gives a deep but short surgical anesthesia. These measures will all contribute to increasing the possibility of maintaining a pregnancy.
20. During anesthesia, the eyes of the mouse will stay wide open. The cornea will quickly dry out, leading to severe pain when the animal recovers. It is therefore highly recommendable to moisten the eyes with a drop of PBS as soon as the mouse has fallen asleep.
21. The normal gestation time in most mouse strains is 19–20 days. Outbred CD1—which is one of the most commonly used mice as recipients for embryo transfer—tends to deliver very reliably on E19. A commonly encountered problem following embryo transfer is low implantation rates and dying/reabsorbing fetuses. If the number of fetuses that develop to term is too low, the size of each conceptus will be larger then normal, which often results in difficulties at birthing. Litter sizes of less then four pups often cause delivery difficulties in recipient females. These pups generally do not survive, and the mother suffers pain unless a caesarean section is performed. The procedure is fairly simple and the survival rate of the sectioned pups very high *(7)*. During the first half of the pregnancy, the weight gain of the mother is moderate, but weight increases dramatically during the last week. By weighing the recipients every other day, it is possible to predict in which cases the litter size could be expected to be small, because the curve of weight gain will reliably indicate litter size and possible problems during pregnancy. A female that has not given birth on E20 should promptly be opened for caesarean section in any case.
22. It is essential to get the pups dry and warm immediately after sectioning. The breathing can be stimulated by wiping them roughly with a soft paper tissue. A common mistake is to be too soft with the pups. Because they have not gone through the physical stress of birth, they usually do not start breathing regularly unless they are turned around, pinched, and gently massaged with cotton tips or paper tissue pieces. Place the pups with a new foster mother when they are completely freed from all blood, regularly breathing, and have gained a normal body temperature.

Acknowledgments

We gratefully acknowledge Dr. Lluis Montoliu (Centro Nacional de Biotecnologia, Madrid, Spain), Dr.Thom Saunders (University of Michigan, USA), and Dr. Youming Zhang (Gene Bridges GmbH) for advice and protocols for BAC DNA purification and microinjection.

Reference

1. Glaser, S., Anastassiadis, K. and Stewart, A.F. Current issues in mouse genome engineering. *Nat Genet* 2005 **37**, 1187–93.
2. Zhang, Y., et al. DNA cloning by homologous recombination in Escherichia coli. *Nat Biotechnol* 2000 **18**, 1314–7.

3. Gordon, J.W., et al., Genetic transformation of mouse embryos by microinjection of purified DNA. *Proc Natl Acad Sci USA*, 1980 **77**(12), 7380–4.
4. Gordon, J.W. and Ruddle, F.H. Integration and stable germ line transmission of genes injected into mouse pronuclei. *Science* 1981, **214**(4526), 1244–6.
5. Lin, T.P. Microinjection of mouse eggs. *Science* 1966, **151**(708), 333–7.
6. Brinster, R.L., et al. Somatic expression of herpes thymidine kinase in mice following injection of a fusion gene into eggs. *Cell* 1981, **27**(1 Pt 2), 223–31.
7. Nagga, A., el al. *Manipulating the Mouse Embryos, a Laboratory Manual.* 2nd ed. Cold Spring Harbor Laboratory Press, New York, 2002.
8. Pinkert, C.A., and H.G. Polites. Transgenic animal production focusing on the mouse model. In *Transgenic Animal Technology*, C.A. Pinkert, Editor. Academic Press, Inc., San Diego, 1994.
9. Peterson, K.R., et al. Transgenic mice containing a 248-kb yeast artificial chromosome carrying the human beta-globin locus display proper developmental control of human globin genes. *Proc Natl Acad Sci USA*, 1993, **90**(16), 7593–7.
10. Schedl, A., et al., A method for the generation of YAC transgenic mice by pronuclear microinjection. *Nucleic Acids Res.* 1993, **21**(20), 4783–7.
11. Sternberg, N.L. Cloning high molecular weight DNA fragments by the bacteriophage P1 system. *Trends Genet.* 1992, **8**(1), 11–6.
12. Smith, D.J., et al. Construction of a panel of transgenic mice containing a contiguous 2-Mb set of YAC/P1 clones from human chromosome 21q22.2. *Genomics* 1995, **27**(3), 425–34.
13. Nielsen, L.B., et al. Human apolipoprotein B transgenic mice generated with 207- and 145-kilobase pair bacterial artificial chromosomes. Evidence that a distant 5´-element confers appropriate transgene expression in the intestine. *J. Biol. Chem.* 1997, **272**(47), 29752–8.
14. Yang, X.W., Model, P. and Heintz, N. Homologous recombination based modification in Escherichia coli and germline transmission in transgenic mice of a bacterial artificial chromosome. *Nat. Biotechnol.* 1997, **15**(9), 859–65.
15. Giraldo, P. and Montoliu, L. Size matters: use of YACs, BACs and PACs in transgenic animals. *Transgenic Res.* 2001, **10**(2), 83–103.
16. Ristevski S. Making better transgenic models: conditional, temporal, and spatial approaches. *Mol. Biotechnol.* 2005, **29**(2), 153–63.
17. Camper, S.A. and Saunders, T.L. Transgenic rescue of mutant phenotypes using large DNA fragments. In *Genetic Manipulation of Receptor Expression and Function*, D. Acilli, Editor. John Whiley & Sons, New York, 2000, p. 1–22.
18. Lacy, E., et al. A foreign beta-globin gene in transgenic mice: integration at abnormal chromosomal positions and expression in inappropriate tissues. *Cell* 1983, **34**(2), 343–58.
19. Wilkie, T.M., Brinster, R.L., and Palmiter, R.D. Germline and somatic mosaicism in transgenic mice. *Dev. Biol.* 1986, **118**(1), 9–18.

20. Kaufmann, R.M., Pham, C.T., and Ley, T.J. Transgenic analysis of a 100-kb human beta-globin cluster-containing DNA fragment propagated as a bacterial artificial chromosome. *Blood* 1999 **94**, 3178–84.
21. Foster, H.L., Small, J.D., and Fox, J.G. *The Mouse in Biomedical Research.* Academic, New York, 1983.
22. Brinster, R.L., et al. Factors affecting the efficiency of introducing foreign DNA into mice by microinjecting eggs. *Proc. Natl. Acad. Sci. U. S. A.* 1985, **82,** 4438–42.
23. Harris, A.W., et al. The E mu-myc transgenic mouse. A model for high-incidence spontaneous lymphoma and leukemia of early B cells. *J. Exp. Med.* 1989, **167**, 353–71.
24. Chisari, F.V., et al., Molecular pathogenesis of hepatocellular carcinoma in hepatitis B virus transgenic mice. *Cell* 1989, **59**(6), 1145–56.
25. Whittingham, D.G. Culture of mouse ova. *J. Reprod. Fertil. Suppl.* 1971, **14**, 7–21.
26. Refferty, K.A. *Methods in Experimental Embryology of the Mouse.* Johns Hopkins Press, Baltimore, MD, 1970.
27. Flaming, D.G. and Brown, K.T. Micropipette puller design: form of the heating filament and effects of filament width on tip length and diameter. *J. Neurosci. Methods* 1982, **6**, 91–102.
28. Montoliu, L., et al., Visualization of large DNA molecules by electron microscopy with polyamines: application to the analysis of yeast endogenous and artificial chromosomes. *J. Mol. Biol.* 1995, **246**(4), 486–92.
29. Bronson, R.A. and McLaren, A. Transfer to the mouse oviduct of eggs with and without the zona pellucida. *J. Reprod. Fertil.* 1970, **22**(1), 129–37.

6

Bioluminescence Imaging to Evaluate Infections and Host Response In Vivo

Pamela Reilly Contag

Summary

The continued prospect of emerging pathogens and recent events including the acceptance of widespread drug resistance and threats of bioterrorism have introduced the necessity be creative in our development of therapies for bacterial infections. Many pathogens have both acute and persistent phases. There is a need to understand these pathogens throughout their entire life cycle within the host and determine the role that the host response including innate immunity plays in the establishment and maintenance of the infection. Contag et al. first suggested in 1995 that a novel whole animal, non-invasive imaging modality may provide more data from which to draw conclusions about infectious disease progression and pathogenicity in the context of a living animal. Here are presented methods for imaging two animal models that represent advances in both following the progression of infectious disease in the host and the response of the host to the pathogen.

Key Words: Infection; bioluminescence; Listeria; host response; imaging; BLI; GFAP; multi-parameter; two wavelength; biofilms; Streptococcus; bacterial meningitis; inflammation; in vivo; biophotonic; biomarker; animal models; non-invasive; pathogenicity; gall bladder; persistent infection; acute infection.

1. Introduction

Animal models of infectious disease have been used to measure the severity of disease caused by pathogens for over a century, since Louis Pasteur tested pathogens in a mouse model of intraperitoneal (IP) sepsis *(1, 2)*. These in vivo models represented the course of acute bacterial disease and adequately supported the discovery and development of broad-spectrum antibiotics that we have used in treatment s the 1940s. Until recently, these models have only been altered by incremental changes over time because infectious disease models that employ death as an endpoint were relatively easy to perform, could replicate the human septic disease, and were predictive of how efficacious newly discovered antibiotics would be against Gram-positive and Gram-negative bacterial infections in humans. The endpoints measured in these models were time to death and bacterial colony counts (or viral plaque counts) from organs or tissues. More sophisticated endpoints involve histopathology or measurement of immune activation of blood cells *(3–5)*.

Since then, conventional animal models to study infections in vivo have been extended to include chronic rather than acute infections, medical device infections, emerging pathogens and the contemplation of alternative therapies such as immunotherapy *(6–10)*. Researchers were lacking more sophisticated models that would allow longitudinal analysis of the disease in the whole animal, the ability to measure changes in the host and pathogen as the pathogen enters the environment of the host, and the focus on specific drug targets for the measure of drug and target interaction. These developments only emerged in the last decade.

Since 1995, the non-invasive imaging modality, bioluminescence imaging (BLI), has been demonstrated to provide more data from which to draw conclusions about infectious disease progression and pathogenicity through imaging these events in the whole living animal *(11–13)*. In vivo BLI has been described over the last several years *(11,13–15)* as a method that allows real-time, non-invasive imaging, longitudinal or temporal analysis, spatial information of the disease course, and internal controls for more precise statistical evaluation. Bioluminescence or biophotonic imaging is a modality that includes a sensitive CCD camera; a system or box to contain the subject in a darkened controlled environment; and software to drive the operation of the system and perform data analysis.

BLI depends on the ability to image light emitted from the essentially opaque or turbid tissues of mammals and is based on the principles of scattering (index of refraction differences between fluid and cellular organelles) and absorption (chromophores that absorb strongly at wavelength under 600 nm) that are characteristic of opaque tissue. In vivo imaging systems detect surface radiance or the diffuse emission of photons from the surface of the body. Surface radiance is dependent on the photon flux per cell, the number of cells, and the migration of the photons through tissue to the surface. For these in vivo measurements, optical probes greater than 600 nm emitted wavelength are preferred for deeper tissue measurements (>1 cm) *(16)*.

The optical probes used for most bioluminescent in vivo imaging studies are the North American Firefly luciferase, Click Beetle luciferase, Renilla luciferase, and the *Photorhabdus luminescens* luciferase *(17–19)*. Here, we use the North American Firefly Luciferase in the vector PGL3 from Promega Corporation and the *P. luminescens* bacterial luciferase *(19)* modified by Francis et al. *(20,21)*. The fundamental difference between the firefly and the bacterial luciferases as it relates to in vivo imaging is that, for the firefly luciferase, the substrate luciferin has to be injected into the animal before imaging, whereas the enzymes to synthesize decanal, the substrate for the bacterial luciferase, are encoded on the cloned bacterial luciferase operon.

The optical probes described above were incorporated into the genomes of various bacterial strains, cell lines, and animals. The optical instrumentation systems were then paired with these sensitive light-emitting pathogens, tumor cells, and transgenic animals for the study of various diseases and therapies *(11,15,16,20,22–25)*.

BLI has been utilized not only for acute infections *(26,27)* but also as a preferred modality to address chronic infectious diseases caused by biofilm formation in and around both anomalous anatomic sites and implanted devices *(28–33)*. We have used BLI including novel tagged pathogens and genetically modified mice to build more sophisticated animal models of infectious disease *(25,26,29,31,34–38)*. Researchers have demonstrated the ability to track pathogens throughout the whole animal, count pathogenic units, test novel small molecules against these pathogens, and thus bring the field to the next set of questions to answer regarding the life cycle of the pathogens in the host, and the role that the host immune response, especially innate immunity, has in influencing the course and cure of that infection.

Many pathogens have an acute phase of infection and a persistent phase. The prospect of drug resistance, emerging pathogens, and bioterrorism have introduced the necessity to track pathogens throughout the entire life cycle within the host and determine the role that the host response plays in the establishment and maintenance of the infection. Here, we describe methods to address these questions in more detail using complex and multiparameter BLI animal models. This knowledge may help to create therapies that may block all stages of the infection cycle, thereby addressing two essential problems; first, removing the animal and human hosts as reservoirs for the further inoculation of the healthy population, and second, limiting the etiology of chronic microbial infection for common diseases of aging including vascular and central nervous system (CNS) diseases and serious acute microbial infection in our aging population.

1.1. Unique Sites of Replication and Persistent Infection of Listeria in a Murine Model

Hardy et al. *(34, 37)* employed whole body BLI to reveal a previously undescribed replication and dissemination pattern of *Listeria monocytogenes* in the murine gall bladder. Listeria is a Gram-positive bacterium and is an foodborne intracellular pathogen that can cause life-threatening systemic disease in immunocompromised individuals. Much Listeria research has concentrated on transmission from food to the host, the intracellular uptake into macrophages, and the intracellular life cycle. Hardy et al., using BLI demonstrated an extracellular replication state and dissemination events in animals, in the lumen of the gallbladder. This is significant because the spread of listeriosis within the host and delivery back into the environment for infection of new hosts depends on the ability of the bacterium to leave the gall bladder. Both food and hormone, cholecystokinin, were shown to induce biliary excretion of *L. monocytogenes* from the murine gallbladder in a short period of time (within 5 min of induction of gallbladder contraction) (*see* **Figs. 1–3**). These studies are also significant

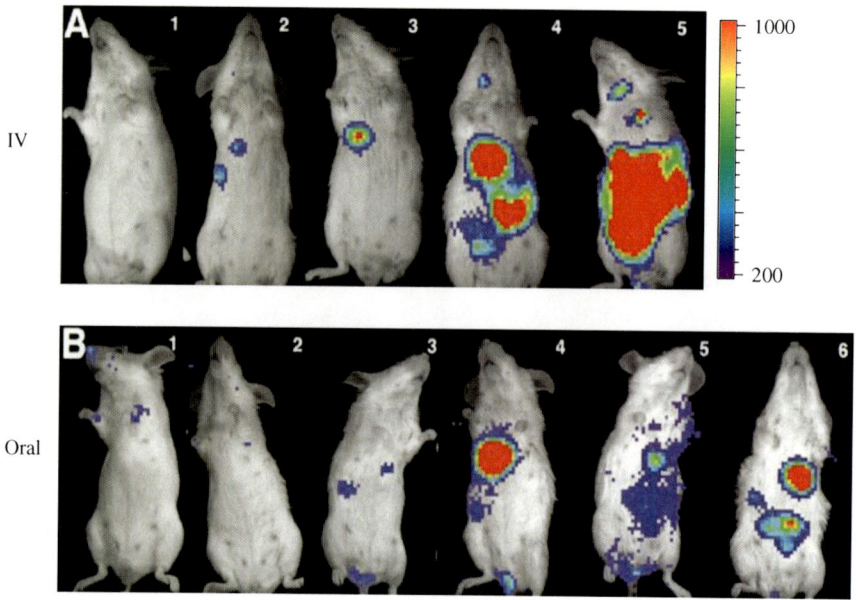

Fig. 1. Bioluminescent imaging of time course of listeriosis in Balb/c mice. (**A**) Time course of lethal intravenous infection shown day 1–5. Luminescent *Listeria monocytogenes* 2C at 4×10^4 cfu per mouse (1 LD50) imaged post-infection. (**B**) Oral infection at 5×10^9 cfu per mouse, shown on day 1–6. Images from Hardy et al. *(34)*.

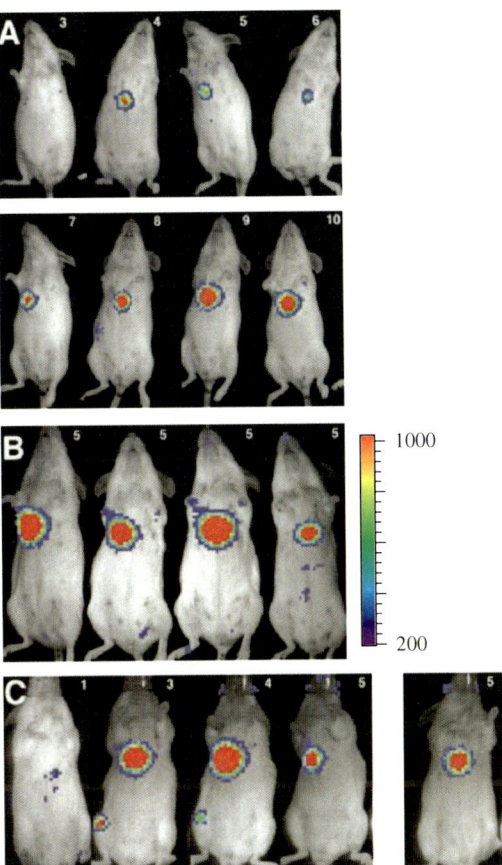

Fig. 2. Replication of *hly* and *hly* int A/B deletion mutants in the gall bladder. **(A)** A luminescent hly deletion of *Listeria monocytogenes* 10403S was inoculated intravenously into Balb/c mice at 10^8 cfu per mouse. The mice were imaged on the indicated days. Two of the five animals displayed the characteristic signal of gall bladder growth. One animal imaged over time post-infection is shown. **(B)** A higher dose (10^9) produced the characteristic signal in all the mice. Four mice are shown on day 5 post-inoculation. An animal displaying a similar signal yielded 3×10^7 cfu from the gall bladder luminal contents. **(C)** Replication of a Δhly ΔinlA.B mutant at inoculation doses of 10^9 and 5×10^8 cfu, left and right panels, respectively.

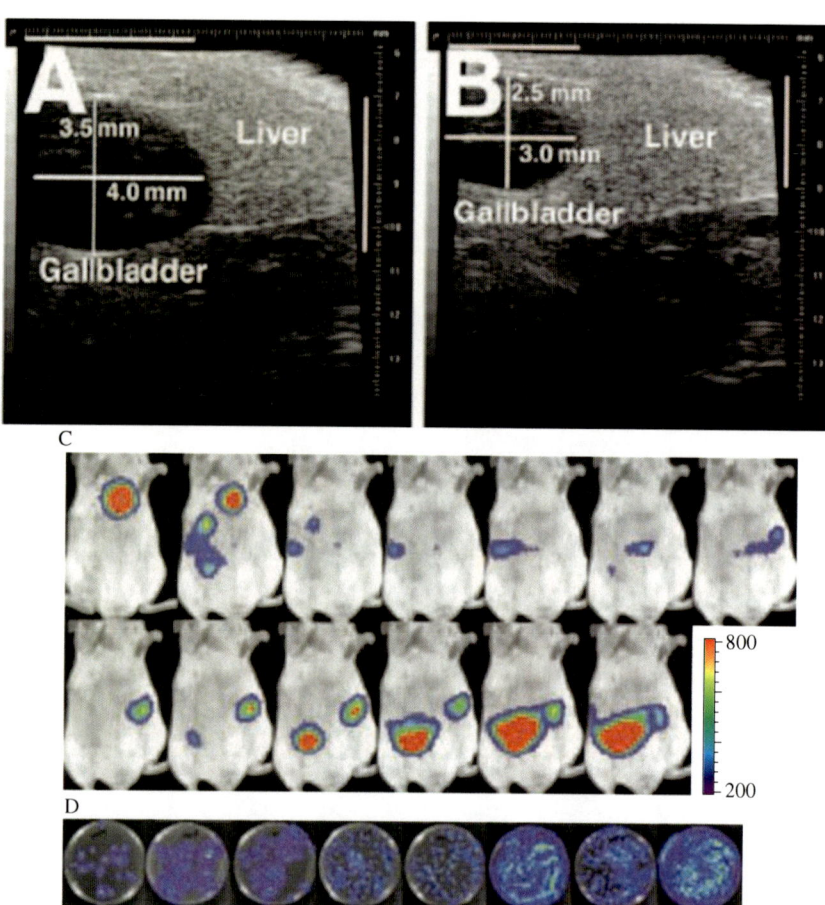

Fig. 3. Cholecystokinin-induced gallbladder contraction and fecal shedding. Normal, uninfected Balb>c mice were starved for 4–6 h and anesthetized, and the gallbladder was imaged using ultrasound. Cholecystokinin octapeptide fragment 26–33 (CCK8) was then injected IP, and images at 3 images per min were acquired to track the gallbladder size. Representative data from one animal are shown. (A) Image before CCK8 administration, (B) image 400 s post-administration, and (C) serial BLI of a representative CD1 mouse infected with virulent *Listeria monocytogenes* strain 2C and injected with CCK8. (D) Efficient expulsion caused an increase in CFU in fecal pellets from mouse C.

to other genera of pathogens, in which the persistent life cycle within the host is not completely understood, such as Mycobacterium.

1.2. Simultaneous BLI Monitoring of Pneumococcal Meningitis and a Biomarker for Neuronal Damage, Glial Fibrillary Acidic Protein

Infection of the brain and spinal cord is one of the most difficult fast onset infectious diseases to diagnose and treat *(41–43)*. In 30% of all patients with meningitis, there are serious sequelae including cognitive impairment. Morbidity and mortality occur in this disease unless proper diagnosis and treatment is begun within 48 h of the first symptoms *(41–44)*.

Cognitive impairment implies neuronal damage and thus simultaneous expression of a biomarker of neuronal damage, and the disease course of bacterial meningitis was investigated. Glial fibrillary acidic protein (GFAP) is an intermediate filament protein expressed in the astrocytes of the CNS and has been used as a surrogate biomarker to monitor neuronal damage because this type of injury is a potent inducer of GFAP. Zhu et al. *(45)* first demonstrated the ability to non-invasively monitor GFAP using bioluminescent imaging. A transgenic mouse was created that carried the luciferase gene under control of the mouse GFAP promoter. Introduction of kainic acid is known to induce neuronal cell death, and in addition, Zhu demonstrated that hippocampal cell death was correlated with luciferase RNA and expression of endogenous GFAP in transgenic GFAP luc mice. Non-invasive bioluminescent imaging of the GFAP luc transgenic mice exhibited a 40-fold induction of GFAP after kainic acid treatment. The experimental model of pneumococcal meningitis was performed using the transgenic GFAP luc mice as the host.

2. Materials

Brain–heart infusion (BHI) media, phosphate-buffered saline (PBS), sterile endotoxin-free water for injection, luciferin, and the anesthetics described are all available from multiple sources. Other than the specific imaging systems including the optical imaging and ultrasound, standard laboratory equipment was used. Pronuclear microinjection and animal breeding services were performed at Xenogen Biosciences in Cranbury, NJ.

2.1. Bacterial Strains

The *L. monocytogenes* strain 10403 *(39)* made bioluminescent by transformation with the plasmid pAUL-A Tn4001 luxABCDE Km^R *(20,21)*, is designated here as strain 2C (*see* **Note 1**). The transposon insert maps to the *fla A* gene by inverse polymerase chain reaction (PCR) *(34)*. Strain 2C (*flaA::lux/kan*

10403S) was used as the donor for transduction of luminescence to a listeriolysin negative strain 10403S Δhly *(40)* and 10403SΔhly *inlA/B* deleted for both listeriolysin and internalins required for intracellular replication, and these transduced strains are designated as 2CΔhly and 2CΔhly $\Delta inlA/B$, respectively.

A clinical isolate of *Streptococcus pneumoniae* A 66.1 (type 3) (Marc Lipsitch, Harvard School of Public Health) was transformed by natural competence using chromosomal DNA from *S. pneumoniae* D39 Tn 4001 *lux* ABCDE Km^R as described by Francis et al. *(20, 21)*. Integration of the lux cassette into the bacterial chromosome had no effect on the bacterial pathogenicity or survival in mice.

2.2. Intensified Charged-Coupled Device Imaging Systems

The system features of the IVIS 100 used in the *L. monocytogenes* study that are critical to the described imaging protocol *(23)* are a 1.0-inch square back-thinned, back-illuminated CCD cooled to –90°C through a closed system. The quantum efficiency of the CCD is 85% at 500–700 nm. The minimum detectable photons for the IVIS 100 system are specified as less than 100 photons/s/cm^2 for a 5-min exposure and binning of 10 (grouping of 10 × 10 pixels) (*see* **Notes 2** and **3**). The animal is photographed in black and white, and the photon image is taken and then superimposed over this black-and-white reference image to orient the viewer and localize the emitted photons. The steps for image collection (*see* **Note 3**) are controlled automatically by LivingImage software (Xenogen Corp., Alameda, CA). The imaging chamber is a light-tight box that includes a heated sample shelf (*see* **Notes 4** and **5**) and gas anesthesia connections (*see* **Note 6**). The stage is controlled by LivingImage software. The heated stage reduces some of the negative effects of anesthesia at body temperature on the animals (*see* **Note 4**). The gas anesthesia system delivers isoflurane gas to a manifold housed in the system imaging chamber, which enables proper dosing per mouse and reduces the amount of handling that would occur with injectable anesthesia.

In vivo bioluminescent imaging of the infected GFAP mouse model (*see* **Subheading 2.4**) is performed with an IVIS Imaging System 200 series (Xenogen Corp.). The features of this system are particularly suited to the two wavelength study of the brain and spinal cord. This instrument is equipped with 17 band pass filters, 20 nm wide with central wavelengths from 420–740 nm. Bacterial and firefly luciferase differ spectrally and require chemically unrelated substrates. Bacterial luciferase has a peak emission wavelength of 489 nm, and firefly luciferase has a peak emission wavelength at 610 nm *(23, 38)*. Their spectral properties are used to differentiate expression of each luciferase. The IVIS imaging system series 200 has an adjustable field of view from 4–26 cm that permits close up views of anatomical sites of one mouse or five mice to be imaged at the same time. In addition, the 200 series system has a custom lens for high resolution at the 4 cm field of view. The system also employs a green laser to project an alignment grid for proper placement of animals *(38)*.

2.3. Ultrasound Imaging of Mouse Gall Bladder for Monitoring Size of Gall Bladder During Listeria Infection

Ultrasound images of cholecystokinin-induced gall bladder contraction are obtained using a VisualSonics model VEVO 660 small animal system (with a 630S probe) as recommended by the manufacturer.

2.4. Mouse Transgenic GFAP-luc Animal Production

Zhu et al. *(45)* have described a mouse model that is the basis for the experimental model below that contains the luciferase gene driven by the murine GFAP promoter in which luciferase expression is followed non-invasively using bioluminescent imaging. An expression construct containing 12 kb of the murine GFAP promoter and the human beta globin intron 2 was cloned into pGL3-basic (Promega) upstream of the luciferase expression cassette *(45)*. The purified transgene was injected into the pronuclei of single-cell fertilized embryos isolated from FVB/N mice. Eight transgenic lines were screened for luciferase expression by BLI at 24 and 48 h following neuronal damage induced by kainic acid treatment (IP injection at 30 mg/kg). One mouse line was chosen and expanded for all subsequent experiments.

2.5. Experimental Mouse Model of Pneumococcal Meningitis

S. pneumoniae A66.1 lux was grown to exponential phase at 37°C as described in **Subheading 3.1**, and various doses of bacteria were introduced through two routes of infection to the GFAP transgenic mice. Female FVB/N-tg GFAP luc mice (22–30 g) were used for the GFAP mouse meningitis experiment to track both infection and inflammation simultaneously using two bioluminescent reporters of different wavelengths *(38)*.

3. Methods

The methods below outline the steps for performing and imaging two animal models of infection. The first model is the real-time tracking and monitoring of *L. monocytogenes* during a murine infection. The second model is simultaneous monitoring of a biomarker for neurodegeneration (GFAP) and streptococcal meningitis.

3.1. Infection of Mice Before Imaging

3.1.1. Infection of Mice with L. monocytogenes

1. Grow *L. monocytogenes* to logarithmic phase in BHI (Difco) is logarithmic phase.
2. Centrifuge bacteria (5 min at 5000 g).
3. Resuspend in PBS to give 5×10^4 or 5×10^9 CFU.
4. Inject intravenously 200 ml into the tail vein of mice.

3.1.2. Infection of Mice with S. pneumoniae

1. A66.1 lux was grown to exponential phase at 37°C in BHI (Difco), supplemented with 10% fetal calf serum, 5% CO_2.
2. Wash in fresh BHI, and the optical density was adjusted to 10^7 CFU/ml, as measured absorbance at 600 nm *(29,38)*.
3. Bacteria were suspended in sterile endotoxin-free saline.
4. Injected intracisternally or intrathecally at several doses.

3.2. Bioluminescence Imaging of Mice to Monitor Listeria Infection

Imaging was performed on an IVIS100 imaging system, described above, Xenogen Corp., as per manufacturer's instructions.

1. Mice were anesthetized with either an IP delivery of Avertin (300 mg/kg) or gas Isoflurane (2–2.5%) (*see* **Note 6**).
2. Mice were then positioned on the heated stage of the IVIS 100 with the anesthesia manifold in place.
3. Mice were imaged for 5 min at a pixel binning of 10.
4. Total photon emission is measured for defined regions of interest and quantified using the LivingImage software package (Xenogen Corp.). Post-imaging mice are kept warm as they are returned to their cages (*see* **Notes 4** and **5**).

3.3. Ultrasound Imaging of Mouse Gall Bladder

1. Mice were anesthetized with isoflurane gas or 300 mg/kg IP avertin.
2. Mice were then prepared for imaging by removal of hair with a shaver and depilatory cream (*see* **Note 7**).
3. Cholecystokinin peptide fragment 26–33 (CCK8) was delivered to the mice through an IP injection.
4. Images were obtained at a rate of 2 or 30 Hz for a total of 300 frames per image sequence with 3–10 acquired sequences per peptide injection *(34)*.

3.4. Ex Vivo Analysis and Bacterial Counts

1. Whole gall bladders were excised for cfu determination.
2. Gall bladders were placed in 500 or 1000 µl of PBS, and luminal contents were released by lancing the organ.

BLI of Infections and Host Response   111

3. The bacteria recovered from the gall bladder were vortexed and plated on 50 μg/ml kanamycin.
4. If required, histopathology is performed on unruptured, fixed gall bladders that are imbedded in paraffin and sliced into 4-μm sections. Microscopy is done after staining sections with hematoxylin/eosin and tissue Gram stain.
5. To measure bioluminescence and optical density of the culture remove 100 μl and 1 ml samples, respectively, and measure photon counts in a Hamamatsu camera (Hamamatsu, Hamamatsu City, Japan) and optical density in a spectrophotometer. Photon counts are then normalized to OD *(34)*.

3.5. Experimental Mouse Model of Pneumococcal Meningitis

1. Mice were anesthetized with 2.5% isoflurane and the inoculation sites on the head and spinal cord.
2. Mice were shaved (*see* **Note 7**).

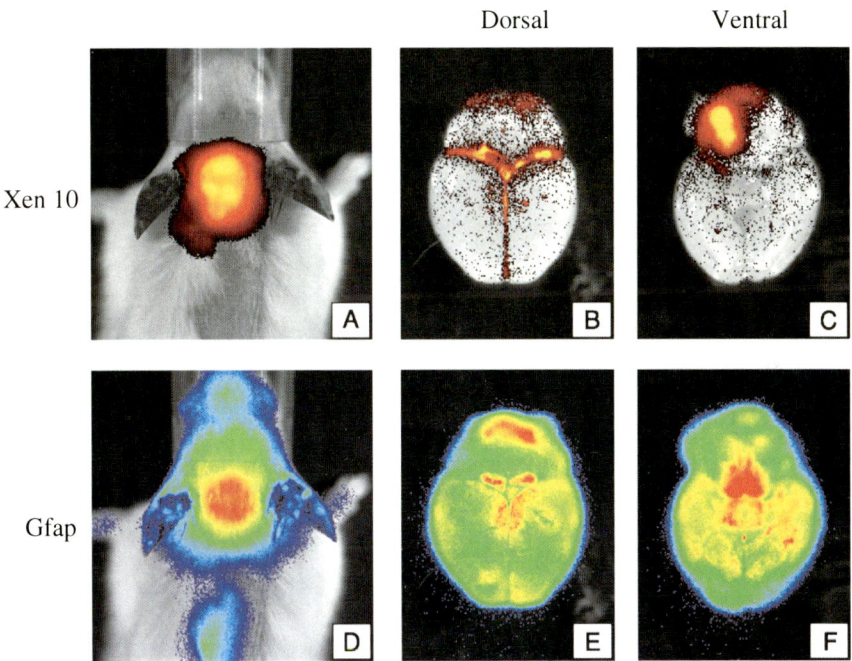

Fig. 4. Serial images of a single mouse after 19 h of untreated meningitis. Panel A and B are in vivo lux (**A**) and luc (**B**) images. Panels **C–F** are ex vivo images of the dorsal and ventral brain as labeled, and panel E shows the ventral brain site of inoculation. The bacterial signal was apparent in regions with trapped cerebral spinal fluid. The GFAP signal was spread throughout the brain with some localized expression, indicating some differing regional expression.

3. A 30.5 ga needle was used to inject 10 μl of a suspension of 10^4 CFU of *S. pneumoniae* A66.1 lux in sterile pyrogen-free saline intracisternally. The uninfected control group received the sterile pyrogen-free saline vehicle. Mice were observed as they recovered from anesthesia for behavior abnormalities; none were observed.
4. At several times, post-infection mice (n = 5) were treated with ceftriaxone (100 mg/kg in 100 μl saline), then every 12 h for 3 days for a total of six treatments (*see* **Fig. 4**) *(38)*.

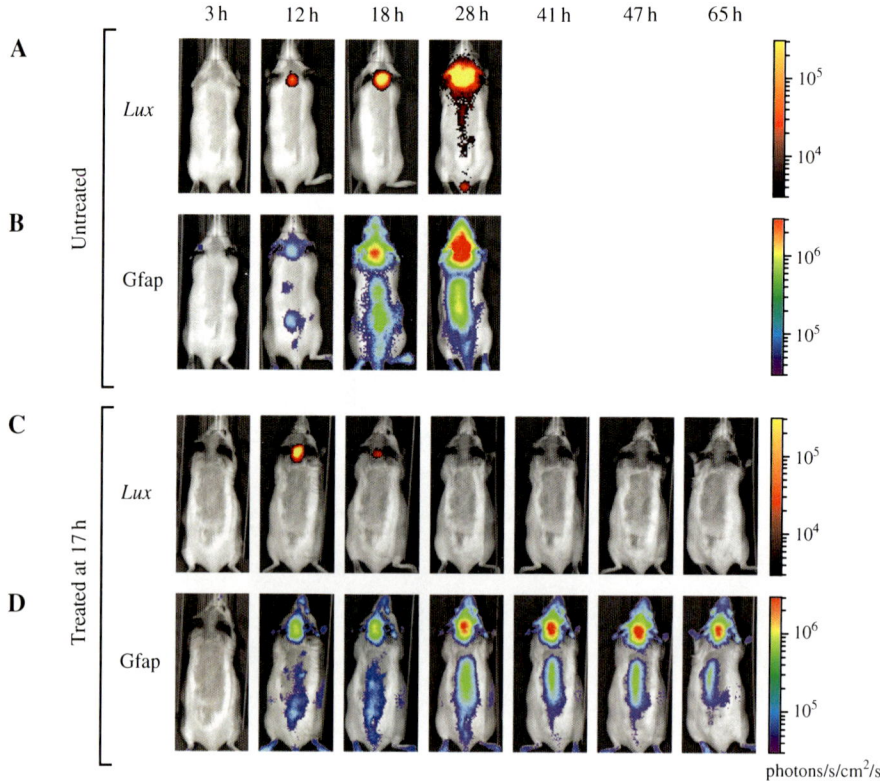

Fig. 5. Panels **A** and **B** are serial images (time of imaging indicated above) of an untreated mouse over 28 h showing lux (*Streptococcus pneumoniae*) and luc (GFAP) expression, respectively. Panels **C** and **D** are serial images of a mouse treated subcutaneously with ceftriaxone 100 mg/kg at 17 h post-infection and every 12 h after for a total of six treatments. Panel **C** shows that this treatment regimen eradicated the infection, however, without a corresponding reduction in Gfap expression (panel **D**). Only mice treated before 11 h post-infection showed a reduction in Gfap expression.

3.6. Bioluminescent Imaging at Two Wavelengths Using an Intensified Charged-Coupled Device Imaging System

1. Before imaging, infected GFAP transgenic luc mice were anesthetized through an IVIS manifold with 2.5% isoflurane. Images were acquired using a 500-nm filter (to image bacterial luciferase) and a 620-nm filter (to image the firefly luciferase) (*see* **Notes 8** and **9**).
2. Additionally, images were taken before and after the injection of luciferin (BioSynth, Staad, Switzerland) (IP injection of 150 mg/kg in 200 µl saline) (*see* **Notes 10** and **11**) *(38)*.

The only light collected before the injection of luciferin is the light emitted from the bacterial luciferase in the genome of the *S. pneumoniae*. Following the injection of luciferin, the light collected is emitted from both the bacterial and the firefly luciferases simultaneously (*see* **Fig. 5**). Those images recorded while the 500 nm filter is in place display only the bacterial luciferase emission because the firefly emission is low in this range. After luciferin injection, those images collected with the 620-nm filter in place are bacterial and firefly luciferases combined. To separate and quantitate the bacterial luciferase expression and host response, subtract the bacterial luciferase contribution from the total using the mathematical tools in the Living Image 2.5 software (Xenogen Corp.) *(38)*.

4. Notes

1. *L. monocytogenes* 2C has an LD 50 that is fourfold higher than the parental strain 104103 *(34)*.
2. When imaging early time points at a camera setting used to achieve high sensitivity, there may be background luminescence emitted from untreated animals. This background may be from several sources. Alfalfa in mouse chow may emit light (mainly fluorescence), various plasticware have background light emission, luciferin alone may be a source of light contamination, and some anesthetics have been demonstrated to cause higher background. It is important to run the proper controls for each possibility.
3. Carefully read the manufacturers recommendations for default and customized imaging settings.
4. A warmed stage and sufficient care immediately post-imaging, especially while the mice are still anesthetized, may reduce any mortality of the mice.
5. All experimental procedures with animals must be performed in accordance with guidelines of the relevant Institutional Animal Care and Use Committee (IACUC). If the IACUC protocol requires, moribund mice be then sacrificed; the time of sacrifice is considered to be the time of death.
6. Isofluorane, a gas anesthetic, requires special user handling to avoid its being released into the laboratory environment.

7. Under certain circumstances such as pigmented skin or fur or a deep location of the signal, wetting or shaving the fur or use of a depilatory cream to remove the fur of the mice may increase the collection of light.
8. Photon emission from bioluminescent S. pneumoniae decreases by 95% as the cells enter stationary phase (20).
9. Mice are observed as they recovered from anesthesia. Only those displaying no behavioral abnormalities were retained for this study.
10. Luciferin as a substrate for the enzymatic production of light from firefly luciferase is administered IP at around 150 mg/ml to 200 mg/kg in 200 µl of sterile saline. A study of luciferin kinetics that demonstrated that luciferin crosses the blood–brain and placental barriers. We have observed that some tissues (highly vascularized) may take up luciferin at a faster rate. It is important to understand the kinetics of luciferin as it relates to each model specifically.
11. Luciferin is dissolved and diluted in water as directed by the manufacturers. Luciferin is diluted in saline before injection.

Acknowledgments

Thank you to Jonathan Hardy, Christopher Contag (Stanford University), Anthony Purchio, and Brad Rice (Xenogen Corporation) for providing excellent data images and suggestions and to Christopher Contag for critical reading of the manuscript. Thank you to Chris Dalesio for his assistance in reformatting the images. Thank you to Kathy Kelsey for excellent administrative help in the completion of the manuscript.

References

1. Miller, J. T., Rahimi, S. Y. & Lee, M. (2005). History of infection control and its contributions to the development and success of brain tumor operations. *Neurosurg Focus* **18**, e4.
2. Thurston, A. J. (2000). Of blood, inflammation and gunshot wounds: the history of the control of sepsis. *Aust N Z J Surg* **70**, 855–61.
3. Zhao, Y. X. & Tarkowski, A. (1995). Impact of interferon-gamma receptor deficiency on experimental Staphylococcus aureus septicemia and arthritis. *J Immunol* **155**, 5736–42.
4. Verdrengh, M. & Tarkowski, A. (1997). Role of neutrophils in experimental septicemia and septic arthritis induced by Staphylococcus aureus. *Infect Immun* **65**, 2517–21.
5. Neely, A. N., Hoover, D. L., Holder, I. A. & Cross, A. S. (1996). Circulating levels of tumour necrosis factor, interleukin 6 and proteolytic activity in a murine model of burn and infection. *Burns* **22**, 524–30.
6. Tissi, L., von Hunolstein, C., Mosci, P., Campanelli, C., Bistoni, F. & Orefici, G. (1995). In vivo efficacy of azithromycin in treatment of systemic infection and

septic arthritis induced by type IV group B Streptococcus strains in mice: comparative study with erythromycin and penicillin G. *Antimicrob Agents Chemother bf* 39, 1938–47.
7. Pier, G. B., Meluleni, G. & Neuger, E. (1992). A murine model of chronic mucosal colonization by Pseudomonas aeruginosa. *Infect Immun* **60**, 4768–76.
8. Naiki, Y., Michelsen, K. S., Schroder, N. W., Alsabeh, R., Slepenkin, A., Zhang, W., Chen, S., Wei, B., Bulut, Y., Wong, M. H., Peterson, E. M. & Arditi, M. (2005). MyD88 is pivotal for the early inflammatory response and subsequent bacterial clearance and survival in a mouse model of Chlamydia pneumoniae pneumonia. *J Biol Chem* **280**, 29242–9.
9. Kelly, B. P., Furney, S. K., Jessen, M. T. & Orme, I. M. (1996). Low-dose aerosol infection model for testing drugs for efficacy against Mycobacterium tuberculosis. *Antimicrob Agents Chemother* **40**, 2809–12.
10. Atrasheuskaya, A. V., Bukin, E. K., Fredeking, T. M. & Ignatyev, G. M. (2004). Protective effect of exogenous recombinant mouse interferon-gamma and tumour necrosis factor-alpha on ectromelia virus infection in susceptible BALB/c mice. *Clin Exp Immunol* **136**, 207–14.
11. Contag, C. H., Contag, P. R., Mullins, J. I., Spilman, S. D., Stevenson, D. K. & Benaron, D. A. (1995). Photonic detection of bacterial pathogens in living hosts. *Mol Microbiol* **18**, 593–603.
12. Contag, P. R. (2002). Whole-animal cellular and molecular imaging to accelerate drug development. *Drug Discov Today* **7**, 555–62.
13. Contag, P. R., Olomu, I. N., Stevenson, D. K. & Contag, C. H. (1998). Bioluminescent indicators in living mammals. *Nat Med* **4**, 245–7.
14. Benaron, D. A., Contag, P. R. & Contag, C. H. (1997). Imaging brain structure and function, infection and gene expression in the body using light. *Philos Trans R Soc Lond B Biol Sci* **352**, 755–61.
15. Contag, C. H., Spilman, S. D., Contag, P. R., Oshiro, M., Eames, B., Dennery, P., Stevenson, D. K. & Benaron, D. A. (1997). Visualizing gene expression in living mammals using a bioluminescent reporter. *Photochem Photobiol* **66**, 523–31.
16. Troy, T., Jekic-McMullen, D., Sambucetti, L. & Rice, B. (2004). Quantitative comparison of the sensitivity of detection of fluorescent and bioluminescent reporters in animal models. *Mol Imaging* **3**, 9–23.
17. de Wet, J. R., Wood, K. V., Helinski, D. R. & DeLuca, M. (1985). Cloning of firefly luciferase cDNA and the expression of active luciferase in Escherichia coli. *Proc Natl Acad Sci USA* **82**, 7870–3.
18. Wilson, T. & Hastings, J. W. (1998). Bioluminescence. *Annu Rev Cell Dev Biol* **14**, 197–230.
19. Ruby, E. G. & Nealson, K. H. (1976). Symbiotic association of Photobacterium fischeri with the marine luminous fish Monocentris japonica; a model of symbiosis based on bacterial studies. *Biol Bull* **151**, 574–86.

20. Francis, K. P., Yu, J., Bellinger-Kawahara, C., Joh, D., Hawkinson, M. J., Xiao, G., Purchio, T. F., Caparon, M. G., Lipsitch, M. & Contag, P. R. (2001). Visualizing pneumococcal infections in the lungs of live mice using bioluminescent Streptococcus pneumoniae transformed with a novel gram-positive lux transposon. *Infect Immun* **69**, 3350–8.
21. Francis, K. P., Joh, D., Bellinger-Kawahara, C., Hawkinson, M. J., Purchio, T. F. & Contag, P. R. (2000). Monitoring bioluminescent Staphylococcus aureus infections in living mice using a novel luxABCDE construct. *Infect Immun* **68**, 3594–600.
22. Zhao, H., Doyle, T. C., Coquoz, O., Kalish, F., Rice, B. W. & Contag, C. H. (2005). Emission spectra of bioluminescent reporters and interaction with mammalian tissue determine the sensitivity of detection in vivo. *J Biomed Opt* **10**, 41210.
23. Rice, B. W., Cable, M. D. & Nelson, M. B. (2001). In vivo imaging of light-emitting probes. *J Biomed Opt* **6**, 432–40.
24. Contag, C. H., Jenkins, D., Contag, P. R. & Negrin, R. S. (2000). Use of reporter genes for optical measurements of neoplastic disease in vivo. *Neoplasia* **2**, 41–52.
25. Zhang, N., Weber, A., Li, B., Lyons, R., Contag, P. R., Purchio, A. F. & West, D. B. (2003). An inducible nitric oxide synthase-luciferase reporter system for in vivo testing of anti-inflammatory compounds in transgenic mice. *J Immunol* **170**, 6307–19.
26. Burns, S. M., Joh, D., Francis, K. P., Shortliffe, L. D., Gruber, C. A., Contag, P. R. & Contag, C. H. (2001). Revealing the spatiotemporal patterns of bacterial infectious diseases using bioluminescent pathogens and whole body imaging. *Contrib Microbiol* **9**, 71–88.
27. Rocchetta, H. L., Boylan, C. J., Foley, J. W., Iversen, P. W., LeTourneau, D. L., McMillian, C. L., Contag, P. R., Jenkins, D. E. & Parr, T. R., Jr. (2001). Validation of a noninvasive, real-time imaging technology using bioluminescent Escherichia coli in the neutropenic mouse thigh model of infection. *Antimicrob Agents Chemother* **45**, 129–37.
28. Kuklin, N. A., Pancari, G. D., Tobery, T. W., Cope, L., Jackson, J., Gill, C., Overbye, K., Francis, K. P., Yu, J., Montgomery, D., Anderson, A. S., McClements, W. & Jansen, K. U. (2003). Real-time monitoring of bacterial infection in vivo: development of bioluminescent staphylococcal foreign-body and deep-thigh-wound mouse infection models. *Antimicrob Agents Chemother* **47**, 2740–8.
29. Kadurugamuwa, J. L., Modi, K., Yu, J., Francis, K. P., Orihuela, C., Tuomanen, E., Purchio, A. F. & Contag, P. R. (2005). Noninvasive monitoring of pneumococcal meningitis and evaluation of treatment efficacy in an experimental mouse model. *Mol Imaging* **4**, 137–42.
30. Kadurugamuwa, J. L., Sin, L. V., Yu, J., Francis, K. P., Purchio, T. F. & Contag, P. R. (2004). Noninvasive optical imaging method to evaluate postantibiotic effects on biofilm infection in vivo. *Antimicrob Agents Chemother* **48**, 2283–7.

31. Kadurugamuwa, J. L., Sin, L., Albert, E., Yu, J., Francis, K., DeBoer, M., Rubin, M., Bellinger-Kawahara, C., Parr, T. R., Jr. & Contag, P. R. (2003). Direct continuous method for monitoring biofilm infection in a mouse model. *Infect Immun* **71**, 882–90.
32. Kadurugamuwa, J. L., Sin, L. V., Yu, J., Francis, K. P., Kimura, R., Purchio, T. & Contag, P. R. (2003). Rapid direct method for monitoring antibiotics in a mouse model of bacterial biofilm infection. *Antimicrob Agents Chemother* **47**, 3130–7.
33. Xiong, T., Zhang, Z., Liu, B. F., Zeng, S., Chen, Y., Chu, J. & Luo, Q. (2005). In vivo optical imaging of human adenoid cystic carcinoma cell metastasis. *Oral Oncol* **41**, 709–15.
34. Hardy, J., Francis, K. P., DeBoer, M., Chu, P., Gibbs, K. & Contag, C. H. (2004). Extracellular replication of Listeria monocytogenes in the murine gall bladder. *Science* **303**, 851–3.
35. Doyle, T. C., Nawotka, K. A., Kawahara, C. B., Francis, K. P. & Contag, P. R. (2006). Visualizing fungal infections in living mice using bioluminescent pathogenic Candida albicans strains transformed with the firefly luciferase gene. *Microb Pathog* **40**, 82–90.
36. Kadurugamuwa, J. L., Modi, K., Yu, J., Francis, K. P., Purchio, T. & Contag, P. R. (2005). Noninvasive biophotonic imaging for monitoring of catheter-associated urinary tract infections and therapy in mice. *Infect Immun* **73**, 3878–87.
37. Hardy, J., Margolis, J. J. & Contag, C. H. (2006). Induced biliary excretion of Listeria monocytogenes. *Infect Immun* **74**, 1819–27.
38. Kadurugamuwa, J. L., Modi, K., Coquoz, O., Rice, B., Smith, S., Contag, P. R. & Purchio, T. (2005). Reduction of astrogliosis by early treatment of pneumococcal meningitis measured by simultaneous imaging, in vivo, of the pathogen and host response. *Infect Immun* **73**, 7836–43.
39. Dussurget, O., Cabanes, D., Dehoux, P., Lecuit, M., Buchrieser, C., Glaser, P. & Cossart, P. (2002). Listeria monocytogenes bile salt hydrolase is a PrfA-regulated virulence factor involved in the intestinal and hepatic phases of listeriosis. *Mol Microbiol* **45**, 1095–106.
40. Lauer, P., Chow, M. Y., Loessner, M. J., Portnoy, D. A. & Calendar, R. (2002). Construction, characterization, and use of two Listeria monocytogenes site-specific phage integration vectors. *J Bacteriol* **184**, 4177–86.
41. Scheld, W. M., Koedel, U., Nathan, B. & Pfister, H. W. (2002). Pathophysiology of bacterial meningitis: mechanism(s) of neuronal injury. *J Infect Dis* **186** (Suppl 2), S225–33.
42. Nau, R. & Bruck, W. (2002). Neuronal injury in bacterial meningitis: mechanisms and implications for therapy. *Trends Neurosci* **25**, 38–45.

43. Kim, T. S. & Perlman, S. (2003). Protection against CTL escape and clinical disease in a murine model of virus persistence. *J Immunol* **171**, 2006–13.
44. van der Flier, M., Geelen, S. P., Kimpen, J. L., Hoepelman, I. M. & Tuomanen, E. I. (2003). Reprogramming the host response in bacterial meningitis: how best to improve outcome? *Clin Microbiol Rev* **16**, 415–29.
45. Zhu, L., Ramboz, S., Hewitt, D., Boring, L., Grass, D. S. & Purchio, A. F. (2004). Non-invasive imaging of GFAP expression after neuronal damage in mice. *Neurosci Lett* **367**, 210–2.

7

Intravital Two-Photon Imaging of Natural Killer Cells and Dendritic Cells in Lymph Nodes

Susanna Celli, Béatrice Breart, and Philippe Bousso

Summary

Two-photon microscopy makes it possible to image in real-time fluorescently labeled cells located in deep tissue environments. We describe a procedure to visualize the behavior of natural killer (NK) cells and dendritic cells (DC) in the lymph nodes of live, anesthetized mice. Intravital two-photon imaging is a powerful tool to study the migration and cell interactions of immune cells such as NK and DC in physiological settings.

Key Words: NK cells; dendritic cells; lymph node; intravital imaging; two-photon microscopy.

1. Introduction

Recent advances in imaging technology provide us with the opportunity to visualize immune cells in real time, at the single cell level and in physiological settings. In particular, two-photon laser-scanning microscopy is a technique of choice to track the behavior of lymph node (LN) cells located up to 100–300 µm below the LN surface *(1–5)*. Initial studies have been performed using explanted LNs maintained under physiological conditions of temperature and oxygen *(6–8)*. Two-photon imaging of lymphocyte behavior in LNs can also be performed in surgically exposed inguinal or popliteal LNs of live, anesthetized mice *(9–11)*. Although more demanding, intravital imaging offers several advantages over imaging of explanted organs, including preservation of vascular and lymphatic flow, innervation, oxygen metabolism and possibly soluble gradients. This approach also has some caveats such as possible biases introduced by the anesthesia and/or the surgical procedure. Intravital imaging of the LN also carries intrinsic technical challenges such as the avoidance of movements of the sample caused by breathing and maintenance of a physiological temperature close to the exposed LN. Here, we describe a procedure for intravital two-photon imaging of the popliteal LN. Some parts of this procedure were adapted from previous studies *(9,10)*.

2. Materials

1. Two-photon laser-scanning microscopy system including an upright microscope (*see* **Note 1**), a Ti:Sapphire laser tuned to 900 nm, a ×20 or ×40 dipping objective with long working distance (for example, Olympus 20× 0.95 NA, Zeiss Achroplan 40× 0.8 NA).
2. Homemade heating platform (*see* **Fig. 1A**). A heating element is glued onto a rectangular shaped piece of duralumin (180 × 140 × 3 mm) using thermal adhesive.

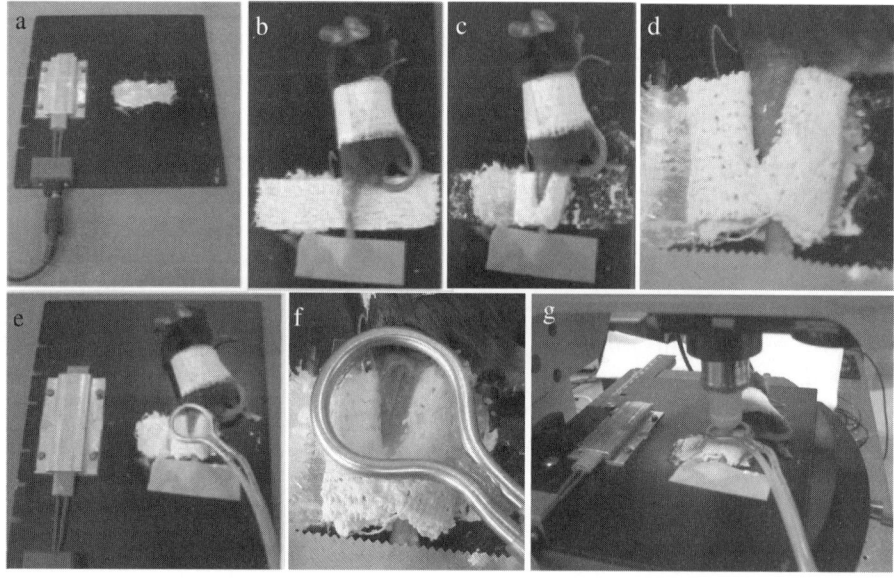

Fig. 1. Surgical exposure of the popliteal lymph node (LN) for intravital imaging. These pictures illustrate some of the important steps during the preparation of the popliteal LN for intravital imaging. (**A**) Custom-designed heating platform. (**B–D**) Preparation of the plaster cast to immobilize the lower hind leg. (**E and F**) A ring-shaped, metallic tube, glued to a coverslip is placed on the top of the popliteal LN. (**G**) After completion of the LN exposure, the heated platform is placed on the microscope stage.

The heating element is set up to heat the platform at 37°C. The platform has been designed to fit in the microscope stage.
3. Sample heating setup: peristaltic pump, inline solution heater (SH-27B Warner Instruments), heater controller (TC-344B Warner Instruments), semi-flexible vinyl temperature probe, tygon tubes, and tube connectors.
4. Ring-shaped [inner diameter (ID) 18 mm] metallic tube [inox, outer diameter (OD) 3 mm, ID 2 mm], cover slip (22 × 22 × 0.13 mm).
5. Dissecting microscope (Leica MZ 12/5) and fiber optics for incident illumination with cold light source (Leica CLS ×150).
6. Anesthesia: Ketamine and Medetomidine HCl (Dormitor), stored at 4°C.
7. Small animal clipper (Harvard apparatus).
8. Microsurgical instruments: Mayo scissor and Adson forceps for the skin, two microdissecting forceps (Dumont n°5) and one microdissecting scissor, microvascular clamp (micro glover curved), needle holder (Halsey standard), 6/0 and 8/0 silk (Ethicon), ophthalmic cautery fine type, polythene catheter OD 0.80, ID 0.40 mm (Portex, England).
9. 5-(and-6)-Carboxyfluorescein diacetate, succinimidyl ester (CFSE), stock solution 10 mM in DMSO (store at −20°C). SNARF, stock solution 10 mM in DMSO (store at −20°C).
10. Phosphate-buffered saline (PBS) buffer.
11. Fetal calf serum (decomplemented).
12. 0.9% sodium chloride solution for injection (Baxter, Maurepas, France).
13. Betadine solution (Asta Medica, Merignal, France).
14. Heparin sodium.
15. Plaster bandages.
16. Surgical tape.
17. Cyanoacrylate glue.
18. Acetic silicone sealant.
19. Pre-warmed distilled water (37°C).
20. Disposable sheets of cellulose fiber (Harvard apparatus).

3. Methods

We describe a procedure to visualize in real time the behavior of innate immune cells, such as dendritic cells (DC) and natural killer (NK) cells, in the LNs of live, anesthetized mice. Motility and cell–cell interactions can be studied using various immunological contexts that will not be detailed here.

3.1. Labeling and Adoptive Transfer of NK Cells and DC

1. Isolate NK cells (from LNs and spleen) and splenic DC using magnetic sorting and/or a fluorescence-activated cell sorter.

2. Resuspend 5×10^6 NK cells and 3×10^6 DC at 1×10^7 cells/ml in PBS with 5 μM CFSE and SNARF dye, respectively. Incubate the cell for 10 min at 37°C. Add 1 volume of fetal calf serum to stop the reaction. Wash twice. Resuspend the cells in PBS.
3. Inject NK cells intravenously in the recipient mice and inject DCs in the footpad. Intravital imaging can be performed at 24 h.

3.2. Anesthesia

Inject the mouse i.p. with ketamine (24 mg/kg) and i.m. with metedomidine HCl (170 μg/kg). Check the level of anesthesia by pinching the tail and by checking body and whisker movements. To maintain a deep level of anesthesia during the procedure (that may last several hours), it is recommended to inject the mouse each hour with a reduced dose of anesthetics.

3.3. Jugular Vein Cannulation (Optional)

Preparation of a vascular access enables the injection of cells, dyes, or drugs during image acquisition (*see* **Note 2**). We chose to cannulate the external jugular vein, which is distant from the site of the popliteal LN.

1. Shave the hair of the neck using a small animal clipper. Place the mouse on its back on the surgical board with its head toward the operator. Maintain the mouse head in extension by gently pulling its anterior teeth with a thin rubber band.
2. Clean the neck area with betadine solution, and perform a lateral skin incision along the neck.
3. Expose the external jugular vein by gently dissociating from the surrounding tissues. Burn and cut all the collateral veins.
4. Place a vascular clamp on jugular vein at the level of the clavicle to prevent air entry during the insertion of the cannula. Tie the other extremity of the jugular with 8/0 silk.
5. Perform a small incision on the superior wall of the jugular vein and insert the cannula filled with heparinized solution.
6. Tie the cannula with 8/0 silk. The vascular clamp is removed, and the blood flow through the cannula is checked. A drop of glue is used to fix the cannula to the surrounding tissues. The skin is closed by silk suture 6/0, once the glue is completely solidified. The cannula is washed regularly with heparinized solution.

3.4. Preparation of the Popliteal LN

For all the following steps, the mouse should be transferred onto the heated platform (*see* **Fig. 1A**). The use of this stage simplifies the management of the mouse during the preparation of the sample and contributes to the overall stability of the intravital setup.

1. Shave the hair of the left lower hind leg using a small animal clipper.
2. Tape the tail of the mouse to the right side of the body using surgical tape. The left lower hind leg is gently pulled out and kept in extension using surgical tape as shown in **Fig. 1B**.
3. Perform a skin incision on the area of the popliteal space, and remove a small circular piece of skin.
4. Carefully expose the popliteal LN by removing some of the fat covering the popliteal cavity (*see* **Note 3**).
5. Once the exposition is completed, cover the LN with small moist gauze. Immobilization of the leg is then achieved by preparing a plaster cast.
6. Cut two rectangular pieces of plaster bandages (80 × 25 mm), quickly immerse them in water, and then let drop the excess of water. Place the plaster bandages one on top of other with the left leg lying in the middle (*see* **Fig. 1B**). Plaster bandages are folded twice to create a block of plaster on each side of the leg (*see* **Fig. 1C**). The height of the plaster blocks should match that of the LN surface (*see* **Note 4**). The two plaster bandages are connected at the ankle level to complete leg immobilization (*see* **Fig. 1D**).
7. Allow the plaster cast to dry (*see* **Note 5**). The shape of the plaster cast is designed to (1) prevent movements of the leg and movements because of breathing by fixing the leg to the heated platform and (2) create a flat surface around the LN on which to place a cover slip (see below).

3.5. Maintenance of the Exposed LN at Physiological Temperature

Keeping the LN at a physiological temperature is critical part of the procedure, because lymphocyte motility and behavior are highly temperature-dependent *(6)*. Several factors contribute to a temperature drop close to the sample, including anesthesia, surgical exposition of the LN, and thermal transfer between the sample and the objective. Although the role of the heated platform is to help maintain the mouse body temperature within physiological range, it is also essential to correct the temperature locally, close to the sample. An approach similar to that previously described by Mempel et al.*(12)* is used with slight modifications.

1. Place and glue a cover slip onto the plaster cast. Only a film of saline solution must remain between the LN and the cover slip. A metallic ring-shaped tube (*see* **Fig. 1F**) should be previously fixed on the top of the cover slip by using silicone sealant.
2. Place the heated platform with the intravital setup onto the stage of the microscope. Sheets of cellulose fibers can be used as a blanket to cover parts of the animal body that are not directly under the objective.

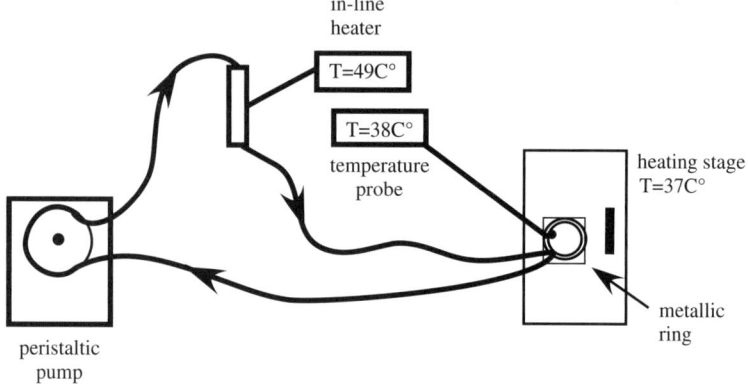

Fig. 2. Schematic representation of the water circuit used to maintain the temperature of lymph node (LN) area within physiological ranges. A peristaltic pump is used to circulate water in an inline heater and then inside a ring-shaped metallic tube glued to a coverslip and placed on the top of the LN. The ring and the coverslip create a small chamber filled with water in which a dipping objective is immersed. Water within the chamber is maintained at 38°C with the help of the warm water circulating in the metallic tube.

3. Connect the extremities of the metallic tube with tygon tubes. Use a peristaltic pump so that distilled water heated with an in-line heater (set to 49°C) circulate into the tube. A schematic representation of the set-up is depicted in **Fig 2**.
4. Add distilled water (pre-warmed at 37°C) in the center of the metallic ring. The ring placed on the coverslip creates a 'reservoir' in which a dipping objective can be immerse.
5. Monitor the temperature of the water in the ring with the help of the temperature probe. The heated liquid circulating inside the metallic tube should maintain the temperature of the water in the ring at 38°C. If necessary, adjust the temperature of the in-line heater accordingly.

3.6. Two-Photon Laser-Scanning Microscopy

1. Tune the Ti:Sapphire laser at 900 nm.
2. Set-up the microscope configuration, filter (505–550 nm band pass) for the detection of CFSE, and filter (570/645) for the SMARF.
3. Check that the dipping objective is properly immersed in the water contained within the metallic ring (*see* **Fig. 1G**). Add more warm water if needed.
3. Using mercury lamp illumination, look through the eyepiece and focus on the surface of the LN.

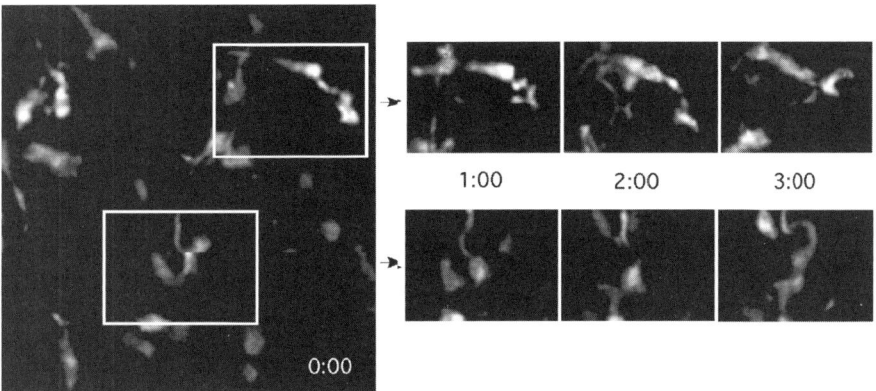

Fig. 3. *In vivo* imaging of dendritic cell (DC) behavior in the popliteal lymph node (LN). Splenic DCs purified from Ubiquitin-GFP mice were injected i.v. At 24 h, the popliteal LN was prepared for intravital imaging. Time frame images show two individual DCs over time. Time is indicated as min:sec.

4. Start scanning using the Ti:Sapphire laser (*see* **Note 6**). Adjust the gain and offset of the detectors. While scanning, move along the x-, y-, and z-axis to locate an area containing a high-enough numbers of fluorescently labeled cells.
5. Once a suitable location is found, acquired Z-stacks of optical sections (typically 10 sections spaced by 4 μm) every 30 s for a total duration of 30–60 min (*see* **Note 7**).
6. Control the level of anesthesia before starting a new image acquisition. If necessary, proceed to an additional injection of anesthetics.
7. Process and analyze the data using 3D-tracking software such as Imaris or Volocity (*see* **Note 8**). Quicktime movies can be generated following a maximum intensity projection of each Z-stack using the publicly available softwares ImageJ and OsiriX. Time frame images of individual DCs *in vivo* are shown in **Fig. 3**.

4. Notes

1. It is recommended to use non-descanned detectors (rather than internal photomultipliers) for optimal signal collection.
2. Intravenous injection of 2 mg tetramethylrhodamine-dextran (10 kDa) can be used to visualize blood vessels.
3. The popliteal LN is always located below the junction of the popliteal vein and one of its left branches.
4. During the preparation of the plaster cast, adjust the height of the plaster blocks to that of the LNs by gently applying a microscope slide over the plaster blocks and mouse leg while the solidification process takes place.
5. This process is rather quick (5 min) on the heated stage.

6. Care should be taken to minimize the laser power used for excitation as excessive power can cause phototoxicity resulting in cell movement arrest *(13)*.
7. At steady state, about half of the LN NK cells are located in the paracortical area of the LN. To ensure that the lack of cell movement is not due to laser-induced photoxocity, it may useful to cotransfer naïve T cells (known to be highly motile, 11 µm/min)
8. 2D cell tracking can also be performed using the publicly available Mtrack2 plugin of ImageJ (http://rsb.info.nih.gov/ij/plugins/index.html).

References

1. Denk, W., Strickler, J. H., and Webb, W. W. (1990). Two-photon laser scanning fluorescence microscopy. *Science* **248,** 73–6.
2. Cahalan, M. D., Parker, I., Wei, S. H., and Miller, M. J. (2002). Two-photon tissue imaging: seeing the immune system in a fresh light. *Nat Rev Immunol* **2,** 872–80.
3. Bousso, P., and Robey, E. A. (2004). Dynamic behavior of T cells and thymocytes in lymphoid organs as revealed by two-photon microscopy. *Immunity* **21,** 349–55.
4. Huang, A. Y., Qi, H., and Germain, R. N. (2004). Illuminating the landscape of in vivo immunity: insights from dynamic in situ imaging of secondary lymphoid tissues. *Immunity* **21,** 331–9.
5. Dustin, M. L. (2004). Stop and go traffic to tune T cell responses. *Immunity* **21,** 305–14.
6. Miller, M. J., Wei, S. H., Parker, I., and Cahalan, M. D. (2002). Two-photon imaging of lymphocyte motility and antigen response in intact lymph node. *Science* **296,** 1869–73.
7. Bousso, P., and Robey, E. (2003). Dynamics of CD8+ T cell priming by dendritic cells in intact lymph nodes. *Nat Immunol* **4,** 579–85.
8. Hugues, S., Fetler, L., Bonifaz, L., Helft, J., Amblard, F., and Amigorena, S. (2004). Distinct T cell dynamics in lymph nodes during the induction of tolerance and immunity. *Nat Immunol* **5,** 1235–42.
9. Miller, M. J., Wei, S. H., Cahalan, M. D., and Parker, I. (2003). Autonomous T cell trafficking examined in vivo with intravital two-photon microscopy. *Proc Natl Acad Sci USA* **100,** 2604–9.
10. Mempel, T. R., Henrickson, S. E., and Von Andrian, U. H. (2004). T-cell priming by dendritic cells in lymph nodes occurs in three distinct phases. *Nature* **427,** 154–9.
11. Bajenoff, M., Breart, B., Huang, A. Y., Qi, H., Cazareth, J., Braud, V. M., Germain, R. N., and Glaichenhaus, N. (2006). Natural killer cell behavior in lymph nodes revealed by static and real-time imaging. *J Exp Med* **203,** 619–31.
12. Mempel, T. R., Scimone, M. L., Mora, J. R., and von Andrian, U. H. (2004). In vivo imaging of leukocyte trafficking in blood vessels and tissues. *Curr Opin Immunol* **16,** 406–17.
13. Shakhar, G., Lindquist, R. L., Skokos, D., Dudziak, D., Huang, J. H., Nussenzweig, M. C., and Dustin, M. L. (2005). Stable T cell-dendritic cell interactions precede the development of both tolerance and immunity in vivo. *Nat Immunol* **6,** 707–14.

8

Dissection of the Antiviral NK Cell Response by MCMV Mutants

Stipan Jonjic, Astrid Krmpotic, Jurica Arapovic, and Ulrich H. Koszinowski

Summary

Our understanding of virus control by natural killer (NK) cells relies mainly on in vitro observations. The significance of these findings for virus control in vivo is not yet fully understood. Complexity is added by the fact that many viruses, particularly herpesviruses, are equipped with sets of genes that, dependent on the genetic background of the host, modify the NK cell response. The advent of recombinant DNA technology and mutagenesis procedures for BAC-cloned viral genomes has made it possible not only to screen for viral proteins with such functions but also to assess their biological relevance. Mutant viruses with gene defects reveal the efficacy and complexity of NK cell control. Here, we describe procedures to assess the NK cell response to mouse cytomegalovirus (MCMV), a prominent virus model for studying NK cell functions in vivo.

Key Words: Mouse cytomegalovirus; NK cells; immune evasion; NKG2D; Ly49H; MULT-1; RAE-1; H60.

1. Introduction

Cells of the innate immune system, including macrophages, dendritic cells (DCs) and natural killer (NK) cells, play a decisive role in the early control of viral infection. DCs are important for the induction of adaptive immunity and for the activation of NK cells *(1)*. NK cells are the effectors of the innate immune system. They recognize and eliminate infected cells *(2)*. The activation of NK cells relies on the integration of a series of signals from both activating and inhibitory receptors. Unlike T and B lymphocyte receptor (TCR and BCR) genes, the genes encoding these NK receptors do not undergo somatic mutation. Activated NK cells kill target cells through granzyme- and perforin-dependent mechanisms. NK cells also promote and coordinate the activity of other innate and adaptive immune effector cells through the release of cytokines *(3)* and, for example, through receptors for immunoglobulin G (IgG) Fc fragments that connect innate and adaptive immune responses *(4)*.

Studies on mouse cytomegalovirus (MCMV) revealed several mechanisms by which viruses modulate the mammalian immune response *(5)*. The MCMV genome has a size of 230 kb and codes for more than 180 genes *(6)*, among which less than 50 genes are essential for virus replication. All other genes are probably involved in the interaction with host functions. Mutagenesis of bacterial artificial chromosome (BAC)-cloned herpesviral genomes in *Escherichia coli* paved the way for more extensive studies on the role of individual viral genes for virus pathogenesis *(7–9)*. MCMV not only interferes with the adaptive immune response but also disrupts the innate immune response mediated by NK cells *(10)*. The principles adopted by MCMV to escape or subvert the NK cell response include (1) selection of escape mutants of viral ligands for activating NK receptors, (2) down-regulation of cellular ligands for activating NK cell receptors, (3) use of virally encoded MHC class I homologues as ligands for inhibitory NK cell receptors, (4) differential modulation of MHC class I expression, and (5) expression of viral Fc receptor to interfere with antibody dependent cellular cytotoxicity (ADCC) (*see* also **Table 1**). In addition, the MCMV genome encodes proteins that affect NK cell function indirectly by regulating apoptosis, by modulating cytokines and chemokines, and by altering the properties of infected DCs and macrophages *(10,11)*.

NK cell activity dictates the susceptibility to primary MCMV infection. The NK gene complex (NKC) encoding the Ly49 family of NK receptors defines the susceptibility of mice to MCMV infection *(12)*. Additional genetic loci have also been implicated in NK-cell-dependent resistance *(13,14)*. Resistant mice show a robust antiviral NK cell response and control virus replication. Susceptible strains fail to mount an effective NK cell response, which leads to increased viral loads and disease *(12)* (*see* **Fig. 1**).

The list of MCMV genes involved in NK cell regulation is still growing. Remarkably, most conventional mouse strains are susceptible to MCMV because of the viral ability to inhibit NK cell activation *(15)* (*see* **Fig. 1**). In many cases, deletion of one viral gene can change the resistance phenotype of a mouse strain. **Table 1** summarizes examples of MCMV deletion mutants

Table 1
NK Cell Regulation by MCMV Genes Revealed by Deletion Mutants

Protein function	Mutant phenotype	Examples	References
Ligand for activating NK receptor	Resistance to NK cells and enhanced virulence	• Ly49H recognition of MCMV m157 in Ly49H$^+$ mice (e.g., C57BL/6)	Arase et al. 2002; Smith et al. 2002; Bubic et al. 2004
		• Ly49P recognition of unknown MCMV protein in context of H-2D^k in Ma/My mice	Desrosiers et al. 2005
		• Recognition of unknown MCMV ligand by unknown NK receptor encoded by Cmv4 in PWK/Pas mice	Adam et al. 2006
Ligand for inhibitory NK receptor	NK cell-dependent attenuation	• Binding of MCMV m144 (viral MHC class I homologue) to unknown NK receptor	Farrell et al. 1997
		• Binding of MCMV m157 to Ly49I in Ly49H$^-$ Ly49I$^+$ mice (e.g., 129/SvJ)	Arase et al. 2002
Inhibitor of cellular ligand for activating NK receptor	NK cell-dependent attenuation	Down-modulation of NKG2D ligands by MCMV proteins m145, m152, m155, m138.	Krmpotic et al. 2002; Lodoen et al. 2003; Lodoen et al. 2004; Hasan et al. 2005; Krmpotic et al. 2005; Lenac et al. 2006
Homolog of chemokine	NK cell-dependent attenuation	m131/m129	Fleming et al. 1999
Viral Fc receptor	Potentially enhanced sensitivity to ADCC	m138 MCMV	Crnkovic-Mertens et al. 1998

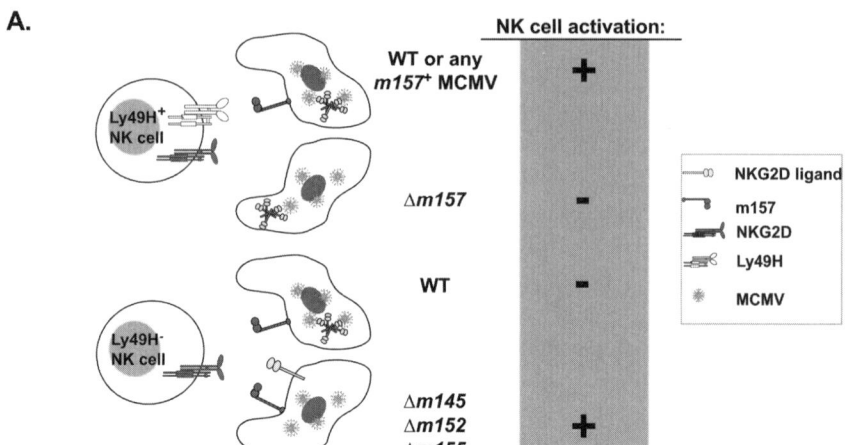

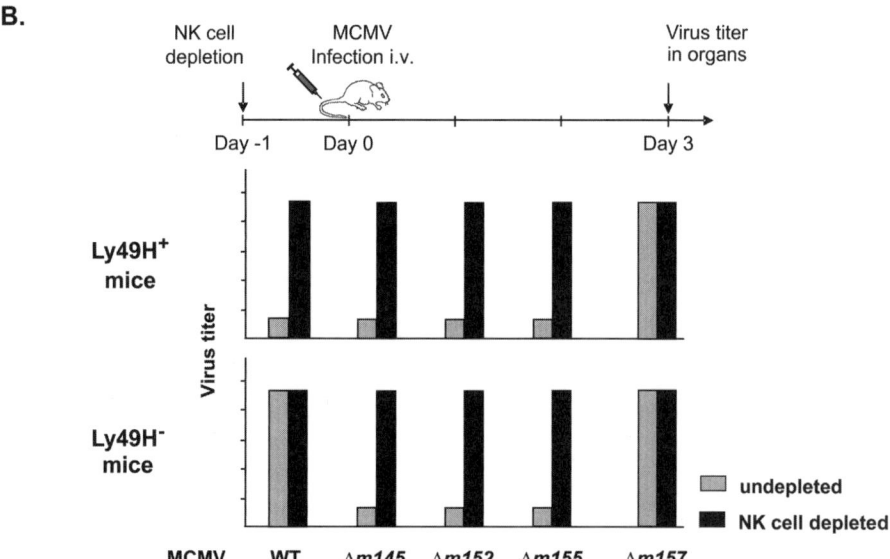

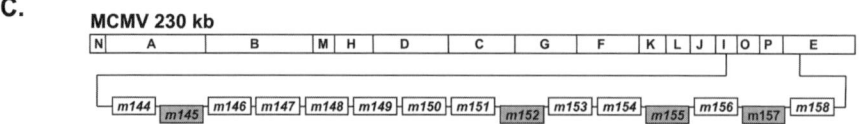

with altered susceptibility to NK cells in vivo. Formal proof for the association of a loss or gain in host resistance to virus infection with the expression of a particular viral gene requires not only deletion mutants but also revertant viruses in which the deleted gene is re-inserted into the viral genome.

If a viral protein regulates NK cells, the difference between mutant and wild-type (WT) virus should be lifted in the NK cell deficient host or after NK cell depletion. One should also bear in mind that a specific phenotype of the viral deletion mutant can be expected only if a particular immune function regulated by the viral protein is preserved. For instance, the MCMV gene *m157* encodes the ligand for the *activating* NK cell receptor Ly49H, and deletion of the gene leads to resistance of the mutant virus to NK cell control in mouse strains that express the Ly49H NK cell receptor *(16)* (*see* **Fig. 1**). If the viral immune modulatory protein is a ligand for an *inhibitory* NK receptor, the deletion mutant is attenuated in an NK-cell-dependent manner *(17)*. The same phenotype is observed when the viral protein down-regulates cellular ligand(s) for activating NK cell receptors *(15,18,19)*. Several MCMV proteins down-modulate cellular ligands for the NKG2D receptor. Deletion of any of these genes leads to NK-cell- and NKG2D-dependent attenuation of the respective mutant virus (*see* **Fig. 1**).

The situation is more complex if viral genes have more than one function, for instance, down-modulation of MHC class I molecules and modulation of ligands for activating NK cell receptors *(15,20)*. In theory, the lack of host MHC class I molecules should render MCMV-infected cells sensitive to NK cell control because of the lack of engagement of inhibitory NK cell receptors. However, this is not the case because the virus simultaneously inhibits the expression of NKG2D ligands. Therefore, cells infected with WT MCMV are

◄──

Fig. 1. Principles of NK cell regulation by MCMV and mutant phenotypes. (**A**) Schematic view of interaction between NK receptors (Ly49H and NKG2D) and their ligands expressed on cells infected with different MCMV mutants. (**B**) The susceptibility of different mutant viruses to NK cell control 3 days after injection into either Ly49H$^+$ or Ly49H$^-$ mice. (**C**) *Hind*IIIE cleavage map of MCMV Smith strain is shown with the expanded region containing most of the *m145* gene family members, several of which are involved in the regulation of NK cell function including *m145*, *m152*, and *m155* responsible for the down-modulation of NKG2D ligands MULT-1, RAE-1, and H60, respectively. The product of *m157* gene serves as a ligand for activating NK receptor Ly49H.

resistant to control by NK cells, with the exception of Ly49H$^+$ NK cells (*see* **Fig. 1**) *(16)*. If, however, a viral protein has multiple functions, the phenotype of the mutant virus in vivo is not predictable, and the best way to investigate its role is by the use of mutant viruses, as described in this chapter.

2. Materials

Standard equipment and material for cell culture, virus preparation, and handling laboratory mice are required (centrifuges, incubators, refrigerators, pipettes, tubes, cell culture dishes, animal hoods, small surgical instruments, etc.). The other major materials are listed below.

1. MCMV Smith strain, ATCC, VR-1399 *(21)*
   BAC-derived MCMV WT virus (Smith strain) *(9)* and mutant virus strains (e.g., Δ*m145* *(19)*, Δ*m152* *(22)*, Δ*m155* *(18)*, Δ*m157* *(16)* and their corresponding revertant virus strains).
   BAC-derived MCMV expressing green fluorescence protein (GFP): WT MCMV-GFP, MCMV-GFPΔ6, MCMV-GFPΔ6S1, MCMV-GFPΔ6S2, MCMV-GFPΔ6S3, MCMV-GFPΔ*m144*, MCMV-GFPΔ*m145*, MCMV-GFPΔ*m146*, MCMV-GFPΔ*m147/m148* *(19)*.
2. Mice: C57BL/6, BALB/c, C57BL/6 RAG1$^{-/-}$.
3. Cell lines: mouse embryonic fibroblasts (MEFs); NIH/3T3, ATCC CRL-1658; BALB/3T3, ATCC CCL-163; M2-10B4, ATCC CRL-1972; IC-21 cells, ATCC TIB-186; SVEC4-10, ATCC CRL-218; B12 *(23)*.
4. Fetal calf serum (FCS) (GIBCO-BRL).
5. Minimal essential medium (MEM) (GIBCO-BRL).
6. Dulbecco's Modified Eagle's Medium (DMEM) (GIBCO-BRL).
7. RPMI-1640 (GIBCO-BRL).
8. Methyl cellulose medium (viscous medium): In 320 mL distilled water add 8.8 g methyl cellulose (4000 centipoise, Sigma). Stir with a large magnetic stirrer. Leave medium for about 4 days at 4°C until methyl cellulose has completely dissolved. Autoclave for 30 min at 121°C. Supplement with 40 mL of 10x MEM, 40 mL FCS, 40,000 U Penicillin, 40 mg Streptomycin, 4 mL 1 M HEPES, and 0.88 g sodium bicarbonate.
9. Phosphate-buffered saline (PBS), (10× stock): 2.0 g KCl, 2.0 g KH$_2$PO$_4$, 80 g NaCl, 11.4 g Na$_2$HPO$_4$ anhydrous. Dissolve Na$_2$HPO$_4$ in 200 mL ddH$_2$O with stirring and heating and dissolve remaining three salts in 700 mL ddH$_2$O; combine both solutions. Adjust pH to 7.4 with 5 M NaOH. Adjust volume to 1 L.
10. Trypsin for cell culture: 0.25% Trypsin plus 1 mM EDTA (GIBCO-BRL).
11. FACS medium: PBS, 10 mM EDTA, 20 mM HEPES (pH 7.2), 2% FCS, 0.1% sodium azide (NaN$_3$).
12. Tris/NH$_4$Cl (erythrocyte lysing buffer): 90 mL of 0.16 M NH$_4$Cl and 10 mL of 0.17 M Tris–HCl, pH 7.65. Adjust to pH 7.2 with HCl.

13. 15% Sucrose/virus suspension buffer (VSB): 50 mM Tris–HCl (pH 7.8), 12 mM KCl, 5 mM Na_2EDTA, 15% (w/v) sucrose.
14. Propidium iodide (Sigma-Aldrich).
15. 2.5% (w/v) paraformaldehyde dissolved in PBS.
16. Triton X-100 (Sigma-Aldrich).
17. Gelatin from cold water fish skin (Sigma-Aldrich).
18. Tissue culture (TC) dishes: Petri dishes 59 cm^2; Petri dishes 156 cm^2; 6-well TC plates; 12-well TC plates; 48-well TC plate; Round bottom 96-well TC plate (BD Falcon, Greiner).
19. Conical tubes 50 mL (BD Falcon).
20. Eppendorf tubes (Eppendorf).
21. Luer slip syringe 2 mL (BectonDickinson).
22. Hypodermic needle 25 G, 0.5 × 19 mm.
23. Insulin syringe 1 mL 29 G, 0.33 × 12.7 mm (BD Micro-Fine).
24. Infrared lamp 200 W.
25. Iris forceps and scissors with straight and curved ends.
26. Ultracentrifuge bottles 25 × 89 mm (Beckman).
27. Ultracentrifuge (Beckman L7-65, Rotor type TI 55.2).
28. Inverted microscope (Olympus).
29. γ-Counter (Berthold).
30. Confocal laser-scanning microscope (Olympus).
31. Mouse anti-mouse NK1.1 mAb, clone PK136, ATCC HB191.
32. Rabbit antiserum to Asialo-GM1 (Wako Chemicals); control rabbit serum.
33. Anti-Ly49C/I mAb, clone 5E6 *(24)*.
34. Rat anti-CD49b-biotin mAb, clone DX5; control rat IgM-biotin Ab (BD Pharmingen).
35. Streptavidin-Phycoerythrin (SA-PE) (BD Pharmingen).
36. Rat anti-mouse NKG2D (e.g., clone 191004, R&D Systems) or Armenian hamster anti-mouse NKG2D, clone C7 *(25)*.
37. Mouse anti-mouse Ly49H mAb unlabeled, clone 3D10 *(26)*.
38. PE-conjugated NKG2D tetramer *(15)*.
39. Rat anti-MULT-1 mAb, clone 1D6 *(19)*.
40. PE-conjugated rat anti-mouse RAE-1αβγ mAb, clone CX1 (BD Pharmingen).
41. PE-conjugated rat IgG2b, isotype control (Southern Biotechnology).
42. Rat anti-mouse H60, clone 205326 (R&D Systems).
43. PE-conjugated goat anti-rat IgG (H/L) (Caltag).
44. Mouse anti-MCMV gp48 mAb, CROMA229 *(27)*.
45. Fluorescein isothiocyanate (FITC)-labeled goat anti-mouse Ig (BD Pharmingen).
46. TRITC-labeled goat anti-rat F(ab)$_2$ (Santa Cruz Biotechnology).
47. Poly (I•C) (Amersham Pharmacia Biotech).

48. Sodium Chromate [^{51}Cr] (Na$_2^{51}$CrO$_4$) (Amersham Pharmacia Biotech). This product is radioactive material, and it should be used only by authorized persons in designated laboratory settings.

3. Methods

3.1. Assessment of NK Cell Response to MCMV Mutants In Vivo

The methods described below outline procedures to assess the NK cell control of MCMV in vivo and to study viral proteins involved in the regulation of NK cell response by the use of viral mutants (*see* **Note 1**). This includes the depletion and/or modulation of NK cells in vivo by the use of specific monoclonal and polyclonal antibodies and the ex vivo testing of NK cell activation, the preparation of virus stocks for use in vivo, the infection of mice by injecting virus or virus-infected cells through different routes and the subsequent testing of the virus titers in organs.

In essence, viral proteins that regulate NK cell functions or that may have this property act in the context of a virus infection. Therefore, virus mutants and controls are studied with respect to NK cell control in vivo. A prerequisite is that mutants and corresponding revertants have WT virus growth kinetics on permissive cells in cell cultures in vitro. This excludes effects of mutant construction on genes that define virus fitness in the absence of NK cells. If the function of the viral gene under study is not known, the virus growth kinetics in an immunocompetent host are determined. Groups of mice are infected with the mutant virus lacking the gene of interest or with the WT control virus. Virus titers in organs of infected animals are determined 3–4 days after infection (*see* **Note 2**). If the mutant grows to a lower or to a higher titer in vivo than the control virus, the principle that causes this alteration has to be explained. The change in virus productivity may be due to an altered susceptibility to NK cells. Therefore, the viruses are studied in mice that differ with respect to NK cell function. For instance, if the virus encodes a gene, which is a ligand for an activating NK receptor, the phenotype of the deletion mutant is lost in mice that lack the respective NK receptor (*see* **Fig. 1**). Therefore, the lack of phenotype in one mouse strain does not exclude a phenotype in another mouse strain that has a different set of NK receptors.

In a similar approach, the contribution of NK cells to latency control and to the resolution of recurrent virus infection should be testable. To date, little is known regarding the contribution of NK cells to long-term immune surveillance during virus latency and reactivation. Depletion of NK cells led however to a high frequency of recurrence when performed in conjunction with depletion of CD4$^+$ or CD8$^+$ T cells *(28)*. This suggests that NK cells may play a role in the control of latent MCMV infection. If this was true, mutant viruses lacking genes that down-modulate NK cell functions should be less prone to cause recurrent infection (*see* **Note 3**).

3.1.1. Infection of Mice

3.1.1.1. INFECTION OF MICE WITH MCMV

1. Inject 6–8-week-old mice (*see* **Note 4**) by the intraperitoneal (i.p.) or the intravenous (i.v.) route with $2–5 \times 10^5$ PFU of TC-derived mutant or control MCMV strains in a volume of 0.5 mL of PBS (or culture medium without FCS) using a hypodermic needle. Intravenous injection should be given in a tail vein after warming the mice under an infrared lamp.
2. Kill the mice with carbon dioxide inhalation 3 or 4 days after infection (*see* **Note 2**), collect organs (spleen, liver, and lungs) in 5-mL tubes containing 2 mL of medium and store them at –20°C until the virus titration (*see* **Subheading 3.1.5.**).

Newborn mice should be infected by injection of 100–500 PFU of TC-derived virus i.p. in 50 µL of PBS (*see* **Note 5**). To avoid the injected fluid leaking through the injection site, insert the hypodermic needle under the skin of the thoracic region and pass it into the peritoneal cavity. Observe this process to avoid liver damage by the needle.

When salivary gland-derived virus (SGV) is used, take a lower dose (5×10^4 PFU to 1×10^5 PFU) because SGV is more virulent than the TC-derived virus, depending on the strain of mice to be used. SGV should be injected i.p. in a volume of 0.5 mL of PBS (or medium without FCS).

3.1.1.2. INFECTION OF MICE BY INTRAVENOUS INJECTION OF INFECTED FIBROBLASTS

Instead with the free virus, mice can be infected by i.v. inoculation of infected MEFs (*see* **Note 6**). This ex vitro/in vivo protocol is suitable for testing the function of NK cells and the significance of various MCMV immunoevasins involved in the modulation of surface ligands for NK cell receptors. Intravenously injected fibroblasts are trapped in lung capillaries, and the analysis of NK cell control can be restricted to this organ *(15)*.

1. Seed MEFs in complete medium (3% MEM) in 59 cm² TC Petri dishes (10^6 cells/Petri dish).
2. Infect MEFs with 2 PFU/cell of TC-derived MCMV mutants or control virus strains and allow virus adsorption for 30 min.
3. Facilitate infection by centrifugation at 800 *g* for 30 min (*see* **Note 7**) and incubate the cells for 12 h at 37°C and 5% CO_2.

4. Collect the cells by trypsinization and wash them twice in complete medium to remove free virus.
5. Resuspend cells in medium without FCS and inject i.v. into recipient mice in a volume of 0.5 mL (10^4–10^5 cells per mouse) (*see* **Note 4**).
6. Kill the mice by carbon dioxide inhalation 3 days later.
7. Collect lungs of infected animals and freeze them at –20°C before virus titration.
8. Determine the virus titers using standard plaque-forming assay (*see* **Subheading 3.1.5.**).

3.1.2. Depletion of NK Cells In Vivo and Blocking of NK Receptors

Attenuation of mutant virus growth in vivo early after infection is an indication that the protein under study might be involved in the regulation of innate immunity. Evidence is provided by depleting NK cells, by blocking NK receptors, or by using mouse strains with specific genetic defects.

1. Groups of infected mice are either subjected to NK cell depletion or serve as controls and receive isotype-matched irrelevant antibodies or control serum.
2. Deplete NK cells in NK1.1$^+$ mice by the injection of anti-NK1.1 mAb PK136 *(29)* (1 mg per mouse in 0.5 mL of PBS i.p.) 6–24 h before infection (*see* **Note 8**).
3. NK1.1$^-$ mice are depleted from NK cells by i.p. injection of rabbit antiserum to asialo-GM1 (Wako Chemicals) at the dose of 25–50 μL diluted in 0.5 mL of PBS, 6–24 h before infection.
4. Depletion of Ly49 subsets of NK cells is achieved by mAbs specific for particular Ly49 receptors (e.g., 150–500 μg/mouse of 5E6 mAb for Ly49C/I$^+$ NK cell subset) *(24)* in 0.5 mL of PBS i.p., 6–24 h before infection.
5. To control the efficacy of NK cell depletion, take an aliquot of splenic tissue from at least one animal from each group subjected to NK cell depletion as well as from control animals. Prepare single-cell suspensions and lyse erythrocytes with lysis buffer. Wash cells twice in FACS medium and adjust to 1×10^6 cells per sample. Stain cells with anti-CD49b-biotin mAb (clone DX5, pan-NK cell marker) followed by SA-PE or control isotype-matched antibody and perform FACS analysis.
6. The function of a particular NK cell receptor can be analyzed through blockage by specific mAbs. For example, rat mAb 191004 (R&D Systems) specifically blocks the NKG2D receptor and mouse 3D10 mAb blocks the Ly49H NK cell receptor *(26)*. To block a particular NK cell receptor in vivo inject i.p. 100 μg/mouse of specific blocking mAb in 0.5 mL of PBS 6–24 h before infection.

3.1.3. Assessment of Mutant MCMV Attenuation by Challenging Infection In Vivo

The consequence of deletion of particular viral gene on virus fitness in vivo can be tested by determining the lethal dose 50 (LD$_{50}$) of the mutant virus.

1. Infect at least 10 mice per group by i.v. or i.p. inoculation of mutant viruses or control virus (*see* **Subheading 3.1.1.1.**) using a range of virus doses (between 10^5 and 5×10^6 PFU) to determine LD_{50} for each virus (*see* **Note 9**).
2. Monitor the mortality rate within each group daily.
3. Calculate the mortality rate and determine LD_{50} for mutant and control viruses.
4. If the mutant viruses differ in their virulence as compared to control viruses, use NK-cell-depleted mice (*see* **Subheading 3.1.2.**) to determine whether the attenuation/virulence phenotype can be linked to NK cells.
5. Alternatively, inject mice with the LD_{50} determined for the control virus (*see* **Note 9**) and daily monitor the mortality within each group.

3.1.4. Preparation of Virus Stocks

The quality of virus stock preparation defines the experimental quality. Therefore, for any comparison, the virus stocks to be compared need to be of similar quality. Virus stocks for in vivo use can either be TC-derived or ex vivo derived from salivary glands of MCMV-infected mice (SGV).

3.1.4.1. Preparation of TC-Derived MCMV Stocks

1. Grow MEFs or another permissive cell line (e.g., NIH/3T3, BALB/3T3, M2-10B4, B12) (*see* **Note 10**) in Petri dishes (156 cm^2) in a complete medium to reach subconfluence.
2. Remove the excess of the medium and infect cells (0.01 PFU of MCMV per cell) in a volume of the medium sufficient to cover the cell monolayer (about 6 mL/Petri dish).
3. Incubate for 4–6 h at 37°C in 5% CO_2 to allow virus adsorption. Avoid carrying out virus adsorption in a large volume of the medium because this could reduce the efficacy of infection.
4. Add a fresh medium and incubate for additional 3–5 days to reach absolute infection, which can be visualized by its cytopathic effect.
5. Collect the virus-containing supernatant and centrifuge at 6,400 g for 20 min to remove cell debris and cell-associated virus (*see* **Note 11**).
6. Pellet the virus from the supernatant by ultracentrifugation at 26,000 g at 4°C for 2–3 h.
7. Decant the supernatant and leave the virus pellet with 200 µL of residual medium on ice for about 12 h.
8. Resuspend the pellet in medium and purify the virus by ultracentrifugation on the sucrose gradient (18 mL 15% sucrose/VSB in Centrifuge Bottles 25 × 89 mm), by using a swing rotor at 72,000 g at 4°C for 2 h.
9. After decanting the supernatant, leave the pellet in 500 µL of 15% sucrose/VSB overnight before resuspending (*see* **Note 11**).
10. Pool supernatants and freeze aliquots at –80°C.

11. Before using the stock in vivo, thaw one aliquot and determine the virus titer (*see* **Subheading 3.1.5.**).

3.1.4.2. Preparation of Salivary Gland-Derived MCMV Stocks

Salivary glands are the most important sites for horizontal spread of the MCMV. Owing to unknown epigenetic factors, virus produced in salivary glands (SGV) is more virulent than virus derived from other tissue or virus produced in cultured cells *(30)*.

1. Inject young (4–6 weeks old) female BALB/c mice with 2×10^5 PFU of TC-derived virus i.p. in 0.5 mL of culture medium without FCS.
2. Kill mice 14 days after infection with carbon dioxide inhalation and collect salivary glands (pool of parotid, greater sublingual, and mandibular glands) (*see* **Note 12**).
3. Prepare organ homogenates in cold PBS or in medium without FCS (3 mL per mouse) on ice using the stainless steel wire mesh.
4. Remove the tissue debris by centrifugation for 5 min at 800 *g* at 4°C.
5. Freeze the aliquots at –80°C.
6. Determine the virus titer (*see* **Subheading 3.1.5.**) and inject mice with 5×10^4 PFU of the first passage of SGV MCMV in 0.5 mL of medium without FCS (*see* **Note 13**).
7. Repeat this procedure 2–3 more times before stock production.

3.1.5. Determination of Virus Titers Using the Standard Plaque-Forming Assay

Titers of infectious MCMV in organ homogenates or virus stocks can be determined by a standard plaque-forming assay on MEFs (or alternatively on NIH/3T3, BALB/3T3, M2-10B4, or B12 cells) (*see* **Note 10**).

1. Seed MEFs in complete medium in 48-well TC plates on the day before titration to reach subconfluence.
2. Thaw mouse organs stored at –20°C.
3. Prepare organ homogenates in 2 mL of complete cold medium (3% MEM) by passing them through a stainless steel wire mesh.
4. Make serial $\log_{10}$ dilutions of each organ homogenate, remove excess of medium from the plates, and add 100 µL of each dilution per well at least in duplicate.
5. Incubate plates for 30 min at 37°C to allow virus adsorption.
6. Centrifuge plates for 30 min at 800 *g*.
7. Incubate plates for 30 min at 37°C before adding 0.5–1 mL of viscous medium (methyl cellulose) per well (*see* **Note 14**).
8. Count virus plaques after 3–4 days using an inverted microscope.
9. Calculate the number of PFU per organ or per gram of tissue (*see* **Note 15**).

3.1.6. Adoptive Transfer of NK Cells

Adoptive cell transfer into MCMV-infected recipient mice is a frequently used approach to study the antiviral capacity of immune cells. Adoptive transfer of NK cells into suckling MCMV-infected mice provided the evidence of the importance of NK cells in the immunosurveillance of MCMV *(31)*. A similar approach can also be used in adult mice. The use of RAG1-deficient or SCID mice as donors of NK cells avoids a contamination by T and B cells.

1. Kill RAG1-deficient mice with carbon dioxide inhalation and aseptically remove spleen (*see* **Note 16**).
2. Prepare single-cell suspension of splenocytes.
3. Lyse the erythrocytes with lysing buffer.
4. Wash the cells twice in the medium without FCS.
5. Adjust the number of cells and transfer them i.v. into syngeneic recipient mice in 0.5 mL of the medium without FCS (in case of suckling mice, inject the cells i.p.). It is recommended that at least three different doses of cells should be tested for each virus.
6. Infect recipient mice with mutant or control viruses 6 h after cell transfer (*see* **Subheading 3.1.1.1.**).
7. Kill the mice with carbon dioxide inhalation 3 days after infection, collect and freeze their organs at −20°C (spleen, lungs, and liver).
8. Determine virus titers in each organ with standard plaque-forming assay (*see* **Subheading 3.1.5.**).

3.2. Assessment of NK Cell Response to MCMV Mutants In Vitro

3.2.1. Cytotoxic ^{51}Cr-Release NK Assay on MCMV-Infected Cells

Cytotoxic ^{51}Cr-release NK assay has been frequently used as a correlate of in vivo susceptibility of virus mutants to NK cell control. Not all permissive cells are however suitable for a ^{51}Cr-release NK cell assay. The cells need to be constitutively susceptible to NK cell lysis and at the same time permissive for MCMV infection. B12 cells (immortalized BALB/c fibroblasts) *(23)* have served to show that viral inhibitors of NKG2D ligands compromise NK cell lysis of infected cells, whereas cells infected with the viruses lacking the NKG2D inhibitor m152 are sensitive to NK cell lysis *(15)*.

1. Prepare effector NK cells from splenocytes of mice pretreated 16 h earlier with 200 µg of poly (I•C) injected i.p.
2. Adjust effector cell number to a cell concentration appropriate for achieving the desired starting E : T ratio. Transfer effector cells to round bottom 96-well TC plates and perform serial twofold dilutions of cells in final volume of 100 µL. Leave empty the rows A and B for spontaneous and maximal release controls.

3. Infect target cells (e.g., B12 cells) with recombinant mutant or control WT MCMV (*see* **Subheading 3.1.1.2.**) or leave cells uninfected.
4. Harvest target cells 12 h after infection by trypsinization and wash cells twice with culture medium without FCS.
5. Resuspend cells in 0.5 mL of medium without serum and add 100 µCi of $Na_2^{51}CrO_4$.
6. Mix gently and incubate for 60 min at 37°C and 5% CO_2. Rotate tube gently every 15 min to mix cells.
7. Wash the cells 4× in complete medium, adjust the number of ^{51}Cr-labeled target cells to $2–5 \times 10^4$ per mL.
8. Add 100 µL of the target cell suspension to prepared 96-well plates containing effector cells. Add 100 µL of complete medium to wells in row A (spontaneous release controls) and 100 µL of 5% Triton X-100 to wells in row B (maximal release controls).
9. Centrifuge plates for 3 min at 200 *g*.
10. Incubate plates at 37°C and 5% CO_2 for 4 h.
11. Harvest 100 µL of supernatant from each well.
12. Measure radioactivity released into the supernatant with a γ-counter and calculate percent of Cr-release in each well.

3.2.2. Detection of MCMV Genes Involved in the Down-Modulation of NKG2D Ligands Using Flow Cytometry and Confocal Microscopy

MCMV encodes a number of proteins that down-modulate NKG2D ligands *(15,18,19,32–34)*. These proteins were originally identified by surface staining of cells infected with various MCMV mutants. As the NKG2D receptor can recognize different ligands, a soluble NKG2D receptor (e.g., NKG2D tetramer, *see* **Note 17**) should be used to locate the viral modulatory gene, whereas specific antibodies should be used to define the ligand regulated by particular viral immunoevasins (*see* **Fig. 2**). To localize individual genes involved in the modulation of NKG2D ligands, the following strategy was used. First, mutant viruses lacking sets of genes were used to define the genomic region encoding proteins involved in the down-modulation of NKG2D ligands. The NKG2D receptor recognizes all three mouse NKG2D ligands, so a soluble receptor (tetramer) is optimal for screening. Use of mutants with smaller deletions, and finally mutants lacking only a single gene led to the identification of the four NKG2D regulators within the MCMV genome known so far. To determine which NKG2D ligand is regulated by a given viral protein, specific monoclonal antibodies were used. One should bear in mind that different mouse strains may express different NKG2D ligands (e.g., C57BL/6 mice do not express H60).

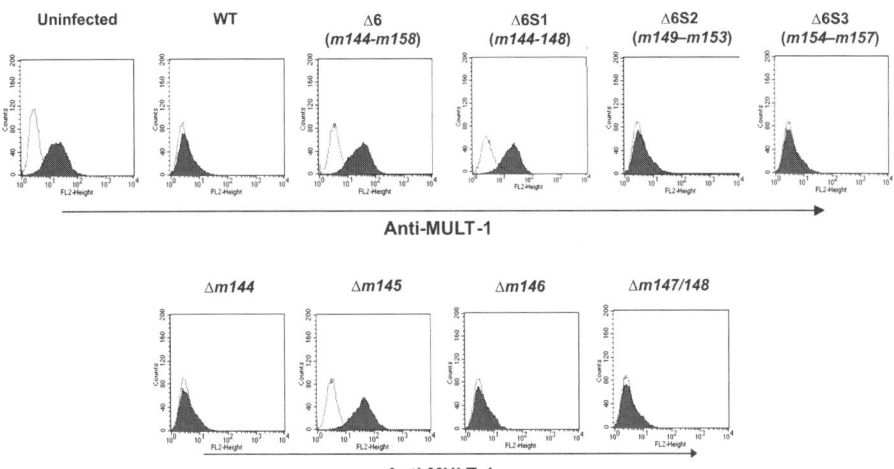

Fig. 2. Flow cytometric screening for MCMV gene involved in the down-modulation of NKG2D ligand MULT-1. To determine the genomic region encoding protein involved in the down-modulation of MULT-1, the SVEC4-10 cells were infected with WT MCMV or mutant virus Δ6 lacking the set of genes (*m144–m158*). Dissection of Δ6 region into three subregions (Δ6S1, Δ6S2, and Δ6S3) revealed that the gene responsible for the viral interaction with the expression of MULT-1 is located within Δ6S1 region. Infection of cells with single deletion mutants for each individual gene from Δ6S1 region identified the *m145* as MCMV gene responsible for down-modulation of MULT-1. Twelve hours p.i. cells were stained with anti-MULT-1 mAb. Reproduced from **ref. 19** by copyright permission of The Rockefeller University Press.

1. Infect permissive cells (e.g., BALB/c and C57BL/6 MEFs, B12, SVEC4-10, NIH/3T3) that express NKG2D ligands with either WT or mutant MCMV (*see* **Subheading 3.1.1.2.**).
2. Harvest the cells 12 h after infection by trypsinization or EDTA (*see* **Note 18**).
3. Resuspend cells in FACS medium. Keep cell suspensions on ice.
4. Stain infected and uninfected control cells with PE-NKG2D tetramer *(15)* (*see* **Note 17**) diluted in FACS medium for 25 min on ice. Use specific mAbs for the staining of individual NKG2D ligands (MULT-1, H60, and RAE-1) rat anti-MULT-1 mAb 1D6 *(19)*, rat PE-anti-RAE-1αβγ mAb CX1 (BD Pharmingen), rat anti-H60, clone 205326 (R&D Systems). For each mAb, use appropriate isotype-matched controls.
5. Wash cells twice with FACS medium.
6. Analyze cells stained with PE-NKG2D tetramer and cells stained with a directly labeled primary antibody by flow cytometry.
7. Incubate cells stained with unlabeled primary antibody with appropriate fluorescence-labeled secondary antibody for 25 min on ice.

8. Wash cells twice with FACS medium and analyze them by flow cytometry.

To distinguish infected from uninfected cells, intracellular staining with the mAb CROMA 229 *(27)*, which recognizes MCMV gp48 can be applied. Alternatively, a recombinant virus expressing GFP can be used.

3.2.3. Confocal Analysis of NKG2D Ligands in MCMV-Infected Cells

1. For confocal analysis, grow infected and uninfected cells on glass coverslips in 12-well plates.
2. Wash cells twice in PBS.
3. Fix cells with 2.5% (w/v) paraformaldehyde for 20 min at room temperature (RT) and permeabilize by 5 min incubation in 0.1% (w/v) Triton X-100.
4. Wash cells twice in PBS.
5. Block unspecific binding with 0.2% (w/v) fish skin gelatine diluted in PBS for 10 min at RT.
6. For visualization of NKG2D ligands, add specific mAb (e.g., rat anti-MULT-1 mAb) or control isotype-matched mAb diluted in PBS. For intracellular localization of NKG2D ligands, use mAbs specific for different cellular compartments and incubate cells for additional 40 min at RT.
7. Wash cells 4× in PBS.
8. Add fluorescence-labeled secondary antibody and incubate for 40 min at RT.
9. Wash cells 4× in PBS.
10. Wash cells briefly in the distilled H_2O.
11. Mount the coverslips using mowiol.
12. Leave from 30 min up to few weeks in dark place.
13. Analyze samples with a confocal laser-scanning microscope.

3.2.4. Monitoring of Surface Resident Ligands for NK Cell Receptors

To study the fate of surface resident ligands for activating NK cell receptors, the assay is based on tagging of the surface ligand with specific mAbs (e.g., rat anti-MULT-1 mAb) and following the fate of antibody-bound surface protein by flow cytometry and confocal microscopy. This assay is also called uptake assay *(35)*.

3.2.4.1. UPTAKE ASSAY FOR SURFACE RESIDENT MULT-1

1. Seed 4×10^5 SVEC4-10 cells per well in 6-well TC plates in 10% DMEM for 6–12 h.
2. Remove excess of medium and infect cells (4 PFU of TC-derived MCMV per cell in 750 µL of 10% DMEM per well).
3. Incubate plates for 30 min at 37°C, 5% CO_2 to allow virus adsorption.
4. Add 3.5 mL of warm medium and centrifuge plates at 800 *g* for 30 min to enhance infection.

5. Incubate cells at 37°C, 5% CO_2 (the incubation period can be from 0 to 4 h).
6. Leave plates on ice for few minutes.
7. Remove medium and wash plates twice with cold PBS.
8. Remove PBS and add rat anti-MULT-1 mAb 1D6 or isotype-matched control mAb in 750 μL of cold PBS per well.
9. Incubate plates for 30 min on ice and remove unbound antibodies by washing twice with cold PBS.
10. Add warm 10% DMEM and incubate plates for periods ranging from 20 min to 24 h.
11. Harvest cells by Trypsin-EDTA and wash cells twice in FACS medium.
12. Add secondary mAbs (goat anti-rat-PE) diluted in 50 μL of FACS medium per sample and incubate for 30 min.
13. Wash cells twice with 1 mL FACS medium.
14. Resuspend cells in 500 μL of FACS medium and analyze the samples by flow cytometry.

4. Notes

1. Frequently used MCMV strains are the strains Smith and K181 *(21,36)*. To study the function of viral immunomodulatory genes in vivo, the virus mutants, as well as their corresponding revertant strains and WT virus, should be derived from the same genome. Conventional mutagenesis protocols for herpesviruses are based on the insertion of marker genes into the viral genome by homologous recombination during infection of cells, which allows disruption or deletion of viral genes *(37,38)*. Generation of viral mutants requires a selection process against the WT virus. The recombinant virus has to be plaque-purified. Recently, this method has been replaced in most labs by using the BAC-cloned viral genome *(7–9,39,40)*. This allows the introduction of any kind of genomic mutation along with the appropriate controls in the bacterial plasmid before reconstitution of the virus mutant.
2. The antiviral activity of NK cell is most prominent during the early phase of infection. Therefore, the virus titers in organs of infected animals should be determined not later than 3–4 days after infection.
3. MCMV latency is usually established about 12 weeks after infection (BALB/c strain, TC-grown virus). For virus reactivation, mice are subjected to depletion of $CD4^+$ and $CD8^+$ T cells, with or without depletion of NK cells. NK cells should be depleted using anti-asialoGM1 antibody or by anti-NK1.1 mAbs (in mice expressing this receptor). Animals should be killed on days 5 to 12 after immunodepletion. Reactivation of virus in several organs is then detectable in a significant proportion but not all mice *(28)*.
4. Laboratory mouse strains differ in their NK-cell-dependent sensitivity to MCMV *(13,41,42)*. This has to be considered when the attenuation of virus mutants and their susceptibility to immune surveillance are assessed.

5. The maturation time and the expression of different NK cell receptors vary (*43,44*). It may take several weeks to reach the maximum of NK cell activity. Therefore, newborn mice are very sensitive to MCMV infection, and infections with virus doses over 100 PFU result in a high mortality rate.
6. Syngeneic primary cell lines have to be used, otherwise (e.g., in the case of injection of non-syngeneic MEFs or tumour cell line) NK cells of recipient mice would respond to transferred cells and reject them independently of infection.
7. Centrifugation of cells after adding the virus enhances the efficacy of the infection about 25-fold (*45*).
8. The selective depletion of NK cells has limitations. Most antibodies used for this purpose are directed against proteins that are expressed also on other cell subsets. The most specific antibody used for NK cell depletion is PK136 that recognizes NK1.1 (*29*). In NK1.1$^-$ mice depletion of NK cells is usually performed by rabbit antiserum to asialo-GM1 (AGM1). This antiserum is not only specific for NK cells, and therefore, the effects cannot be strictly ascribed to NK cells. NK cell receptors can be blocked by specific mAbs without depleting the cells. This approach is used for blocking the activation receptors Ly49H and NKG2D. Owing to potential side effects of immunoglobulins, it is important to use matched control antibodies when assessing the effect of blocking or cytolytic mAbs. The antibody preparation can influence the in vivo result. The most important thing is that the antibody be free of LPS (endotoxins) and other components that may induce unwanted responses. Therefore, it is advantageous to use purified antibodies. Sterile filtrated ammonium-sulfate precipitates of mAbs, dialyzed against PBS, are frequently used and give satisfactory results.
9. Viral LD_{50} depends on several conditions. First, mouse strains differ in their sensitivity to MCMV. Second, for the infection, TC-derived virus or SGV can be used. SGV is much more virulent. Third, the LD_{50} increases with age. Newborn mice infected with doses over 100 PFU of TC-derived WT MCMV have a high mortality rate.
10. MEFs are the cells of choice for the preparation of MCMV stocks for in vivo use, as compared to immortalized cell lines. As host cell components are present in virus preparations, in vivo experiments should be conducted with virus grown on cells syngeneic to the animals used in experiments.
11. Cell-associated virus can be prepared and used for in vitro studies. Cell-associated virus may however be less suitable for in vivo studies than virus released into the supernatant. Sucrose/VSB may be toxic for newborn mice and PBS should be used to resuspend the virus pellet. To prevent bacterial contamination of virus stock, filtration through 0.45 μm sterile filters either before or after gradient ultracentrifugation may be advisable. Note that virus is lost with each preparatory step.

12. For preparation of SGV, cervical lymph nodes are removed, and the surrounding fatty tissue is cleared. Salivary glands are easily accessible by forceps. Usually submandibular and sublingual glands are collectively removed, whereas parotid glands should be removed individually. Using curved forceps, all three pairs of the salivary glands can be removed in one step.
13. SGV induces a surplus of inflammatory cytokines in vivo *(46)*. This may influence the pattern of the early immune response and may cause difficulties in assessing the effects of single viral gene products on the innate immune response. To obtain a high SGV titer, it is recommended to deplete $CD4^+$ T lymphocytes.
14. Viscous medium is added to avoid horizontal spread of virus particles. Because of its viscosity, this medium cannot be added by using a pipette. Instead, a sterile beaker can be used.
15. The same procedure can be used for the determination of the virus titer in virus stocks. Centrifugal enhancement (about 25-fold) should be avoided.
16. There are many variations in protocols used for the isolation of NK cells. Instead of using RAG1-deficient mice or SCID mice, splenocytes can be obtained from conventional laboratory mice and NK cell population enriched by the incubation with a cocktail of monoclonal antibodies against CD4, CD8, CD24, I-E, and I-A molecules, followed by complement mediated lysis. Another option for NK cell enrichment is the positive selection of NK cells with magnetic beads.
17. NKG2D tetramers used were generated as described by Krmpotic et al.*(15)*. The extracellular domain of the NKG2D molecule was fused to an NH_2-terminal biotinylation tag50-specific biotinylation site cloned into a *pET3a* expression vector (Novagen). Recombinant NKG2D proteins were generated as insoluble proteins following transduction of this *pET3a-NKG2D* in *E. coli* strain BL21 (DE3). Refolded NKG2D was purified by gel filtration (Superdex 200HR, Pharmacia) as a single peak corresponding to the predicted size (34 kD) of homodimeric molecules. The protein was further biotinylated with the biotin ligase BirA and subsequently tetramerized by addition of SA-PE (Molecular Probes) at a molar ratio of 4:1. Tetramers were purified by gel filtration over a Superdex 200HR column and stored at 2 mg/mL at 4°C in PBS (pH 8.0) containing 0.02% sodium azide, 1 g/mL of pepstatin, 1 g/mL of leupeptin, and 0.5 mM EDTA.
18. Several proteins (e.g., MCMV fcr-1) are sensitive to trypsin treatment. It may be necessary to use EDTA for harvesting infected cells.

Acknowledgments

U.H. Koszinowski is supported by the Deutsche Forschungsgemeinschaft through SFB 455. S. Jonjic is supported by Croatian Ministry of Science Grant 0062004 and FP6 Marie Curie Research Training grant 019248, and A. Krmpotic is supported by an International Research Scholars grant from the Howard Hughes Medical Institute and by Croatian Ministry of Science Grant 0062007.

Reference

1. Reis e Sousa, C. (2004) Activation of dendritic cells: translating innate into adaptive immunity. *Curr. Opin. Immunol.* **16**, 21–5.
2. French, A. R., and Yokoyama, W. M. (2003) Natural killer cells and viral infections. *Curr. Opin. Immunol.* **15**, 45–51.
3. Biron, C. A. (2001) Interferons alpha and beta as immune regulators–a new look. *Immunity* **14**, 661–4.
4. Merrill, J. E., Ullberg, M., and Jondal, M. (1981) Influence of IgG and IgM receptor triggering on human natural killer cell cytotoxicity measured on the level of the single effector cell. *Eur. J. Immunol.* **11**, 536–41.
5. Alcami, A., and Koszinowski, U. H. (2000) Viral mechanisms of immune evasion. *Immunol. Today* **21**, 447–55.
6. Rawlinson, W. D., Farrell, H. E., and Barrell, B. G. (1996) Analysis of the complete DNA sequence of murine cytomegalovirus. *J. Virol.* **70**, 8833–49.
7. Messerle, M., Crnkovic, I., Hammerschmidt, W., Ziegler, H., and Koszinowski, U. H. (1997) Cloning and mutagenesis of a herpesvirus genome as an infectious bacterial artificial chromosome. *Proc. Natl. Acad. Sci. U. S. A.* **94**, 14759–63.
8. Wagner, M., and Koszinowski, U. H. (2004) Mutagenesis of viral BACs with linear PCR fragments (ET recombination). *Methods Mol. Biol.* **256**, 257–68.
9. Wagner, M., Jonjic, S., Koszinowski, U. H., and Messerle, M. (1999) Systematic excision of vector sequences from the BAC-cloned herpesvirus genome during virus reconstitution. *J. Virol.* **73**, 7056–60.
10. Jonjic, S., Bubic, I., and Krmpotic, A. (2006) Innate immunity to cytomegaloviruses, in *Cytomegaloviruses: Molecular biology and immunology* (Reddehase, M. J., ed.), Caister Academic Press, Wymondham, Norfolk, UK, pp. 285–319.
11. Orange, J. S., Fassett, M. S., Koopman, L. A., Boyson, J. E., and Strominger, J. L. (2002) Viral evasion of natural killer cells. *Nat. Immunol.* **3**, 1006–12.
12. Scalzo, A. A. (2002) Successful control of viruses by NK cells–a balance of opposing forces? *Trends Microbiol.* **10**, 470–4.
13. Desrosiers, M. P., Kielczewska, A., Loredo-Osti, J. C., Adam, S. G., Makrigiannis, A. P., Lemieux, S., Pham, T., Lodoen, M. B., Morgan, K., Lanier, L. L., and Vidal, S. M. (2005) Epistasis between mouse Klra and major histocompatibility complex class I loci is associated with a new mechanism of natural killer cell-mediated innate resistance to cytomegalovirus infection. *Nat. Genet.* **37**, 593–9.
14. Adam, S. G., Caraux, A., Fodil-Cornu, N., Loredo-Osti, J. C., Lesjean-Pottier, S., Jaubert, J., Bubic, I., Jonjic, S., Guenet, J. L., Vidal, S. M., and Colucci, F. (2006) Cmv4, a new locus linked to the NK cell gene complex, controls innate resistance to cytomegalovirus in wild-derived mice. *J. Immunol.* **176**, 5478–85.
15. Krmpotic, A., Busch, D. H., Bubic, I., Gebhardt, F., Hengel, H., Hasan, M., Scalzo, A. A., Koszinowski, U. H., and Jonjic, S. (2002) MCMV glycoprotein

gp40 confers virus resistance to CD8+ T cells and NK cells in vivo. *Nat. Immunol.* **3**, 529–35.
16. Bubic, I., Wagner, M., Krmpotic, A., Saulig, T., Kim, S., Yokoyama, W. M., Jonjic, S., and Koszinowski, U. H. (2004) Gain of virulence caused by loss of a gene in murine cytomegalovirus. *J. Virol.* **78**, 7536–44.
17. Farrell, H. E., Vally, H., Lynch, D. M., Fleming, P., Shellam, G. R., Scalzo, A. A., and Davis-Poynter, N. J. (1997) Inhibition of natural killer cells by a cytomegalovirus MHC class I homologue in vivo. *Nature* **386**, 510–4.
18. Hasan, M., Krmpotic, A., Ruzsics, Z., Bubic, I., Lenac, T., Halenius, A., Loewendorf, A., Messerle, M., Hengel, H., Jonjic, S., and Koszinowski, U. H. (2005) Selective down-regulation of the NKG2D ligand H60 by mouse cytomegalovirus m155 glycoprotein. *J. Virol.* **79**, 2920–30.
19. Krmpotic, A., Hasan, M., Loewendorf, A., Saulig, T., Halenius, A., Lenac, T., Polic, B., Bubic, I., Kriegeskorte, A., Pernjak-Pugel, E., Messerle, M., Hengel, H., Busch, D. H., Koszinowski, U. H., and Jonjic, S. (2005) NK cell activation through the NKG2D ligand MULT-1 is selectively prevented by the glycoprotein encoded by mouse cytomegalovirus gene m145. *J. Exp. Med.* **201**, 211–20.
20. Ziegler, H., Thäle, R., Lucin, P., Muranyi, W., Flohr, T., Hengel, H., Farrell, H., Rawlinson, W., and Koszinowski, U. H. (1997) A mouse cytomegalovirus glycoprotein retains MHC class I complexes in the ERGIC/cis-Golgi compartments. *Immunity* **6**, 57–66.
21. Smith, M. G. (1954) Propagation of salivary gland virus of the mouse in tissue cultures. *Proc. Soc. Exp. Biol. Med.* **86**, 435–40.
22. Krmpotic, A., Messerle, M., Crnkovic-Mertens, I., Polic, B., Jonjic, S., and Koszinowski, U. H. (1999) The immunoevasive function encoded by the mouse cytomegalovirus gene m152 protects the virus against T cell control in vivo. *J. Exp. Med.* **190**, 1285–96.
23. Del Val, M., Schlicht, H. J., Ruppert, T., Reddehase, M. J., and Koszinowski, U. H. (1991) Efficient processing of an antigenic sequence for presentation by MHC class I molecules depends on its neighboring residues in the protein. *Cell* **66**, 1145–53.
24. Sentman, C. L., Hackett, J., Jr., Kumar, V., and Bennett, M. (1989) Identification of a subset of murine natural killer cells that mediates rejection of Hh-1d but not Hh-1b bone marrow grafts. *J. Exp. Med.* **170**, 191–202.
25. Ho, E. L., Carayannopoulos, L. N., Poursine-Laurent, J., Kinder, J., Plougastel, B., Smith, H. R., and Yokoyama, W. M. (2002) Costimulation of multiple NK cell activation receptors by NKG2D. *J. Immunol.* **169**, 3667–75.
26. Smith, H. R., Chuang, H. H., Wang, L. L., Salcedo, M., Heusel, J. W., and Yokoyama, W. M. (2000) Nonstochastic coexpression of activation receptors on murine natural killer cells. *J. Exp. Med.* **191**, 1341–54.
27. Reusch, U., Muranyi, W., Lucin, P., Burgert, H. G., Hengel, H., and Koszinowski, U. H. (1999) A cytomegalovirus glycoprotein re-routes MHC class I complexes to lysosomes for degradation. *EMBO J.* **18**, 1081–91.

28. Polic, B., Hengel, H., Krmpotic, A., Trgovcich, J., Pavic, I., Luccaronin, P., Jonjic, S., and Koszinowski, U. H. (1998) Hierarchical and redundant lymphocyte subset control precludes cytomegalovirus replication during latent infection. *J. Exp. Med.* **188**, 1047–54.
29. Koo, G. C., and Peppard, J. R. (1984) Establishment of monoclonal anti-Nk-1.1 antibody. *Hybridoma* **3**, 301–3.
30. Osborn, J. E., and Walker, D. L. (1971) Virulence and Attenuation of Murine Cytomegalovirus. *Infect. Immun.* **3**, 228–236.
31. Bukowski, J. F., Warner, J. F., Dennert, G., and Welsh, R. M. (1985) Adoptive transfer studies demonstrating the antiviral effect of natural killer cells in vivo. *J. Exp. Med.* **161**, 40–52.
32. Lodoen, M., Ogasawara, K., Hamerman, J. A., Arase, H., Houchins, J. P., Mocarski, E. S., and Lanier, L. L. (2003) NKG2D-mediated natural killer cell protection against cytomegalovirus is impaired by viral gp40 modulation of retinoic acid early inducible 1 gene molecules. *J. Exp. Med.* **197**, 1245–53.
33. Lodoen, M. B., Abenes, G., Umamoto, S., Houchins, J. P., Liu, F., and Lanier, L. L. (2004) The cytomegalovirus m155 gene product subverts natural killer cell antiviral protection by disruption of H60-NKG2D interactions. *J. Exp. Med.* **200**, 1075–81.
34. Lenac, T., Budt, M., Arapovic, J., Hasan, M., Zimmermann, A., Simic, H., Krmpotic, A., Messerle, M., Ruzsics, Z., Koszinowski, U. H., Hengel, H., and Jonjic, S. (2006) The herpesviral Fc receptor fcr-1 down-regulates the NKG2D ligands MULT-1 and H60. *J. Exp. Med.* **203**, 1843–50.
35. Mansouri, M., Bartee, E., Gouveia, K., Hovey Nerenberg, B. T., Barrett, J., Thomas, L., Thomas, G., McFadden, G., and Fruh, K. (2003) The PHD/LAP-domain protein M153R of myxomavirus is a ubiquitin ligase that induces the rapid internalization and lysosomal destruction of CD4. *J. Virol.* **77**, 1427–40.
36. Hudson, J. B., Walker, D. G., and Altamirano, M. (1988) Analysis in vitro of two biologically distinct strains of murine cytomegalovirus. *Arch. Virol.* **102**, 289–95.
37. Mocarski, E. S., Jr., and Kemble, G. W. (1996) Recombinant cytomegaloviruses for study of replication and pathogenesis. *Intervirology* **39**, 320–30.
38. Spaete, R. R., and Mocarski, E. S. (1987) Insertion and deletion mutagenesis of the human cytomegalovirus genome. *Proc. Natl. Acad. Sci. U. S. A.* **84**, 7213–7.
39. Borst, E. M., Crnkovic-Mertens, I., and Messerle, M. (2004) Cloning of beta-herpesvirus genomes as bacterial artificial chromosomes. *Methods Mol. Biol.* **256**, 221–39.
40. Borst, E. M., Posfai, G., Pogoda, F., and Messerle, M. (2004) Mutagenesis of herpesvirus BACs by allele replacement. *Methods Mol. Biol.* **256**, 269–79.
41. Scalzo, A. A., Fitzgerald, N. A., Wallace, C. R., Gibbons, A. E., Smart, Y. C., Burton, R. C., and Shellam, G. R. (1992) The effect of the Cmv-1 resistance gene, which is linked to the natural killer cell gene complex, is mediated by natural killer cells. *J. Immunol.* **149**, 581–9.

42. Scalzo, A. A., Fitzgerald, N. A., Simmons, A., La Vista, A. B., and Shellam, G. R. (1990) Cmv-1, a genetic locus that controls murine cytomegalovirus replication in the spleen. *J. Exp. Med.* **171**, 1469–83.
43. Williams, N. S., Kubota, A., Bennett, M., Kumar, V., and Takei, F. (2000) Clonal analysis of NK cell development from bone marrow progenitors in vitro: orderly acquisition of receptor gene expression. *Eur. J. Immunol.* **30**, 2074–82.
44. Fraser, K. P., Gays, F., Robinson, J. H., van Beneden, K., Leclercq, G., Vance, R. E., Raulet, D. H., and Brooks, C. G. (2002) NK cells developing in vitro from fetal mouse progenitors express at least one member of the Ly49 family that is acquired in a time-dependent and stochastic manner independently of CD94 and NKG2. *Eur. J. Immunol.* **32**, 868–78.
45. Hudson, J. B. (1988) Further studies on the mechanism of centrifugal enhancement of cytomegalovirus infectivity. *J. Virol. Methods* **19**, 97–108.
46. Trgovcich, J., Stimac, D., Polic, B., Krmpotic, A., Pernjak-Pugel, E., Tomac, J., Hasan, M., Wraber, B., and Jonjic, S. (2000) Immune responses and cytokine induction in the development of severe hepatitis during acute infections with murine cytomegalovirus. *Arch. Virol.* **145**, 2601–18.
47. Arase, H., Mocarski, E. S., Campbell, A. E., Hill, A. B., and Lanier, L. L. (2002) Direct recognition of cytomegalovirus by activating and inhibitory NK cell receptors. *Science* **296**, 1323–6.
48. Smith, H. R., Heusel, J. W., Mehta, I. K., Kim, S., Dorner, B. G., Naidenko, O. V., Iizuka, K., Furukawa, H., Beckman, D. L., Pingel, J. T., Scalzo, A. A., Fremont, D. H., and Yokoyama, W. M. (2002) Recognition of a virus-encoded ligand by a natural killer cell activation receptor. *Proc. Natl. Acad. Sci. U. S. A.* **99**, 8826–31.
49. Fleming, P., Davis-Poynter, N., Degli-Esposti, M., Densley, E., Papadimitriou, J., Shellam, G., and Farrell, H. (1999) The murine cytomegalovirus chemokine homolog, m131/129, is a determinant of viral pathogenicity. *J. Virol.* **73**, 6800–9.
50. Crnkovic-Mertens, I., Messerle, M., Milotic, I., Szepan, U., Kucic, N., Krmpotic, A., Jonjic, S., and Koszinowski, U. H. (1998) Virus attenuation after deletion of the cytomegalovirus Fc receptor gene is not due to antibody control. *J. Virol.* **72**, 1377–82.

9

Analyzing Antibody–Fc-Receptor Interactions

Falk Nimmerjahn and Jeffrey V. Ravetch

Summary

Cellular receptors for immunoglobulins (Fc-receptors; FcR) are central mediators of antibody-triggered effector functions. Immune complex (IC) binding to FcRs results in a variety of reactions such as the release of inflammatory mediators, antibody dependent cellular cytotoxicity (ADCC) and phagocytosis of ICs. Analyzing antibody–FcR (Ab–FcR) interactions in vitro is essential to determine the effector mechanisms, binding characteristics and affinity parameters that will impact and predict antibody activity in vivo. The methods described in this chapter include the generation of ICs and soluble FcR variants, as well as ELISA and FACS-based assays to study Ab–FcR interactions.

Key Words: Antibody; Fc-receptor; immune complex; ELISA; FACS; cellular receptors; effector functions; ADCC; antibody mutant.

1. Introduction

Antibodies have shown promising results in human therapy of cancer and viral infections *(1–3)*. Studies in mice and results from human clinical trials have demonstrated that cytotoxic antibodies mediate their effector functions to a great extent by crosslinking cellular receptors for immunoglobulins (Fc-receptors; FcR) *(4–9)*. There is strong interest in generating antibodies with preferential binding to activating FcRs and in analyzing the outcome of immune complex (IC)-triggered crosslinking of FcRs on cells of the innate and adaptive immune system *(3, 10–13)*. This chapter will provide protocols for the most important techniques and the generation of essential tools to study the interaction of antibodies with FcRs.

From: *Methods in Molecular Biology, vol. 415: Innate Immunity*
Edited by: J. Ewbank and E. Vivier © Humana Press Inc., Totowa, NJ

Interactions between antibodies and their receptors or effector pathways range in affinity over several orders of magnitude (10^6–10^9 M^{-1}), which necessitates various assay systems to study these interactions on a functional level. The most commonly used experimental systems are enzyme-linked immunosorbent assays (ELISA)-, fluorescent activated cell sorting (FACS)-, or surface plasmon resonance (SPR)-based assays. These assays have a different sensitivity and therefore provide different levels of information. The use of soluble FcRs has greatly simplified the design of quantitative assays, such as ELISA or SPR analysis *(14–16)*. With the exception of the high-affinity FcγRI, all other FcRs bind monomeric antibodies with medium or low affinity *(9)*, which precludes the use of ELISA or FACS analysis as assay systems. To study this group of receptors the use of antigen–antibody complexes (ICs) is essential. In addition, IC-mediated activation of effector cells provides important information about the outcome of FcR triggering on effector cells. The way these reagents are made can significantly influence the experimental outcome, and it is crucial to adhere to certain rules to obtain consistent results. Moreover, the design of ELISA-based antibody–FcR (Ab–FcR) interaction analysis is very sensitive with respect to buffers with different salt concentrations and washing solutions. Therefore, the central aim of this chapter is to provide information on how these experiments can be done in an easy-to-use and reproducible way. With these basic tools in hand, Ab-FcR-related questions can be addressed in a broad variety of experimental systems.

2. Materials
2.1. Production of Immune Complexes

1. Antibodies: Mouse anti-TNP antibody isotypes (IgG1, IgG2a, and IgM) are available from BD-Pharmingen (San Diego, CA). IgG2b and IgG3 anti-TNP antibodies were kindly provided by Brigitta Heyman (University of Uppsala, Sweden).
2. BSA-TNP (Sigma, St. Louis, MO) was biotinylated using a DSB-X biotinylation kit (Molecular probes, Eugene, OR).
3. Crosslinking anti-mouse F(ab)$_2$ fragments [e.g. phycoerythrin (PE) labeled] (Jackson ImmunoResearch, West Grove, PA).

2.2. Cell Culture

1. Dulbecco's Modified Eagle's Medium (DMEM) (Invitrogen, Carlsbad, CA), supplemented with 10% heat-inactivated fetal bovine serum (FBS, Hyclone, Logan, UT) and penicillin, streptomycin, and glutamate (Pen/Strep/Glu from Invitrogen; use 1:100); this medium is referred to as "culture medium." For protein production, FBS is replaced using Nutridoma SP (1:100; Roche, Indianapolis, IN); this medium is referred to as "production medium."

2. Dulbecco's phosphate-buffered saline (DPBS) and trypsin/ethylenediaminetetraacetic acid (EDTA) (1×) solution (0.25 or 0.05%) (Invitrogen).

2.3. Transient Transfection of 293T cells

1. Double-distilled water (18.2 MΩ-cm) should be autoclaved followed by filtration using a 0.2-μm filter (e.g., Stericup from Millipore, Billerica, MA).
2. 2× HBS buffer: dissolve 11.9 g of 4-(2-hydroxyethyl)-1-piperazineethanesulfonic acid (HEPES), 16.4 g of NaCl, and 0.21 g of Na_2HPO_4 in 1 L of double-distilled water. Adjust the pH to exactly 7.05; filter the solution through a 0.2-μm filter and store in aliquots at –20°C.
3. 2.5 M $CaCl_2$: Dissolve the appropriate amount of $CaCl_2$ in double-distilled water, filter the solution through a 0.2-μm filter, and store it at 4°C.
4. 100 mM Chloroquine: Dissolve the appropriate amount of chloroquine in double-distilled water, filter the solution through a 0.2-μm filter, and store it in aliquots at –20°C.
5. DNA: prepare the DNA with any available preparation kit (e.g., Maxi-prep kits from Qiagen, Carlsbad, CA, or Invitrogen); dissolve the DNA in 1× TE buffer (10 mM Tris–HCl pH 7.5, 1 mM EDTA pH 8.0) or double-distilled water and store at 4°C.
6. Sodium azide: Make a 10% solution (w/v) in double-distilled water and filter with a 0.2-μm filter.
7. Protease inhibitor cocktail (e.g., Complete Mini, Roche).

2.4. ELISA with Soluble FcRs and ICs

1. Anti-Flag 96-well plates (Sigma-Aldrich, St. Louis, MO).
2. Horseradish peroxidase (HRP)-coupled streptavidin (Roche).
3. 3, 3′, 5, 25′-tetramethylbenidine (TMB) substrate solution (KPL, Gaithersburg, MA).
4. Stop solution: 6% ortho-phosphoric acid: dilute 85% ortho-phosphoric acid (Fluka BioChemika/Sigma-Aldrich Buchs, Switzerland) with double-distilled water to obtain a 6% solution.

2.5. Immune Complex Binding to FcR-Expressing Cells

1. 96-well V-bottom plates (Nunc, Rochester, NY).
2. FACS buffer: DPBS supplemented with 2% heat-inactivated FBS (incubate FBS for 30 min at 56°C).
3. FcR-blocking antibodies: FcγRIIB/III block (clone 2.4G2, BD-Pharmingen), FcγRIV block (9E9) (8,15).
4. Streptavidin–POD conjugate (Roche).

3. Methods

First, we will start by providing a detailed description of two basic techniques needed to study Ab–FcR interactions: the production of soluble FcRs and the generation of IC. Different FcRs show a distinct pattern of glycosylation, and it has been suggested recently that these sugar side chains are involved in recognizing non-fucosylated antibodies with high affinity *(16)*. Thus, it might be important to use mouse or human cell lines to preserve the FcR glycosylation status. Therefore, we will provide a protocol for production of soluble FcRs in 293T cells. Second, we will give examples how these reagents can be used to gain insight into Ab–FcR interactions, such as ICs binding to cellular FcRs or to soluble FcRs immobilized to 96-well plates. The latter assay is a quick and easy-to-use quality control step before ICs are used for other functional in vitro or in vivo experiments.

3.1. Generation of Soluble Fc-Receptors by Transient Transfection of 293T Cells

1. Soluble FcR are valuable tools to assess quickly the functionality of IC preparations or the affinity of Ab–FcR interactions. A scheme of which portions of the FcRs are used and where affinity tags are added to allow optimal activity of the soluble proteins is shown in **Fig. 1**. An easy way to generate a large quantity of these molecules is by transient transfection of 293T cells, which will be the focus of this section.

A

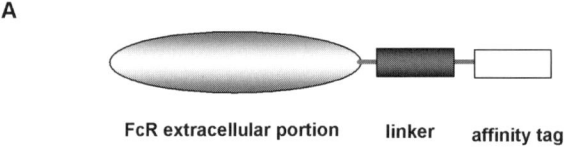

FcR extracellular portion   linker   affinity tag

B

FcγR	Linker	Affinity tag
FcγRI (AA 1-297) FcγRIIB (AA 1-211) FcγRIII (AA 1-216) FcγRIV (AA 1-201)	GGGGSGGGG	3xFlag (DYKDHDGDYKD HDIDYKDDDDK)

Fig. 1. Design of soluble Fc-receptor (FcR) variants. (**A**) Cartoon of the different domains of a soluble FcR molecule, consisting of the extracellular portion of the FcR, a short linker, and an affinity tag to simplify purification. (**B**) Detailed description which portions of the different mouse FcRs are present in the soluble molecules. In addition, the amino acid composition of the linker sequence (one letter amino acid code) and the affinity tag are indicated.

2. Grow 293T cells in a 150-mm culture dish (e.g., Corning, Corning, NY) in culture medium. Typically, a 1:10 split (using trypsin to detach the cells) every 2–3 days will keep the cells at an optimal density. On the day of transfection, the cells should reach about 70–80% confluence. For well-growing cells, this takes about 1.5 days after a 1:10 split.
3. On the day of transfection, warm up the culture medium to 37°C. Remove the old medium and add 24 ml of fresh, warm medium per 150-mm plate.
4. Prepare the transfection reagents and let them adjust to room temperature.
5. Add 3 ml of 2× HBS buffer to one 13-ml plastic tube (e.g., Sarstedt tube 13 ml, 100 × 16 mm) and label it "HBS."
6. Add 2.6 ml of water to another 13-ml tube, followed by 60 µg of DNA, 300 µl of 2.5 M $CaCl_2$, and 10 µl of 100 mM chloroquine; mix well by flicking the tube and label it "DNA" (see **Note 1**).
7. Bubble air through the HBS solution with a 2-ml plastic tip attached to a pipette aid; simultaneously, using a 1-ml pipette, add the DNA solution dropwise to the HBS tube. Continue to bubble air for 30 s; the solution should become slightly cloudy (see **Note 2**).
8. Carefully distribute the solution drop-wise onto the 150-mm culture dish with the 293T cells and incubate for 7–9 h at 37°C (see **Note 3**).
9. Remove the culture medium containing the transfection mix and replace with 20–24 ml of production medium.
10. Harvest the culture supernatant containing the soluble FcR molecules after 5–6 days, spin down cells and debris by centrifugation (600 g; 5 min), and filter the supernatant through 0.2-µm filter (see **Note 4**).
11. This supernatant can be used directly for ELISA (see **Subheading 3.3.**).
12. If the supernatant is not used immediately, it should be stored at 4°C and sodium azide [final concentration of 0.005% (v/v)] and protease inhibitors (e.g., one tablet of Complete Mini Protease Inhibitor Cocktail (Roche) per 100 ml of culture supernatant) should be added (see **Note 5**).

3.2. Generation of Immune Complexes

General considerations: Monomeric antibody binding to the majority of FcRs is virtually undetectable by FACS or ELISA assays (see **Fig. 2A**). The only Ab–FcR interaction that can be analyzed with these methods is the binding of IgG2a to the high-affinity FcγRI (see **Fig. 2A** and **B**). For all other FcRs, the use of IC is a prerequisite (see **Fig. 3**). There are several commercially available antibody/antigen combinations that can be used to study the interaction of ICs with FcRs. The easy access to monoclonal trinitrophenyl (TNP)-specific antibodies in combination with the respective polyvalent antigens (TNP-coupled BSA or OVA) makes it the system of choice for many in vitro and in vivo applications *(17)*; in general, any antibody/antigen combination can be used

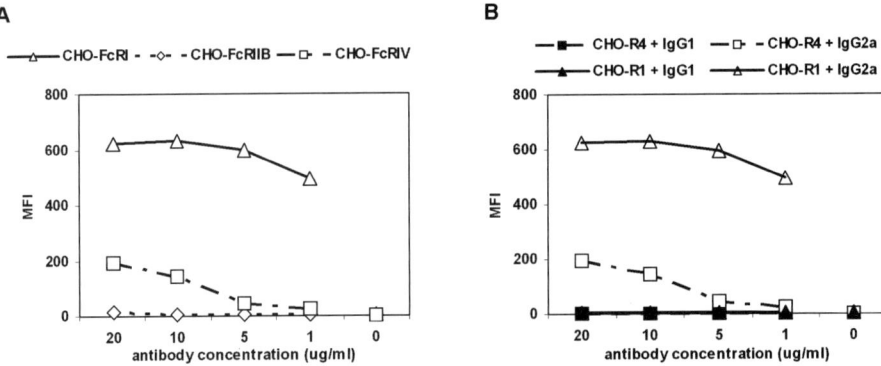

Fig. 2. Binding of monomeric antibodies to cellular FcRs. (**A**) Chinese hamster ovarian (CHO) cells expressing mouse FcγRI, FcγRIIB, or FcγRIV were incubated for 1 h with the indicated amounts of a monomeric IgG2a antibody at 4°C with gentle shaking. Cell bound antibody was detected by incubation with a PE-labeled mouse IgG-specific F(ab)$_2$ fragment. After washing, cells were analyzed on a FACS calibur and the mean fluorescence intensity (MFI) was plotted against the antibody concentration (μg/ml). Whereas there is significant binding to FcγRI-transfected cells only low level or no binding could be observed to FcγRIV- or FcγRIIB-transfected cells. (**B**) Specificity of monomeric antibody binding to CHO-FcγRI- and FcγRIV-transfected cells. Cells were incubated with the indicated amounts of monomeric mouse IgG1 or IgG2a antibodies, and bound antibody was detected as before. In contrast to significant binding of IgG2a, no IgG1 binding to either of these receptors was observed consistent with previous observations *(15)*.

(*see* **Notes 6** and **7**). If a polyvalent antigen is not available, ICs can be generated by using secondary crosslinking F(ab)$_2$ fragments combined with the primary antibodies. In this chapter, we will describe both the generation of TNP-ICs (see point **3.2.1**) and the use of F(ab)$_2$ fragments to crosslink primary antibodies (*see* **Subheading 3.2.2.**). The suggested volumes correspond to an IC preparation that can be used for 20 different samples in ELISA- or FACS-based systems and can be scaled up or down depending on the individual experiment.

3.2.1. TNP–Immune Complexes

1. Warm DPBS to 37°C.
2. Spin antibody preparations for 5 min at 20,000 g in a tabletop centrifuge to remove aggregates in the preparation.
3. Combine 10 μg of anti-TNP antibody and 10 μg of biotinylated TNP-BSA in an Eppendorf tube and add DPBS to 1 ml. As a control, set up another tube with 10 μg of biotinylated TNP-BSA without the antibody (antigen-only control) (*see* **Note 8**).

A ELISA with TNP-ICs and immobilized soluble FcRs

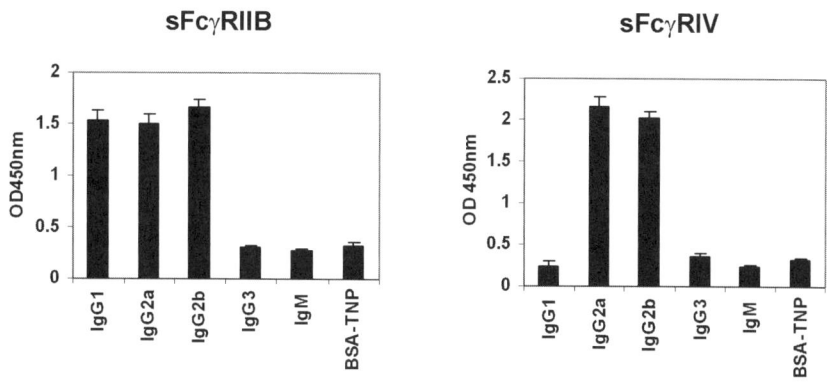

B FACS analysis of IC binding to cellular FcRs

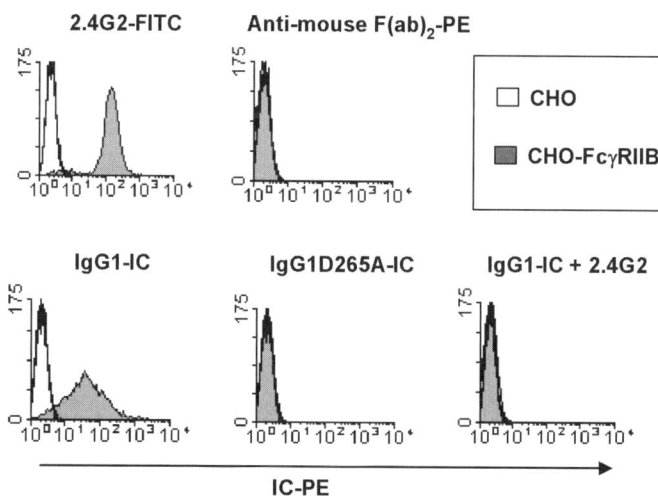

Fig. 3. Analyzing antibody–FcR interactions with immune complexes. (**A**) Flag-tagged soluble FcγRIIB or FcγRIV were captured on anti-Flag 96-well plates (Sigma). ICs were generated by using biotinylated BSA-TNP in combination with TNP-specific IgG1, IgG2a, IgG2b, IgG3, and IgM antibodies. Bound ICs were detected by incubation with horseradish peroxidase-coupled streptavidin. (**B**) IC binding to FcR expressing cells. ICs were generated by co-incubation of mouse IgG1 or an IgG1 mutant that

4. Mix well by pipetting and incubate with gentle shaking (not vortexing) on an orbital shaker (e.g., Nutator; Becton Dickinson, Franklin Lakes, NJ) for 1 h at 37°C.
5. Check that no visible precipitates have formed by flicking the tube.
6. *Optional*: Removal of unincorporated antibody and antigen (*see* **Note 9**)—Transfer the ICs to a gel filtration column (e.g., Ultra Free 300-kDa column, Millipore) and centrifuge in a swing-out rotor at 1700 g until the sample is concentrated to about 200 µl. Unbound antibody and antigen will pass through the filter, whereas ICs will stay in the upper compartment. Add 1.5 ml of DPBS to the column and repeat this concentration twice. Finally, recover the sample from the column and adjust the volume with DPBS to 1 ml. The ICs are now ready to use in ELISA- or FACS-based assays (50 µl per 1×10^5 cells or ELISA well).

3.2.2. Immune Complexes with F(ab)$_2$ Fragments

1. Warm DPBS to 37°C.
2. Spin antibody preparations for 5 min at 20,000 g in a tabletop centrifuge to remove aggregates in the preparation.
3. Combine 10 µg of the primary antibody and 1 µg of the fluorochrome-labeled (e.g., PE) F(ab)$_2$ fragment specific for the primary antibody and adjust the volume to 1 ml with DPBS. As a control, set up another tube containing only 1 µg of fluorochrome-labeled (e.g., PE) F(ab)$_2$ fragment specific for the primary antibody.
4. Mix well by pipetting and incubate with gentle shaking (not vortexing) on an orbital shaker (e.g., Nutator; Becton Dickinson) for 1 h at 37°C.
5. Check that no visible precipitates have formed by flicking the tube; the ICs are ready to use for FACS-based assays as exemplified in **Fig. 3A**.

3.3. ELISA with Soluble FcRs and Immune Complexes

ELISA-based methods provide a quick and easy-to-use system for evaluating the interaction of ICs with FcRs in vitro. Moreover, this assay can be used as a quality control step before IC preparations are used for in vivo experiments or for in vitro stimulation of effector cells (*see* **Note 7**). We have optimized ELISA-based assays with soluble FcRs for Flag-tagged receptors. The availability of precoated and blocked anti-Flag 96-well plates (Sigma) makes this procedure

◀――――――――――――――――――――

Fig. 3. is deficient in binding to FcRs (IgG1-D265A) with a PE-labeled mouse IgG-specific F(ab)$_2$ fragment CHO cells (open histograms) or CHO-FcγRIIB cells (filled histograms) were analyzed for FcγRIIB expression (2.4G2-FITC), binding of the anti-mouse F(ab)$_2$-PE fragment (background control) and binding of IgG1 or IgG1-D265A ICs. In addition, IC binding to FcγRIIB was blocked by co-incubation with 10 µg/ml of 2.4G2 (IgG1 – IC + 2.4G2) as another means of demonstrating specificity.

very quick and straightforward. Moreover, it is not necessary to purify the soluble receptors from the culture supernatant as the FcRs will be captured with high specificity on the 96-well plates.

1. Add 150 µl of culture supernatant containing the soluble FcRs and incubate for 3 h at room temperature or overnight at 4°C (*see* **Note 10**).
2. Wash wells three times with DPBS using either a squeeze bottle or an automated washer (*see* **Note 11**).
3. Add 50 µl of the IC preparation generated according to the protocol in **Subheading 3.2** and incubate for 2 h at room temperature.
4. Wash three times with DPBS and add horseradish-coupled streptavidin (e.g., POD-SAv from Roche used at 1:10,000 in DPBS).
5. Incubate for 30 min at room temperature, wash three times with DPBS, and detect bound ICs by adding TMB substrate solution (100 µl/well).
6. Stop the reaction by adding 100 µl of Stop-solution.
7. A typical result obtained with biotinylated TNP-BSA and different TNP-specific immunoglobulin G (IgG) isotypes is shown in **Fig. 3**.

3.4. Immune Complex Binding to FcR-Expressing Cells

1. In this protocol, we will use Chinese hamster ovarian (CHO) cells that have been stably transfected with individual FcRs. This protocol is by no means restricted to transfected cells and can be used for studying IC binding to primary cells as well.
2. The staining procedure will be performed in V-bottom 96-well plates. For cells that have to be detached first, use 0.05% trypsin or only EDTA instead of 0.25% as the FcRs are sensitive to higher trypsin concentrations.
3. Distribute 1×10^5 cells per well of a V-bottom 96-well plate and spin at 350 g for 5 min.
4. Loosen the cell pellet by vortexing briefly and incubate on ice for 5 min; all the following steps have to be carried out on ice or at 4°C.
5. Add 50 µl of the ICs generated according to the protocol in **Subheading 3.2** per well.
6. *Optional*: Add FcR-blocking antibodies at 10 µg/ml to block IC binding to FcRs (e.g., 2.4G2 to block FcγRIIB and FcγRIII or 9E9 to block FcγRIV).
7. Incubate for 1 h with gentle shaking (level 2–3 on a 1–10 scale) on a horizontal shaker at 4°C.
8. Wash twice by pelleting the cells by centrifugation (350 g) followed by gentle resuspension in 200 µl of cold DPBS per well.
9. If the ICs have been generated by using a crosslinking fluorochrome-labeled F(ab)$_2$ fragment, re-suspend the cells in 300 µl of FACS buffer and proceed to FACS analysis. If ICs have been generated with biotinylated polyvalent antigens (e.g., bio-TNP-BSA), re-suspend the cells in 100 µl of FACS buffer and add a fluorochrome-labeled streptavidin conjugate (e.g., SAv-PE or SAv-FITC from BD-Pharmingen).

10. Incubate for 15 min at 4°C in the dark.
11. Wash the cells three times as in step 7.
12. Resuspend the cells in 300 µl FACS buffer and proceed to FACS analysis.

4. Notes

1. In our experience, the best results are obtained if the DNA is added first, followed by addition of $CaCl_2$ and chloroquine; mix well before further use.
2. An easy way to control if calcium–DNA crystals have formed is to look at the cells under a microscope (×20 magnification) after the crystals have settled down (1–2 h post-transfection).
3. As the transfection mix contains chloroquine, prolonged incubation times will lead to cytotoxicity and result in lower protein yields. Typical transfection efficiencies with this cell type and method range between 80–95%. Whereas lipofection methods can achieve a similar transfection efficiency, they usually lead to a much lower level of protein production.
4. To increase cell viability and protein yield, the medium can be harvested after 2–3 days and be replaced with fresh production medium for another 3 days.
5. The maximum storage time at 4°C is about 1 week. Longer storage times result in a significant loss in activity. The expected concentration of soluble FcRs in the supernatant is typically in the range of 2–5 µg/ml.
6. For other antibody/antigen combinations [e.g., keyhole limpet hemocyanin (KLH) or ovalbumin (OVA)], it is important to consider that the use of polyclonal antibodies might result in a very high order of crosslinking and lead to insolubility of the resulting ICs.
7. If ICs are to be used to study effector cell activation and cytokine release, it is essential to test all reagents for lipopolysaccharide (LPS) contaminations. This includes commercially available antibody preparations that frequently contain significant amounts of LPS. These contaminants have to be removed before use with primary effector cells such as monocytes, macrophages, or dendritic cells (e.g. use Detoxi-Gel from Pierce, Rockford, IL).
8. Antibody and antigen should be in DPBS; if they are in a different buffer, dialyze them against DPBS first. The presence of low levels of sodium azide does not pose a problem.
9. As the affinity of monomeric antibodies to FcR (with the exception of binding of IgG2a to the high-affinity FcγRI) is very low, unbound antibody does not usually result in significant background binding or triggering of FcRs as long as antibody aggregates have been removed by high-speed centrifugation. The purity of ICs can be verified by SDS–PAGE analysis under non-denaturing conditions.
10. The culture supernatant usually contains saturating amounts of soluble FcRs. The amount of bound FcRs can be evaluated by using FcR-specific antibodies (e.g., 2.4G2 for FcγRIIB and III or 9E9 for FcγRIV). Alternatively, the amount of FcRs in the supernatant can be evaluated by direct coating of the supernatant

to ELISA plates (e.g., Nunc Maxisorb, Rochester, New York, USA) overnight followed by blocking and detection with an affinity-tag-specific antibody (e.g. anti-Flag antibody from Sigma).
11. The use of water as a washing solution is not recommended as this frequently leads to a loss of signal intensity; for the same reason, the use of 0.05% Tween 20 is not recommended.

Acknowledgments

We thank Brigitta Heyman for kindly providing anti-TNP antibodies. The stably transfected CHO cell lines were generated by Pierre Bruhns. We thank Eleanor Clowney, Andrew Kim, and Diana Dudziak for critically reading the manuscript. This work was supported in part by grants form the NIH to J.V.R; F.N. is funded by the Bavarian Genome Research Network (Baygene).

References

1. Glennie, M. J., and van de Winkel, J. G. (2003) Renaissance of cancer therapeutic antibodies. *Drug Discov Today* **8,** 503–10.
2. Waldmann, T. A. (2003) Immunotherapy: past, present and future. *Nat Med* **9,** 269–77.
3. Sanz, L., Blanco, B., and Alvarez-Vallina, L. (2004) Antibodies and gene therapy: teaching old 'magic bullets' new tricks. *Trends Immunol* **25,** 85–91.
4. Clynes, R., and Ravetch, J. V. (1995) Cytotoxic antibodies trigger inflammation through Fc receptors *Immunity* **3,** 21–6.
5. Cartron, G., Dacheux, L., Salles, G., Solal-Celigny, P., Bardos, P., Colombat, P., and Watier, H. (2002) Therapeutic activity of humanized anti-CD20 monoclonal antibody and polymorphism in IgG Fc receptor FcgammaRIIIa gene *Blood* **99,** 754–8.
6. Weng, W. K., and Levy, R. (2003) Two immunoglobulin G fragment C receptor polymorphisms independently predict response to rituximab in patients with follicular lymphoma. *J Clin Oncol* **21,** 3940–7.
7. Weng, W. K., Czerwinski, D., Timmerman, J., Hsu, F. J., and Levy, R. (2004) Clinical outcome of lymphoma patients after idiotype vaccination is correlated with humoral immune response and immunoglobulin G Fc receptor genotype. *J Clin Oncol* **22,** 4717–24.
8. Nimmerjahn, F., and Ravetch, J. V. (2005) Divergent immunoglobulin g subclass activity through selective Fc receptor binding. *Science* **310,** 1510–2.
9. Nimmerjahn, F., and Ravetch, J. V. (2006) Fcgamma receptors: old friends and new family members. *Immunity* **24,** 19–28.
10. Shields, R. L., Namenuk, A. K., Hong, K., Meng, Y. G., Rae, J., Briggs, J., Xie, D., Lai, J., Stadlen, A., Li, B., Fox, J. A., and Presta, L. G. (2001) High resolution mapping of the binding site on human IgG1 for Fc gamma RI, Fc gamma RII, Fc

gamma RIII, and FcRn and design of IgG1 variants with improved binding to the Fc gamma R. *J Biol Chem* **276,** 6591–604.
11. Shields, R. L., Lai, J., Keck, R., O'Connell, L. Y., Hong, K., Meng, Y. G., Weikert, S. H., Presta, L. G., Namenuk, A. K., Rae, J., Briggs, J., Xie, D., Stadlen, A., Li, B., and Fox, J. A. (2002) Lack of fucose on human IgG1 N-linked oligosaccharide improves binding to human Fcgamma RIII and antibody-dependent cellular toxicity. *J Biol Chem* **277,** 26733–40.
12. Shinkawa, T., Nakamura, K., Yamane, N., Shoji-Hosaka, E., Kanda, Y., Sakurada, M., Uchida, K., Anazawa, H., Satoh, M., Yamasaki, M., Hanai, N., and Shitara, K. (2003) The absence of fucose but not the presence of galactose or bisecting N-acetylglucosamine of human IgG1 complex-type oligosaccharides shows the critical role of enhancing antibody-dependent cellular cytotoxicity. *J Biol Chem* **278,** 3466–73.
13. Schuster, M., Umana, P., Ferrara, C., Brunker, P., Gerdes, C., Waxenecker, G., Wiederkum, S., Schwager, C., Loibner, H., Himmler, G., and Mudde, G. C. (2005) Improved effector functions of a therapeutic monoclonal Lewis Y-specific antibody by glycoform engineering. *Cancer Res* **65,** 7934–41.
14. Maenaka, K., van der Merwe, P. A., Stuart, D. I., Jones, E. Y., and Sondermann, P. (2001) The human low affinity Fcgamma receptors IIa, IIb, and III bind IgG with fast kinetics and distinct thermodynamic properties. *J Biol Chem* **276,** 44898–904.
15. Nimmerjahn, F., Bruhns, P., Horiuchi, K., and Ravetch, J. V. (2005) FcgammaRIV: a novel FcR with distinct IgG subclass specificity. *Immunity* **23,** 41–51.
16. Ferrara, C., Stuart, F., Sondermann, P., Brunker, P., and Umana, P. (2006) The carbohydrate at FcgammaRIIIa Asn-162. An element required for high affinity binding to non-fucosylated IgG glycoforms. *J Biol Chem* **281,** 5032–6.
17. Getahun, A., and Heyman, B. (2006) How antibodies act as natural adjuvants. *Immunol Lett* **104,** 38–45.

10

The Isolation and Identification of Murine Dendritic Cell Populations from Lymphoid Tissues and Their Production in Culture

David Vremec and Ken Shortman

Summary

Dendritic cells (DC) are widely regarded as the most potent cellular inducers of the adaptive immune response; so, immunotherapy through DC manipulation is a promising option in the future fight against many human ailments. We have developed a method of isolating DC from the mouse that involves efficient extraction from tissues, followed by the selection of the lightest density cells, then depletion of non-DC through a cocktail of monoclonal antibodies and anti-immunoglobulin magnetic beads. Finally, purification and segregation into DC subtypes is achieved by immunofluorescent labeling and sorting. This has demonstrated a network of DC populations differing in surface phenotype and function. We can now produce larger numbers of many of these DC subpopulations from their precursors using bone marrow cultures supplemented with fms-like tyrosine kinase 3 ligand (Flt3L). The culture-generated DC can be aligned with the populations directly isolated from tissues. Combining the in vivo and in vitro systems will make study of murine DC and their alignment to their human counterparts an easier process.

Key Words: Dendritic cell; DC isolation; DC subpopulations; flow cytometry; Flt3 ligand.

1. Introduction

Dendritic cells (DC) play a pivotal role in the immune system. In steady state, in the absence of pathogens or inflammation, the DC system is believed to be involved in the maintenance of self-tolerance through the presentation of self-antigens to the developing T-cell system in the thymus and to mature

T-cells in the periphery. DC in the periphery also act as sentinels for pathogen invasion. On encounter with microbial products or other "danger" signals, the DC are activated to an immunogenic state where their presentation of foreign antigens to T cells initiates an effector T-cell response *(1)*. This DC activation or maturation process involves a series of changes including reduction of antigen uptake, upregulation of surface major histocompatibility complex II (MHC II) and upregulation of co-stimulatory molecules *(2)*. It is therefore important that any procedure for isolation of DC conserves their initial maturation status and does not inadvertently activate the DC. As isolated DC undergo a form of spontaneous activation in culture, some earlier procedures, which depended on the selective adherence properties of DC on prolonged culture, induced DC maturation and changes in DC surface markers *(3)*. Although our procedure involves collagenase digestion, this is carried out for a brief time at room temperature and does not appear to activate the DC. Thus, the surface MHC II and the status of costimulatory molecules on the DC obtained with this method are identical with that obtained (with lower yield) when the DC are extracted without incubation with collagenase by mechanical disruption of tissue at 4°C (unpublished observation).

Over the last decade, it has become clear that, although all DC share many features related to their common roles in antigen processing and T-cell activation, they are a heterogeneous group of cells with important differences in function, including differences in the recognition of microbial products and in cytokine production *(4,5)*. Our separation procedures aim to extract efficiently all DC subtypes from tissues, to enrich them without significant loss or bias, then to separate them into functionally distinct subtypes based on surface marker differences. Collagenase digestion at room temperature is used to extract all DC from tissues without loss of surface markers. The brief exposure to ethylene-diamine tetraacetic acid (EDTA) and mild shear force serve to disassociate the DC from any complexes with T cells. Because DC are a relatively rare cell type, it is difficult and expensive to obtain adequate purity and segregation by direct fluorescence-activated cell sorting on the disassociated tissue; pre-enrichment is advisable. DC are of especially low buoyant density (presumably because of the continuous uptake of extracellular fluids by pinocytosis and endocytosis), so a simple selection of the 3–5% lightest density cells provides an effective enrichment with good recovery. It also eliminates damaged cells and erythrocytes. We use negative selection of non-DC lineage cells with immunomagnetic beads for further pre-enrichment, but care must be taken to avoid loss of any DC subtypes bearing particular markers found more commonly on T cells, B cells, or macrophages. Positive selection of DC on immunomagnetic beads or immunomagnetic particles based on CD11c expression is an alternative approach at this stage, although this does not provide the same level of purity as staining for the same marker followed by fluorescence-activated cell sorting.

Separation of the different DC subtypes can be achieved by multicolor labeling and sorting using a range of surface markers. CD11c is used as a marker to segregate all DC from most impurities, although it should be noted that some other cells, particularly certain natural killer (NK) cells, also express CD11c and need to be depleted using different markers. The combination of CD11c and MHC II can be used for clearer definition of conventional DC (cDC) *(3)*.

Plasmacytoid pre-DC (pDC), the natural type 1 interferon (IFN)-producing cells also enriched in the DC preparation, can readily be distinguished from conventional DC by staining with anti-CD11c and anti-CD45RA or CD45R. pDC are CD11c^{int} CD45RA^{hi} whereas cDC are CD11c^{hi} CD45RA$^-$ *(6)*. When activated to a more dendritic phenotype, however, the pDC shift to a lower expression of CD45RA and move out of the sorting gate that is effective for quiescent pDC. In this case, a marker that is upregulated on activation, such as CD69 *(7)*, or a pDC marker that is not lost upon activation, such as Ly6C *(8)* or 120G8 *(9)*, could be used to segregate the activated pDC. Separation of the cDC into discrete subsets in tissues such as spleen is readily accomplished by staining for CD4 and CD8α, providing in the spleen three distinct subtypes, CD4$^+$8$^-$, CD4$^-$8$^-$, and CD4$^-$8$^+$. CD4$^-$CD8$^+$ cDC are the major producers of IL-12p70 and cause polarization of T cells to a Th1 response *(10)*. They are also the most efficient at cross-presenting antigen *(11)*. CD4$^+$CD8$^-$ cDC produce the largest amounts of the inflammatory cytokines MIP-1α, MIP-1β, and Rantes *(12)*, whereas the CD4$^-$CD8$^-$ cDC fraction are the major producers of IFN-γ *(10)*. Note, however, that in some tissues, pickup of CD4 or CD8αβ from T cells can result in false low-level staining for these markers; thus, in thymus, the major cDC population is CD4$^-$ CD8αα$^+$, but may also show some staining for CD4 or CD8β because of such pickup *(13)*. Other markers may be used because the CD8$^+$ DC are CD24^{hi}CD205^{hi}CD11b^{lo}CD172a(Sirpα)^{lo} while the CD8$^-$ are the converse, CD24^{lo}CD205^{lo}CD11b^{int}CD172a(Sirpα)^{hi} *(14)*. The cDC populations in the lymph nodes are more complex, with at least five subpopulations delineated. They include the three populations found in spleen as well as two migratory DC populations. Both mesenteric and skin-draining lymph nodes possess a CD4$^-$CD8$^{-/lo}$CD205^{int} population, referred to as interstitial DC, that have migrated through the afferent lymph from the gut and dermis, respectively. The subcutaneous lymph nodes also contain a $CD4^- CD8^{lo}$ $CD205^{hi}$ population, representing the mature form of Langerhans cells that have migrated from the epidermis *(15)*.

A deterrent facing those studying DC is the relatively small number that can be efficiently isolated from tissues and the difficulties in studying those cells in vivo. This problem has been avoided using culture methods for generating DC from bone marrow cells or monocytes stimulated by Granulocyte-macrophage colony stimulating factor (GM-CSF) and interleukin-4 (IL-4) *(16)*. These probably correspond, however, to a type of DC only generated in vivo under conditions of inflammation and differ from the steady-state DC found in normal laboratory mice. Recently, large numbers of both pDC and cDC have been produced by culture of bone marrow cells with fms-like tyrosine kinase3 ligand (Flt-3L) *(17,18)*. Our laboratory has recently demonstrated that although the cDC generated in these cultures lack the markers CD4 and CD8α, they can nevertheless be divided into the functional equivalents of $CD8α^-$ and $CD8α^+$ DC using the markers CD11b, CD24, and CD172a *(14)*. Thus, large numbers of the equivalents of the steady-state mouse spleen DC populations can now be produced in culture.

An important task for the future will be to compare and align the DC subtypes isolated from human lymphoid tissues with those obtained from the mouse. The appropriate markers will differ, because human DC lack CD8 and all express CD4; however, other markers, such as CD11b, have already been shown to segregate human cDC subsets *(19)*. The isolation procedures here outlined for mouse DC can readily be modified for human DC isolation (with appropriate changes to human osmolarity medium and corresponding changes of DC buoyant density) *(19)*. In addition, the Flt3L culture system works for human DC production *(20,21)*.

2. Materials

2.1. Organ Removal

1. FCS: Fetal calf serum. Aliquot and store at –20°C.
2. RPMI-FCS: RPMI-1640 is modified to mouse osmolarity (308 mOsm/kg) and additional pH 7.2 Hepes buffering is included to reduce dependence on CO_2 concentration. FCS is added to a final concentration of 2%. After adjusting to approximately pH 7 with CO_2, filter sterilize, and store in 100-ml bottles at 4°C.

2.2. Digestion and Release of DC

1. Enzyme digestion mix: Dissolve collagenase (Type III, Worthington Biochemicals, Freehold, NJ) at 7 mg/ml and DNase I (Boehringer Mannheim, Mannheim,

Germany) at 0.14 mg/ml in RPMI-FCS. Ensure the collagenase used is free of trysin-like proteases (*see* **Note 1**). Aliquot into 1 ml amounts and store as a 7× stock solution frozen at −20°C. Dilute each 1 ml aliquot with 6 ml RPMI-FCS just before use. Filter sterilize if required.
2. EDTA solution: 0.1 M EDTA disodium salt adjusted to pH 7.2. Filter to sterilize and store at 4°C.
3. EDTA-FCS: Add 1 ml of 0.1 M EDTA per 10 ml FCS before use.

2.3. Selection of Light Density Cells

1. BSS-EDTA: contains 150 mM NaCl and 3.75 mM KCl (no Ca^{2+} or Mg^{2+}) and 5 mM EDTA and is adjusted to pH 7.2 and mouse osmolarity (308 mOsm/kg). Filter to sterilize and store in 100-ml bottles at 4°C.
2. BSS-EDTA-FCS: Add EDTA-FCS to BSS-EDTA to a final concentration of 2%.
3. Nycodenz-EDTA: Nycodenz AG powder (Nycomed Pharma AS, Oslo, Norway) is prepared as a 0.372 M stock solution in water, then diluted and adjusted to the appropriate density (at 4°C) and osmolarity (308 mOsm/kg) using BSS-EDTA (for spleen, 1.077 g/cm^3, for thymus, 1.076 g/cm^3, for lymph nodes, 1.082 g/cm^3). Use a 25-ml weighing bottle to determine density at 4°C. Filter to sterilize and store in sealed 10 ml aliquots at −20°C. Thaw at room temperature, mix thoroughly, and cool to 4°C before use (*see* **Note 2**).
4. Polypropylene tubes: 14-ml polypropylene round bottom tubes (Becton Dickinson Labware, Franklin Lakes, NJ).

2.4. Depletion of Non-DC Lineages

1. Monoclonal antibody depletion cocktail: Add together pre-titrated amounts of rat monoclonal antibodies specific for non-DC lineage cells and dilute to the appropriate volume with BSS-EDTA-FCS (*see* **Note 3**). The monoclonal antibodies added can be varied depending on the DC populations required (*see* **Table 1**). Filter to sterilize, aliquot and store at −20°C.
2. Immunomagnetic beads: Either BioMag goat anti-rat immunoglobulin G (IgG)-coated beads (Qiagen, Clifton Hill, Australia) or Dynabead sheep anti-rat IgG-coated beads (Dynal, Oslo, Norway).
3. Spiral rotator: Spiramix 10 (Denley, Billingshurst, England).

2.5. Bone Marrow Cultures

1. Red Cell Lysis Medium (RCLM): 0.168 M NH_4Cl. Filter sterilize and store at 4°C.
2. Flt3L: recombinant murine Flt-3L purified from the tissue culture supernatant produced by a CHO cell line transfected with a soluble murine Flt3L-FLAG peptide construct in the mammalian expression plasmid pEF-BOS (N. Nicola, WEHI, Australia). Alternatively use recombinant human Flt3L.

Table 1
The Contents of the Monoclonal Antibody Depletion Cocktail Vary Depending on Which DC Populations are Required and the Tissue from Which They Are Being Isolated

Clone	Specificity	Spleen and Lymph Node cDC only	Spleen and Lymph Node cDC + pDC	Thymus cDC only	Thymus cDC + pDC
KT3-1.1	CD3ε	+	+	+	+
T24/31.7	CD90	+	+	+	+
TER119	erythroid	+	+	+	+
1A8	Ly6G	+	+	+	–
RA36B2	CD45R	+	–	+	+
ID3	CD19	–	+	–	+
M1/70	CD11b	–	–	+	+
F4/80	F4/80Ag	–	–	+	+
GK1.5	CD4	–	–	+	–

All monoclonal antibodies used are derived from rat, which enables a simple depletion with anti-rat IgG-coated magnetic beads.

3. Flt3L DC culture medium: Modified RPMI-1640, iso-osmotic with mouse serum, with additional Hepes buffering at pH 7.2, is supplemented with 10% FCS, 50 µM 2-mercaptoethanol, and 2 mM L-glutamine, filter sterilized and stored frozen at −70°C. It is further supplemented with 300 ng/ml murine Flt3L just before use (*see* **Note 4**).

2.6. Immunofluorescent Staining and Flow Cytometric Analysis

1. All monoclonal antibodies (*see* **Table 2**) were purified from hybridoma culture supernatant using Protein G sepharose (Amersham Biosciences, Castle Hill, Australia) and conjugated to fluorochromes in-house (*see* **Note 5**). All were titrated to determine saturating levels.
2. Fluorochrome conjugated mAb: Conjugate N418, M1/69 and M1/70 to FITC (Molecular Probes Inc., Eugene, OR), 14.8, M1/69 and NLDC145 to phycoerythrin [(PE) Prozyme, San Leandro, CA], 14.8 and YTS169.4 to allophycocyanin [(APC) Prozyme], N418, YTS169.4 and GK1.5 to AlexaFluor594 (Molecular Probes Inc.) and P84 to biotin-XX succinimidyl ester (Molecular Probes Inc.) following the manufacturer's instructions (*see* **Note 5**). Add FCS to 1% and NaN$_3$ to a final concentration of 10 mM (*see* **Note 6**). Aliquot and store stocks of FITC, AlexaFluor594 and biotin conjugates at −70°C. Stocks of PE and APC conjugates (*see* **Note 7**) and working stocks of FITC, AlexaFluor594 and biotin conjugates are stored at 4°C, protected from light. Dilute to final concentration just before use.
3. PE-Streptavidin (BD PharMingen, San Jose, CA) is titrated to determine saturating levels and stored protected from the light at 4°C. Dilute just before use.
4. Propidium iodide (PI): Prepare a 100 µg/ml PI (Calbiochem, La Jolla, CA) stock solution in normal saline. Aliquot and store at 4°C protected from light (*see* **Note 8**).
5. BSS-EDTA-FCS-PI: Dilute the PI stock in BSS-EDTA-FCS to a final working concentration of 500 ng/ml before adding to cells.

Table 2
Four-Color Flow Cytometric Analysis Using Monoclonal Antibodies Conjugated to Fluorochromes is Used to Define Distinct DC Subpopulations in Spleen, Thymus, and Lymph Node

Clone	N418	14.8	YTS169.4	GK1.5	NLDC145	M1/69	M1/70	P84
Specificity	CD11c	CD45RA	CD8	CD4	CD205	CD24	CD11b	CD172a
Conjugates	FITC, Alexa594	PE, APC	APC, Alexa594	Alexa594	PE	FITC, PE	FITC	Biotin

Their use enables separation of at least four populations in spleen, three in thymus, and six in lymph node.

3. Methods
3.1. Removal of Tissue

1. Remove spleens, or thymus or appropriate pooled lymph nodes from eight mice and place into cold RPMI-FCS (*see* **Note 9**). Take care to minimize contamination of the organs with any fat or connective tissue.

3.2. Digestion and Release of DC

1. The enzyme digestion mix is prepared slightly ahead of time and allowed to warm to room temperature.
2. Strip away any residual fat (*see* **Note 10**) from the tissue using two needles and transfer the organ to a small plastic petri dish containing 7 ml of the enzyme digestion mix. Use a sharp pair of scissors or a single-sided razor blade to cut the tissue into very small fragments (*see* **Note 11**). Transfer the mix containing the fragments to a 10-ml polypropylene tube using a wide bore pasteur pipette and digest with frequent mixing, (use the same pipette) for 20–25 min at room temperature (~22°C) (*see* **Note 12**).
3. Add 600 µl of EDTA solution to the digestion mix and continue mixing for a further 5 min (*see* **Note 13**).
4. Remove any remaining tissue fragments by running the digestion mix through a sieve and then make the volume up to 9 ml with RPMI-FCS. Underlay with 1 ml of cold FCS-EDTA and recover the cells by centrifugation (*see* **Note 14**).

3.3. Selection of Light Density Cells

1. Thaw two 10 ml aliquots of Nycodenz-EDTA of appropriate density at room temperature, mix well and maintain them at 4°C until required (*see* **Note 15**).
2. Transfer 5 ml of the Nycodenz-EDTA into the bottom of each of two polypropylene tubes.
3. Resuspend the cell pellet in the remaining 10 ml of Nycodenz-EDTA (*see* **Note 16**). Gently layer 5 ml of the cell suspension over the Nycodenz-EDTA already in each of the two tubes. Layer 1–2 ml of EDTA-FCS over the cell suspension. Gently mix the interface by inserting the tip of a pasteur pipette, swirling and removing it (*see* **Note 17**).
4. Centrifuge in a swing-out head, refrigerated (4°C) centrifuge for 10 min at 1700 *g*.
5. Use a pasteur pipette to collect the light density fraction in the upper zones (*see* **Note 18**). Discard the bottom 4 ml and the cell pellet. Transfer the light density fraction to a 50-ml tube, dilute up to 50 ml with BSS-EDTA, mix and centrifuge to pellet the cells (*see* **Note 19**). Resuspend the cells in 5 ml BSS-EDTA-FCS.

3.4. Immunomagnetic Bead Depletion of Non-DC Lineage Cells

1. Count the recovered cells (*see* **Note 20**).
2. Add the appropriate monoclonal antibody depletion cocktail to the cell pellet at 10 μl/10^6 cells, resuspend, and incubate at 4°C for 30 min.
3. During the incubation, the appropriate number of magnetic beads is transferred to a 5-ml polypropylene tube and washed three to four times with BSS-EDTA-FCS (*see* **Note 21**). To wash, simply dilute beads with the BSS-EDTA-FCS, place the tube into the magnet, allow the beads to move to the side of the tube, and then remove the supernatant. After the final wash, pellet the beads at the bottom of the tube in a small amount of BSS-EDTA-FCS and keep at 4°C until required.
4. Dilute the cells to 9 ml with BSS-EDTA-FCS and underlay with 1 ml of FCS-EDTA. Centrifuge the cells through the underlay and remove the supernatant from the top, leaving the FCS layer over the cells (*see* **Note 14**). Allow the tube to sit at 4°C for a minute while media runs down the wall of the tube. Remove the media and the FCS (*see* **Note 22**), then resuspend the cells in 400–500 μL of BSS-EDTA-FCS.
5. Remove the liquid remaining on the magnetic bead pellet and transfer the cells to it to produce a bead and cell slurry (*see* **Note 23**). Seal the tube and mix continuously for 20 min at 4°C by rotating at a 30° angle on a spiral mixer (*see* **Note 24**).
6. Dilute the bead cell slurry up to 4 ml with BSS-EDTA-FCS, mix very gently and attach the tube to the magnet for 1–2 min.
7. Recover the supernatant containing unbound DC with a pasteur pipette and transfer to a second 5-ml polypropylene tube. Discard the tube containing magnetic beads bound to non-DC. Reattach the tube to the magnet for a further 1–2 min to ensure removal of any residual beads and transfer the supernatant to a 10-ml tube (*see* **Note 25**).
8. Layer 1 ml of FCS-EDTA under the cell suspension and centrifuge to recover the DC-enriched cells. Resuspend the cells in 2 ml of BSS-EDTA-FCS and count. Hold cells at 4°C until immunofluorescent labeling and flow cytometric analysis.

3.5. Flt3L Bone Marrow Cultures

1. Remove the femur and tibia from the desired number of mice and place in RPMI-FCS at 4°C (*see* **Note 26**).
2. Flush the bone marrow from the bones, using a 21G needle attached to a syringe filled with RPMI-FCS, to create a single-cell suspension. Underlay with FCS and centrifuge.
3. Remove the supernatant and gently resuspend the cells in 2–3 ml of RCLM. Expose the cells to RCLM for a maximum of 30 s (*see* **Note 27**). Immediately dilute the cells with RPMI-FCS, pass them into a new tube through a sieve to remove clumps, underlay with FCS, and wash by centrifugation. Repeat the washing step twice. Resuspend the cells and count.
4. Centrifuge the cells and resuspend the pellet at 1.5×10^6/ml in Flt3L culture medium. Culture the cells for 8–10 days at 37°C in 10% CO_2 in air (*see* **Note 28**).

5. Harvest the cells after culture by gently washing the flasks several times with cold BSS-EDTA-FCS (*see* **Note 29**) and centrifuge to pellet. Resuspend the cells in BSS-EDTA-FCS and count to determine recovery. Maintain at 4°C until immunofluorescent labeling and flow cytometric analysis.

3.6. Immunofluorescent Staining and Flow Cytometric Analysis

1. Prepare a cocktail of pre-titrated fluorochrome-conjugated monoclonal antibodies (*see* **Table 2**) at the appropriate concentration just before use (*see* **Note 30**).
2. Centrifuge the cells and remove the supernatant.
3. Add 10 µl of the fluorochrome-conjugated antibody cocktail per 10^6 cells and incubate at 4°C for 30 min.
4. Resuspend up to a larger volume with BSS-EDTA-FCS and underlay with FCS-EDTA. Centrifuge to wash the cells (*see* **Note 14**).
5. Remove the supernatant leaving the FCS-EDTA layer above the cells and centrifuge again for 30 s to allow any remaining media to wash down the wall of the tube. Remove the remaining media and the FCS-EDTA layer.
6. If a biotinylated antibody was included in the first stage, add a second stage of pre-titrated PE-streptavidin at 10 µl per 10^6 cells. Incubate and wash as for the first stage.
7. Resuspend cells in BSS-EDTA-FCS-PI and maintain cells at 4°C until ready for flow cytometry (*see* **Note 31**).
8. Sort the DC using a FACSVantageSEDiVa (BD Biosciences, San Jose, CA) or analyze using a FACStar Plus (BD Biosciences) (*see* **Note 32**). Both flow cytometers should be set up with a 488 nm emitting argon ion laser for detection of FITC, PE, and PI and a 605 nm emitting dye laser, adjusted to emit a wavelength of 597 nm, to detect APC and AlexaFluor594 and the appropriate dichroic filters (BD Biosciences) (*see* **Table 3**). Select DC on the basis of high forward and side light

Table 3
Excitation and Emission Wavelengths of the Fluorochromes Used and the Appropriate Dichroic Filters

	Fluorochrome	Maximum absorption (nm)	Maximum emission (nm)	Dichroic filter
Argon ion laser (488 nm)	FITC	490	525	530-DF30
	PE	565	575	575-DF26
	PI	536	620	630-DF22
				675-DF20 DiVa
Dye laser (597 nm)	APC	650	660	660-DF20 FACStar
	AlexaFluor594	585	610	630-DF22

scatter, excluding dead cells with high PI fluorescence. Identify $CD11c^{int}CD45RA^{hi}$ pDC and $CD11c^{hi}CD45RA^-$ cDC (*see* **Note 33**). CD45R (B220) is an alternative marker to CD45RA.
9. Appropriate software is used to detail the DC subpopulations as shown in **Figs. 1 and 2**.

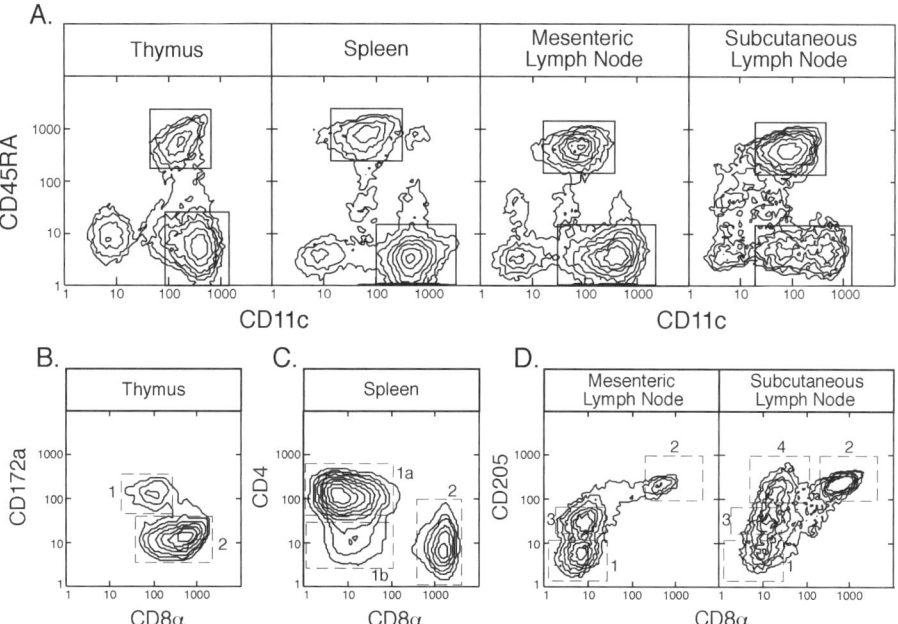

Fig. 1. Separation of mouse DC into discrete subpopulations. (**A**) Dendritic cells from thymus, spleen, and lymph node separated into $CD11c^{int}CD45RA^+$ plasmacytoid pre-DC (pDC) and $CD11c^+CD45RA^-$ conventional DC (cDC). pDC isolated from thymus express relatively more CD11c than those of spleen and lymph node, whereas some cDC from lymph node express lower levels of CD11c. (**B**) cDC from thymus can readily be divided into a minor $CD8\alpha^-CD172a^+$ (1) and a major $CD8\alpha^+CD172a^-$ (2) population. (**C**) Populations 1 and 2 are also represented in the spleen but in different proportions. The major $CD8\alpha^-$ population can be further segregated into $CD4^+$ (1a) and $CD4^-$ (1b) populations. (**D**) Populations 1 and 2 also occur in the lymph node. Both mesenteric and subcutaneous lymph nodes also contain a $CD8\alpha^{-/lo}CD205^{lo}$ (3) population corresponding to interstitial DC. Subcutaneous lymph node contain an extra population that is $CD11c^{lo}CD8\alpha^{lo}CD205^{hi}$ (4) that are the mature version of the Langerhans cell.

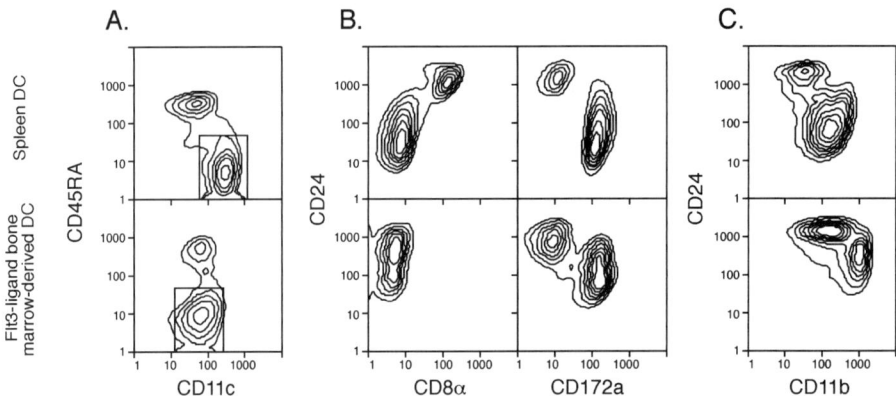

Fig. 2. A comparison of surface marker expression between freshly isolated spleen DC and DC generated from Flt3L bone marrow cultures. It is possible to separate two distinct cDC subpopulations in both cases: a $CD172a^{-/lo}CD11b^{lo}CD24^{hi}$ population corresponding to the spleen $CD8\alpha^+$ fraction and a $CD172a^{hi}CD11b^{hi}CD24^{lo}$ population that corresponds to the spleen $CD8\alpha^-$ fraction. **(A)** pDC and cDC can be separated on the basis of differential expression of CD45RA and CD11c in both spleen and Flt3L bone marrow generated DC. **(B)** $CD8\alpha$ is not expressed by Flt3L generated DC, but CD172a can be used as an alternative marker to distinguish this population. **(C)** Although both subsets are found in each case, their proportions vary. As would be expected, there are proportionally more $CD24^{hi}$ ($CD8\alpha^+$ equivalent) DC in the Flt3L-supplemented cultures.

4. Notes

1. Each individual batch of collagenase should be tested to ensure the absence of trypsin-like proteases. Trypsin and similar proteases can strip molecules from the cell surface and alter the apparent phenotype of cells after digestion. We screen the collagenase by incubating with thymus cells and checking for any loss of CD4 and CD8 from the surface of thymocytes by flow cytometry.
2. The appropriate density of Nycodenz for isolation of DC from various organs was calculated for separation at pH 7.2, 308 mOsm/kg and at 4°C. All these factors affect the buoyant density of cells. As density is temperature dependent, it is paramount when preparing and using Nycodenz that the Nycodenz is at 4°C and the centrifuge used is set to 4°C. Nycodenz of higher density will reduce the purity of the light density fraction, whereas a lower density will decrease the yield. As a rough guide, the light density fraction for spleen DC enrichment should represent 3–5% of spleen cells.
3. Monoclonal antibodies are titrated by surface staining of cells and analysis by flow cytometry. A fluorochrome conjugated anti-rat Ig second stage is used for

titrating unconjugated rat mAb. A concentration resulting in surface saturation of the antigen is considered adequate for efficient depletion.
4. Flt3L should be titrated before use by small-scale bone marrow culture. Suboptimal levels will vastly reduce DC yield. Optimal levels can result in DC recoveries of up to 130% of the starting number of bone marrow cells.
5. In most cases, commercial equivalents of the antibodies are available in a purified or pre-conjugated form.
6. Sodium azide is extremely toxic, and all care should be taken in its use.
7. Phycobiliproteins are extremely sensitive to freezing and thawing. Do not freeze PE or APC.
8. PI is toxic, and care should be taken when it is used. If functional studies are to be undertaken ensure cells are washed post sorting.
9. Smaller or larger numbers of organs (we have worked in a range from a single inguinal lymph node up to 64 spleens) are accommodated by decreasing or increasing all volumes and amounts listed proportionally.
10. Residual fat can prove toxic to cells and will not degrade during the digestion causing clumping; so, it is beneficial to clean organs as much as possible at this stage.
11. The organs should be cut into very small fragments to increase the surface area available to the enzymes. This ensures adequate digestion and maximizes cell yield.
12. Adequate digestion time can vary depending on how well the organs were cut up initially and how well the digesting tissue was mixed during the digestion. If care is taken, 20–25 min is sufficient to digest all but the outer capsule of lymph node and the pulpy tissue from spleen.
13. EDTA is added to all media after digestion to release lymphoid cells bound to the DC and to stop re-formation of multicellular complexes that include DC. Failure to do so may cause loss of DC during purification and also loss of purity if lymphocytes are carried through the purification process attached to DC.
14. All centrifugation steps are performed at 1000 g for 7 min at 4°C unless stated otherwise. Layering FCS under the sample to produce a zonal centrifugation step increases the efficiency of separation from smaller particles and soluble material in the upper supernatant and so avoids the need for repetitive "washing" of the cells.
15. Nycodenz must be mixed well before aliquoting and before use to ensure a solution of uniform density. A most common immediate and obvious result of a failure to do this is the inability to layer the Nycodenz containing cells over the Nycodenz at the bottom of the tube. Poor DC separation will result if mixing is incomplete.
16. Do not overload the density separation, or efficiency will be lost. Use no more than four organs ($\sim 10^9$ cells) per 10 ml of Nycodenz medium.

17. Very sharp interfaces will reduce the efficiency of the density separation, even though it looks "cleaner" after centrifugation. It is better to disrupt the interfaces slightly using the pipette, to create a density gradient.
18. After centrifugation, a distinct band of light density cells will be present at the interface zone between the FCS and the Nycodenz, and the dense cells will have formed a pellet. A gradient of cells of intermediate density will, however, be found between them. When recovering the light density cells, concentrate on the band at the interface but collect all cells down to the 4 ml mark.
19. Adequate dilution and mixing of the cells in Nycodenz medium with EDTA-BSS is essential to recover the light density cells as a pellet during centrifugation.
20. Count all cells, including any remaining erythrocytes, to calculate volumes of mAb depletion cocktail and beads that are required in subsequent steps.
21. For optimal efficiency, use Dynabeads at a 5:1 bead to cell ratio. This is recommended for thymus and lymph node DC purification. For optimal economy when large numbers of cells must be depleted, such as the case with spleen DC, BioMag beads at a 10:1 bead to cell ratio are recommended. Alternatively, a two-stage depletion can be performed to optimize efficiency but reduce cost by initially using BioMag beads at a 5:1 bead to cell ratio, followed by Dynabeads at a 3:1 bead to cell ratio on the reduced number of cells.
22. Centrifuging through a layer of FCS separates cells from unbound (excess) mAb; so, it is important to remove all the supernatant after washing. Excess mAb will compete for binding to the beads and will decrease efficiency of the depletion.
23. Producing the bead cell slurry maximizes bead and cell contact, but care is required to reduce the risk of evaporation.
24. Rotation at a 30° angle is designed to keep the bead cell slurry at the bottom of the tube.
25. DC recovery can be increased by washing the beads and attaching the tube to the magnet a second time; this will, however, lead to reduced purity.
26. We routinely recover $4-7 \times 10^7$ cells per mouse using these bones.
27. RCLM is eventually toxic to all cells, so its addition and mixing should be very gentle, and then, it should be washed away as quickly as possible.
28. Maximum yield of the DC subpopulations can also vary with culture time, so an optimal culture time should be determined for each batch of Flt3L.
29. This washing step should be sufficient to remove any slightly adherent DC from the plastic but leave behind any more adherent macrophages.
30. All immunostaining steps should be performed at 4°C.
31. PI is used to exclude dead cells during flow cytometric analysis.
32. Any other flow cytometer with appropriate lasers and optical set-up can be used.
33. pDC isolated from thymus have higher surface levels of CD11c than those isolated from spleen or lymph node. The cDC generated in the Flt3L bone marrow cultures have lower levels of CD11c than found on freshly isolated cDC and are equivalent to the levels on the pDC developing alongside them.

References

1. Steinman, R.M. (1991) The dendritic cell system and its role in immunogenicity. *Annu. Rev. Immunol.* 9, 271–296.
2. Wilson, N.S., El-Sukkari, D., Belz, G.T., Smith, C.M., Steptoe, R.J., Heath, W.R., Shortman, K. and Villadangos, J.A. (2003) Most lymphoid organ dendritic cell types are phenotypically and functionally immature. *Blood* 102, 2187–2194.
3. Vremec, D and Shortman, K. (1997) Dendritic cell subtypes in mouse lymphoid organs: cross-correlation of surface markers, changes on incubation and differences between thymus, spleen and lymph nodes. *J. Immunol.* 159, 565–573.
4. Shortman, K. and Caux, C. (1997) Dendritic cell development: multiple pathways to natures adjuvant. *Stem Cells* 15, 409–419.
5. Shortman, K. and Liu, Y.J. (2002) Mouse and human dendritic cell subtypes. *Nat. Rev. Immunol.* 2, 151–161.
6. O'Keeffe, M., Hochrein, H., Vremec, D., Caminschi, I., Miller, J.L., Anders, E.M., Wu, L., Lahoud, M.H., Henri, S., Scott, B., Hertzog, P., Tatarczuch, L. and Shortman, K. (2002) Mouse plasmacytoid cells: long-lived cells, heterogeneous in surface phenotype and function, that differentiate into CD8$^+$ dendritic cells only after microbial stimulus. *J. Exp. Med.* 196, 1307–1319.
7. Yasuda, K., Yu, P., Kirschning, C.J., Schlatter, B., Schmitz, F., Heit, A., Bauer, S., Hochrein, H. and Wagner, H. (2005) Endosomal translocation of vertebrate DNA activates dendritic cells via TLR9-dependent and -independent pathways. *J. Immunol.* 174, 6129–6136.
8. Asselin-Paturel, C., Boonstra, A., Dalod, M., Durand, I., Yessaad, N., Dezutter-Dambuyant, C., Vicari, A., O'Garra, A., Biron, C., Briere, F. and Trinchieri, G. (2001) Mouse type I IFN-producing cells are immature APCs with plasmacytoid morphology. *Nat. Immunol.* 2, 1144–1150.
9. Asselin-Paturel, C., Brizard, G., Pin, J.-J., Briere, F. and Trinchieri, G. (2003) Mouse strain differences in plasmacytoid dendritic cell frequency and function revealed by a novel monoclonal antibody. *J. Immunol.* 171, 6466–6477.
10. Hochrein, H., Shortman, K., Vremec, D., Scott, B., Hertzog, P. and O'Keeffe, M. (2001) Differential production of IL-12, IFN-a and IFN-γ by mouse dendritic cell subsets. *J. Immunol.* 166, 5448–5455.
11. Pooley, J.L., Heath, W.R. and Shortman, K. (2001) Cutting edge: intravenous soluble antigen is presented to CD4 T cells by CD8$^-$ dendritic cells, but cross-presented to CD8 T cells by CD8$^+$ dendritic cells. *J. Immunol.* 166, 5327–5330.
12. Proietto, A.I., O'Keeffe, M., Gartlan, K., Wright, M.D., Shortman, K. Wu, L. and Lahoud, M.H. (2004) Differential production of inflammatory chemokines by murine dendritic cell subsets. *Immunobiology* 209, 163–172.
13. Vremec, D., Pooley, J., Hochrein, H., Wu, L. and Shortman, K. (2000) CD4 and CD8 expression by dendritic cell subtypes in mouse thymus and spleen. *J. Immunol.* 164, 2978–2986.

14. Naik, S.H., Proietto, A.I., Wilson, N.S., Dakic, A., Schnorrer, P., Fuchsberger, M., Lahoud, M.H., O'Keeffe, M., Shao, Q.X., Chen, W.F., Villadangos, J.A., Shortman, K. and Wu, L. (2005) Cutting edge: generation of splenic $CD8^+$ and $CD8^-$ dendritic cell equivalents in Fms-like tyrosine kinase 3 ligand bone marrow cultures. *J. Immunol.* 174, 6592–6597.
15. Henri, S., Vremec, D., Kamath, A., Waithman, J., Williams, S., Benoist, C., Burnham, K., Saeland, S., Handman, E. and Shortman, K. (2001) The dendritic cell populations of mouse lymph nodes. *J. Immunol.* 167, 741–748.
16. Sallusto, F. and Lanzavecchia, A. (1994) Efficient presentation of soluble antigen by cultured human dendritic cells is maintained by granulocyte/macrophage colony-stimulating factor plus interleukin-4 and downregulated by tumor necrosis factor-a. *J. Exp. Med.* 179, 1109–1118.
17. Brasel, K., De Smedt, T., Smith, J.L. and Maliszewski, C.R. (2000) Generation of murine dendritic cells from flt3-ligand-supplemented bone marrow cultures. *Blood* 96, 3029–3039.
18. Gilliet, M., Boonstra, A., Paturel, C., Antonenko, S., Xu, X.L., Trinchieri, G., O'Garra, A. and Liu, Y.J. (2002) The development of murine plasmacytoid dendritic cell precursors is differentially regulated by FLT-3-ligand and granulocyte/macrophage colony-stimulating factor. *J. Exp. Med.* 195, 953–958.
19. Vandenabeele, S., Hochrein, H., Mavaddat, N., Winkel, K. and Shortman, K. (2001) Human thymus contains 2 distinct dendritic cell populations. *Blood* 97, 1733–1741.
20. Arrighi, J-F., Hauser, C., Chapuis, B., Zubler, R.H. and Kindler, V. (1999) Long-term culture of human $CD34^+$ progenitors with FLT3-ligand, thrombopoietin, and stem cell factor induces extensive amplification of a $CD34^-$ $CD14^-$ and a $CD34^-CD14^+$ dendritic cell precursor. *Blood* 93, 2244–2252.
21. Chen, W., Antonenko, S., Sederstrom, J.M., Liang, X., Chan, A.S.H., Kanzler, H., Blom, B., Blazar, B.R. and Liu, Y-J. (2004) Thrombopoietin cooperates with FLT3-ligand in the generation of plasmacytoid dendritic cell precursors from human hematopoietic progenitors. *Blood* 103, 2547–2553.

11

Analysis of Individual Natural Killer Cell Responses

Wayne M. Yokoyama and Sungjin Kim

Summary

Typical assays for natural killer (NK) cell function assess the responses of entire NK cell populations. It is now possible to determine the responses of individual NK cells. Herein, two representative assays are described along with examples of how they have helped clarify current understanding of NK cell biology.

Key Words: NK cells; interferon; BrdU; licensing; cytokine.

1. Introduction

After B and T cells, natural killer (NK) cells comprise the third major lymphocyte population *(1,2)* and can be distinguished from other lymphocytes by the absence of B- and T-cell antigen receptors, that is, sIg and T-cell receptor (TCR), respectively. Indeed, they are present in *scid* and Rag-1- or Rag-2- deficient mice, indicating that they do not require gene rearrangement for their development or functions. They typically display a large, granular morphology and are found in peripheral lymphoid tissues and blood. NK cells more closely resemble T cells in effector function, such as cytolytic machinery and cytokine production, and thus, it is not surprising that NK and T cells may share a common progenitor cell. On the contrary, a phenotypic description of NK cells is rooted in discrimination from $CD3^-$ cells. In C57BL/6 mice, this phenotype is based on NK1.1 (Nkrp1c or Klrb1c) expression, that is, NK cells are $NK1.1^+$ $CD3^-$ cells *(2)*.

NK cells appear to be important participants in tumor surveillance because they lyse tumor cells *(3)* and have been used for immunotherapy *(4)*. NK cells also participate in the host response to infections *(5)*, as dramatically illustrated in clinical cases of selective NK cell deficiency associated with recurrent, severe herpes viral infections *(6)*. Mice are more susceptible to certain viral infections [murine cytomegalovirus (MCMV), vaccinia virus, or mouse hepatitis virus] if they have defective NK cell effector function *(7)*, are depleted of NK cells by antibody administration *(8–10)*, or are genetically deficient in NK cell number *(11–13)*. Interestingly, the antibody depletion studies further showed that if anti-NK cell antibody was given to wild-type mice later in the infection, there was no untoward effect *(9)*. Thus, NK cells are significant in early, innate immunity to viral infections.

NK cells provide host defense by killing targets, such as infected cells or tumors, through the exocytosis of preformed granules containing perforin and serine proteases (granzymes) that mediate target apoptosis *(14)*. Targets can also directly trigger NK cell cytokine production. In addition, NK cells respond to inflammatory cytokines produced by other cells, such as interleukin-12 (IL-12), that stimulates their production of interferon-γ (IFN-γ) *(15)*. Other cytokines, such as IL-2, IL-15, and type I interferons will stimulate NK cells directly or indirectly through NK/T cells and plasmacytoid dendritic cells (pDCs), resulting in NK cell proliferation, IFN-γ production, and enhanced killing capacity *(15–21)*. Thus, NK cells can respond directly to targets and can respond directly or indirectly to cytokines.

1.1. Assessment of NK Cell Responses in Heterogeneous Effector Populations

Traditionally, NK cell functions are determined in "bulk" assays, where the functional response of an entire NK cell population is assessed. For example, in target killing assays, heterogeneous populations of NK cells are added in vitro to ^{51}Cr-labeled targets, and killing is determined by release of ^{51}Cr into the cell-free supernatant. In this assay, it is theoretically possible that only a few effector cells do most of the killing, whereas the rest of the effector cells are not activated. Alternatively, many NK cells could be triggered. In other words, target-killing assays essentially report the combined responses of individual NK cells.

Similarly, in response to several stimuli, NK cells produce cytokines, such as IFN-γ, tumor necrosis factor-α (TNF-α), and granulocyte-macrophage colony-stimulating factor (GM-CSF). This response can be readily assessed by ELISA assays of supernatants. Akin to the ^{51}Cr-release target-killing assay, however, assessments of cytokine production in this way represent the responses of the entire NK cell population. Likewise, it is important to note that this limitation also applies to traditional NK cell signal transduction assays such as immunoprecipitation or Western blot analysis for phosphorylated proteins from stimulated, heterogeneous NK cell populations. Finally, NK cell proliferation can be assessed by incorporation of ^{3}H-thymidine, but again, this assay represents the response of the entire population, not of individual NK cells.

1.2. Assessment of Individual NK Cell Responses

It will be helpful to study individual NK cell responses to advance our understanding of NK cell functions. To get around the limitations of the aforementioned assays, various modifications can be employed, such as antibody-mediated depletion of NK cell subsets or purification of certain NK cell subsets, before the assay is performed. Alternatively, flow cytometry-based assays of individual NK cell responses can be done on unpurified cell populations as markers can be used at the assay stage to identify cells of interest. In this way, purification steps can be avoided as well as the inherent problems associated with manipulation of cells before assay, such as inconsistent purification and activation (or inhibition) of cellular responses caused by the purification steps alone. Additionally, investigators can avoid artifacts in the functional assays from antibodies used for purification. At the assay stage, responding individual NK cells can be further characterized in multi-parameter flow cytometric analysis for informative markers. Finally, protocols to directly measure individual NK cell responses have the potential benefit of requiring fewer input cells. Thus, recent advances now provide the NK cell biologist with tools to study individual NK cell responses.

In the following sections, we will provide two examples of how assays for analysis of individual NK cell responses have recently helped advance our understanding of NK cells. Borrowing liberally from our own work, we will provide the context for understanding the advance and a description of the protocols themselves in each section. One assay involves the ex vivo triggering of naïve NK cells with a monoclonal antibody (mAb) against an activation receptor on all NK cells, and the other assesses proliferation of NK cells during the course of a viral infection. In both cases, the response of individual NK cells is assessed by flow cytometry. While these assays may not totally replace the familiar population-based assays of target killing and ELISAs for cytokine production, they provide new complementary tools that are increasingly used in the literature.

1.2.1. Licensing of NK Cells by Host Major Histocompatibility Complex Class I

The defining feature of NK cells is their intrinsic capacity to react to cellular targets *(22)*. Both cytolysis and cytokine production are NK cell effector functions that are initiated by target recognition and are regulated by ubiquitously expressed major histocompatibility complex (MHC) class I molecules. Targets with normal expression of MHC class I are more resistant to NK cells than those lacking MHC class I. These findings form the basis for Kärre's "missing-self" hypothesis, postulating that NK cells survey tissues for normal expression of MHC class I, which prevents NK cell activation *(23)*.

Missing self is now explained by MHC class I-specific, NK cell inhibitory receptors, first shown for mouse Ly49A *(24)*. The inhibitory NK cell receptors fall into two general structural types (reviewed in **ref.** *(25)*). Whereas human killer immunoglobulin (Ig)-like receptors (KIRs) are type I integral membrane proteins with Ig-like domains, human and rodent CD94/NKG2A and rodent Ly49 have type II orientation and are disulfide-linked dimers with C-type lectin-like domains. Despite their structural differences, the inhibitory receptors have several shared features. In addition to relatively restricted expression on NK cells (and NK/T cells for the C-type lectins), the receptors belong to families of highly related molecules (except for CD94 which is the invariant partner chain for NKG2 molecules). Importantly, all inhibitory receptors known to date contain immunoreceptor tyrosine-based inhibitory motifs (ITIMs) in their cytoplasmic domains *(26)*. Upon receptor cross-linking and subsequent ITIM phosphorylation, the ITIMs recruit cytoplasmic tyrosine phosphatases such as SH_2-domain-containing protein tyrosine phosphate-1 (SHP-1). This recruitment then presumably leads to dephosphorylation of phosphorylated molecules, thereby blocking cellular activation. These general mechanisms are now appreciated as being applicable to a large number of other inhibitory receptors that are expressed on a wide variety of hematopoietic cells *(26)*.

NK cell recognition and killing of targets also involves activation receptors, compatible with a two-receptor model for the mechanism of NK cell activation *(27)*. Although inhibitory receptor effects tend to dominate over activation receptor functions, triggering of NK cell effector responses is determined by the engagement (or not) of activation and/or inhibitory receptors on the NK cell by their target cell ligands and integration of signals transduced by such receptors.

A critical issue in NK cell development relates to how NK cells acquire tolerance to self. Whereas the missing-self hypothesis implies that there should be overt NK cell auto-reactivity in MHC class I-deficient hosts, NK cells are not spontaneously overactive in humans or mice with deficiencies in MHC class I expression *(28–34)*. Indeed, the opposite was found, that is, NK cells from MHC class I-deficient mice demonstrate poor killing of MHC class I-deficient targets, even though they appear normal in number, tissue distribution, and expression levels of activation receptors *(28–32)*. They cannot reject MHC class I-deficient bone marrow (BM) grafts that are rejected by normal NK cells. Moreover, NK cell-mediated rejection of MHC class I-sufficient BM grafts depends on which host MHC class I allele is present *(35,36)*. Thus, NK cell functions are also influenced by the host MHC class I environment, not just at the level of effector–target cell interactions.

NK cells from MHC-deficient mice and humans are defective in target killing *(28,30)*. These defects could be due to effects on NK cell subpopulations, or altered expression or function of activation or inhibitory receptors *(37,38)*, issues that are difficult to assess with cellular targets because the universe of NK cell receptors involved in target recognition remains incompletely defined. Moreover, the target-killing assay does not allow assessment of individual NK cell responses.

Recently, we used a target cell-free system to activate individual resting NK cells through NK1.1 (Nkrp1c or Klrb1c), an activation receptor expressed by all mature and developing $CD3^-$ NK cells in C57BL/6 (B6) mice *(39)*. By intracellular staining of individual NK cells for IFN-γ, we showed that anti-NK1.1 cross-linking led to stimulation of large numbers of individual NK cells from wt B6 but not $\beta 2m^{-/-}$ or $K^{b-/-}$ $D^{b-/-}$ mice *(40)*. Similar effects were observed with stimulation through other activation receptors and by tumor targets. This effect was an NK cell autonomous effect. Thus, the intrinsic defects of NK cells from MHC class Ia-deficient hosts suggest that NK cell functional maturation requires specific interaction with host MHC class Ia molecules. We termed this process MHC class I-dependent "licensing" to distinguish it from MHC class I-dependent "education" that implies different events occurring during T-cell development.

The ability to trigger all NK cells and analyze responding cells by flow cytometry aided identification of cells that were selectively licensed by self-MHC. In particular, $Ly49C^+$ NK cells appeared to be selectively licensed by $H2K^b$ in mice bearing only a single MHC class I molecule. In contrast, $Ly49A^+$ NK cells could be licensed by $H2D^d$ as revealed by MHC-congenic, recombinant congenic, and transgenic mice. Finally, retroviral gene transfer into hematopoietic stem cells revealed that licensing requires the cytoplasmic ITIM in the self-MHC-specific receptor for licensing. Thus, licensing occurs as a result of engagement and signaling through self-MHC-specific receptors that were ironically identified as inhibitory receptors in effector responses. Analysis of this process was aided by study of individual NK cell responses.

1.2.2. NK Cell Responses During Viral Infection

In mice, the critical role of NK cells in host defense against viruses has been especially evident in infections with MCMV *(16,18,41)*. Although NK cell responses to cytokines are important in viral control *(15)*, recent studies indicate that the murine NK cell activation receptor, Ly49H, is required for resistance to MCMV infection in vivo. A genome-wide scan had identified the autosomal dominant *Cmv1* resistance gene as being responsible for the genetically determined resistance of certain strains of mice to MCMV *(42)*. Abundant genetic mapping data indicated that *Cmv1* maps to the NK gene complex (NKC) *(43–45)*, which contains clusters of genes encoding the lectin-like NK cell receptors *(46)*. These studies implicated NK cell involvement in the resistant phenotype *(46)*. Indeed, when NK cells were depleted with the anti-NK1.1 mAb, MCMV-resistant C57BL/6 mice became susceptible *(47)*.

The successful isolation of the resistant *Cmv1* allele depended on genetic evidence that the BXD-8 recombinant inbred mouse strain, derived from MCMV-resistant C57BL/6 and susceptible DBA/2 inbred strains, appeared to have inherited the entire NKC and flanking genomic segments from its resistant C57BL/6 progenitor but displayed the susceptible phenotype *(47)*. These mice were subsequently found to have a specific deletion in *Ly49h* *(41,48)*. Furthermore, when resistant C57BL/6 mice were injected with an anti-Ly49H mAb, they became susceptible as measured by viral titers in the spleen and lethality *(41,49)*. Finally, susceptible BALB/c mice become resistant when *Ly49h* was reconstituted by transgenesis *(50)*. These studies established that *Ly49h* is responsible for genetic resistance to MCMV and suggested that Ly49H is an activation receptor that recognizes MCMV-infected cells and activates NK cells to resist infection.

The activation receptor hypothesis was confirmed by a study showing that a DAP12 mutant (Tyr in ITAMs mutated to Phe) mouse could not resist MCMV *(51)*. This is consistent with predictions from previous in vitro signaling studies showing that Ly49H signals through DAP12 *(41,52)*. The ligand for Ly49H during MCMV infection was independently identified by two groups *(53,54)* and established to be the open reading frame, m157, encoded by the virus itself. Thus, Ly49H specifically recognizes MCMV-infected cells and specifically triggers NK cell activation to resist infection.

In addition to target killing, NK cell activation receptors can also trigger cytokine production in vitro *(55,56)*. Indeed, when co-incubated in vitro with

MCMV-infected macrophages or cell lines transfected with m157, Ly49H$^+$ NK cells selectively produce cytokines and chemokines in a coordinate manner as determined by intracellular staining for IFN-γ and lymphotactin/ATAC, MIP-1α, MIP-1β, and RANTES *(53, 57)*. In addition, NK cells produce the same cytokines/chemokines when stimulated with other cytokines, such as IL-2, IL-15, and IL-18, regardless of Ly49H expression. Thus, staining for individual NK cell responses can distinguish between "specific" and "non-specific" NK cell responses as related to expression of Ly49H.

These in vitro findings are relevant to the relative "non-specific" stimulation of NK cells found during early MCMV infection in vivo where IFN-γ production was not confined to the Ly49H$^+$ NK cell subset *(58)*. Moreover, direct assessment of NK cell proliferation by FACS analysis of in vivo acute bromo-deoxyuridine (BrdU) incorporation *(58)* confirmed that infection stimulates NK cell proliferation *(17, 59)* and also demonstrated that early (days 1–2 post-infection) in vivo NK cell proliferation was nonselective with respect to Ly49H expression. This proliferation resembled the cytokine-driven "bystander proliferation" observed in T cells in response to viral infections or stimulation with type I interferons *(60)*, suggesting that the initial phase of viral-induced NK cell proliferation represents a nonspecific response to pro-inflammatory cytokines.

On the contrary, the nonselective proliferation phase was followed by a period of preferential proliferation of Ly49H$^+$ NK cells peaking at days 4–6 of MCMV infection *(58)* and accumulation of Ly49H$^+$ NK cells. Specific proliferation and accumulation of NK cells was abrogated with anti-Ly49H antibody, supporting the hypothesis that Ly49H recognition of MCMV-infected cells stimulates selective proliferation of Ly49H$^+$ NK cells at a later phase. In addition, an initial phase of nonspecific NK cell proliferation was seen with vaccinia infection, but the later specific proliferation of Ly49H$^+$ NK cells was absent. Thus, initial virus-specific NK cell responses may be masked by generic cytokine responses and only later are detectable. These studies were aided by the capacity to analyze individual NK cell responses, such as IFN-γ production, outlined in **Subheading 1.1.1**, and acute BrdU incorporation.

2. Materials

2.1. NK Cell Activation Assay for Licensing Studies

1. Mice (e.g., 2 ~ 4 months old B10.D2: cat. no. 000463; Jackson Laboratory, Bar Harbor, Maine, USA).
2. Dulbecco's phosphate-buffered saline (DPBS: cat. no. 21-031-CM; Mediatech, Herndon, Virginia, USA).

3. Polystyrene 6-well tissue culture plate.
4. Complete RPMI media containing 10% fetal calf serum (FCS) (RPMI-10).
5. 60 × 15 mm petri dish.
6. 10-ml syringe.
7. Cell strainer (cat. no. 352350; BD Falcon, San Jose, California, USA).
8. 15-ml tube (cat. no. 14-959-49D; Fisher, Pittsburgh, Peansylvania, USA) and 5-ml tube (cat. no. 60818-383; VWR, Batavia, Illinois, USA).
9. RBC lysis buffer: Mix 10 ml of 0.17 M Tris–HCl (pH 7.65) and 90 ml of 0.16 M NH_4Cl and adjust to pH 7.2 with HCl (can be stored up to several months at room temperature, Batavia, Illinais, USA).
10. Culture supernatant of 2.4G2 hybridoma (anti-FcγRIII/II, cat. no. HB-197; ATCC, Manassas, Virginia, USA).
11. Antibodies and second-step reagents (anti-NK1.1: cat. no. 553162, anti-Ly49A-biotin: cat. no. 552014, anti-NK1.1-APC: cat. no. 550627, anti-CD3-PE-Cy7: cat. no. 552774, anti-IFNγ-Alexa Fluor 488: cat. no. 557724, Strepavidin-PE: cat. no. 554061, all from BD Biosciences, San Jose, California, USA).
12. Flow cytometry buffer; PBS containing 1% FCS and 0.09% NaN_3.
13. BD Cytofix/Cytoperm Kit (cat. no. 555028; BD Biosciences).

2.2. In Vivo BrdU Incorporation Assays

1. Mice (e.g., 2 ~ 4 months old B10.D2: cat. no. 000463; Jackson Laboratory).
2. RPMI media containing 5% FCS (RPMI-5).
3. 60 × 15 mm petri dish.
4. 10-ml syringe.
5. Cell strainer (cat. no. 352350; BD Falcon).
6. 15-ml tube (cat. no. 14-959-49D; Fisher) and 5-ml tube (cat. no. 60818-383; VWR).
7. RBC lysis buffer: Mix 10 ml of 0.17 M Tris–HCl (pH 7.65) and 90 ml of 0.16 M NH_4Cl, and adjust to pH 7.2 with HCl (can be stored up to several months at room temperature).
8. Culture supernatant of 2.4G2 hybridoma (anti-FcγRIII/II, cat. no. HB-197; ATCC).
9. Antibodies (anti-NK1.1: cat. no. 553162, anti-Ly49A-biotin: cat. no. 552014, anti-NK1.1-APC: cat. no. 550627, anti-CD3-PE-Cy7: cat. no. 552774, Strepavidin-PE: cat. no. 554061, all from BD Biosciences).
10. Flow cytometry buffer; PBS containing 1% FCS and 0.09% NaN_3.
11. FITC BrdU Flow Kit (cat. no. 559619; BD Biosciences).

3. Methods

3.1. NK Cell Activation Assay for Licensing Studies

In this method, freshly isolated NK cells are stimulated with an immobilized mAb (PK136) against NK1.1 (Nkrp1c or Klrb1c), an activation receptor expressed by all immature and mature NK cells in C57BL/6 and C57BL/10 mice (*see* **Note 1**). Brefeldin A is added to prevent secretion of IFN-γ to enable detection in individual cells by anti-IFN-γ staining and flow cytometry.

3.1.1. Antibody Coating

1. Dilute purified mAb PK136 (anti-NK1.1) to 0.5–2 μg/ml in DPBS.
2. Add 1 ml to each well of polystyrene 6-well tissue culture plates.
3. Incubate plates at 37°C in CO_2 incubator for 1 h 20 min.
4. Rinse plates three times with DPBS (*see* **Note 2**).

3.1.2. Preparation of Cell Suspension from Spleen

1. For each mouse, place one 60 × 15 mm petri dish containing 5 ml of RPMI-10 on ice.
2. Euthanize a mouse.
3. Remove spleen and place it in the petri dish.
4. Using the plunger of a 10-ml syringe, grind spleen against a cell strainer in a circular motion until mostly fibrous debris remains.
5. Transfer cell suspension into a 15-ml polypropylene centrifuge tube, centrifuge at 300 *g* for 5 min at 4°C, and discard supernatant.
6. Resuspend pellet in 1 ml of RBC lysis buffer by pipetting up and down with P-1000 and brief vortexing.
7. Incubate for 5 min at room temperature.
8. Fill the tube with cold RPMI-10, centrifuge at 300 *g* for 5 min at 4°C, and discard supernatant.
9. Resuspend pellet in 10 ml of RPMI-10, count cells, centrifuge at 300 *g* for 5 min at 4°C, and discard supernatant.
10. Resuspend cells at 1×10^7/ml in cold RPMI-10.

3.1.3. Stimulation

1. Discard DPBS from antibody-coated plate and add 1 ml (1×10^7 cells) of cell suspension to each well.
2. Incubate for 45 min at 37°C in CO_2 incubator.
3. Add brefeldin A (comes as a 1000x stock in Cytofix/Cytoperm Kit) to final 1x concentration as follows: thaw brefeldin A stock and dilute to 100x working stock in prewarmed RPMI-10, and add 10 μl of 100x working stock into each well and mix by gentle swirling.
4. Incubate for additional 6-8 h in 37°C CO_2 incubator.
5. Stop stimulation by adding 5ml of cold flow cytometry buffer (*see* **Note 3**).

3.1.4. Staining of Cell Surface Antigens

1. Harvest cells from well by pipetting up and down vigorously about 30 times using a 5 ml pipette.

2. Transfer cell suspension into a 15-ml centrifuge tube, centrifuge at 300 g for 5 min at 4°C, and discard supernatant.
3. Resuspend pellet in 200 μl of culture media of 2.4G2 hybridoma to block Fc receptor.
4. Stain cells ($2 \sim 10 \times 10^6$) at $2 \sim 5 \times 10^7$/ml for cell surface antigens with specific antibodies in 5-ml tube.

   a. First, incubate cells with biotinylated-antibody against NK cell antigen that is expressed by a subset of NK cells (e.g., anti-Ly49A-biotin) for 30 min on ice.
   b. Wash once with flow cytometry buffer; vortex briefly, fill tube with flow cytometry buffer, centrifuge at 300 g for 5 min and pour off supernatant.
   c. Incubate with a mixture of streptavidin-phycoerythrin (PE), anti-NK1.1-APC, anti-CD3-PE-Cy7 for 30 min on ice (*see* **Notes 4–8**).

5. Vortex briefly, fill tube with flow cytometry buffer, centrifuge at 300 g for 5 min at 4°C and pour off supernatant.
6. Wash one more time as in **step 5** and vortex thoroughly.

3.1.5. Staining of Intracellular Antigen

1. Add to each tube 250 μl of Cytofix/Cytoperm solution (provided in Cytofix/Cytoperm Kit) with P-1000, mix thoroughly by pipetting up and down, and vortex.
2. Incubate for 20 min on ice (vortex halfway through and at the end).
3. Add 2 ml of Perm/Wash solution (provided in Cytofix/Cytoperm Kit), vortex thoroughly, and incubate about 10 min on ice. (*see* **Notes 9** and **10**).
4. Centrifuge at 300 g for 5 min at 4°C and pour off supernatant.
5. Vortex briefly, add 2 ml of Perm/Wash solution, vortex thoroughly, centrifuge at 300 g for 5 min at 4°C, and pour off supernatant.
6. Add to each tube 50 μl of anti-cytokine antibody (e.g., anti-IFNγ-Alexa Fluor488 at 5 μg/ml in Perm/Wash buffer) with P-200, mix by pipetting up and down, and vortex thoroughly.
7. Incubate for 30 min on ice (vortex halfway through and at the end).
8. Wash two times with Perm/Wash solution as in **step 5**.

3.1.6. Flow Cytometry

1. Add to each tube 200 μl of flow cytometry buffer (*see* **Note 9**).
2. Use flow cytometry to gate on the CD3⁻ NK1.1⁺ cells as representing the NK cell population.
3. Determine the % of NK cells that stain for IFN-γ. Also, determine the % of NK cells bearing the marker of interest that produce IFN-γ.

3.2. In Vivo BrdU Incorporation Assay

In this method, the acute (3 h) BrdU incorporation assay is described because phenotype of NK cells may be altered in long-term BrdU incorporation studies (*see* **Note 11**). NK cells normally proliferate poorly in the periphery without stimuli. However, their proliferation is dramatically enhanced in response to various stimuli, such as viral infections.

3.2.1. Injection of BrdU into Mice

1. Inject i.p. 2 mg of BrdU (200 μl of 10 mg/ml in PBS, provided in BrdU Flow Kit) into mice (*see* **Note 12**).
2. the mice back into a cage for 3 h.

3.2.2. Preparation of Cell Suspension from Spleen

1. For each mouse, place one 60 × 15 mm petri dish containing 5 ml of RPMI-5 on ice.
2. Euthanize a mouse.
3. Remove spleen and place it in the petri dish.
4. Using the plunger of a 10-ml syringe, grind spleen against a cell strainer in a circular motion until mostly fibrous debris remains.
5. Transfer cell suspension into a 15-ml centrifuge tube, centrifuge at 300 g for 5 min at 4°C, and discard supernatant.
6. Resuspend pellet in 1 ml of RBC lysis buffer by pipetting up and down with P-1000 and brief vortexing.
7. Incubate 5 min at room temperature.
8. Fill the tube with cold RPMI-5, centrifuge at 300 g for 5 min at 4°C, discard supernatant.
9. Resuspend cells at 4×10^7/ml in culture media of 2.4G2 hybridoma to block Fc receptor.

3.2.3. Staining of Cell Surface Antigens

1. Stain cells ($2 \sim 10 \times 10^6$) at $2 \sim 5 \times 10^7$/ml for their cell surface antigens and wash as in **steps 4–6** in **Subheading 3.1.4**.

3.2.4. BrdU Staining

1. Add to each tube 250 μl of Cytofix/Cytoperm solution (provided in BrdU Flow Kit) with P-1000, mix thoroughly by pipetting up and down, and vortex.
2. Incubate for 20 min on ice (vortex halfway through and at the end).
3. Add 2ml of Perm/Wash solution (provided in BrdU Flow Kit), vortex thoroughly, and incubate about 10 min on ice (*see* **Notes 9** and **10**).
4. Centrifuge at 300 g for 5 min at 4°C and pour off supernatant.

5. Vortex briefly, add 2 ml of Perm/Wash solution, vortex thoroughly, spin, and pour off supernatant.
6. Add to each tube 125 µl of Cytoperm Plus buffer (provided in BrdU Flow Kit) with P-200, mix by pipetting up and down, and vortex thoroughly.
7. Incubate for 10 min on ice.
8. Add 2 ml of Perm/Wash solution, vortex thoroughly, spin, and pour off supernatant.
9. Add to each tube 125 µl of Cytofix/Cytoperm solution with P-200, mix by pipetting up and down, and vortex thoroughly.
10. Incubate for 5 min on ice.
11. Wash twice with Perm/Wash solution as in **step 5**.
12. Add to each tube 100 µl of diluted DNAse (300 µg/ml) (provided in BrdU Flow Kit) with P-200, mix by pipetting up and down, and vortex thoroughly.
13. Incubate for 1 h 30 min at 37°C (vortex halfway through and at the end).
14. Wash twice with Perm/Wash solution as in **step 5**.
15. Add to each tube 50 µl of anti-BrdU-FITC (50× diluted anti-BrdU-FITC, provided in FITC BrdU Flow Kit) in Perm/Wash solution and vortex thoroughly.
16. Incubate for 30 min at room temperature (vortex halfway through and at the end).
17. Wash twice with Perm/Wash solution as in **step 5**.

3.2.5. Flow Cytometry

1. Add to each tube 300 µl of flow cytometry buffer (*see* **Note 9**).
2. Use flow cytometry to gate on the $CD3^-$ $NK1.1^+$ cells as representing the NK cell population.
3. Determine the % of NK cells that stain for BrdU. Also, determine the % of NK cells bearing the marker of interest that incorporate BrdU.

4. Notes

1. It is important to note that the PK136 mAb may not be appropriate if mouse strains other than those on the C57BL/6 or C57BL/10 (B6 or B10) background are used. In other mouse strains, mAb PK136 may recognize other Nkpr1 isoforms, such as Nkrp1b, an inhibitory receptor in SJL/J or Swiss.NIH mice *(61,62)*. Alternative stimuli include mAbs against other activation receptors on all mouse NK cells in all strains, such as CD16 (FcγRIII, mAb 2.4G2), or on mouse NK cell subsets in B6 mice, such as Ly49D or Ly49H.
2. Do not let the wells dry before adding cells. If more time is needed, keep plates with DPBS filled at room temperature, and discard DPBS just before adding cells into the wells.
3. The plates can then be stored at 4°C overnight.
4. If the NK cell antigen is expressed at high levels and a specific antibody is available as PE-conjugated form, skip **steps a** and **b** and stain together with antibodies in **step c**.

5. Saturating concentration of each antibody should be predetermined before use. Final concentration at 5 µg/ml of mAb can be used in most cases.
6. The choice of antibodies is dictated by the nature of the experiment. At the minimum, staining for cell surface expression of CD3 and NK1.1 should be performed.
7. The use of immobilized anti-NK1.1 to stimulate does not generally impair staining with anti-NK1.1.
8. It may be of interest to stain for other cell surface markers to determine if they correlate with IFN-γ production.
9. Following this step, the cells can be stored overnight at 4°C in the dark.
10. If more Perm/Wash solution is needed, it can be replaced with flow cytometry buffer containing 0.1% (w/v) saponin (cat. no. S7900; Sigma).
11. To measure proliferation of an immune cell subset in vivo, long-term BrdU incorporation assays are commonly used whereby BrdU is added to the drinking water for several days. However, one should be cautious in interpreting data obtained from the long-term assays of NK cell proliferation, because individual NK cells or subsets of NK cells may differ in their survival and trafficking, an issue that requires further studies. Moreover, the phenotype of NK cells could be altered during a long-term BrdU-labeling period inasmuch as BrdU is a mutagen. Thus, to avoid these complications, acute BrdU incorporation assays are preferable, particularly when measuring in vivo proliferation of subsets of NK cells in response to infection or other stimuli.

    This staining protocol is a modified protocol from Instruction Manual of BrdU Flow Kits (BD Pharmingen™). If more BrdU solution is needed, it can be made with BrdU powder (cat. no. B5002; Sigma) by dissolving the powder in DPBS at 10 mg/ml. The BrdU solution needs to be aliquoted and stored at −80°C. Thawed solution should be used within hours.
12. Typically, we use mice that were infected with MCMV 2 days or 5 days ago.

References

1. Trinchieri, G. (1989) Biology of natural killer cells. *Adv Immunol* **47**, 187–376.
2. Yokoyama, W. M. (1999) Natural killer cells, Chapter 17. In *Fundamental Immunology* (Paul, W. E., Ed.), pp. 575–603, Lippincott-Raven, New York.
3. Herberman, R. (1982) *NK Cells and Other Natural Effector Cells*, Academic Press, New York.
4. Rosenberg, S. A., Lotze, M. T., Muul, L. M., Leitman, S., Chang, A. E., Ettinghausen, S. E., Matory, Y. L., Skibber, J. M., Shiloni, E., Vetto, J. T., Seipp, C. A., Simpson, C., and Reichert, C. M. (1985) Observations on the systemic administration of autologous lymphokine-activated killer cells and recombinant interleukin-2 to patients with metastatic cancer. *N Engl J Med* **313**, 1485–92.
5. Bancroft, G. J. (1993) The role of natural killer cells in innate resistance to infection. *Curr Opin Immunol* **5**, 503–10.

6. Biron, C. A., Byron, K. S., and Sullivan, J. L. (1989) Severe herpesvirus infections in an adolescent without natural killer cells. *N Engl J Med* **320**, 1731–5.
7. Shellam, G. R., Allan, J. E., Papadimitriou, J. M., and Bancroft, G. J. (1981) Increased susceptibility to cytomegalovirus infection in beige mutant mice. *Proc Natl Acad Sci USA* **78**, 5104–08.
8. Bukowski, J. F., Woda, B. A., Habu, S., Okumura, K., and Welsh, R. M. (1983) Natural killer cell depletion enhances virus synthesis and virus-induced hepatitis in vivo. *J Immunol* **131**, 1531–8.
9. Bukowski, J. F., Woda, B. A., and Welsh, R. M. (1984) Pathogenesis of murine cytomegalovirus infection in natural killer cell-depleted mice. *J Virol* **52**, 119–28.
10. Welsh, R. M., Dundon, P. L., Eynon, E. E., Brubaker, J. O., Koo, G. C., and O'Donnell, C. L. (1990) Demonstration of the antiviral role of natural killer cells in vivo with a natural killer cell-specific monoclonal antibody (NK 1.1). *Nat Immun Cell Growth Regul* **9**, 112–20.
11. Orange, J. S., and Biron, C. A. (1996) An absolute and restricted requirement for IL-12 in natural killer cell IFN-gamma production and antiviral defense. Studies of natural killer and T cell responses in contrasting viral infections. *J Immunol* **156**, 1138–42.
12. Loh, J., Chu, D. T., O'Guin, A. K., Yokoyama, W. M., and Virgin, H. W. t. (2005) Natural killer cells utilize both perforin and gamma interferon to regulate murine cytomegalovirus infection in the spleen and liver. *J Virol* **79**, 661–7.
13. Kennedy, M. K., Glaccum, M., Brown, S. N., Butz, E. A., Viney, J. L., Embers, M., Matsuki, N., Charrier, K., Sedger, L., Willis, C. R., Brasel, K., Morrissey, P. J., Stocking, K., Schuh, J. C., Joyce, S., and Peschon, J. J. (2000) Reversible defects in natural killer and memory CD8 T cell lineages in interleukin 15-deficient mice. *J Exp Med* **191**, 771–80.
14. Henkart, P. A. (1994) Lymphocyte-mediated cytotoxicity: two pathways and multiple effector molecules. *Immunity* **1**, 343–46.
15. Biron, C. A., Nguyen, K. B., Pien, G. C., Cousens, L. P., and Salazar-Mather, T. P. (1999) Natural killer cells in antiviral defense: function and regulation by innate cytokines. *Annu Rev Immunol* **17**, 189–220.
16. Grimm, E. A., Robb, R. J., Roth, J. A., Neckers, L. M., Lachman, L. B., Wilson, D. J., and Rosenberg, S. A. (1983) Lymphokine-activated killer cell phenomenon. III. Evidence that IL-2 is sufficient for direct activation of peripheral blood lymphocytes into lymphokine-activated killer cells. *J Exp Med* **158**, 1356–61.
17. Orange, J. S., and Biron, C. A. (1996) Characterization of early IL-12, IFN-alphabeta, and TNF effects on antiviral state and NK cell responses during murine cytomegalovirus infection. *J Immunol* **156**, 4746–56.
18. Carnaud, C., Lee, D., Donnars, O., Park, S. H., Beavis, A., Koezuka, Y., and Bendelac, A. (1999) Cutting edge: cross-talk between cells of the innate immune system: NKT cells rapidly activate NK cells. *J Immunol* **163**, 4647–50.

19. Nguyen, K. B., Salazar-Mather, T. P., Dalod, M. Y., Van Deusen, J. B., Wei, X. Q., Liew, F. Y., Caligiuri, M. A., Durbin, J. E., and Biron, C. A. (2002) Coordinated and distinct roles for IFN-alphabeta, IL-12, and IL-15 regulation of NK cell responses to viral infection. *J Immunol* **169**, 4279–87.
20. Andrews, D. M., Scalzo, A. A., Yokoyama, W. M., Smyth, M. J., and Degli-Esposti, M. A. (2003) Functional interactions between dendritic cells and NK cells during viral infection. *Nat Immunol* **4**, 175–81.
21. Krug, A., French, A. R., Barchet, W., Fischer, J. A. A., Dzionek, A., Pingel, J. T., Orihuela, M. M., Akira, S., Yokoyama, W. M., and Colonna, M. (2004) TLR9-dependent recognition of MCMV by IPC and DC generates coordinated cytokine responses that activate antiviral NK cell function. *Immunity* **21**, 107–19.
22. Lanier, L. L., Phillips, J. H., Hackett, J., Jr., Tutt, M., and Kumar, V. (1986) Natural killer cells: definition of a cell type rather than a function. *J Immunol* **137**, 2735–9.
23. Ljunggren, H. G., and Karre, K. (1990) In search of the 'missing self': MHC molecules and NK cell recognition. *Immunol Today* **11**, 237–44.
24. Karlhofer, F. M., Ribaudo, R. K., and Yokoyama, W. M. (1992) MHC class I alloantigen specificity of Ly-49+ IL-2-activated natural killer cells. *Nature* **358**, 66–70.
25. Yokoyama, W. M. (1997) What goes up must come down: the emerging spectrum of inhibitory receptors. *J Exp Med* 186, 1803–08.
26. Long, E. O. (1999) Regulation of immune responses through inhibitory receptors. *Annu Rev Immunol* **17**, 875–904.
27. Yokoyama, W. M. (1995) Natural killer cell receptors. *Curr Opin Immunol* **7**, 110–20.
28. Furukawa, H., Yabe, T., Watanabe, K., Miyamoto, R., Miki, A., Akaza, T., Tadokoro, K., Tohma, S., Inoue, T., Yamamoto, K., and Juji, T. (1999) Tolerance of NK and LAK activity for HLA class I-deficient targets in a TAP1-deficient patient (bare lymphocyte syndrome type I). *Hum Immunol* 60, 32–40.
29. Vitale, M., Zimmer, J., Castriconi, R., Hanau, D., Donato, L., Bottino, C., Moretta, L., de la Salle, H., and Moretta, A. (2002) Analysis of natural killer cells in TAP2-deficient patients: expression of functional triggering receptors and evidence for the existence of inhibitory receptor(s) that prevent lysis of normal autologous cells. *Blood* **99**, 1723–9.
30. Bix, M., Liao, N. S., Zijlstra, M., Loring, J., Jaenisch, R., and Raulet, D. (1991) Rejection of class I MHC-deficient haemopoietic cells by irradiated MHC-matched mice. *Nature* **349**, 329–31.
31. Hoglund, P., Ohlen, C., Carbone, E., Franksson, L., Ljunggren, H. G., Latour, A., Koller, B., and Kärre, K. (1991) Recognition of beta 2-microglobulin- negative (beta 2m-) T-cell blasts by natural killer cells from normal but not from beta 2m- mice: nonresponsiveness controlled by beta 2m- bone marrow in chimeric mice. *Proc Natl Acad Sci USA* **88**, 10332–36.

32. Liao, N. S., Bix, M., Zijlstra, M., Jaenisch, R., and Raulet, D. (1991) MHC class I deficiency: susceptibility to natural killer (NK) cells and impaired NK activity. *Science* **253**, 199–202.
33. Ljunggren, H. G., Van Kaer, L., Ploegh, H. L., and Tonegawa, S. (1994) Altered natural killer cell repertoire in Tap-1 mutant mice. *Proc Natl Acad Sci USA* **91**, 6520–4.
34. Dorfman, J. R., Zerrahn, J., Coles, M. C., and Raulet, D. H. (1997) The basis for self-tolerance of natural killer cells in beta2-microglobulin- and TAP-1- mice. *J Immunol* **159**, 5219–25.
35. Ohlen, C., Kling, G., Hoglund, P., Hansson, M., Scangos, G., Bieberich, C., Jay, G., and Karre, K. (1989) Prevention of allogeneic bone marrow graft rejection by H-2 transgene in donor mice. *Science* **246**, 666–8.
36. Bennett, M., Yu, Y. Y., Stoneman, E., Rembecki, R. M., Mathew, P. A., Lindahl, K. F., and Kumar, V. (1995) Hybrid resistance: 'negative' and 'positive' signaling of murine natural killer cells. *Semin Immunol* **7**, 121–7.
37. Raulet, D. H., Vance, R. E., and McMahon, C. W. (2001) Regulation of the natural killer cell receptor repertoire. *Annu Rev Immunol* **19**, 291–330.
38. Sentman, C. L., Olsson, M. Y., and Karre, K. (1995) Missing self recognition by natural killer cells in MHC class I transgenic mice. A 'receptor calibration' model for how effector cells adapt to self. *Semin Immunol* 7, 109–19.
39. Kim, S., Iizuka, K., Kang, H. S., Dokun, A., French, A. R., Greco, S., and Yokoyama, W. M. (2002) In vivo developmental stages in murine natural killer cell maturation. *Nat Immunol* **3**, 523–8.
40. Kim, S., Poursine-Laurent, J., Truscott, S. M., Lybarger, L., Song, Y. J., Yang, L., French, A. R., Sunwoo, J. B., Lemieux, S., Hansen, T. H., and Yokoyama, W. M. (2005) Licensing of natural killer cells by host major histocompatibility complex class I molecules. *Nature* **436**, 709–13.
41. Brown, M. G., Dokun, A. O., Heusel, J. W., Smith, H. R., Beckman, D. L., Blattenberger, E. A., Dubbelde, C. E., Stone, L. R., Scalzo, A. A., and Yokoyama, W. M. (2001) Vital involvement of a natural killer cell activation receptor in resistance to viral infection. *Science* **292**, 934–7.
42. Scalzo, A. A., Fitzgerald, N. A., Simmons, A., La Vista, A. B., and Shellam, G. R. (1990) Cmv-1, a genetic locus that controls murine cytomegalovirus replication in the spleen. *J Exp Med* **171**, 1469–83.
43. Scalzo, A. A., Lyons, P. A., Fitzgerald, N. A., Forbes, C. A., Yokoyama, W. M., and Shellam, G. R. (1995) Genetic mapping of Cmv1 in the region of mouse chromosome 6 encoding the NK gene complex-associated loci Ly49 and musNKR-P1. *Genomics* **27**, 435–41.
44. Forbes, C. A., Brown, M. G., Cho, R., Shellam, G. R., Yokoyama, W. M., and Scalzo, A. A. (1997) The Cmv1 host resistance locus is closely linked to the Ly49 multigene family within the natural killer cell gene complex on mouse chromosome 6. *Genomics* **41**, 406–13.

45. Depatie, C., Muise, E., Lepage, P., Gros, P., and Vidal, S. M. (1997) High-resolution linkage map in the proximity of the host resistance locus CMV1. *Genomics* **39**, 154–63.
46. Brown, M. G., Scalzo, A. A., Matsumoto, K., and Yokoyama, W. M. (1997) The natural killer gene complex-a genetic basis for understanding natural killer cell function and innate immunity. *Immunol Rev* **155**, 53–65.
47. Scalzo, A. A., Fitzgerald, N. A., Wallace, C. R., Gibbons, A. E., Smart, Y. C., Burton, R. C., and Shellam, G. R. (1992) The effect of the Cmv-1 resistance gene, which is linked to the natural killer cell gene complex, is mediated by natural killer cells. *J Immunol* **149**, 581–9.
48. Lee, S. H., Girard, S., Macina, D., Busa, M., Zafer, A., Belouchi, A., Gros, P., and Vidal, S. M. (2001) Susceptibility to mouse cytomegalovirus is associated with deletion of an activating natural killer cell receptor of the C-type lectin superfamily. *Nat Genet* **28**, 42–5.
49. Daniels, K. A., Devora, G., Lai, W. C., O'Donnell, C. L., Bennett, M., and Welsh, R. M. (2001) Murine cytomegalovirus is regulated by a discrete subset of natural killer cells reactive with monoclonal antibody to ly49h. *J Exp Med* **194**, 29–44.
50. Lee, S. H., Zafer, A., de Repentigny, Y., Kothary, R., Tremblay, M. L., Gros, P., Duplay, P., Webb, J. R., and Vidal, S. M. (2003) Transgenic expression of the activating natural killer receptor Ly49H confers resistance to cytomegalovirus in genetically susceptible mice. *J Exp Med* **197**, 515–26.
51. Sjolin, H., Tomasello, E., Mousavi-Jazi, M., Bartolazzi, A., Karre, K., Vivier, E., and Cerboni, C. (2002) Pivotal role of KARAP/DAP12 adaptor molecule in the natural killer cell-mediated resistance to murine cytomegalovirus infection. *J Exp Med* 195, 825–34.
52. Smith, K. M., Wu, J., Bakker, A. B., Phillips, J. H., and Lanier, L. L. (1998) Cutting edge: Ly-49D and Ly-49H associate with mouse DAP12 and form activating receptors. *J Immunol* **161**, 7–10.
53. Smith, H. R., Heusel, J. W., Mehta, I. K., Kim, S., Dorner, B. G., Naidenko, O. V., Iizuka, K., Furukawa, H., Beckman, D. L., Pingel, J. T., Scalzo, A. A., Fremont, D. H., and Yokoyama, W. M. (2002) Recognition of a virus-encoded ligand by a natural killer cell activation receptor. *Proc Natl Acad Sci USA* **99**, 8826–31.
54. Arase, H., Mocarski, E. S., Campbell, A. E., Hill, A. B., and Lanier, L. L. (2002) Direct recognition of cytomegalovirus by activating and inhibitory NK cell receptors. *Science* **296**, 1323–6.
55. Cuturi, M. C., Anegon, I., Sherman, F., Loudon, R., Clark, S. C., Perussia, B., and Trinchieri, G. (1989) Production of hematopoietic colony stimulating factors by human natural killer cells. *J Exp Med* **169**, 569.
56. Kim, S., and Yokoyama, W. M. (1998) NK cell granule exocytosis and cytokine production inhibited by Ly-49A engagement. *Cell Immunol* **183**, 106–12.

57. Dorner, B. G., Smith, H. R. C., French, A. R., Kim, S., Poursine-Laurent, J., Beckman, D. L., Pingel, J. T., Kroczek, R. A., and Yokoyama, W. M. (2004) Coordinate expression of cytokines and chemokines by natural killer cells during murine cytomegalovirus infection. *J Immunol* **172**, 3119–31.
58. Dokun, A. O., Kim, S., Smith, H. R., Kang, H. S., Chu, D. T., and Yokoyama, W. M. (2001) Specific and nonspecific NK cell activation during virus infection. *Nat Immunol* **2**, 951–6.
59. Biron, C. A., Sonnenfeld, G., and Welsh, R. M. (1984) Interferon induces natural killer cell blastogenesis in vivo. *J Leukoc Biol* **35**, 31–7.
60. Tough, D. F., Borrow, P., and Sprent, J. (1996) Induction of bystander T cell proliferation by viruses and type I interferon in vivo. *Science* **272**, 1947–50.
61. Kung, S. K., Su, R. C., Shannon, J., and Miller, R. G. (1999) The NKR-P1B gene product is an inhibitory receptor on SJL/J NK cells. *J Immunol* **162**, 5876–87.
62. Carlyle, J. R., Martin, A., Mehra, A., Attisano, L., Tsui, F. W., and Zuniga-Pflucker, J. C. (1999) Mouse NKR-P1B, a novel NK1.1 antigen with inhibitory function. *J Immunol* **162**, 5917–23.

12

Isolation and Analysis of Human Natural Killer Cell Subsets

Guido Ferlazzo

Summary

Natural killer (NK) cells were originally defined as mediators of spontaneous cytotoxicity against virus-infected and tumor cells. In human peripheral blood, the majority of NK cells can mediate cell lysis mainly through perforin and granzymes. It has been recently shown, however, that the majority of NK cells in human secondary lymphoid organs are primarily immunoregulatory by secreting cytokines immediately after activation and do not express perforin and granzymes. Because lymph nodes (LN) harbor 40% and peripheral blood only 2% of all lymphocytes in humans, this newly characterized NK cell compartment in LN and related tissues probably outnumbers perforin$^+$cytolytic NK cells in our body. Although human NK cell biology has so far mainly studied peripheral blood NK cells, we have lately focused on human NK cells harbored in lymphoid tissues and identified procedures for their optimal isolation as well as their phenotypic and functional characterization.

Key Words: Natural killer cells; secondary lymphoid organs; killer Ig-like receptors; lymph nodes; lymphoid cell development; interleukin-2; flow cytometry; fluorescence in situ hybridization; human.

1. Introduction

For many years, natural killer (NK) cells were considered to be a homogeneous lymphocyte population with excellent cytotoxic capability. Nowadays, NK cells rather appear to comprise various subsets that differ in function and in part in phenotype. Human peripheral blood mononuclear cells (PBMNC) contain around 10% of NK cells *(1)*. The majority ($\geq$95%) belongs to the

$CD56^{dim}$ $CD16^+$ cytolytic NK subset *(2–4)*. These cells carry homing markers for inflamed peripheral sites and carry perforin to rapidly mediate cytotoxicity. The minor NK cell subset in blood (≤5%) is $CD56^{bright}CD16^-$ *(2–4)*. These NK cells lack perforin (or display low level of it) but secrete large amounts of interferon γ (IFNγ) and tumor necrosis factor β (TNFβ) upon activation and are superior to $CD56^{dim}$ NK cells in this latter function *(3,4)*. $CD56^{bright}$ $CD16^-$ cells proliferate more vigorously than their $CD56^{dim}CD16^+$ counterparts. Consistent with this feature, they uniquely express the high-affinity receptor for interleukin-2 (IL-2) (CD25), the alpha chain of IL-7 receptor and CD161, the receptor for the stem cell factor, also named c-kit. In addition, they display homing markers for secondary lymphoid organs, namely CCR7 and CD62L *(2)*.

Notably, the major histocompatibility complex (MHC) class I allele-specific killer immunoglobulin (Ig)-like receptors (KIR) are expressed on subsets of $CD56^{dim}CD16^+$ cytolytic NK cells, whereas the immunoregulatory $CD56^{bright}CD16^-$ NK subset expresses uniformly CD94/NKG2A and lacks KIRs *(3)*.

Recent reports have shown that a substantial amount of human NK cells homes to secondary lymphoid organs. These account for around 5% of mononuclear cells in uninflamed lymph nodes (LN) and 0.4–1% in inflamed tonsils and LN *(5,6)*. These NK cells constitute a remarkable pool of innate effector cells, because LN harbor 40% of all lymphocytes, whereas peripheral blood contains only 2% of all lymphocytes *(7,8)*. Therefore, LN NK cells are in physiological conditions 10 times more abundant than blood NK cells.

As might be anticipated from CCR7 and CD62L expression on $CD56^{bright}$ NK cells in blood, the $CD56^{bright}$ NK subset is enriched in all secondary lymphoid organs analyzed so far (LN, tonsils, and spleen) *(6)*.

Likewise, $CD56^{bright}$ NK cells of peripheral blood, secondary lymphoid tissue NK cells are perforin low/negative and show extremely poor cytolytic activity. Perforin and cytotoxicity can however be promptly upregulated by IL-2 on secondary lymphoid organ NK cells, and at the same time, they acquire the expression of CD16, as well as KIRs *(6)*. Therefore, activation converts secondary lymphoid organ NK cells into cytotoxic effectors analogous to blood $CD56^{dim}CD16^+$ NK cells.

The ontogenic link between the two main NK cell subsets is controversial. Thus, $CD56^{bright}$ NK cells have been proposed either as precursors of $CD56^{dim}$ cells or as derived from $CD56^{dim}$ cells *(9–11)*. Indeed, because it is not known yet whether mice have NK cell subsets equivalent to $CD56^{bright}$ and $CD56^{dim}$, data regarding human NK cell subset development have been limited, and at present, the developmental relationship between cytokine secreting CD56^{bright} and cytolytic CD56^{dim} NK cells is still unknown as are the sites of terminal NK cell differentiation.

To ascertain putative difference in the lifespan of distinct NK cell subsets, and therefore on their developmental relationship, we have evaluated their specific telomere length by flow-FISH (a quantitative fluorescence in situ hybridization recently adapted to flow cytometry). We found that both peripheral blood CD56^{bright} and LN NK cells display longer telomeres than peripheral blood CD56^{dim} NK cells.

Finally, taking into account the recently formulated hypothesis of LN as possible sites of NK cell ontogeny, we analyzed for the first time the features of NK cells directly isolated from human efferent lymph and from either inflamed or not inflamed LN. We observed that NK cells isolated from human thoracic duct efferent lymph, but not NK cells resident in autologous non-inflamed LN, express CD16 and KIR. Conversely, in vivo expression of KIR and CD16 in LN NK cells occurs during LN paracortical/follicular hyperplasia, characterized by extensive lymphocyte proliferation and therefore abundant cytokine production.

Our data imply a scenario in which KIR and CD16 acquisition represents a key maturation step from CD56^{bright} into CD56^{dim} NK cells. This differentiation can take place in LN during inflammation. Next, NK cells would leave the LN through the efferent lymphatics and recirculate in peripheral blood.

Many open questions still remain to be addressed, and this chapter presents methods that can be applied to the isolation and analysis of human NK cell subsets.

2. Materials

2.1. NK Cell Isolation and Culture

1. 100-μm nylon cell strainer (BD Labware, Mountain View, CA).
2. Ficoll–Hypaque (Amersham Pharmacia, Piscataway, NJ, USA).
3. NK Cell Isolation Kit II (Miltenyi Biotec, Gladbach, Germany).
4. RPMI-1640 medium.
5. Fetal calf serum (Cambrex, Verviers, Belgium).
6. rIL-2 (Proleukin, Chiron, Milan, Italy).
7. IL-12 (Peprotech, London, UK).
8. Monensin (Sigma, Milan, Italy) or GolgiStop (containing monensin) (BD Pharmingen, San Diego, CA).
9. DNase (Genentech Inc., San Francisco, CA).
10. Buffer: phosphate-buffered saline (PBS) pH7.2, supplemented with 0.5% bovine serum albumin (BSA) and 2 mM ethylenediaminetetraacetic acid (EDTA).

2.2. mAbs, Flow Cytometry, and NK Cell Phenotypic Analysis

1. Allophycocyanin- or fluorescein isothiocyanate (FITC)-conjugated anti-CD3, allophycocyanin-conjugated CD62L, phycoerythrine (PE)- or CyChrome-conjugated anti-CD56, FITC-conjugated anti-CD16, anti-CD69, anti-HLA-DR, anti-CD8, anti-CCR7, anti-CD117, allophycocyanin-conjugated anti-IFN-γ, anti-perforin, anti-granzymeA-, and anti-granzymeB were purchased from BD PharMingen.
2. PE-conjugated anti-NKp30, NKp46, NKp44, and anti-CD127 from Immunotech-Coulter (Marseille, France).
3. CO202 mAb (anti-CD48), XA141 (anti-CD158a/p58.1 and anti-CD158h/p50.1), Y249 (anti-CD158b1/p58.2 and anti-CD158j/p50.2), AZZ158 (anti-CD158e1/e2/p70 and anti-CD158k/p140), Z176 (anti-p75-AIRM1), Z199 (anti-CD159a/NKG2A), ECM217/1 (anti-NKG2D), CO54 (anti-2B4), and ON56 [anti-NK-T-B Ag (NTBA)] were produced in A. Moretta's or L. Moretta's laboratories (University of Genova, Italy and Istituto Giannina Gaslini, Genova, Italy).
4. Human γ globulin (human therapy grade from Biotest srl, Milan, Italy).
5. FITC- or PE-conjugated isotype-specific goat antimouse Abs (Southern Biotechnology Associates, Birmingham, AL).
6. Paraformaldehyde (1% solution in PBS).
7. Saponin (Sigma) (0.1% in PBS).

2.3. NK Cell Cytolytic Activity by Flow Cytometry

1. PKH26 Red Fluorescent Cell Linker Kit (Sigma-Aldrich, St. Louis, MO).
2. TO-PRO-3 (Molecular Probes, Eugene, OR).
3. FACS permeabilizing Solution, Becton Dickinson, CA. Dilute an aliquot of the 10× concentrate 1:10 with distilled H_2O before use.
4. Anti-CD107a PE-conjugated (BD Pharmingen).
5. Monensin.
6. NK cell target cell line (e.g., K562, 721.221, CCRF-CEM, FO1).
7. Buffer: PBS pH7.2, supplemented with 0.5% BSA and 2 mM EDTA.

2.4. Proliferation Assay by CFSE Dilution

1. 5,6-Carboxyfluorescein diacetate succinimidyl ester (CFSE) dilution.
2. BSA.
3. PBS.

2.5. Analysis of Telomere Length

1. Telomere PNA Kit/FITC for Flow Cytometry (DakoCytomation, Denmark).
2. 1301 cell line (ATCC) (control cells).
3. PBS.
4. Pure water.

3. Methods

During this study, it was necessary to isolate human NK cells either from peripheral blood or from secondary lymphoid tissues, that is, lymph, LN, tonsils, and spleen.

Although secondary lymphoid organs can harbor, as a whole, large quantities of NK cells, LN, and tonsils usually contain NK cells at a lower percentage than blood (*see* **Fig. 1**). Therefore, in addition to the classical procedures employed for cell isolation from solid tissues, LN and tonsil NK cell isolation requires extra attention to collect an amount of NK cells suitable for their characterization or for functional experiments or sorting. In addition, the purity of the NK cells is critical in most of the functional experiments: in the absence of flow cytometry sorting facilities, this latter issue often calls for specific modifications of commercial kits for magnetic cell sorting.

With regard to functional assay with rare NK cells, we found that classical cytotoxicity assay by Cr^{51} release could be replaced with a cytofluorimetric assay, which can determine the percentage of target cell lysis employing a lower number of effector cells. As an alternative method to detect indirectly NK cell

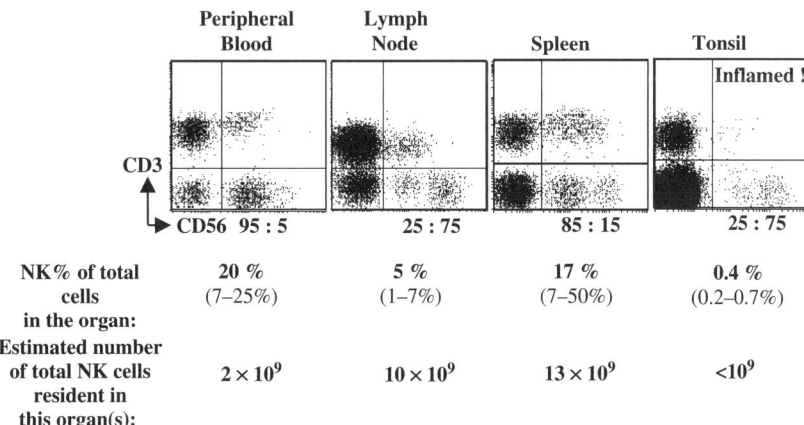

Fig. 1. Natural killer (NK) cell content in secondary lymphoid organs. NK cells are enriched in human secondary lymphoid organs. Single mononuclear cell suspensions were obtained from peripheral blood, lymph nodes, spleens, and tonsils. Cells were then analyzed by flow cytometry using specific mAb against CD3 (allophycocyanin conjugated) and CD56 (PE conjugated). Numbers indicate respectively the median value of NK cell percentages in all the experiments (ranges are shown in brackets) and the estimated number of total NK cells.

cytolytic activity, we found it useful to evaluate the rate of degranulation in the presence of target cells. The degranulation can be easily detected by flow cytometry staining of the cell surface with anti-CD107a monoclonal antibodies.

Finally, to determine the developmental relationship of different NK cell subsets, we have been measuring their telomere length. The measurement of telomere length has been widely employed for assessing the proliferative history of distinct cell subsets, including naïve and memory T cells *(12–16)*. In most normal somatic cells, telomere sequences are lost during replication, and therefore, telomere length inversely correlates with cell age. In our study, we employed flow-FISH because (1) this technique allows the assessment of telomere length with greater sensitivity than traditional methodologies *(15,16)* and (2) because it is possible to perform the analysis with as few as 5×10^4 NK cells, as it might be critical in case of a small number of NK cells isolated from tissues.

3.1. NK Cell Isolation and Culture

3.1.1. Isolation from Peripheral Blood

Human lymphocytes can be isolated from peripheral blood by density centrifugation over a step gradient consisting of a mixture of the carbohydrate polymer Ficoll™ and the dense iodine-containing compound metrizamide. This yields a population of mononuclear cells at the interface that has been depleted of red blood cells and most polymorphonuclear leukocytes or granulocytes. The resulting population, called PBMNC, consists mainly of lymphocytes and monocytes. Although this population is readily accessible, it is not necessarily representative of the lymphoid system, as only recirculating lymphocytes can be isolated from blood.

1. Collect peripheral blood in a tube containing anticoagulant or use defibrinated blood.
2. Dilute the blood with an equal volume of PBS.
3. Layer two-thirds of the diluted blood over one-third of Ficoll™ using a pipette with as little mixing as possible at the interface. As Ficoll™ is of greater density than cell suspension, a distinct interface will be formed.
4. At room temperature, centrifuge for 30 min at 800 *g* without break.
5. After centrifugation, there will be a well-defined lymphocyte layer at the interface. Using a pipette, carefully remove the cells from the interface and transfer to a new centrifuge tube.
6. Dilute the transferred cells with PBS to reduce the density of the solution. Centrifuge at 800 *g* for 10 min to pellet the lymphocytes, then discard the supernatant (*see* **Note 1**).

7. Wash PBMNC from the interface twice in PBS.
8. Resuspend the cell pellet in 40 μl of buffer per 10^7 total cells and add 10 μl of Biotin-Antibody Cocktail per 10^7 total cells (*see* **Notes 2** and **3**).
9. After 10 min of incubation at 4°C cells, wash with buffer by adding 10–20× labeling volume and then centrifuge at 300 *g* for 10 min (not required in Miltenyi protocol).
10. Completely pipette off the supernatant and add 80 μl of buffer per 10^7 total cells, 20 μl of Anti-Biotin Microbeads per 10^7 total cells and 50 μl of anti-CD3 Microbeads per 10^8 total cells. The use of anti-CD3 Microbeads is not required in the Miltenyi protocol, but it prevents unwanted CD3 contamination in the NK cell fraction; as contamination may range between 2 and 7%, addition of anti-CD3 Microbeads should be confined to experiments requiring high level of NK cell purity (*see* **Note 4**).
11. Incubate cells for 15 min at 4°C and then wash as described above. Resuspend up to 10^8 cells in 500 μl of buffer and perform magnetic separation.

3.1.2. Isolation from Lymph Node

1. Remove adipose tissue from resected LN with scalpels and scissors.
2. Dissociate LN by extensively mincing the tissues with scissors. It is important to keep the sample wet by adding small amounts of PBS during the mincing procedure (*see* **Note 5**).
3. Add DNAse to PBS at 100 U/ml. DNAse helps to minimize clumping and optimize recovery (the procedure does not affect cell-surface phenotype) (*see* **Note 6**).
4. Collect dissociated cells and filter through a 100-μm nylon cell strainer to exclude undissociated fragments. During filtering, press the sample through the strainer by a 5-ml syringe plunger. Debris and dead cells are then eliminated using a Ficoll–Hypaque discontinuous gradient as in **Subheading 3.1.1** (without dilution of cells suspension).
5. Extensively wash the single-cell suspension obtained at the gradient interface.
6. Total LN mononuclear cells can be now analyzed by flow cytometry or used for functional experiments that do not foresee NK cell sorting.

Alternatively, NK cells can be negatively selected by the NK Cell Isolation Kit II as described above (*see* **Subheading 3.1.1.**). Evaluate the percentage of NK cells in the isolated population using anti-CD3 and anti-CD56 mAbs. Because the kit has been developed for isolation of human NK cells from peripheral blood and NK cells are rarer in human LN, consider replicating the steps from 1 to 6 of the procedure described in **Subheading 3.1.1.** The replicated procedure will avoid contamination of other leukocytes because LN and blood contain different percentages of the distinct leukocyte subsets.

3.1.3. Isolation from Tonsils

Proceed as in **Subheading 2.1.2.** A larger number of mononuclear cells can be isolated compared with LN, but NK cells are usually represented in lower percentage, probably because of extensive B-cell proliferation in inflamed tonsils (the surgical resection consequential to an inflammatory process is usually the source of these specimens). Owing to the large number of B cells, the replication of steps 1–6 is mandatory for NK cell sorting with a magnetic separation kit.

3.1.4. Isolation from Spleen

Proceed as in **Subheading 3.1.2** regarding tissue dissociation and as in **Subheading 3.1.1** for magnetic separation of NK cells, because the leukocyte subset distribution resembles peripheral blood distribution.

3.1.5. Isolation from Lymph

Given 2 ml of lymph collected during surgical resection of esophageal cancer involving the thoracic duct,

1. Dilute lymph with PBS 1:4 and centrifuge at 300 g for 10 min.
2. Resuspend cell pellet in buffer and isolate mononuclear cells by density gradient centrifugation on a Ficoll–Hypaque gradient as described in **Subheading 3.1.1**.

In most cases, a limited amount of lymph will be available. As NK cells represent only 0.2% total mononuclear cells in lymph, only limited NK cell analysis can in general be performed.

3.2. NK Cell Phenotypic Analysis

3.2.1. Direct Immunofluorescence

Add 1 mg/ml human γ-globulin to cell suspension to block non-specific Fc-receptor binding (*see* **Note 7**). Label cells with fluorochrome-conjugated mAb for 30 min at 4°C, wash and analyze by flow cytometry.

3.2.2. Indirect Immunofluorescence

Saturate cell non-specific binding sites with human γ-globulin (*see* **Note 7**), add the relevant mAb and incubate for 30 min at 4°C.

After extensive washings, add FITC- or PE-conjugated isotype-specific goat anti-mouse antibodies and incubate for 30 min at 4°C. Negative controls will include directly labeled or unlabeled isotype-matched irrelevant mAbs.

3.2.3. Intracytoplasmic Staining

Lytic granules and cytokines are detected by intracytoplasmic staining and flow cytometry. For cytokines detection, cells are incubated in culture medium alone or in culture medium supplemented with 10 ng/ml PMA and 500 ng/ml ionomycin, at 37°C for 6 h. After the first hour, add monensin to a final concentration of 2 mM. Alternatively, cells can be stimulated by cytokines (e.g., IL-2, IL-12, IL-18) for 24 h and monensin added for the last 5 h.

Cells are then washed, fixed, permeabilized, and stained with relevant mAbs as described below:

1. Resuspend the cell suspension in cold PBS (no BSA) plus 1% paraformaldehyde (*see* **Subheading 2.2**.), vortexing cells for 5 min during fixation.
2. Wash twice with PBS.
3. Resuspend cells in buffer with 0.1% saponin to permeabilize the cells.
4. Centrifuge at 300 *g* for 10 min.
5. Gently resuspend the cell pellet.
6. Saturate cell non-specific binding sites with human γ-globulin (*see* **Note 7**), add the relevant mAb, and incubate for 30 min at 4°C.
7. Centrifuge at 300 *g* for 10 min.
8. Wash twice in buffer with 0.1% saponin.
9. Centrifuge at 300 *g* for 10 min and resuspend in buffer for flow cytometry analysis.

3.3. Evaluation of NK Cell Cytotoxicity by Flow Cytometry

3.3.1. Cytotoxic Assay by Flow Cytometry

To evaluate the cytolytic activity, we have been using either the NK cell-sensitive lymphoblastoid cell line 721.221, which does not express surface HLA class I molecules, or the T-lymphoblastoid cell line CCRF-CEM. The latter was used as target because of its expression of UL16-binding protein (ULBP) 1 and ULBP2, two ligands of NKG2D, the only activating receptor expressed by freshly isolated LN or tonsil NK cells. Other NK-sensitive cell lines can be successfully employed (as long as they do not adhere strongly to cell culture plastic).

The assay employs two fluorescent reagents: PKH26 and TO-PRO-3. PKH26 inserts into plasma membranes because of its lipophilic aliphatic residue without damaging the cells. TO-PRO-3 is a membrane-impermeable DNA stain.

Cytotoxicity assays are performed as follows:

1. Staining of target cells with the dye molecule PKH26:
    a. Wash extensively 10^6 target cells in plain PBS and remove as much buffer as possible after centrifugation at 350 *g* for 5 min (*see* **Note 8**).

b. In a 15-ml conical tube, resuspend target cells in Diluent C (provided in PKH26 kit) to a density of 10^7 cells/ml (concentrations from 5×10^5 to 10^7 have worked), do not vortex (see **Note 9**).
c. Dilute PKH26 1:1000 in Diluent C (see **Note 10**).
d. Add 1 volume stain to 1 volume cell suspension and mix immediately by inversion.
e. Incubate for 5 min at RT, agitate occasionally.
f. Wash with 10 ml of 10% serum supplemented RPMI twice (serum blocks the reaction).

2. Resuspend cells in culture medium to desired density. Use at least 1000 target cells per sample.
3. Incubate target cells with secondary lymphoid organ or peripheral blood NK cells at appropriate NK cell/target ratios, as in a classical Cr^{51} release assay. In contrast to radioactive assays, duplicate/triplicate measurements can be avoided if different effector/target ratios are performed, because the methodology presents less experimental variability.
4. Spin at 300 g for 3 min at the beginning of the incubation to ensure close contact of target and effector cells.
5. Incubate for 2–4 h depending on effector and target cell type. Differently from Cr^{51} release assay, effector/target incubation can be carried out for longer periods without increasing background of activity (we tested safely cytolytic activity after 24 h). Include two tubes with target cells only. Incubation can be performed directly in FACS tubes if sample number allows. Otherwise, incubate in V-bottom 96-μwell plates and transfer to FACS tubes at the end of incubation period.
6. Centrifuge FACS tubes at 350 g for 5 min. Target cell tube 1: Background—discard supernatant and resuspend cells in 500 μl buffer.

    Target cell tube 2: Maximum lysis—resuspend one tube with target cells in 500 μl 1 × BD permeabilizing solution and incubate for 10 min at RT. Wash once with staining buffer, resuspend in 500 μl of buffer.

    Samples: discard supernatant and resuspend cells in 500 μl of buffer.
7. Analysis: to all tubes add 10 μl of a 10 μM TO-PRO-3 solution immediately before FACS analysis. Analyze samples by flow cytometry.

PKH26 is excited at 551 nm, and TO-PRO-3, using the He-Ne laser, at 633 nm. Thus, target cells are easily gated by PKH26 red fluorescence, while undamaged target cells can be distinguished from lysed cells by their ability to exclude the DNA stain after the coincubation with effector cells. Moreover, the assay allows for surface staining of heterogeneous target cell populations. Background and maximum TO-PRO-3 stainings were obtained by incubation with medium and detergent, respectively (see **Fig. 2**).

Human NK Cell Subsets

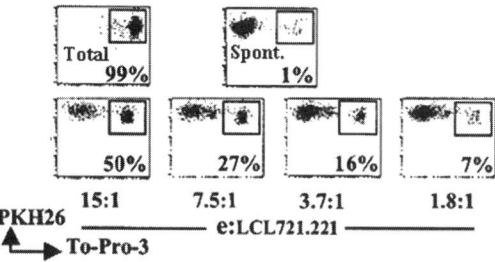

Fig. 2. Evaluation of NK cell cytotoxicity by flow cytometry. The indicated E:T ratios (e:LCL721.221) were calculated using the number of NK cells added to the cytotoxicity assays. Numbers inside the dot plots indicate the percentage of lysed cells (in the small quadrant) at the indicated E:T ratios. Background (Spont.) and maximum TO-PRO-3 stainings (Total) were obtained by incubation with medium and detergent, respectively.

3.3.2. CD107a Assay

CD107a assay is an alternative method to evaluate the cytotoxic properties of NK cells "ex vivo" and in particular when a limited number of cells can be employed for the study.

Target cell killing by NK cells mainly relies on degranulation, which leads to the release of lytic granules containing perforin and granzymes. During this process, lysosomal membranes and plasma membranes merge, and vesicle membrane proteins become visible on the cell surface. Among these vesicle membrane protein, lysosomal-associated membrane protein-1 (CD107a) can be used to study the regulation of cytolytic activity of NK cells. Recently, a correlation between CD107a surface expression and effector cell cytotoxicity was established *(17, 18)*, and we could confirm in a series of experiments a strong correlation between NK cell cytotoxic activity and CD107a expression. Also, we were able to determine NK cell activation of cytotoxic function in response to target cells on a single-cell level, which allows to analyze the activity of the distinct NK cell subsets independently.

Whole PBMNC or purified NK cells can be used as effectors in CD107a assay. To detect spontaneous degranulation, a control sample without target cells must be included.

1. Plate in a 96-μwells NK cells and targets at effector/target (E/T) ratio of 1:1 (2×10^5 effector cells : 2×10^5 target cells in a volume of 200 μl of complete medium).
2. In each well containing the E/T suspension or E alone, add 8 μl of PE-conjugated anti-CD107a before incubation.

3. Incubate at 37°C for 4 h (*see* **Note 11**).
4. After the first hour, add 5 μl of 2 mM monensin to inhibit cell secretion.
5. At the end of coincubation, wash in buffer and stain with mAbs (anti-CD3 and anti-CD56) for flow cytometric analysis. Surface expression of CD107a can be assessed separately in NK cell subsets.

3.4. Proliferation Assay by CFSE Dilution

1. Wash cells in PBS + 0.1% BSA.
2. Resuspend cells at $1–2 \times 10^7$ cells/ml.
3. Label cells by adding 1 μl of the 5 mM CFSE stock/ml (5 μM final).
4. Incubate in a 37°C water bath for 10 min. Gently mix after 5 min.
5. Add PBS + 0.1% BSA to the top of the tube and centrifuge.
6. Wash cells in PBS + 0.1% BSA and once in the medium in which cells will be used.

Cells can be cultured according to the experimental design, and cell proliferation will be evaluated by flow cytometry, analyzing the dilution CFSE fluorescence (FL-1) occurring upon cell divisions.

3.5. Analysis of Telomere Length

The Telomere PNA Kit/FITC for flow cytometry provides a convenient method for the detection of the telomeric sequences in nucleated hematopoietic cells from vertebrates.

This kit uses PNA probes, which are superior to DNA probes in terms of sensitivity and specificity. The method is optimal for the estimation of telomere length, as the fluorescence intensity of the cells is directly correlated to the length of the telomeres *(15,19)*.

It is recommended that all samples are run in duplicate and that the mean values of the duplicate determinations are used for data analysis.

3.5.1. Procedure

3.5.1.1. DAY 1

1. Wash sample cells and 1301 cells in PBS.
2. Count cells in Bürker chamber (*see* **Note 12**).
3. For each sample, mix 10^6 test cells and 10^6 control cells (1301 cell line) (*see* **Note 13**). We have successfully analyzed telomere length starting from as few as 2×10^5 NK cells. 1301 control cells can be maintained at 10^6.
4. Add PBS to a total of 6 ml. Divide the mixture into four 1.5-ml aliquots (1×10^6 cells) and place in 1.7-ml microcentrifuge tubes, labeled A, B, C, and D.
5. Centrifuge cells at 500 *g* for 5 min.
6. Remove as much as possible of supernatant by suction/aspiration.

7. Add 300 μl of Hybridization Solution, Vial 1 of the kit, to each of the two tubes A and B (controls) and 300 μl of Telomere PNA Probe/FITC in hybridization solution, Vial 2 of the kit, to each of the other two tubes C and D.
8. Mix by using a vortex mixer.
9. Place the four tubes in a pre-warmed heating block adjusted to 82°C for 10 min (*see* **Note 14**).
10. Mix by using a vortex mixer and place the tubes in the dark at RT overnight.

3.5.1.2. DAY 2

1. Add 1 mL of wash solution (Vial 3 of the kit diluted 1:10) to each of the four tubes.
2. Mix by using a vortex mixer.
3. Place the tubes in a pre-warmed heating block adjusted to 40°C for 10 min.
4. Mix by using a vortex mixer and centrifuge tubes at 500 g for 5 min.
5. Gently decant the supernatant into appropriate waste container.
6. Add 1 ml of wash solution (Vial 3 of the kit diluted 1:10) to each microcentrifuge tube.
7. Mix by using a vortex mixer.
8. Place the tubes in a pre-warmed heating block adjusted to 40°C for 10 min.
9. Mix by using a vortex mixer and centrifuge tubes at 500 g for 5 min.
10. Gently decant the supernatant.

3.5.2. DNA Staining

1. Add 0.5 ml of DNA-staining solution (Vial 4 of the kit diluted 1:10) to each of the four tubes.
2. Mix by using a vortex mixer. Transfer the four cell suspensions A, B, C, and D to four flow cytometry tubes.
3. Mix the samples by using a vortex mixer just before the transfer to minimize cell loss caused by adherence of cells to tube walls.
4. Leave tubes in the dark at 2–8°C for 2–3 h (*see* **Note 15**).

3.5.3. Analysis

Analyze samples on a flow cytometer using logarithmic scale FL1-H for probe fluorescence and linear scale FL3-H for DNA staining (*see* **Fig. 3**).

Propidium iodide staining allows that flow cytometric analysis is performed gating on G0/1-cells. The relative telomere length (RTL) value is calculated as the ratio between the telomere signal of each sample and the control cell (1301 cell line) with correction for the DNA index of G0/1 cells. This correction is performed to standardize the number of telomere ends per cell and thereby telomere length per chromosome.

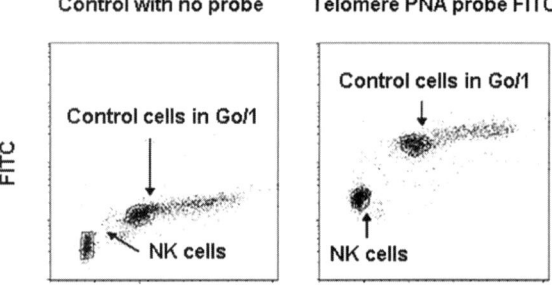

Fig. 3. Analysis of telomere length by Flow-FISH. Left panel: dot plot of FL1-height (Telomere PNA probe FITC) versus FL3-height (Propidium Iodide) of cells without Telomere PNA Probe. Gates are set around cells in the G0/1-phase for both sample cells (NK cells isolated from lymph node) and control cells (1301 cell line). Right panel: dot plot of FL1-height (Telomere PNA probe FITC) versus FL3-height (Propidium Iodide) of cells hybridized with Telomere PNA Probe/FITC. Gates are set as above. Statistical data on these cells, together with the DNA index of the cells determined by traditional DNA measurement, are then used for calculation of RTL of the sample cells compared to the control cells in the following way: RTL = (mean FL1 sample cells with probe − mean FL1 sample cells ADD × 100 without probe) × DNA index of control cells × 100 / (mean FL1 control cells with probe − mean FL1 sample cells with probe) DNA index sample cells. In the example, the figures were RTL=$(20.3 - 4.4) \times 2 \times 100 / (202.5 - 10.9) \times 1 = 16.6\%$ The above RTL value indicates that the average telomere fluorescence per chromosome/genome in the sample cells (normal human mononuclear cells isolated from blood) is about 16.6% of the telomere fluorescence per chromosome/genome in the control cells (1301 cell line).

4. Notes

1. Platelets can be present among PBMNC of the interface after centrifugation. An excess of platelets can interfere with NK cell functional assay or during culture. Contaminating platelets may be removed at this time.

   a. Add 0.5 ml of PBS and 50 µl of thrombin and then gently resuspend the pellet.
   b. Make up to 10 ml with PBS and mix thoroughly.
   c. Centrifuge cells very slowly (1 g) for 3 min; the platelets will clump and settle to the bottom.
   d. Draw off the supernatant (containing the cells) and place in another tube, leaving the aggregated platelets behind.
   Alternatively,
   e. Centrifuge cells (60 g) for 1 min to pellet the cells.

f. Pour off supernatant and resuspend the resulting cells in PBS.
g. Repeat two more times. One will notice that each time this is done, the resulting supernatant will become more clear as the platelets are being removed.

2. Use pre-cooled solution. This will prevent non-specific cell labeling.
3. When working with less than 10^7 cells, use same volumes as indicated.
4. We are aware that Miltenyi Biotec is preparing a new NK cell isolation kit, which is supposed to bypass the needing of anti-CD3 beads to avoid contamination. It should soon be commercially available.
5. We do not recommend enzymatic digestion for NK cell isolation from secondary lymphoid tissues. Enzymatic digestion with hyaluronidase and collagenase, usually required for cell isolation from other solid tissues (e.g., cancer specimens), does not improve the final amount of NK cells harvested.
6. PBS additioned with DNAse should be used during mechanic dissociation and, in particular, when cells are resuspended to be centrifuged. Centrifugation in the absence of DNase may result in extensive cell clumping after cell pellet resuspension.
7. Avoid the use of human γ-globulin when staining for CD16.
8. For an optimal PKH26 staining, all serum should be carefully eliminated from the cell suspension.
9. Depending on the available number of cells, the staining can be also performed on flow cytometry 5-ml tubes or in V-bottom 96-μwells (for fewer cells).
10. PKH26 is provided in ethanol solution: work fast and avoid evaporation.
11. Optimal expression of CD107a occurs between 3 and 6 h post stimulation.
12. To achieve a good reproducibility, it is important that the initial cell count for the control cells is accurate. If cells are counted in a Bürker chamber, we recommend that 64 fields should be counted.
13. 1301 cell line should be preferred as control cells as they have very long telomeres. 1301 is a sub-line of the EBVgenome negative T-cell leukemia line, CCRF-CEM.
14. Denaturation should be performed at minimum 80°C and maximum 84°C. Check the temperature of the heating block carefully. Denaturing temperatures below 75°C impair results seriously.
15. The 2–3 h DNA staining time in the protocol should be regarded as the minimum time required. If it is more convenient to stain the samples for longer (up to 24 h), then this can be done without affecting the results if samples are kept in the dark at 2–8°C.

References

1. Robertson, M. J., and J. Ritz. (1990). Biology and clinical relevance of human natural killer cells. *Blood* **76**:2421.
2. Campbell, J. J., S. Qin, D. Unutmaz, D. Soler, K. E. Murphy, M. R. Hodge, L. Wu, and E. C. Butcher. (2001). Unique subpopulations of CD56+ NK and NK-T

peripheral blood lymphocytes identified by chemokine receptor expression repertoire. *J. Immunol.* **166**:6477.
3. Jacobs, R., G. Hintzen, A. Kemper, K. Beul, S. Kempf, G. Behrens, K. W. Sykora, and R. E. Schmidt. (2001). CD56bright cells differ in their KIR repertoire and cytotoxic features from CD56dim NK cells. *Eur. J. Immunol.* **31**:3121.
4. Cooper, M. A., T. A. Fehniger, S. C. Turner, K. S. Chen, B. A. Ghaheri, T. Ghayur, W. E. Carson, and M. A. Caligiuri. (2001). Human natural killer cells: a unique innate immunoregulatory role for the CD56bright subset. *Blood* **97**:3146.
5. Fehniger, T. A., M. A. Cooper, G. J. Nuovo, M. Cella, F. Facchetti, M. Colonna, and M. A. Caligiuri. (2003). CD56bright natural killer cells are present in human lymph nodes and are activated by T cell-derived IL-2: a potential new link between adaptive and innate immunity. *Blood* **101**:3052.
6. Ferlazzo, G., D. Thomas, S.-L. Lin, K. Goodman, B. Morandi, W. A. Muller, A. Moretta, and C. Munz. (2003). The abundant NK cells in human secondary lymphoid tissues require activation to express killer cell Ig-like receptors and become cytolytic. *J. Immunol.* **172**:1455.
7. Westermann, J., and R. Pabst. (1992). Distribution of lymphocyte subsets and natural killer cells in the human body. *Clin. Invest.* **70**:539.
8. Trepel, F. (1974). Number and distribution of lymphocytes in man: a critical analysis. *Klin. Wochenschr.* **52**:511.
9. Freud, A. G., B. Becknell, S. Roychowdhury, H. C. Mao, A. K. Ferketich, G. J. Nuovo, T. L. Hughes, T. B. Marburger, J. Sung, R. A. Baiocchi, M. Guimond, and M. A. Caligiuri. (2005). A human CD34(+) subset resides in lymph nodes and differentiates into CD56bright natural killer cells. *Immunity* **22**:295–304.
10. Loza, M. J., and B. Perussia. (2004). The IL-12 signature: NK cell terminal CD56+high stage and effector functions. *J. Immunol.* **172**:88–96.
11. Mailliard, R. B., S. M. Alber, H. Shen, S. C. Watkins, J. M. Kirkwood, R. B. Herberman, and P. Kalinski. (2005). IL-18-induced CD83+CCR7+ NK helper cells. *J. Exp. Med.* **202**:941–953.
12. Hastie, N. D., M. Dempster, M. G. Dunlop, A. M. Thompson, D. K. Green, and R. C. Allshire. (1990). Telomere reduction in human colorectal carcinoma and with ageing. *Nature* **346**:866–868.
13. Figueroa, R., H. Lindenmaier, M. Hergenhahn, K. V. Nielsen, and P. Boukamp. (2000). Telomere erosion varies during *in vitro* aging of normal human fibroblasts from young and adult donors. *Cancer Res.* **60**:2770–2774.
14. Weng, N. P., B. L. Levine, C. H. June, and R. J. Hodes. (1995). Human naive and memory T lymphocytes differ in telomeric length and replicative potential. *Proc. Natl. Acad. Sci. U. S. A.* **92**:11091–11094.
15. Rufer, N., W. Dragowska, G. Thornbury, E. Roosnek, and P. M. Lansdorp. (1998). Telomere length dynamics in human lymphocyte subpopulations measured by flow cytometry. *Nat. Biotechnol.* **8**:743–747.

16. Serakinci, N., and J. Koch. (1999). Detection and sizing of telomeric repeat DNA in situ. A performance comparison of PRINS and PNA assays. *Nat. Biotechnol.* **17**:200–201.
17. Betts, M. R., J. M. Brenchley, D. A. Price, S. C. De Rosa, D. C. Douek, M. Roederer, and R. A. Koup. (2003). Sensitive and viable identification of antigen-specific CD8+ T cells by a flow cytometric assay for degranulation. *J. Immunol. Methods* **281**:65–78.
18. Penack, O., C. Gentilini, L. Fischer, A. M. Asemissen, C. Scheibenbogen, E. Thiel, and L. Uharek. (2005). CD56dimCD16neg cells are responsible for natural cytotoxicity against tumor targets. *Leukemia* **19**:835–840.
19. Hultdin, M., E. Grönlund, K.-F. Norrback, E. Eriksson-Lindström, T. Just, and G. Roos. (1998). Telomere analysis by fluorescence *in situ* hybridization and flow cytometry. *Nucleic Acids Res.* **26**:3651–3656.

13

Innate Immune Function of Eosinophils

From Antiparasite to Antitumor Cells

Fanny Legrand, Gaetane Woerly, Virginie Driss, and Monique Capron

Summary

Eosinophils are multifunctional leukocytes classically described as being involved in helminth parasitic infections and allergic diseases. Previously restricted to an exclusive role in the release of cytotoxic mediators, they are now also considered to be immunoregulatory cells and potential effectors in innate immune responses. Eosinophils are mainly found in tissues, so specific procedures are needed for their isolation from venous blood and for functional assays. Murine models are very useful for the dissection of eosinophil physiology in vivo. But murine eosinophils significantly differ from human ones. A complete understanding of eosinophil biology therefore requires comparative study of eosinophils from different mammalian species. We summarize here the main experimental protocols used to study human, mouse, and rat eosinophil biology. We focus on technical improvements of existing methods that optimize purification and in vitro functional studies of eosinophils.

Key Words: Eosinophil purification; hypodense; flow cytometry; activation markers; mediator release; cytotoxicity; heterogeneity; animal models; cell lines.

1. Introduction

Eosinophils have long been known to be involved in helminth parasitic infections and allergic diseases. Indeed, they have been characterized until recently by this functional duality. Today, eosinophils are considered rather as really multifunctional leukocytes, involved in inflammatory processes as well as in modulation of innate and adaptive immunity *(1)*. Eosinophils express membrane receptors involved in adaptive immune responses like

Fc receptors, class II major histocompatibility complex (MHC), or co-stimulatory molecules including CD86 and CD28 that allow interactions with lymphocytes and antigen-presenting cells *(2)*. They also display an array of receptors involved in innate immunity, including lectin-type receptors and protease-activated receptors; Toll-like receptor (TLR)-specific mRNAs have also been detected *(3)*. Triggering eosinophils by engagement of receptors for cytokines, immunoglobulins, or complement can lead to the secretion of numerous cytokines [interleukin (IL)-2, IL-4, IL-5, IL-10, IL-12, IL-13, IL-16, IL-18, and transforming growth factor (TGF)α/β], chemokines (RANTES, eotaxin), and lipid mediators (platelet-activating factor, leukotriene C4). We have demonstrated that there is a specific release of Th2 and Th1 immunoregulatory cytokines and cationic proteins, dependent on the nature of stimulus *(4,5)*. Furthermore, eosinophils store in their specific granules great quantities of cytokines and cationic proteins, known to be highly cytotoxic mediators *(6,7)*. Eosinophil granules contain a crystalloid core composed of major basic protein (MBP) and a matrix containing eosinophil cationic protein (ECP), eosinophil-derived neurotoxin (EDN), and eosinophil peroxidase (EPO) *(8)*. They can also generate reactive oxygen species (ROS). This high cytotoxic potential can lead to deleterious effects in inflammatory and allergic processes, but it can also be deployed effectively to combat parasitic targets. We have in particular demonstrated that eosinophils are able to kill schistosomula, the young larvae of the helminth parasite *Schistosoma mansoni*, through an IgE-dependent antibody-dependent cellular cytotoxicity (ADCC) mechanism *(9,10)*.

Eosinophils are able to integrate the different signals of their environment. They can respond to danger signals, including bacteria *(11)* or signals derived from stressed and necrotic cells *(12)*. Moreover, eosinophil numbers have been documented to be elevated in peripheral blood and/or to infiltrate the tissues in some malignant disorders *(13,14)*. Several studies have convincingly shown that tissue or blood eosinophilia was correlated with a significantly better prognosis *(15)*. A putative tumoricidal activity of eosinophils is supported by recent in vitro *(16)* and animal studies *(17,18)*. Not only efficient during helminthic infections, recent experimental results implicate the eosinophil cytotoxic potential in antitumor immunity. Eosinophils are also able to induce apoptosis of human tumor cells, and a new method has been developed to study this tumoricidal activity (submitted data).

Many arguments favor a primordial role for eosinophils in innate immune responses: the great storage potential of their granules, their prominent tissue localization at the interface with the external environment, their capacity for recognition and integration of danger signals, and their high and specific cytotoxic content, which can be rapidly and selectively mobilized through finely regulated degranulation pathways.

In this chapter, we describe methods that can be needed to functionally characterize eosinophils. It is clear that we still have much to learn about this remarkable type of cell.

2. Materials
2.1. Isolation Protocols

1. Dextran solution: 4.5% (w/v) dextran T500 (DxT500, Amersham Pharmacia Biotech, Uppsala, Sweden) in normal saline.
2. Tyrode's gel: 0.1% (w/v) D-Glucose (Merck, Darmstadt, Germany), 0.1% (w/v) $NaHCO_3$, 0.02% (w/v) KCl, 0.8% (w/v) NaCl, 0.005% (w/v) Na_2HPO_4, 0.1% (w/v) gelatin, pH 7.5–7.8. First dissolve all the components except the gelatin in distilled water (smaller volume than desired final solution). Dissolve then the gelatin in a small volume of distilled water while heating slightly (50–60°C). Maintain the prepared medium at a warm temperature and add the dissolved gelatin (avoid solidification of the gelatin). Complete with water to the desired volume and sterilize the solution while heating slightly by filtering successively through cellulose acetate filters of 0.45-μm and 0.22-μm pore size. This solution can be stored at 4°C for several months. Just before use, add 0.003% (w/v) DNase I grade II from bovine pancreas (Boehringer Mannheim, Mannheim, Germany) to avoid DNA aggregates.
3. Metrizamide (Sigma-Aldrich, St. Louis, MO): Prepare a stock solution of metrizamide 30% (w/v) in Tyrode's gel. Sterilize through a 0.22-μm filter and store at –20°C in the dark (*see* **Note 1**). This solution is stable for very long periods of time. Prepare working solutions by dilution in Tyrode's gel (as mentioned in **Table 1**). In order to be faster during the purification, we recommend preparing these gradients solutions 1 day before purification and storing them at 4°C.
4. Phosphate-buffered saline (PBS) 1×: always without calcium and magnesium.
5. Eosinophil purity: The cytocentrifuge (Shandon Cytospin 2) uses a unique sample chamber assembly that incorporates a plastic sample chamber, a filter card to act as a blotter, a microscope slide, and a unique stainless steel slide clip. Perform May-Grünwald Giemsa (MGG) coloration using RAL 555 (Rieux, France).
6. Percoll solution (d = 1.082 g/mL): In a graduated cylinder, add nine parts (v/v) of Percoll (d = 1.13 g/mL, Amersham Biosciences) to one part (v/v) of 9% (w/v) NaCl to have a solution adjusted to approximately 340 mOsm/kg H_2O. Transfer this 90% Percoll solution in a bottle and adjust density with physiological saline (0.9% NaCl). Calculate the required volume of normal saline (V) using the formula: $V = 1.7 \times V_o - V_{90}$ with V_o = volume of undiluted Percoll required; V_{90} = volume of total 90% Percoll (*see* **Note 2**). The solution is stable at 4°C for several weeks. As all the preparation should be made under a laminar flow hood in sterile conditions, filtering is not necessary.

Table 1
Preparation of Metrizamide Solutions

Number of gradients		2		4		6	
Total volume (mL)		28		56		84	
Prepared volume (mL)		30		60		90	
Percentage	Density (g/mL)	Stock (mL)	Gel (mL)	Stock (mL)	Gel (mL)	Stock (mL)	Gel (mL)
18	1.105	3.0	2.0	6.0	4.0	9.0	6.0
20	1.115	3.35	1.65	6.7	3.3	10.05	4.95
22	1.125	3.65	1.35	7.3	2.7	10.95	4.05
23	1.130	3.85	1.15	7.7	2.3	11.55	3.45
24	1.135	4.0	1.0	8.0	2.0	12.0	3.0
25	1.140	4.15	0.85	8.3	1.7	12.45	2.55
Total		22	8	44	16	66	24

Stock: solution of metrizamide 30%; gel: Tyrode's gel.

7. Spinal needle (Terumo Corp., Tokyo, Japan).
8. Lysis buffer (isotonic ammonium chloride solution): 155 mM NH$_4$Cl, 10 mM NaHCO$_3$, and 0.1 mM ethylenediaminetetraacetic acid (EDTA), pH 7.4. Sterile filtered through 0.45-µm and 0.22-µm filters. Stable for months at 4°C.
9. Anti-human CD16 and streptavidin-coated magnetic beads, CS and LD columns, and MACS system are from Miltenyi Biotec, Bergisch Gladbach, Germany.
10. Fraction V bovine serum albumin (BSA) (Eurobio, Les Ulis, France): 10% (w/v) stock solution in PBS, filter sterilized and stored at 4°C.
11. Biotinylated anti-mouse antibodies (anti-CD19, anti-CD90, and anti-CD8α) and anti-rat antibodies (anti-CD45R, anti-mononuclear phagocytes and anti-CD8α): Pharmingen, San Diego, CA.
12. Nylon filter (90 µm): PA 90/35 PW WH (SAATItech, Appiano Gentile, CO, Italy).

2.2. Cell Culture

1. The eosinophil culture medium (complete medium): RPMI 1640 supplemented with 10% heat-inactivated fetal calf serum (FCS), 25 mM HEPES, 2 mM L-glutamine, 10 mM sodium pyruvate, and 10 µg/mL of gentamycin. Use 100 IU/mL penicillin and 100 µg/mL streptomycin to replace gentamycin.
2. Recombinant human IL-5 (rhIL-5) (Peprotech, Rocky Hill, NJ): Reconstitute in water to a concentration of 100 µg/mL. Make small frozen aliquots. Prepare working solution at 5 µg/mL by dilution in complete medium. Store at 4°C. Stable for a few months.

3. Peroxidase cyanide buffer: EPO is specifically detected by a technique based on the resistance of EPO enzyme to cyanide (*see* **Note 3**). Dissolve one tablet (10 mg) of the peroxidase substrate 3,3′-diaminobenzidine tetrahydrochloride (DAB) (Sigma-Aldrich, St. Louis, MO) in 15 mL of 0.1 M Tris–HCl, pH 7.6, containing 0.4 mg/mL potassium cyanide. This solution is stored at 4°C for 1 month. Just before use, add 3 μL of 3% (w/w) H_2O_2 solution to 1 mL of this buffer for a single use.

2.3. Analysis of Specific Markers and Activation Markers

2.3.1. Flow Cytometry Analysis

1. Saponin (Sigma): 10% (w/v) stock solution in PBS, filter sterilized and stored at 4°C.
2. Paraformaldehyde (Sigma): Prepare a 4% (w/v) solution in PBS. In order to dissolve it, add concentrated NaOH to increase the pH (up to 10–11) until the solution is completely clear and then readjust the pH to 7.5 with concentrated HCl. The solution is aliquoted and can be stored for a long time at −20°C and for 1 month at 4°C.
3. Fluorescence-activated cell sorting (FACS) membrane buffer (FACS buffer): 1% (w/v) BSA, 0.05% NaN_3 in PBS.
4. FACS intracellular buffer (IC buffer): 1% (w/v) BSA, 0.1% (w/v) saponin in PBS.
5. Antibodies: anti-mouse FcγR (2.4.G2), anti-human MBP, and anti-human EPO mAbs are from Pharmingen. The anti-human ECP (EG2) is from Pharmacia.

2.3.2. Immunocytochemistry

1. Tris-buffered saline (TBS): 0.05 M Tris–HCl, 154 mM NaCl, pH 7.6.
2. Dako pen, the APAAP detection system (rabbit anti-mouse immunoglobulins and mouse APAAP complex), New Fuchsin kit, and Mayer's hematoxylin (ready-to-use aqueous solution) are from Dako (Glostrup, Denmark).
3. Immumount (Shandon, Pittsburgh, PA).
4. Stratalinker UV 2400 (Stratagene, La Jolla,CA).
5. Human serum albumin (HSA) (Sigma): 10% (w/v) stock solution in PBS, filter sterilized and stored at 4°C.
6. Fluoromount G (Southern Biotechnology Assoc., Birmingham, AL).

2.4. Functional Tests

2.4.1. Mediator Release Assays by Chemiluminescence

1. White 96-well cell culture plate with flat bottom and individual lid (Nunc, Roskilde, Denmark).
2. RPMI 1640 without phenol red (Cambrex Bio Science, Rockland, ME) (*see* **Note 4**).

3. Luminol (Sigma) is dissolved at 5 mg/mL in dimethyl sulfoxide (DMSO) and stored in single use aliquots: volume of 15 µL for measurement of ROS and 80 µL for measurement of EPO. Aliquots are stored at –20°C in the dark.
4. ROS buffer: 15 mM Tris–HCl, 150 mM NaCl, pH 7.4.
5. EPO buffer: 15 mM Tris–HCl, pH 6.0.
6. D-Luciferin (Roche Diagnostics, Basel, Switzerland): 0.16 mM in EPO buffer. Twenty-microliter aliquots are stored at –20°C for single use (*see* **Note 5**).
7. Stimulus: rhIL-5 at a final concentration of 20 ng/mL.
8. Luminometer: Victor^{2TM} Wallac (PerkinElmer, Shelton, CT) allows the measurement of the intensity of light released by a chemical reaction (chemiluminescence). Data analysis is performed with Wallac 1420^{TM} software.

2.4.2. Cytotoxicity Assays

1. Hank's medium and Modified Eagle's Medium (MEM) (Cambrex).
2. 96-well flat-bottomed microtiter plates (Nunc, Roskilde, Denmark).
3. PKH-26 red fluorescent cell linker kit (Sigma): Store at room temperature or refrigerate. Keep dye stock tightly capped to prevent evaporation.
4. Fluorescein isothiocyanate (FITC)-conjugated annexin V (Pharmingen).
5. Binding buffer (1×): 10 mM HEPES/NaOH, pH 7.4, 140 mM NaCl, 2.5 mM CaCl$_2$. Sterilize through 0.45-µm filter. Store at 4°C.
6. Colorectal adenocarcinoma cell line, Colo-205 (ATCC LGC Promochem, Middlesex, UK).

3. Methods

Eosinophils, which are mainly found in tissues, only represent 1–5% of circulating leukocytes in normal donors. Specific experimental procedures are thus needed for eosinophil purification from peripheral blood. Two main protocols of purification can be used to isolate eosinophils. The first to be described, using metrizamide discontinuous gradients, is based on the specific cell density of eosinophils and allows one to obtain separately eosinophils with normal density (normodense) and with lower density (hypodense), the latter representing the activated phenotype. The second technique, more efficient regarding the eosinophil purity and the most commonly used today, consists of blood centrifugation on a Percoll gradient followed by a negative immunomagnetic selection by anti-CD16-coated microbeads and the MACS system. Even if eosinophils are end-stage cells, they can be maintained overnight in culture. This rest, after the stress of purification, allows turnover of membrane receptors. Eosinophils are characterized by their phenotypical and functional heterogeneity, strongly dependent upon eosinophil donors. Studying phenotype is possible by immunofluorescent staining using two efficient and complementary techniques like flow cytometry (FCM) and immunocytochemistry

(ICC). Indeed, techniques such as enzymatic or fluorescent ICC analysis can be used to confirm the membrane or intracellular expression of the proteins of interest observed by FCM analysis. Moreover, ICC on cytospin preparations can complement results obtained by FCM by localizing the storage sites of the molecules studied (granular, cytoplasmic, or membrane-like staining).

To understand the effector roles of eosinophils, it is necessary to characterize the functional properties of these cells. We describe here two functional assays: mediator release and cytotoxicity assays. Release of several cytotoxic mediators, either neo-synthesized (e.g., ROS) or preformed as granule cationic proteins (e.g., EPO), can be measured by chemiluminescence. After activation, release of ROS generated by NADPH-oxidase can be measured by luminol-dependent chemiluminescence. The ROS produced by eosinophil activation oxidize luminol, and this oxidation results in an excited state of luminol, which fluoresces and releases photons as it decays to a lower energy level. Released photons are then collected by a luminometer (*see* **Note 6**). EPO, which constitutes ~25% of the total protein mass of specific granules, catalyzes the oxidation reactions to form highly ROS and reactive nitrogen metabolites. As EPO is released during degranulation by activated eosinophils, it can be considered to be a marker of eosinophil activation and degranulation. EPO release can also be measured by chemiluminescence. The cytotoxicity assays presented here are based on protocols measuring in vitro beneficial cytotoxic properties of eosinophils toward two different targets, helminthic parasite larvae and tumor cell lines. First, parasite killing by purified human eosinophils is an IgE-dependent ADCC mechanism and seems restricted to hypodense eosinophils present only in highly hypereosinophilic patients. Concerning antitumor activity of eosinophils, several in vitro arguments allow one to postulate that eosinophils are also able to use effectively their cytotoxic potential toward human tumor cell lines by inducing their apoptosis (submitted data).

To complete our description of methods for eosinophil study, we describe here isolation protocols for mouse and rat eosinophil purification. The main experimental procedures detailed for human eosinophils—immunostaining (FCM and ICC) and mediator release and cytotoxicity experiments toward parasite targets—can be used with murine eosinophils. Protocol modifications are then mentioned in notes if needed. Although murine models are advantageous to study eosinophil physiology in vivo, several differences observed between murine and human eosinophils have to be seriously considered *(19)*. First, mouse eosinophils are poorly granulated, in contrast to rat and human eosinophils (see **Fig. 1c** and **Fig. 2**), and therefore granule release is more

Table 2
Different Expression of Ig Receptors in Human, Mouse, and Rat Eosinophils (18,19)

Receptors	Ligand	Human	Mouse	Rat
FcεRI	IgE	Yes[a]	No[b]	Yes
CD89	IgA	Yes	No	mRNA
pIgR	IgA/IgM	Yes	No	No
FcαµR	IgA/IgM	No	mRNA	No
Tf-R	Tf/IgA	Yes (Tf-R1)	Yes (Tf-R1, -R2)	No

Tf, transferrin.
[a] Only expressed on activated cells.
[b] In human IL-5/FcεRIα transgenic mouse, eosinophils express FcεRI with a distribution similar to human cells.

difficult to induce. Second, differences in IgE and IgA membrane receptor expression have been described, as summarized in **Table 2** *(20)*.

3.1. Isolation Protocols

3.1.1. Purification of Human Eosinophils

3.1.1.1. BLOOD SAMPLING

1. Collect human venous blood from subjects in vacutainers containing heparin or sodium citrate as anticoagulant. Be careful of taking a minimum of 80–100 mL of venous blood if it is a normal donor in order to have enough purified eosinophils. The donors should not have received any corticosteroids or histamine antagonists at the time of the study. It is better they not received such therapy in the preceding weeks.
2. Maintain fresh blood at room temperature and use it as soon as possible (maximum 3–4 h).

3.1.1.2. METRIZAMIDE DISCONTINUOUS GRADIENTS

1. Maintain all the solutions used before erythrocyte lysis at room temperature. Perform dextran sedimentation by adding five volumes of heparinized blood to one volume of the dextran solution under a laminar flow hood at room temperature. When the level of sedimentation is stabilized, collect the leukocyte rich supernatant, wash it in Tyrode's gel, and centrifuge at $500 \times g$ for 8 min (*see* **Note 7**).
2. Remove the contaminating erythrocytes by hypotonic lysis by adding one volume of 0.2% (w/v) ice-cold NaCl solution to the cell pellet for 10 s while mixing. Restore the physiological osmolarity by addition of one volume of 1.6% (w/v) NaCl

solution. Centrifuge immediately at 4°C for 8 min at 350 × g. Repeat once the lysis step.
3. Wash the cells with Tyrode's gel containing DNase (see **Note 8**). Centrifuge at 4°C for 8 min at 700 × g. Resuspend the preparation of mixed leukocytes at 25 × 10^6 to 50 × 10^6 cells per mL in Tyrode's gel.
4. Prepare the discontinuous metrizamide gradient (this step can be performed during dextran sedimentation) by using metrizamide solutions prepared 1 day before as mentioned in **Subheading 2.1., step 3**. With a transfer pipet carefully layer 2 mL of each metrizamide solution (from 1.140 to 1.105 g/mL density) in a 15 mL tube (see **Note 9**) and then layer on the top 2 mL of leukocytes suspension in Tyrode's gel. The details of the metrizamide gradient are presented in the table in **Fig. 1**.
5. Centrifuge at 1200 × g for 45 min at 20°C (without brake) and recover the different layers of the metrizamide gradient. Wash the cell suspensions three times with PBS [centrifugation respectively at 700 × g (see **Note 10**), 550 × g, and 450 × g].
6. Determine the eosinophil purity: cytocentrifuge the cell suspension for 2 min at 10 × g (prepare a dilution in PBS to have 10^5 cells/100 µL/cytospin) and then stain cytospin with RAL 555 coloration. Typical results are presented in **Fig. 1**. One can see morphological changes such as cytoplasmic vacuoles and reduced numbers and size of granules in activated eosinophils, called "hypodense." We recommend counting a minimum of 300 cells per cytospin for a good estimation. The contaminating cells are mainly neutrophils (see **Note 11**).

3.1.1.3. CD16-Coated Beads and MACS System

1. The first step consists of separation of polynuclear and mononuclear cells using a Percoll gradient (see **Note 12**). Distribute fresh blood in 50-mL tubes (maximum 15 mL per tube) and dilute with PBS (1:1 v/v). After mixing, add gently 15 mL Percoll solution (d = 1.082 g/mL) at the bottom of the tube using a spinal needle and centrifuge at room temperature, 700 × g for 20 min without brake.
2. After centrifugation, the pellet contains red blood cells and granulocytes and the thin layer at the interface of Percoll and plasma contains mononuclear cells. After removal of excess plasma, collect carefully the peripheral blood mononuclear cell (PBMC) layer. Remove all the Percoll and transfer the pellet to new 50-mL tubes (no more than 10 mL pellet volume per tube) (see **Note 13**).
3. Perform a hypotonic lysis to remove the contaminating erythrocytes by adding 40 mL of ice-cold lysis buffer directly onto the remaining pellet, mix vigorously, and leave on ice for 15 min. After centrifugation at 500 × g, 4°C for 15 min, pool all the pellets in one tube and repeat a second lysis step for 10 min on ice (see **Note 14**).
4. Resuspend the granulocytes in cold PBS, count them, and determine the percentage of neutrophils and eosinophils by staining cytocentrifuged preparation with RAL 555 coloration. We recommend counting a minimum of 300 cells per cytospin for a good estimation.

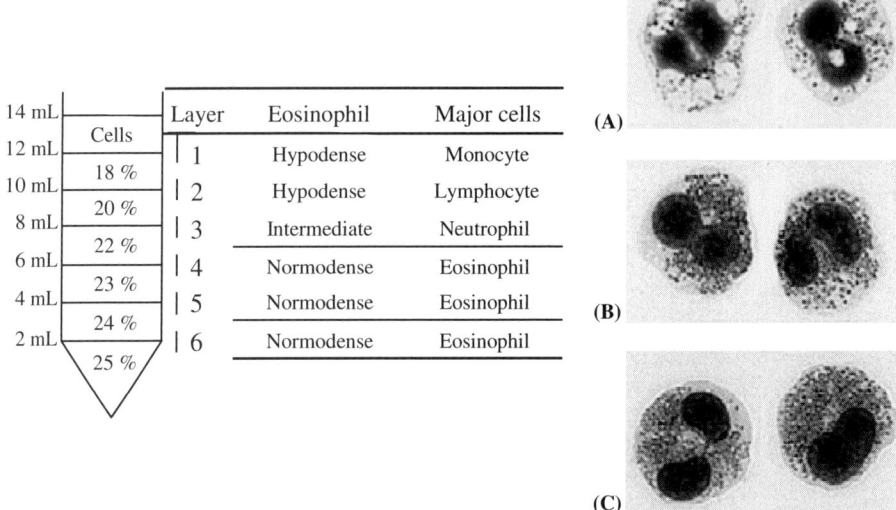

Fig. 1. Human eosinophils purified on metrizamide gradients. Cells are centrifuged on microscope slides and stained with RAL 555 as described in **Subheading 3**. Notice the bilobated nucleus and numerous orange granules. **(A)** Layer 2: hypodense eosinophils, **(B)** Layer 3: eosinophils with intermediate density, **(C)** Layer 4: normodense eosinophils (× 100). **(D)** Representation of the metrizamide gradient layers and cellular composition.

5. The second part of the purification procedure consists of a negative selection of eosinophils using the MACS system. Resuspend granulocytes in 500 µL ice-cold filtered PBS–BSA 0.5% and incubate with anti-CD16-coated magnetic beads (80 µL beads for 100×10^6 polynuclear neutrophils) for 30 min in the fridge (less cold than on ice) with intermittent gently mixing (*see* **Note 15**).
6. Equilibrate the column with cold PBS–BSA 0.5% (it is recommended to perform this step during incubation with the magnetic beads). For CS column, proceed as follows: Flush gently the column from the side with 10 mL of buffer to remove air bubbles, and then add 15 mL buffer at the top. Just before adding the granulocyte suspension (in 4 mL), add 3 mL of ice-cold PBS–BSA 0.5%. Eosinophils are eluted following passage through the field of a permanent magnet. After reaching 25 mL of eluate, flush the column again from the side and continue to collect eluate until 50 mL. Centrifuge the eluted cells at $350 \times g$, 4°C, and count them in PBS.
7. Determine the purity of eosinophils on cytocentrifuged preparations after RAL 555 coloration. A purity >98% is usually obtained. One really needs to be careful during all the procedure to avoid contaminating cells. Contaminating cells are mainly lymphocytes (*see* **Note 16**).

3.1.2. Purification of Mouse Eosinophils

Mouse eosinophils are purified from IL-5 transgenic mice by negative selection using the MACS system *(21)*. These transgenic mice have the advantage to exhibit massive eosinophilia in blood and different organs like spleen, bone marrow, and peritoneal exudate.

1. To obtain peritoneal cells, flush the peritoneal cavity with ice-cold PBS. Perform gentle dissociation of the spleen in ice-cold PBS to obtain splenocytes. Isolate bone marrow cells from femur and tibia of mice by flushing the bone marrow cavities with ice-cold PBS.
2. Perform a hypotonic lysis to remove erythrocytes by adding ice-cold lysis buffer on the pellet (5 mL for spleen cells and 2–3 mL for bone marrow cells). Mix and leave for 10 min (spleen cells) or 5 min (bone marrow cells) on ice (*see* **Note 17**). Centrifuge for 10 min at 350 × g, 4°C.
3. Remove aggregates by filtering the cell suspensions onto a 90-µm nylon filter. Place a piece of sterile filter on the top of a 50-mL tube. Prewet the filter with two drops of PBS and then gently layer the cells and let them go through. Centrifuge for 10 min at 350 × g, 4°C.
4. Resuspend the cells at 1 × 10^8 cells/mL in cold PBS containing 0.5% BSA and 2 mM EDTA, and incubate for 30 min at 4°C with biotinylated anti-CD19, anti-CD90 (Thy1.2), and anti-CD8α (Ly-2). These antibodies will remove B cells, macrophages, and T lymphocytes, respectively.
5. After washing with PBS, resuspend the cells at 1 × 10^8 cells/mL in PBS + 2 mM EDTA and incubate for 15 min at 4°C with streptavidin magnetic beads.
6. Equilibrate the LD column with cold PBS + 0.5% BSA + 2 mM EDTA.
7. Collect the eosinophils by passage through the field of a permanent magnet. Collect the eluate and centrifuge at 350 × g, 4°C for 10 min.
8. Count the cells and estimate the cell purity on cytospin preparation (*see* **Subheading 3.1.1.**). Typical results are presented in **Fig. 2**. Notice the morphological changes of mouse eosinophils compared with human eosinophils, such as low granularity and circular nucleus. Cell purity is usually 95–97%.

3.1.3. Purification of Rat Eosinophils

Rat eosinophils are obtained from the peritoneal cavity of LouM rats by negative selection using the MACS system *(22)*. In order to induce a cell influx in the peritoneal cavity, we injected rats intraperitoneally with 30 mL of 0.9% (v/w) NaCl 48 h before killing.

1. To obtain peritoneal cells, flush the peritoneal cavity with 40 mL of ice-cold PBS.
2. Remove aggregates by filtering the cell suspensions onto a 90-µm nylon filter, as described in **Subheading 3.1.2**.

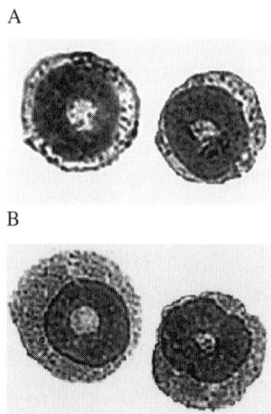

Fig. 2. Murine eosinophils stained with RAL 555 as described in **Subheading 3**. Note the circular nucleus of a mouse (**A**) and a rat (**B**) eosinophil. Notice the low granularity (dark spots) in mouse eosinophils compared with human and rat ones (× 100).

3. Perform a hypotonic lysis to remove erythrocytes by adding 5 mL ice-cold lysis buffer on the pellet, mix and leave for 5 min on ice. Centrifuge for 10 min at 350 × g, 4°C.
4. Resuspend the cells at 2 × 10^6 cells/mL in culture medium and culture for 90 min at 37°C in order to allow macrophages to adhere to the surface of the culture flask.
5. Collect supernatants and centrifuge at 350 × g, 4°C for 10 min.
6. Resuspend the cells at 1 × 10^8 cells/mL in PBS containing 0.5% BSA and 2 mM EDTA.
7. Incubate the cells for 30 min at 4°C with biotinylated anti-CD45R, anti-mononuclear phagocytes (1C7), and anti-CD8α. These antibodies will remove B cells, macrophages, and T cells.
8. After washing with PBS, resuspend the cells at 1 × 10^8 cells/mL in PBS + 2 mM EDTA and incubate for 15 min at 4°C with streptavidin magnetic beads.
9. Collect eosinophils, count them, and estimate the cell purity as described in **Subheading 3.1.2., steps 7** and **8**. The cell morphology of rat eosinophils is presented in **Fig. 2**. Cell purity obtained with rat eosinophils is usually 85–90% (*see* **Note 18**).

3.2. Culture Conditions

3.2.1. Eosinophils

1. Purified human eosinophils can be cultured overnight at 2 × 10^6 per mL in complete medium. Increased adhesion of activated eosinophils and mortality of these end-stage cells explain a usual output of 50–75% (*see* **Note 19**).

2. IL-5 (1–2.5 ng/mL) can be added to the culture medium to improve eosinophil survival. Higher concentrations of IL-5 could activate eosinophils and induce apoptosis (not recommended).
3. As the purification protocol can generate some cellular stress and increase the basal activation state of eosinophils, overnight culture with complete medium is recommended and allows turnover of membrane receptors for phenotype study and eosinophils to rest before functional assays.
4. Purified mouse and rat eosinophils are cultured up to 4 days in culture medium supplemented with 2.5 ng/mL rhIL-5.

3.2.2. Eosinophilic Cell Lines

Three eosinophilic cell lines are available:

1. YY-1: The YY-1 cell line was kindly provided by Dr. J. Chihara (Akita University, Akita, Japan) (23). This eosinophilic cell line is derived from the promyelocytic leukemia cell line HL-60. Cells are maintained in culture in complete medium and induced to further differentiate into eosinophil-like cells using 0.3 mM butyric acid as shown in **Fig. 3**.
2. AML14.3D10: The AML14.3D10 cell line was kindly provided by Dr. Cassandra Paul (Wright State University, Dayton, OH) (24). This cell line is a fully differentiated eosinophilic myelocyte subline of the AML14 cell line that continues to proliferate and maintain a differentiated eosinophil phenotype without cytokine supplementation. It displays many of the characteristics of mature peripheral blood eosinophils, including constitutive expression of the granule cationic proteins (MBP, EPO, EDN, and ECP) and Charcot–Leyden crystal (CLC) protein as shown in **Fig. 3**. AML14.3D10 cells are maintained in culture in complete medium containing 5×10^{-5} M β-mercaptoethanol.
3. Eol-1: The Eol-1 cell line was kindly provided by Dr. J. Chihara (Akita, Japan) (25). This cell line is derived from the blood of a 33-year-old man with acute eosinophilic leukemia following hypereosinophilic syndrome. Although the Eol-1 cells do not show visible granules, they express ECP and the fusion protein FIP1L1–PDGFRα, which has been recently shown to be associated to the myeloproliferative form of the hypereosinophilic syndrome and to chronic eosinophilic leukemia (26).
4. YY-1 and AML14.3D10 cell lines are cultured at 37°C in 5% CO_2, up to a maximum density of 10^6 cells/mL. Cultures are diluted in fresh medium twice a week. Fresh cultures are started from frozen stocks after a maximum of 30 passages to minimize genetic drift and phenotypic changes. The Eol-1 cell line is maintained in culture in complete medium at 37°C in 5% CO_2 and is diluted in fresh medium twice a week.

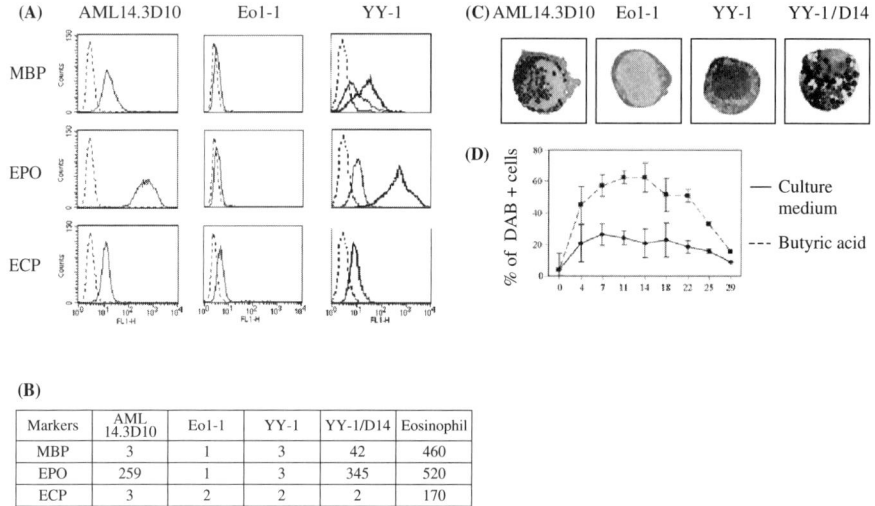

Fig. 3. Analysis by flow cytometry (**A, B**) and by immunocytochemistry (**C, D**) of cationic proteins contained in the eosinophilic cell lines, AML14.3D10, Eol-1, and YY-1. (**A**) Fluorescein isothiocyanate (FITC)-conjugated anti-cationic protein (thin line) or isotype-matched antibody (dashed line). Staining obtained with YY-1 cells differentiated for 14 days under butyric acid (YY-1/D14) is represented in bold line. (**B**) Mean fluorescence intensity values of staining obtained for the cationic proteins in the eosinophilic cell lines in comparison with blood peripheral eosinophils. (**C**) Eosinophil peroxidase (EPO) detection using 3,3´-diaminobenzidine tetrahydrochloride (DAB) staining of cytospin (× 100). (**D**) Kinetic of YY-1 differentiation by butyric acid as determined by the percentage of DAB+ cells.

3.3. Analysis of Specific Markers and Activation Markers

3.3.1. Flow Cytometry Analysis

Eosinophils are highly granular cells. Forward scatted/side scatter (FSC/SSC) criteria are represented by a dot plot in **Fig. 4A**.

1. Membrane staining: To avoid the turnover of membrane receptors, perform the whole experiment on ice (*see* **Note 20**). After two washes in PBS, resuspend the eosinophils at 4×10^6/mL in FACS buffer. Add 50 µL of the cell suspension in round-bottom 96-well plates (2×10^5 cells/well). For direct staining, incubate with the labeled antibody 30–45 min in the dark and then wash the cells twice with FACS buffer (200 µL/well) (*see* **Note 21**). At the end, resuspend the cellular pellet in 100 µL PBS, transfer samples in appropriate FACS tubes, and add 100 µL of

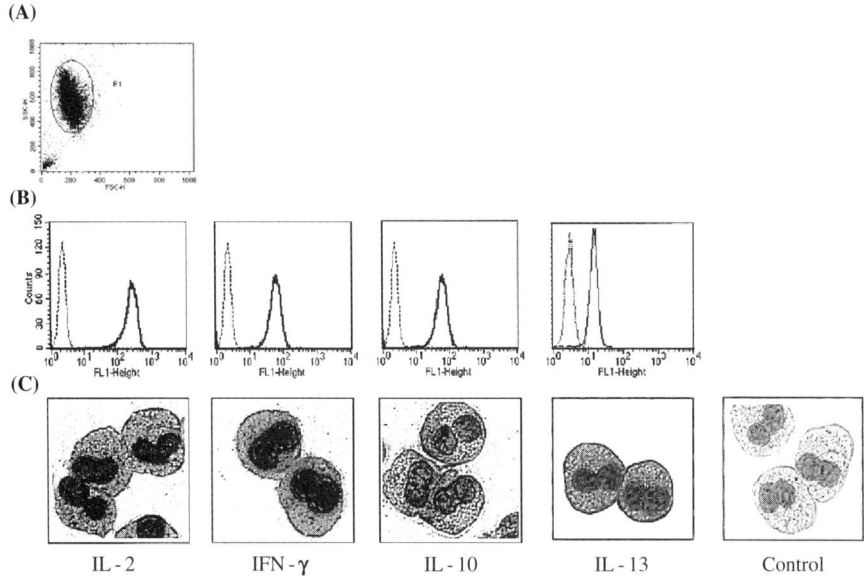

Fig. 4. Flow cytometry and immunocytochemistry (ICC) analysis of eosinophils. (**A**) Dot plot representation of freshly purified eosinophils. (**B, C**) Intracellular staining of Th1 and Th2 cytokines analyzed by immunofluorescence cytometry and ICC. (**B**) Fluorescein isothiocyanate (FITC)-conjugated anti-cytokine (bold line) or isotype-matched antibody (dashed line). Dot plot and histograms are obtained using a FACSCalibur equipped with the CellQuest software. (**C**) ICC staining on cytospin eosinophils revealed using APAAP detection system and New Fuchsin coloration (× 100). (**B**) and (**C**) are reproduced from **ref. *(2)*** (IL-2, IFN-γ, IL-10) and **ref. *(5)*** (IL-13).

FACS buffer. Staining with directly-labeled antibody is highly recommended and preferred to indirect staining (*see* **Note 22**). For indirect staining, incubate the cells with the primary antibody for 30 min, wash twice with FACS buffer (200 μL/well), and then add the secondary antibody for 20 min in the dark. If necessary, non-specific binding of secondary antibody can be previously blocked by the addition of serum (10%) of the same species directly on the cell pellet during 10 min at 4°C before incubation with the labeled antibody. Analyze the samples within 2 h (*see* **Note 22**).

2. Intracellular staining: After fixation of the cells, the experiment can be performed at room temperature. To fix the cells, resuspend eosinophils at 4×10^6/mL in PBS and then add the same volume of 4% paraformaldehyde for 10 min at 4°C. After washing in PBS and centrifugation for 7 min at $350 \times g$, permeabilize the cells (4×10^6/mL) for 10 min in IC buffer. Add directly 50 μL (2×10^5) of the cell suspension

to round-bottom 96-well plates. For direct staining, first block non-specific binding by incubating the cells for 10 min with 10% of serum from the species of your antibody and then add the labeled antibody for 30 min in the dark. For indirect staining, incubate the cells with the primary antibody for 30 min, wash twice with IC buffer, block non-specific binding with serum (10%), and incubate with the secondary antibody for 20 min in the dark (*see* **Note 23**). Wash the cells twice with IC buffer (200 μL) and once with PBS in order to remove saponin. Resuspend the pellet in 200 μL PBS + 0.5% BSA and analyze within 2 h (*see* **Note 24**).
3. Acquisition step: A total of 10^4 events are usually acquired for each sample. Analyze on a FACSCalibur^{TM} using the CellQuest^{TM} software (Becton Dickinson, Mountain View, CA) (*see* **Note 25** and **Fig. 4**).

3.3.2. Immunocytochemistry

For the two procedures described below, perform cytospins from freshly purified eosinophils as described above or from eosinophils cultured overnight in complete medium with rhIL-5 (2.5 ng/mL).

3.3.2.1. Enzymatic Immunocytochemistry

The cytospin preparations presented in **Fig. 4** are processed for using the APAAP method, followed by counterstaining with hematoxylin.

1. Encircle eosinophil spot with Dako pen to provide a barrier to liquids. Perform each step of incubation by adding around 100 μL of solution on the spot, and after each incubation or washing step, tap off excess liquid (*see* **Note 26**).
2. Fix in cold acetone/methanol (1:1 v/v) for 10 min and rehydrate in TBS for 10 min (*see* **Note 27**).
3. each incubation step, wash cytospins for 3 × 10 min in TBS containing 0.1% BSA (*see* **Notes 28** and **29**).
4. After blocking unspecific sites with TBS + 3% BSA for 30 min, incubate cytospins with unlabeled mouse anti-human cytokine or isotype control monoclonal antibodies at 40 μg/mL in TBS + 3% BSA overnight at 4°C.
5. Incubate the slides with rabbit anti-mouse immunoglobulins (1:25) in TBS + 3% BSA for 1 h at room temperature, followed by incubation with the APAAP complex (1:40) for 1 h (*see* **Note 30**).
6. After an additional wash for 2 × 10 min in TBS, the reaction is developed with New Fuchsin substrate (*see* **Note 31**).
7. Counterstain slides with Mayer's hematoxylin for 4 min (nuclear blue staining) and apply coverslips with Immumount.

3.3.2.2. Immunofluorescence

1. After **step 1** (*see* **Subheading 3.3.2.1.**), fix eosinophil cytospins in 4% paraformaldehyde for 10 min and rehydrate in 0.05 *M* PBS, pH 7.4 (*see* **Note 32**).

2. Quench endogenous fluorescence by exposition to UV (3 × 2 min at 0.12 joules) using a Stratalinker, followed by incubation with 0.1 M glycine for 4 min and incubation with 50 mM NH$_4$Cl pH 7.4, for 15 min. After each incubation step, wash cytospins for 10 min in PBS under soft agitation (*see* **Note 33**).
3. To prevent non-specific binding, incubate the slides for 30 min with PBS + 3% BSA supplemented with 5% HSA. After one additional wash in PBS, incubate cytospins with primary antibodies overnight in the fridge in a wet chamber.
4. After washing 2 × 10 min in PBS + 0.1% BSA, incubate the slides with secondary antibodies in PBS + 3% BSA + 5% HSA for 2 h and then with streptavidin–Texas red in PBS + 3% BSA for 1 h.
5. Perform a final step of washes with PBS + 0.1% BSA + 0.02% Tween 20 for 2 × 10 min and with PBS for 10 min. Cytospins are mounted with Fluoromount G and analyzed by fluorescence microscopy using an oil immersion objective. Texas red fluorophores require excitation at 543 nm; select the fluorescence emissions wavelengths pass bands between 581 and 621 nm in accordance with the emission spectral analysis.

3.4. Functional Tests

3.4.1. Mediator Release Assays by Chemiluminescence

3.4.1.1. Ros Production

1. Pre-incubate purified eosinophils (3 × 10^5 cells per well) in 25 µL RPMI 1640 without phenol red in white 96-well plate for 30 min at 37°C in 5% CO_2 (*see* **Note 34**).
2. Program the luminometer and prewarm it at 37°C.
3. Five minutes prior measurement, prepare luminol solution (0.015 mg/mL) with one 15-µL aliquot in 3 mL ROS buffer.
4. Add 25 µL of stimulus solution to the 25 µL of eosinophil suspension immediately before running measurements in the Victor^{2TM} Wallac luminometer. Measure ROS production immediately by the addition of 50 µL of luminol (in injector 1) (*see* **Note 35**). Perform kinetics at 37°C over 60-min time period, and for each time-point, count chemiluminescence for 5 s. Results are expressed in counts per second (CPS).

3.4.1.2. Measurement of EPO for Assessment of Degranulation

1. Incubate purified eosinophils (2 × 10^5) in 50 µL RPMI 1640 without phenol red either alone or with 50 µL stimulus (e.g., rhIL-5 at a final concentration of 20 ng/mL) in 96-well plates at 37°C in 5% CO_2 (final volume of 100 µL).
2. After 2 h stimulation, centrifuge the plate 10 min at 300 × g and transfer 50 µL of supernatant in a white 96-well plate. Keep the plate on ice.
3. Prepare the following solutions and keep them on ice in the dark:

a. Luminol (80-µL aliquot) and D-luciferin (20-µL aliquot) in 4 mL EPO buffer.
b. H_2O_2 solution: 15 µL of 30% (w/w) H_2O_2 solution (stored in the dark at 4°C) in 3 mL EPO buffer.

These solutions should be used immediately and cannot be stored.
4. Program the luminometer.
5. Add the luminol/D-luciferin solution in injector 1 and H_2O_2 solution in injector 2 (see **Note 36**). Start running the protocol: 100 µL luminol/D-luciferin and 50 µL H_2O_2 will be added per well (see **Note 37**) to the supernatants. Only one measure is taken and the results are expressed in CPS.

3.4.2. Cytotoxicity Assays

3.4.2.1. Cytotoxic Activity of Eosinophils Toward Parasitic Targets

1. *Schistosoma mansoni* schistosomula are the targets of the cytotoxicity assays and their obtention is based on the transformation of the infective larvae, the cercariae, into schistosomula after skin penetration, which is enhanced by light illumination. Schistosomula are prepared according to the skin-penetration procedure: under sterile conditions, shave with caution the abdomen of an anesthetized normal albino OF1 mouse and cut off the shaved skin. Degrease the inner leaflet of this skin sample with a gauze to facilitate the transpenetration of the cercariae, the invading form of *S. mansoni*. Fill the bottom part of a glass collector with Hank's medium. Place then the thin skin sample on the top of the collector (without air bubbles underneath) and assemble the upper part of the chamber. Fix the system with pliers and put the cercaria suspension on the skin. Protect the top of the collector from the light to allow cercariae to be attracted by the light at the bottom and to cross the skin barrier. Incubate the bottom collector for 2 h in a water bath at 37°C. During skin penetration, cercariae mature to become larvae called schistosomula. After separation of the two parts of the collector, recover then the larvae-containing medium and centrifuge it at $100 \times g$ for 3 min. Keep schistosomula overnight in 96-well flat-bottomed plate at 37°C, 5% CO_2, in MEM with 1% normal serum before use in the cytotoxicity assay. They can be kept longer in medium culture. Only schistosomula obtained 3 h after skin penetration are susceptible targets for cytotoxicity (see **Note 38**).
2. Because eosinophils need to be activated by antibodies to become cytotoxic in this ADCC procedure, use unheated serum samples from *S. mansoni*-infected patients (IgE-rich) as a source of IgE antibodies (see **Note 39**).
3. For IgE-mediated eosinophil-dependent cytotoxicity, add 50 µL of the schistosomula suspension containing 50 targets (1000/mL) in MEM supplemented with 1% heat-inactivated FCS in flat-bottomed microtiter plates. Then add 50 µL of unheated serum from *S. mansoni*-infected patients or normal subjects as controls at a final dilution of 1:32 together with 100 µL of effector cells at an effector/target ratio of 5000/1 (see **Note 40**) (*9*).

Innate Immune Function of Eosinophils

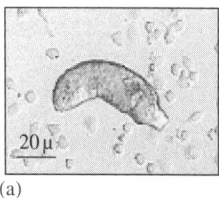
(a)

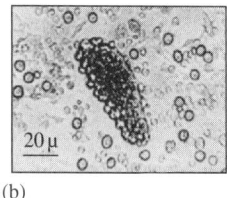
(b)

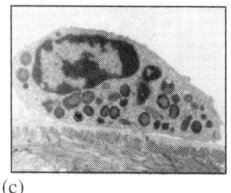
(c)

Fig. 5. Eosinophil cytotoxicity toward *Schistosoma mansoni* schistosomula. (**A**) Living schistosomule. (**B**) After 48-h coculture, eosinophils firmly adhered to schistosomula. Photographs by light microscopy (× 100). (**C**) Adhesion of an hypodense eosinophil to schistosomula by electron microscopy.

4. Evaluate the percentage of schistosomula with adherent eosinophils and the percentage of cytotoxicity (dead schistosomula) by optical microscopic examination after 24- and 48-h culture at 37°C, 5% CO_2. After 24 h, eosinophil adherence to schistosomula can be easily observed and quantified. After 48 h, the percentage of dead schistosomula can be counted (*see* **Fig. 5**). No additional staining is required to differentiate dead or alive schistosomula, because live parasites are transparent and dead larvae are becoming black and are surrounded by numerous adherent eosinophils.

3.4.2.2. Cytotoxic Activity of Eosinophils Toward Tumor Cell Lines

1. Determine eosinophil-mediated cytotoxicity against tumor cell lines by using PKH-26 and Annexin V-FITC stainings (*see* **Note 41**).
2. After overnight culture in complete medium, test for eosinophils' (effector cells; E) cytotoxicity against carcinoma cell lines (target cells; T). Culture eosinophils in 96-well U-bottom plates at 2×10^6 per mL (4×10^5 eosinophils per well).
3. Stain target cells with 10 μM PKH-26 according to the manufacturer's instructions. Perform all steps at room temperature. Place 2×10^6 cells in a conical bottom polypropylene tube and wash once in PBS. After centrifugation for 5 min at $350 \times g$, carefully aspirate the supernatant leaving no more than 25 μL on the cell pellet. Resuspend the cells at 10^7 per mL in diluent C (supplied with the kit), pipetting to ensure complete dispersion. Do not vortex. Prior to staining, prepare a 20 μM PKH-26 dye solution using diluent C. Add an equal volume of this PKH-26 solution to the cell suspension and immediately mix by pipetting. Incubate for 5 min. Stop the reaction by adding an equal volume of FCS. After 1 min incubation, dilute with 10 mL PBS. Centrifuge at $400 \times g$ for 10 min (*see* **Note 42**). After three washes, resuspend stained cells in complete medium and add the target cells (1.6×10^4 per well) to the plate containing eosinophils (E:T ratio of 25:1).
4. After 6-h coculture, measure apoptotic cell death by FCM using FITC-conjugated Annexin V. Recover the cells from the plate and wash the samples twice with cold

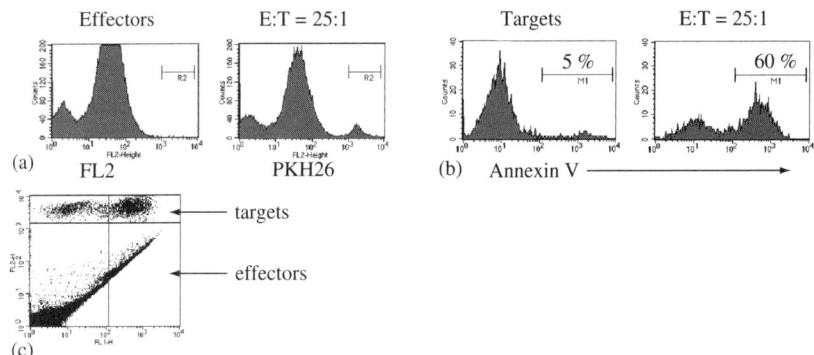

Fig. 6. Eosinophil tumoricidal activity after 6-h coculture with Colo-205. (**A**) FL2 fluorescence of effectors (eosinophils) and effectors plus targets cells. A high PKH26 staining of targets allows to gate them on R2 (FL2 high cells). (**B**) Annexin V–fluorescein isothiocyanate (FITC) staining (FL1) on PKH26+ cells, gated on R2. Eosinophil coculture induces apoptosis of Colo-205 (55%). (**C**) Dot plot representation, not gated. Observe the good separation of the cell populations. Histograms and dot plot are obtained using a FACSCalibur equipped with the CellQuest software.

PBS. Resuspend cells in binding buffer at 2×10^6/mL and incubate with Annexin V-FITC (5 µL per test) for 15 min at room temperature in the dark (*see* **Note 43**). Analyze by FCM as soon as possible (within 1 h).
5. Acquisition step: A total of 5000 PKH-26$^+$ cells are usually acquired per experimental condition and analyzed on a FACSCalibur™ using the CellQuest™ software. Eosinophil-mediated cytotoxicity against tumor cells is determined by evaluation of the number of PKH-26$^+$ Annexin V$^+$ cells. A representative experiment performed with a colon carcinoma cell line is shown in **Fig. 6**.

4. Notes

1. Metrizamide is not autoclavable; therefore the solution has to be sterilized by filtration. This is an important point because metrizamide solutions are susceptible to bacterial degradation.
2. Percoll has to be used at room temperature to avoid errors of measurement caused by volume retraction. The coefficient 1.7 used in the formula comes from the following equation: $(d0 - 1)/(d - 0.1 d10 - 0.9)$, where $d0$ = density of Percoll undiluted (1.13 g/mL); d = desired density of final working solution (1.082 g/mL); and $d10$ = density of 1.5 M NaCl (1.058).
3. Detection of EPO by ICC using DAB staining and by FCM are two complementary protocols: DAB solution allows one to detect only mature EPO whereas anti-EPO

antibody used for FCM can detect mature and immature forms of the eosinophil peroxidase.
4. The use of RPMI 1640 without phenol red, rather than PBS, allows an optimal activation of eosinophils. Neither FCS nor BSA should be present in the medium because these proteins give a high background in chemiluminescence measurements.
5. Contrary to D-luciferin obtained from Sigma, we have found D-luciferin from Roche Diagnostics to be excellent for measurement of eosinophil peroxidase release.
6. This protocol has the advantage of detecting general synthesis of ROS (superoxide, oxygen radicals, and peroxide), and of not being restricted to superoxide anion generation as are methods based on cytochrome c *(27)*.
7. Dextran sedimentation step allows removing of erythrocytes, platelets, and some small lymphocytes. This step stands generally for around 20–30 min.
8. The use of DNase is important to remove all dead cells from the cell suspension before performing the metrizamide separation.
9. Use polystyrene tube (transparent) and not polypropylene tube.
10. Metrizamide is a slightly hypertonic medium and can alter the cells, so high-speed centrifugation is needed to remove all the metrizamide.
11. There is a cell density variation with the physiological state of the eosinophils *(28)*. In normal donors, eosinophils are found mainly in layer 4 and a few in layer 3. By contrast, in eosinophilic donors, hypodense eosinophils are principally found in layers 1 and 2.
12. Percoll solution has to be used at room temperature for good separation of polynuclear and mononuclear cells. Ficoll solution (d = 1.077 g/mL) has also been tested and compared with Percoll (d = 1.082 g/mL). Ficoll is less efficient because more contaminating lymphocytes are found in the polynuclear cell pellets.
13. To avoid lymphocyte or monocyte contamination, we recommend very careful removal of the PBMC layer without touching the tube wall.
14. It has been observed that the type of buffer used for erythrocyte lysis can influence some functional studies, as shown in mouse *(1)*. For example, compared with NaCl lysis, the use of ammonium chloride buffer can interfere with mouse eosinophil antigen-presentation activity, because ammonium chloride is an inhibitor of lysosome acidification needed for antigen presentation.
15. Anti-human CD16-coated magnetic beads are used for negative selection to separate neutrophils from eosinophils, which do not usually express CD16. A subpopulation of $CD16^+$ eosinophils can, however, be found in some patients *(29)*. The proportion of this minor population can differ between patients.
16. Compared with metrizamide, Percoll (not hypertonic) is less toxic for the cells. Percoll gradients and negative selection by anti-CD16 is routinely used because this procedure allows one to increase the yield and the degree of purity and is faster than metrizamide gradients. It leads, however, to the obtention of the whole

eosinophil population and does not allow the comparison of normodense and hypodense activated eosinophils.

17. For peritoneal cells, no lysing step is necessary as no erythrocytes are found in the peritoneal exudate.
18. For higher cell purity (95–98%), rat eosinophils can be obtained by sorting peritoneal cells according to their typical forward and side scatter.
19. We have noticed that culture yield is increased when RPMI without phenol red is used when making complete medium.
20. During all the procedure, do not vortex eosinophils. Just shake tubes or tap plates to avoid stress and activation of the cells.
21. For mouse eosinophils, it is absolutely necessary to block Fc receptors by anti-Fcγ receptor antibody (2.4.G2) before primary antibody incubation.
22. We recommend using directly-labeled monoclonal antibodies for eosinophil membrane receptor detection as we have noticed that weakly expressed receptors are usually not detected by indirect staining *(21)*.
23. Do not fix the eosinophil samples with paraformaldehyde for further analysis, rather leave them on ice and analyze them within 2 h. We have observed staining alterations after fixation.
24. An important step is blocking of non-specific binding by serum pre-incubation particularly for intracellular staining. Although a blocking step is not really necessary for membrane staining on freshly purified eosinophils, one do need to use blocking after overnight culture, otherwise you might observe a second fluorescent peak.
25. It could be interesting to compare membrane and intracellular staining on freshly purified eosinophils and after overnight culture. This will allow one to observe whether expression or increased expression of molecule of interest at the plasma membrane correlates with decreased expression of the intracellular pool, or whether culture allows neo-synthesis of the protein with, or without, membrane expression. Eosinophils are able to express several receptors or cytokines and to store them in intracellular pools. Very often intracellular expression is higher than membrane expression *(21)*.
26. The acquisition step has to be performed the day of the experiment because eosinophils, contrary to lymphocytes, cannot be fixed for overnight analyses.
27. Avoid drying the spot during all the procedure. If you have to suspend the protocol for a moment, add TBS onto the cells. Ready-to-use degreased microscope glass slides are needed.
28. By performing fixation in acetone/methanol, one can keep the cytospin preparations for 1 year at $-20°C$. For this procedure, let dry the cytospins in a hood after fixation and freeze individually in aluminium. When needed, let thaw at room temperature for 30 min and then rehydrate in 0.05 M TBS, pH 7.4, for 15 min.
29. TBS is a suitable diluent for this protocol. Phosphate buffers must be avoided as phosphate is a substrate for alkaline phosphatase and, therefore, competes with the

chromogenic substrate. Additionally, phosphate in the dilution buffer for APAAP will form an insoluble precipitate with magnesium.

30. Perform each step of washing in a bath containing around 200 mL of TBS + 0.1% BSA under gentle agitation.
31. The APAAP method consists of using soluble immune complexes of calf intestinal alkaline phosphatase and anti-alkaline phosphatase. This method is preferred for staining eosinophils, as this leukocyte cell type is rich in endogenous peroxidase and furthermore it adds an amplification step (alkaline phosphatase–anti-alkaline phosphatase).
32. One can increase the intensity of the reaction by performing a second round of APAAP reaction. It is the case for interferon-γ and IL-10 detection, presented in **Fig. 4C**.
33. Contrary to the intracellular procedure, after fixation in paraformaldehyde for membrane staining, one cannot freeze the cytospins. The fixed cytospin preparations can, however, be stored for 1 week at 4°C in a humid chamber.
34. Unstained human eosinophils exhibit marked autofluorescence in comparison with other leukocytes due to granule-associated fluorescent substances. Eosinophil endogenous fluorescence is more pronounced in green than in red wavelengths. Ultraviolet exposition and glycine and NH_4Cl incubations are required for quenching the autofluorescence. Add 100 μL of PBS during UV exposition to avoid drying spot.
35. We noted that, before activation, a pre-incubation (30 min) of eosinophils with serum-free medium effectively decreases cellular stress and background.
36. Use of injectors allows one to perform a homogenous and synchronous addition of luminol, D-luciferin, or H_2O_2 solutions directly in the wells of the plate and to start measurements rapidly.
37. To avoid the degradation of EPO in supernatants and the loss of signal, the measure of chemiluminescence must be taken immediately. Keep supernatants and solutions on ice until measurement. Here, the released peroxidase catalyzes the oxidation of luminol by H_2O_2, which gives the excited molecule 3-aminophthalate. As the reactant H_2O_2 is added in excess, photons collected are proportional to the quantity of catalyst, which is EPO. The reaction is amplified by D-luciferin.
38. Mouse strain: OF1. Avoid taking young mice with too thin skin. Be careful during skin cleaning to avoid making holes in sample that could allow passive passage of cercaria. Use first sterile gauze soaked with culture medium to put out skin muscles and fatty tissues and then wipe with dry cotton.
39. Unheated sera from infected patients have to be used to allow evaluation of IgE-dependent cytotoxicity (heat-labile).
40. Effector cells are hypodense eosinophils, purified by metrizamide discontinuous gradients. The cytotoxic activity of eosinophils purified from hypereosinophilic patients is increased when compared with eosinophils from normal donors. Moreover, there is a heterogeneity of eosinophil cytotoxic function among

different subpopulations of eosinophils purified from the same patient; hypodense eosinophils are more cytotoxic than normodense eosinophils (28).

41. Annexin V-FITC is used to quantitatively determine the percentage of cells undergoing apoptosis, by detecting the phosphatidylserine (PS) exposed on the cell membrane of apoptotic cells. In the early phases of apoptosis, PS translocates from the inner leaflet of the plasma membrane to the outer leaflet.

42. PKH26 staining is a primordial parameter for this experiment. High and homogenous PKH26 staining allows a good discrimination of the two cell populations and consequently a quantification of the percentage of apoptotic targets, by gating on R2. As one can see in **Fig. 6A and B**, eosinophil fluorescence in FL2 channel does not interfere with FL2 fluorescence of stained targets.

43. Because of the Ca^{2+}-dependent affinity of Annexin V for PS, the cell staining with FITC-conjugated Annexin V has to be performed in a Ca^{2+}-enriched binding buffer.

Acknowledgments

The authors thank Sylvie Loiseau, Jean Paul Papin, and Christelle Faveeuw for providing helpful technical comments during the preparation of the manuscript. Results concerning eosinophilic cell lines are unpublished data.

References

1. Rothenberg, M. E., and Hogan, S. P. (2006) The eosinophil. *Annu Rev Immunol* **24**, 147–74.
2. Woerly, G., Roger, N., Loiseau, S., Dombrowicz, D., Capron, A., and Capron, M. (1999) Expression of CD28 and CD86 by human eosinophils and role in the secretion of type 1 cytokines (interleukin 2 and interferon gamma): inhibition by immunoglobulin a complexes. *J Exp Med* **190**, 487–95.
3. Nagase, H., Okugawa, S., Ota, Y., Yamaguchi, M., Tomizawa, H., Matsushima, K., et al. (2003) Expression and function of Toll-like receptors in eosinophils: activation by Toll-like receptor 7 ligand. *J Immunol* **171**, 3977–82.
4. Woerly, G., Roger, N., Loiseau, S., and Capron, M. (1999) Expression of Th1 and Th2 immunoregulatory cytokines by human eosinophils. *Int Arch Allergy Immunol* **118**, 95–7.
5. Woerly, G., Lacy, P., Younes, A. B., Roger, N., Loiseau, S., Moqbel, R., et al. (2002) Human eosinophils express and release IL-13 following CD28-dependent activation. *J Leukoc Biol* **72**, 769–79.
6. Yazdanbakhsh, M., Tai, P. C., Spry, C. J., Gleich, G. J., and Roos, D. (1987) Synergism between eosinophil cationic protein and oxygen metabolites in killing of schistosomula of Schistosoma mansoni. *J Immunol* **138**, 3443–7.
7. Young, J. D., Peterson, C. G., Venge, P., and Cohn, Z. A. (1986) Mechanism of membrane damage mediated by human eosinophil cationic protein. *Nature* **321**, 613–6.

8. Decot, V., and Capron, M. (2006) [Eosinophils: structure and functions]. *Presse Med* **35,** 113–24.
9. Gounni, A. S., Lamkhioued, B., Ochiai, K., Tanaka, Y., Delaporte, E., Capron, A., et al. (1994) High-affinity IgE receptor on eosinophils is involved in defence against parasites. *Nature* **367,** 183–6.
10. Nutten, S., Papin, J. P., Woerly, G., Dunne, D. W., MacGregor, J., Trottein, F., et al. (1999) Selectin and Lewis(x) are required as co-receptors in antibody-dependent cell-mediated cytotoxicity of human eosinophils to Schistosoma mansoni schistosomula. *Eur J Immunol* **29,** 799–808.
11. Svensson, L., and Wenneras, C. (2005) Human eosinophils selectively recognize and become activated by bacteria belonging to different taxonomic groups. *Microbes Infect* **7,** 720–8.
12. Stenfeldt, A. L., and Wenneras, C. (2004) Danger signals derived from stressed and necrotic epithelial cells activate human eosinophils. *Immunology* **112,** 605–14.
13. Ionescu, M. A., Rivet, J., Daneshpouy, M., Briere, J., Morel, P., and Janin, A. (2005) In situ eosinophil activation in 26 primary cutaneous T-cell lymphomas with blood eosinophilia. *J Am Acad Dermatol* **52,** 32–9.
14. Munitz, A., and Levi-Schaffer, F. (2004) Eosinophils: 'new' roles for 'old' cells. *Allergy* **59,** 268–75.
15. Fernandez-Acenero, M. J., Galindo-Gallego, M., Sanz, J., and Aljama, A. (2000) Prognostic influence of tumor-associated eosinophilic infiltrate in colorectal carcinoma. *Cancer* **88,** 1544–8.
16. Munitz, A., Bachelet, I., Fraenkel, S., Katz, G., Mandelboim, O., Simon, H. U., et al. (2005) 2B4 (CD244) is expressed and functional on human eosinophils. *J Immunol* **174,** 110–8.
17. Cormier, S. A., Taranova, A. G., Bedient, C., Nguyen, T., Protheroe, C., Pero, R., et al. (2006) Pivotal advance: eosinophil infiltration of solid tumors is an early and persistent inflammatory host response. *J Leukoc Biol* **79,** 1131–9.
18. Mattes, J., Hulett, M., Xie, W., Hogan, S., Rothenberg, M. E., Foster, P., et al. (2003) Immunotherapy of cytotoxic T cell-resistant tumors by T helper 2 cells: an eotaxin and STAT6-dependent process. *J Exp Med* **197,** 387–93.
19. Dombrowicz, D., and Capron, M. (2001) Eosinophils, allergy and parasites. *Curr Opin Immunol* **13,** 716–20.
20. Decot, V., Woerly, G., Loyens, M., Loiseau, S., Quatannens, B., Capron, M., et al. (2005) Heterogeneity of expression of IgA receptors by human, mouse, and rat eosinophils. *J Immunol* **174,** 628–35.
21. Kayaba, H., Dombrowicz, D., Woerly, G., Papin, J. P., Loiseau, S., and Capron, M. (2001) Human eosinophils and human high affinity IgE receptor transgenic mouse eosinophils express low levels of high affinity IgE receptor, but release IL-10 upon receptor activation. *J Immunol* **167,** 995–1003.

22. Dombrowicz, D., Quatannens, B., Papin, J. P., Capron, A., and Capron, M. (2000) Expression of a functional Fc epsilon RI on rat eosinophils and macrophages. *J Immunol* **165,** 1266–71.
23. Honda, K., Yamada, Y., Cui, C., Saito, N., Kayaba, H., Kobayashi, Y., et al. (1999) Effect of eotaxin on the generation of reactive oxygen species from eosinophil cell line, YY-1. *Int Arch Allergy Immunol* **120 Suppl 1,** 48–50.
24. Baumann, M. A., and Paul, C. C. (1998) The AML14 and AML14.3D10 cell lines: a long-overdue model for the study of eosinophils and more. *Stem Cells* **16,** 16–24.
25. Mayumi, M. (1992) EoL-1, a human eosinophilic cell line. *Leuk Lymphoma* **7,** 243–50.
26. Griffin, J. H., Leung, J., Bruner, R. J., Caligiuri, M. A., and Briesewitz, R. (2003) Discovery of a fusion kinase in EOL-1 cells and idiopathic hypereosinophilic syndrome. *Proc Natl Acad Sci USA* **100,** 7830–5.
27. Suzuki, M., Kato, M., Hanaka, H., Izumi, T., and Morikawa, A. (2003) Actin assembly is a crucial factor for superoxide anion generation from adherent human eosinophils. *J Allergy Clin Immunol* **112,** 126–33.
28. Prin, L., Capron, M., Tonnel, A. B., Bletry, O., and Capron, A. (1983) Heterogeneity of human peripheral blood eosinophils: variability in cell density and cytotoxic ability in relation to the level and the origin of hypereosinophilia. *Int Arch Allergy Appl Immunol* **72,** 336–46.
29. Davoine, F., Labonte, I., Ferland, C., Mazer, B., Chakir, J., and Laviolette, M. (2004) Role and modulation of CD16 expression on eosinophils by cytokines and immune complexes. *Int Arch Allergy Immunol* **134,** 165–72.

14

Ex Vivo and In Vitro Primary Mast Cells

Michel Arock, Alexandra Le Nours, Odile Malbec, and Marc Daëron

Summary

Mast cells are cells of the innate immune system whose biological responses are markedly modulated by effector molecules of adaptive immunity, i.e., antibodies. They thus contribute to anti-infectious defense but also to antibody-dependent inflammatory responses. They are especially well known as inducers of allergic reactions. They are widely distributed in most tissues, but in low numbers. They are not readily purified, and with a poor yield. For these reasons, means to generate large numbers of homogenous non-transformed mast cells have been developed. We describe here (1) fractionation methods suitable for purifying mouse or rat peritoneal mast cells and for purifying human mast cells of various origins, and (2) conditions for generating pure cultured mast cell populations from mouse, rat, and human tissues.

Key Words: Mouse; rat; human; mast cells; purification; culture.

1. Introduction

Like many other cells of the myeloid lineage, mast cells stand at the interface between innate and adaptive immunity. They contribute to anti-infectious defense when pathogens interact directly with specific receptors *(1)* or indirectly through the binding of bacterial products to Toll-like receptors (TLRs) *(2,3)*. They initiate the allergic reaction when their high-affinity immunoglobulin (Ig)E receptors (FcεRI) *(4)*, which are constitutively occupied by IgE antibodies, are aggregated by a multivalent allergen *(5)*. They also initiate IgG-dependent inflammatory responses when immune complexes aggregate their low-affinity IgG receptors (FcγRIIIA in mice) in murine models of autoimmune diseases *(6)*. Mast cell biology has therefore been actively investigated, in mice, rats, and humans.

Mast cells are widely distributed among tissues, in virtually all organs *(7)*. In all cases, however, mast cells represent a minor population. Sources of homogenous mast cells have therefore been searched for and used for experimental studies. Methods based on differential sedimentation in high-density medium have proved useful to obtain reasonably pure ex vivo mature tissue mast cells *(8–10)*. One such method is described here. The major limitation of these techniques is their poor yield, making biochemical analyses difficult. For this reason, transformed mast cell lines of various origins have been extensively used. The most popular has been the rat basophilic leukemia (RBL) cell line, which was understood to be a mucosal-type immature rat mast cell line *(11,12)* . RBL cells have been invaluable for the biochemical characterization and the cloning of FcεRI subunits *(13)*, as well as the dissection of proximal signaling following FcεRI aggregation *(14)*. RBL has also been widely used for transfection of cDNAs and for the mutagenesis analysis of the functional properties of a variety of membrane receptors. A few mouse and human tumor mast cell lines have been described, often lacking FcεRI *(15–20)*. In any case, tumor mast cells are not suitable for addressing issues such as differentiation and proliferation. Cultured mast cells derived from various sources of precursor cells permit such studies. Mouse *(21)* or rat *(22)* bone marrow-derived mast cells (BMMCs) and human mast cells (HMCs) derived from hematopoietic stem cells *(23)* have been instrumental in analyzing growth factor requirements and intracellular signaling leading to mast cell proliferation and activation. Considerable numbers of pure primary mouse mast cells can indeed be generated and kept in culture for several months, enabling in vitro biochemical analyses and in vivo studies, especially by reconstituting mast cell-deficient mice with cultured mast cells derived from genetically modified mice *(24)*.

2. Materials
2.1. Mouse Mast Cells
2.1.1. Ex Vivo Mouse Mast Cells

1. 6–9 week-old mice.
2. Metrizamide.
3. Tris-A–EDTA buffer: 0.025 M Tris–HCl, 0.12 M NaCl, 0.005 M KCl, 0.01 M EDTA, and 0.5 mg/ml human serum albumin.
4. Siliconized 10-ml glass tubes.
5. Toluidine blue solution: 0.5% toluidine blue and 30% ethanol, pH 1.

2.1.2. Cultured Mouse Mast Cells

2.1.2.1. MOUSE BMMCs

1. 6–9 week-old mice.
2. RPMI 1640 medium supplemented with 10% heat-inactivated fetal calf serum, penicillin (100 IU/ml), and streptomycin (100 µg/ml) (complete RPMI).
3. Recombinant mouse interleukin (IL)-3 (*see* **Note 1**).
4. Wright–Giemsa stain.
5. Mota's fixative prepared as follows: add 4 g of lead acetate (basic) to 50 ml of distilled deionized water; stir at slow speed and add 2–4 ml of glacial acetic acid to dissolve the lead acetate and make the solution clear; add 50 ml of absolute ethanol; keep tightly closed and store at room temperature; prepare fresh every 1–2 month.

2.2. Rat Mast Cells

2.2.1. Ex Vivo Rat Mast Cells

1. 10–15 week-old rats.
2. Metrizamide.
3. Tris-A–EDTA buffer: 0.025 M Tris–HCl, 0.12 M NaCl, 0.005 M KCl, 0.01 M EDTA, and 0.5 mg/ml human serum albumin.
4. Siliconized 10-ml glass tubes.
5. Toluidine blue solution containing 0.5% toluidine blue and 30% ethanol, pH 1.

2.2.2. Cultured Rat Mast Cells

1. 10–15 week-old rats.
2. Iscove's modified Dulbecco's medium (IMDM) supplemented with 5×10^{-5} M 2-mercaptoethanol, penicillin (100 IU/ml), streptomycin (l00 µg/ml), 20% heat-inactivated fetal calf serum, and 2–10 ng/ml of recombinant rat IL-3 (rRIL-3) (*see* **Note 2**) (complete IMDM).
3. Hayem's solution.

2.3. HMCs

2.3.1. Ex Vivo HMCs

1. Tyrode buffer: 137 mM NaCl, 2.7 mM KCl, 0.36 mM Na_2HPO_4, and 5.55 mM glucose.
2. TE buffer: Tyrode buffer containing 2 mM EDTA.
3. HEPES buffer: 20 mM HEPES, 125 mM NaCl, 5 mM KCl, and 0.5 mM glucose.
4. HA buffer: HEPES buffer plus 0.25 mg/ml bovine serum albumin (BSA).
5. Phosphate-buffered saline (PBS) (without $MgCl_2/CaCl_2$).
6. Turk's staining solution containing 0.01% crystal violet in 3% acetic acid.
7. Percoll gradients.
8. MACS C1 column.

2.3.2. Cultured HMCs Derived from Hematopoietic Progenitors

1. Serum-free mast cell culture medium (SFCM): IMDM supplemented with L-glutamine (2×10^{-3} mol/l), penicillin (100 IU/l), streptomycin (l00 µg/l), 7.5×10^{-5} M β-mercaptoethanol, BSA (2×10^{-4} mol/l), iron-saturated human transferrin (5×10^{-7} mol/l), insulin (1.7×10^{-6} mol/l), 80 ng/ml recombinant human stem cell factor (rhSCF) (*see* **Note 3**), 50 ng/ml recombinant human IL-6 (rhIL-6), and 2 ng/ml rhIL-3 (first week only).
2. Magnetic cell separator MiniMACS, MidiMACS, VarioMACS, or SuperMACS.
3. MACS column(s) type MS+/RS+, LS+/VS+, or XS+ (plus RS+, VS+, or XS+ column adapter).
4. Pre-separation filters.
5. Ficoll-Paque® (d:1.077 g/ml) and PBS supplemented with 2 mM EDTA or 0.6% anticoagulants citrate dextrose-formula A (6% ACD-A: 22.3 g/l glucose, 22 g/l sodium citrate, and 8 g/l citric acid in H_2O).
6. Buffer: PBS pH 7.2, supplemented with 0.5% BSA and 2 mM EDTA or 0.6% ACD-A. Degas buffer by applying vacuum.
7. FcR blocking reagent.
8. CD34 microbeads.
9. Fluorochrome-conjugated CD34 antibody [e.g., CD34-phycoerythrin (CD34-PE)] and fluorochrome-conjugated CD45 antibody [e.g., CD45-fluorescein isothiocyanate (CD45-FITC)] for control of CD34 progenitor cell isolation.
10. PBS containing 0.1% human AB serum.
11. PBS containing 0.1% NaN_3 and 2% paraformaldehyde.

3. Methods
3.1. Mouse Mast Cells
3.1.1. Ex Vivo Mouse Mast Cells

Mouse mast cells can be purified from cells recovered by peritoneal washing, among which they represent 3–4% of total cells.

1. Lethally anesthetized 6–9 week-old mice are injected intraperitoneally with 2 ml Tris-A–EDTA buffer. Following careful massage of the abdomen, the abdominal cavity is opened and the fluid is recovered with a Pasteur pipette.
2. Cells are washed once and resuspended at 1×10^7 cells/ml in the same buffer at room temperature.
3. Aliquots of 2 ml peritoneal cell suspension are gently layered over 1.5 ml of 22.5% metrizamide dissolved in Tris-A–EDTA buffer, in siliconized tubes, and centrifuged at $270 \times g$ for 10 min at room temperature. Most cells sediment at the interface above the metrizamide, whereas mast cells sediment at the bottom of the tube.
4. The upper layer, the interface (containing cells), and most of the metrizamide solution are very carefully discarded, leaving 50–100 µl above the small mast cell

pellet (*see* **Note 4**). Pelleted cells are gently resuspended and transferred into new tubes.
5. Under these conditions, 1×10^5 mast cells (>95% pure) can be recovered per mouse.

3.1.2. Cultured Mouse Mast Cells

3.1.2.1. MOUSE BMMCs

Bone marrow (BM) cells are harvested from the femurs of 6–9 week-old mice.

1. After lethal anesthesia of the animals, femurs are carefully dissected to eliminate any muscle fragment.
2. One extremity is cut with sterile scissors, and the needle of a 10-ml syringe previously filled with sterile RPMI is inserted in the medullar canal. BM cells are flushed out and collected in a sterile 15-ml tube.
3. After sedimentation of aggregates, cells in suspension are collected in another 15-ml tube and centrifuged, and the equivalent of cells from two femurs are resuspended in 60 ml complete OptiMEM supplemented with 1 ng/ml recombinant murine IL-3.
4. The cell suspension is transferred into culture flasks and placed at 37°C in a humidified CO2 incubator.
5. Every 3–4 days (i.e., twice a week), non-adherent cells are centrifuged, resuspended at 3×10^5 viable nucleated cells/ml in fresh IL-3-containing culture medium, and placed in new culture flasks.
6. After 3 weeks, cultures consist of virtually pure (>90%) FcεRI$^+$, Kit$^+$ mast cells as judged by staining and immunofluorescence.
7. Mouse BMMCs (mBMMCs) thus obtained keep growing in IL-3-containing medium for at least 2–3 months (*see* **Notes 5** and **6**).

3.1.3. Identifying and Phenotyping Mouse Mast Cells

3.1.3.1. WRIGHT–GIEMSA STAINING

1. Cells are resuspended at 2×10^5 cells/ml in RPMI.
2. 100 µl of the suspension, mounted onto clean slides, is placed in cytospin chambers. Slides are spun at $14 \times g$ for 5 min.
3. They are air-dried and stained with Wright–Giemsa before they are mounted in one to two drops Permount under a cover slip.

3.1.3.2. TOLUIDINE BLUE STAINING

1. Cytospin slides prepared as above are fixed by adding several drops of Mota's fixative to cover the cells for 10 min. NB: Mota's evaporates quickly, so additional drops must be added once or twice to prevent crystal formation.

2. Fixative is removed by slowly running water down, not directly on cells, and slides are dried.
3. Cells are stained for 20 min with two to three drops of acidic toluidine blue.
4. Stain is removed by running water down the slide.
5. Slides are air-dried and mounted in one to two drops Permount under a cover slip.

3.1.3.3. ANALYSIS OF THE EXPRESSION OF FcεRI AND KIT BY IMMUNOFLUORESCENCE

1. Aliquots of 5×10^5 cells are incubated in 50 µl complete RPMI containing 10 µ/mp mouse IgE or rat anti-mouse Kit mAb (ACK2) for 1 h at 0°C in Eppendorf tubes.
2. They are washed twice in PBS and incubated for 30 min at 0°C in 50 µl complete RPMI containing an appropriate dilution of $F(ab')_2$ fragments of FITC-labeled rat anti-mouse immunoglobulin (for mouse IgE) or FITC-labeled mouse anti-rat immunoglobulin (for ACK2). Negative controls are incubated with FITC-labeled secondary $F(ab')_2$ only.
3. Cells are washed once and resuspended in 500 µl PBS.
4. Fluorescence is analyzed by flow cytometry.

3.2. Rat Mast Cells

3.2.1. Ex Vivo Rat Mast Cells

The same procedure as that described for mouse mast cells can be applied to rat peritoneal cells. Because of the body weight of rats, mast cells can be obtained in relatively high numbers.

3.2.2. Cultured Rat Mast Cells

BM cells are obtained from outbred male Wistar rats, 10–15 week-old (weight 225–250 g), with the same method as for mouse mast BM cells except for the following points.

1. Complete IMDM is used as culture medium.
2. Mononuclear cell counts are performed by dilution of the cell suspension (1:10) in Hayem's solution and numeration in a hemocytometer; cell suspension is therefore adjusted to 2.5×10^5 viable nucleated BM cells/ml in complete IMDM.
3. A final concentration of 2.5×10^5 viable nucleated BM cells/ml is suspended in complete IMDM in 175-cm^2 flasks that are kept at 37°C in a humidified incubator with 5% CO_2 in air.
4. These cultures are refed and restimulated with fresh complete medium containing rRIL-3 every 4–6 days.
5. Percentage of mast cells is determined by differential counts (400 cells) on Wright–Giemsa and toluidine blue-stained cytocentrifuge preparations.

6. Rat BMMCs (rBMMCs) proliferate rapidly initially, representing 85% of cells after approximately 15 days. After this, the percentage of BMMCs increases slowly to almost 100% before cell viability declines around the fifth week (*see* **Notes 7** and **8**).

3.3. HMCs

3.3.1. Ex Vivo HMCs

Mast cells with different phenotypes can be purified from various human tissues, including lung, intestine, uterus, skin, or foreskin. The general protocol comprises a first step of enzymatic treatment of the desired tissue, in order to obtain a monocellular suspension. This enzymatic treatment is followed by a first step of purification by centrifugation over Percoll gradient, and mast cells are further enriched by separation from contaminating cells using anti-c-kit immunomagnetic beads (*see* **Note 9**).

3.3.1.1. Tissue Processing

HMCs are isolated under sterile conditions by a four-step enzymatic tissue dispersion method.

1. Macroscopically normal human tissue is obtained from surgical specimens.
2. The tissue is placed in TE buffer at 4°C immediately after resection until dissection is started.
3. If present in human intestine, the mucosa is separated mechanically from the submucosa/muscular layers. Always for intestinal specimens, mucus is removed by incubation with acetylcysteine at 1 mg/ml, and epithelial cells are detached with 5 mM EDTA.
4. Whatever the nature of the specimen used, the tissue is enzymatically digested by a four-step incubation (each for 30 min) with four enzymes (3 mg/ml pronase corresponding to 21 U/ml, 0.75 mg/ml chymopapain corresponding to 0.39 U/ml, 1.5 mg/ml collagenaseD corresponding to 0.405 U/ml, and 0.15 mg/ml elastase corresponding to 15.75 U/ml). During the first digestion step, the tissue (mucosa, in the case of an intestinal specimen) is chopped finely with scissors. For the third and fourth digestion steps, the incubation buffer is supplemented with DNase at 15 µg/ml corresponding to 15 U/ml.
5. The cells freed after the last two digestion steps are separated from tissue fragments by filtration through a polyamide Nybolt filter (pore size, 300 µm), washed, pooled, and counted after staining with Turk's solution.
6. The viability of cells is measured by dye exclusion using trypan blue staining.

3.3.1.2. Enrichment of Mast Cells

Cells are fractionated on discontinuous Percoll gradients and further enriched in mast cells by separation from contaminating cells using immunomagnetic beads.

1. 5 ml of cell suspension resuspended in HA buffer are layered carefully over 20 ml of Percoll solution (density = 1.037 g/ml) in a 50 ml polypropylene tube.
2. After centrifugation (500 × g, 20°C, 15 min), the cell sediment is harvested, washed twice with HA buffer, and counted.
3. For immunomagnetic separation, cells are resuspended in 1 ml of HA buffer containing 1 mg/ml albumin and 5 µg/ml mAbYB5.B8 (directed against human c-kit) and incubated for 30 min at 4°C with gentle rolling.
4. The cells are then washed in HA buffer and resuspended in 500 µl of HA buffer containing 1 mg/ml albumin.
5. Finally, the cell suspension is incubated with a goat anti-mouse IgG antibody coupled to paramagnetic beads for 30 min at 4°C. During incubation, the tubes are gently rolled.
6. After washing in HA buffer, mast cells are enriched by magnetic separation of the cells using a MACS C1 column placed in magnetic field.
7. After separation, cells are counted, and washed in culture medium without antibiotics.

3.3.2. Cultured HMCs Derived from Hematopoietic Progenitors

Pure populations of HMCs are obtained by long-term culture of human normal hematopoietic progenitors ($CD34^+$ cells) in the continuous presence of stem cell factor (SCF). Sources of these $CD34^+$ cells are mainly BM, cord blood, or peripheral blood. Usually, mononuclear cells from BM and cord blood contain 0.5–1% of $CD34^+$ cells, whereas peripheral blood contains <0.1% of $CD34^+$ cells. Hematopoietic progenitor cells can be rapidly and efficiently enriched to a purity of about 85–98% using positive selection of magnetically labeled $CD34^+$ cells. Mononuclear cells from peripheral blood (PBMCs), cord blood, or BM are obtained by density gradient centrifugation over Ficoll-Paque®. For immunomagnetic separation, $CD34^+$ hematopoietic progenitor cells are magnetically labeled using MACS CD34 MicroBeads. The magnetically labeled cells are enriched on positive selection columns in the magnetic field (*see* **Note 10**).

3.3.2.1. PURIFICATION OF $CD34^+$ PROGENITORS FROM HUMAN CORD BLOOD, BM, OR PERIPHERAL BLOOD CELLS

3.3.2.1.1. Sampling of Cord Blood Cells.

1. Dilute anticoagulated (preservative-free heparin sodium, 1.0 ml for 50 ml of sample) cord blood 1:4 with PBS.

3.3.2.1.2. Sampling of BM Cells.

1. Collect BM in 50-ml tubes containing 5 ml PBS containing 2 mM EDTA or 0.6% ACD-A or 200 U/ml heparin and store at 4°C if the cells cannot be processed immediately.

2. For release of the cells from aggregates, dilute in 10× excess of RPMI 1640 medium containing 0.02% collagenase B and 100 U/ml DNase and shake gently at room temperature for 45 min.
3. Pass cells through 30-µm nylon mesh or filter. Wet filter with buffer before use.

3.3.2.1.3. Sampling of Peripheral Blood Cells.

1. Start with fresh human blood treated with an anticoagulant, e.g., heparin, citrate, ACD-A or citrate phosphate dextrose (CPD) or leukocyte-rich buffy coat not older than 8 h. Dilute cells with two to four volumes of PBS containing 2 mM EDTA or 0.6% ACD-A.
2. Obtention of a mononuclear cell suspension: Carefully layer 35 ml of diluted cell suspension over 15 ml of Ficoll-Paque®. Centrifuge for 35 min at $400 \times g$ at 20°C in a swinging-bucket rotor (without brake). Aspirate the upper layer leaving the mononuclear cell layer undisturbed at the interphase. Carefully collect interphase cells and wash twice in PBS containing 2 mM EDTA or 0.6% ACD-A. Centrifuge for 10 min at $300 \times g$ at 20°C. Resuspend cell pellet in a minimal volume of buffer. Mononuclear cell counts are performed by dilution of the cell suspension (1:100) in Hayem's solution and numeration in a hemocytometer. Adjust the cell suspension to 300 µl of buffer per 10^8 total cells. For $<10^8$ total cells, use 300 µl. Proceed to magnetic labeling.
3. Magnetic labeling of CD34$^+$ progenitor cells: Add 100 µl FcR blocking reagent per 10^8 total cells to the cell suspension to inhibit unspecific or Fc-receptor-mediated binding of CD34 MicroBeads to non-target cells. Label cells by adding 100 µl CD34 MicroBeads per 10^8 total cells, mix well, and incubate for 30 min at 6–12°C. Wash cells carefully and resuspend in appropriate amount of buffer (MS+/RS+ column: 500–1000 µl; LS+/VS+ column: 1–10 ml, maximum 2×10^8 cells per ml). Proceed to magnetic separation.
4. Magnetic separation of $<2 \times 10^9$ mononuclear cells: Choose a column type (MS+/RS+ or LS+/VS+) according to the number of total unseparated cells and place it (with column adapter) in the magnetic field of the MACS separator. Fill and rinse with buffer (MS+/RS+: 500 µl; LS+/VS+: 3 ml). Pass mononuclear cells through 30-µm nylon mesh or pre-separation filter to remove clumps. Wet filter with buffer before use. Apply cells to the column, allow cells to pass through the column, and wash with buffer (MS+/RS+: 3×500 µl; LS+/VS+: 3×3 ml). Remove column from separator, place column on a suitable tube, and pipette buffer on top of column (MS+/RS+: 1 ml; LS+/VS+: 5 ml). Firmly flush out retained cells with pressure using the plunger supplied with the column. Repeat magnetic separation step: apply the eluted cells to a new prefilled positive selection column (for $<10^7$ CD34$^+$ cells: MS+/RS+; for $<10^8$ CD34$^+$ cells: LS+/VS+), wash, and elute retained cells in buffer (MS+/RS+: 500 µl; LS+/VS+: 2.5 ml).
5. Magnetic separation of 2×10^9 to 2×10^{10} mononuclear cells: Assemble XS+ column and place it in the column holder of the SuperMACS using XS+ column

adapter. Turn three-way stopcock to position "fill." Fill the column from the bottom with buffer from the syringe until the buffer reaches the syringe cylinder. Turn the three-way stopcock to position "run" and rinse column by filling from the top with buffer. Allow buffer to run into the column. Then, add more buffer. Rinse with 50 ml of buffer. Close three-way stopcock; leave the syringe attached during separation, except when refilling with buffer. Move column in the magnetic field of the SuperMACS by turning the handle. Pass cells through 30-μm nylon mesh or filter to remove cell clumps. Apply cells into the syringe cylinder that is set up on the XS+ column and turn three-way stopcock to position "run." Allow the cells to pass through the column. Remove flow resistor and wash with 4 × 30 ml buffer. Close three-way stopcock and remove column out of the magnetic field of the SuperMACS by turning the XS+ adapter handle backward. Detach syringe from the three-way stopcock, fill with buffer, and attach to port A of the XS+ column. Elute retained cells with 20 ml buffer using the syringe. Repeat magnetic separation step: apply the eluted cells to a new prefilled XS+ column or VS+ column, wash, and elute retained cells in buffer.

6. Evaluation of hematopoietic progenitor cell purity: Evaluate the purity of the isolated hematopoietic progenitor cells by flow cytometry or fluorescence microscopy. Fluorescent staining of $CD34^+$ cells can be accomplished by direct immunofluorescent staining using an antibody recognizing an epitope different from that recognized by the CD34 monoclonal antibody QBEND/10 (e.g., CD34-PE). For optimal discrimination of $CD34^+$ cells from other leukocytes, counterstain cells with an antibody against CD45 (e.g., CD45-FITC). $CD34^+$ cells express CD45 at a lower level as compared with lymphocytes. Use the antibodies at appropriate concentrations recommended by the manufacturers. Typically, staining for 5 min at 6–12°C should be sufficient. After fluorescence staining, wash and resuspend the cells in buffer.

3.3.2.2. CULTURE CONDITIONS

1. $CD34^+$ cells are cultured at 37°C in a humidified atmosphere containing 5% CO_2 at a starting density of 1×10^5 cells/ml in SFCM.
2. $CD34^+$ cells may initially proliferate to generate 100 times or more the starting number of cells if rhSCF is combined with rhIL-3 and rhIL-6. Luxurious growth is observed during the first 2–3 weeks, although debris will begin to accumulate because of non-mast cell lineage apoptosis and necrosis.
3. Adherent macrophages also may start increasing in number by 2 weeks. Cultures are checked weekly, and non-adherent mast cell-committed progenitors are separated from adherent cells and debris. Non-adherent cells are gently pipetted and removed to a new flask, or non-adherent cells and supernatant are centrifuged at 150 × g for 5 min, and cells are resuspended in complete media with growth factors every 5–7 days.

4. The number and percentage of mast cells in the cultures are assessed weekly using Wright–Giemsa and acidic toluidine blue (see **Subheading 3.1.3.**) staining of cytocentrifuged samples. Usually, 10^6 CD34$^+$ cells at 8–10 weeks may give rise to 5×10^7 to 15×10^7 HHPMC with <5% contamination with other cell types.
5. Immunostaining for tryptase can be also used to further identify mast cells in the cultures. Briefly, cytocentrifuged preparations are fixed in cold acetone for 30 min, air-dried, and then stored at –20°C. Fixed samples are then treated with mouse IgG1 anti-tryptase monoclonal antibodies and finally submitted to a Fast-Red immunostaining.
6. After >10 weeks in culture, >95% of the cells are identified as mast cells by their morphological features, the presence of metachromatic granules, and their positivity for tryptase. These cells can then used for subsequent experiments as pure normal HMCs (see **Notes 11** and **12**).
7. Because CD34$^+$-derived mast cells usually do not express spontaneously a significant amount of FcεRI at their membrane, they can be incubated for 5 additional days with 2–10 μg/ml human myeloma IgE and 20 ng/ml rhIL-4 to enhance the expression of FcεRI on their membrane.

3.3.2.3. PHENOTYPING HMCS FOR KIT AND FcεRI EXPRESSION

1. For flow cytometric analysis, 5×10^5 cells are incubated for 60 min at 4°C with the monoclonal receptor antibodies [mouse IgG against the α-chain of the high-affinity IgE receptor (FcεRI) or against Kit (YB5.B8)] in 50 μl PBS containing 0.1% human AB serum.
2. After two washes with PBS, cells are incubated for 45 min at 4°C with a 1:20 dilution of anti-mouse IgG conjugated with FITC.
3. Following two washes, cells are suspended in 400 μl PBS containing 0.1% NaN$_3$, fixed with 2% paraformaldehyde solution, and analyzed on an EPICS Profile flow cytometer (Coulter, Krefeld, Germany).
4. Negative controls are done in the absence of antibody, with IgG and isotype-matched desmin antibody D33 (DAKO, Trapper, France).
5. For flow cytometric analysis, Elite Workstation software (Coulter) is used.

4. Notes

1. Recombinant IL-3 can be replaced by 4–20% conditioned medium made of culture supernatant of cells stably transfected with cDNA encoding IL-3.
2. rRIL-3 can be replaced by 5–20% conditioned medium made of culture supernatant of rat mononuclear splenocytes (10^6 cells/ml) stimulated with 1 mg/l Concanavalin A for 5–7 days.
3. rhSCF can be replaced by 5–20% conditioned medium made of culture supernatant of cells stably transfected with cDNA encoding rhSCF.
4. Because mast cells are a minor population amongst peritoneal cells, even a minimal contamination will drastically alter mast cell purity. For this reason, all cells at

the interface must be discarded. This is why siliconized tubes are used. Carefully wiping tube walls before collecting mast cell pellets may also improve purity.
5. Although primary BM cells can be kept frozen in liquid nitrogen before they are cultured, if frozen, mBMMCs quickly die upon thawing.
6. Mast cell numbers are calculated by determining the percentage of acidic toluidine blue positive cells out of total Wright–Giemsa-positive cells. Toluidine blue should stain mast cells red-purple (metachromatic staining) and the background blue (orthochromatic staining). Metachromasia, tissue elements staining a color different from the dye solution, is due to the pH, dye concentration, and temperature of the basic dye. Blue or violet dyes will show a red color shift, and red dyes will show a yellow color shift with metachromatic tissue elements.
7. Rat BM cells can be kept frozen in liquid nitrogen before they are cultured, but it is impossible to keep rBMMCs frozen.
8. Mast cell numbers are calculated exactly as described above for mBMMC cultures.
9. Isolated HMCs cannot be frozen using classical techniques.
10. Once purified, $CD34^+$ cells can be kept frozen in liquid nitrogen before they are cultured, whereas it is impossible to keep mature HMC frozen.
11. Isolated human mast cells can be maintained in culture for up to 3 months in the continuous presence of rhSCF. Mast cell purity increases during culture and reaches nearly 100%. During the first week of culture, mast cell numbers decrease, but after that time they start to proliferate. Cultured mast cells do not change their histamine content, phenotype, or morphology. They are even more responsive toward IgE-dependent stimulation, which causes the release of high amounts of histamine, leukotrienes, and cytokines such as tumor necrosis factor (TNF)-α and IL-5.
12. Once pure, HMC cultures stop growing at 8–10 weeks. They nevertheless stay alive without any significant changes in their phenotype for another 2–5 months provided that they are maintained in SCF-containing medium.

References

1. Arock, M., Ross, E., Lai-Kuen, R., Averlant, G., Gao, Z. and Abraham, S. N. (1998) Phagocytic and tumor necrosis factor alpha response of human mast cells following exposure to gram-negative and gram-positive bacteria. *Infect Immun* **66**, 6030–6034.
2. Varadaradjalou, S., Feger, F., Thieblemont, N., Hamouda, N. B., Pleau, J. M., Dy, M. and Arock, M. (2003) Toll-like receptor 2 (TLR2) and TLR4 differentially activate human mast cells. *Eur J Immunol* **33**, 899–906.
3. Majewska, M. and Szczepanik, M. (2006) [The role of Toll-like receptors (TLR) in innate and adaptive immune responses and their function in immune response regulation]. *Postepy Hig Med Dosw (Online)* **60**, 52–63.
4. Metzger, H. (1991) The high affinity receptor for IgE on mast cells. *Clin Exp Allergy* **21**, 269–279.

5. Metzger, H. (1992) Transmembrane signaling: the joy of aggregation. *J Immunol* **149**, 1477–1487.
6. Gregory, G. D. and Brown, M. A. (2006) Mast cells in allergy and autoimmunity: implications for adaptive immunity. *Methods Mol Biol* **315**, 35–50.
7. Krishnaswamy, G., Ajitawi, O. and Chi, D. S. (2006) The human mast cell: an overview. *Methods Mol Biol* **315**, 13–34.
8. Salari, H., Takei, F., Miller, R. and Chan-Yeung, M. (1987) Novel technique for isolation of human lung mast cells. *J Immunol Methods* **100**, 91–97.
9. Hachisuka, H., Kusuhara, M., Higuchi, M., Okubo, K. and Sasai, Y. (1988) Purification of rat cutaneous mast cells with Percoll density centrifugation. *Arch Dermatol Res* **280**, 358–362.
10. He, D., Esquenazi-Behar, S., Soter, N. A. and Lim, H. W. (1990) Mast-cell heterogeneity: functional comparison of purified mouse cutaneous and peritoneal mast cells. *J Invest Dermatol* **95**, 178–185.
11. Eccleston, E., Leonard, B. J., Lowe, J. S. and Welford, H. J. (1973) Basophilic leukaemia in the albino rat and a demonstration of the basopoietin. *Nat New Biol* **244**, 73–76.
12. Seldin, D. C., Adelman, S., Austen, K. F., Stevens, R. L., Hein, A., Caulfield, J. P. and Woodbury, R. G. (1985) Homology of the rat basophilic leukemia cell and the rat mucosal mast cell. *Proc Natl Acad Sci USA* **82**, 3871–3875.
13. Blank, U., Ra, C., Miller, L., White, K., Metzger, H. and Kinet, J. P. (1989) Complete structure and expression in transfected cells of high affinity IgE receptor. *Nature* **337**, 187–189.
14. Turner, H. and Kinet, J. P. (1999) Signalling through the high-affinity IgE receptor Fc epsilonRI. *Nature* **402**, B24–B30.
15. Dunn, T. B. and Potter, M. (1957) A transplantable mast-cell neoplasm in the mouse. *J Natl Cancer Inst* **18**, 587–601.
16. Barsumian, E. L., McGivney, A., Basciano, L. K. and Siraganian, R. P. (1985) Establishment of four mouse mastocytoma cell lines. *Cell Immunol* **90**, 131–141.
17. Butterfield, J. H., Weiler, D., Dewald, G. and Gleich, G. J. (1988) Establishment of an immature mast cell line from a patient with mast cell leukemia. *Leuk Res* **12**, 345–355.
18. Kirshenbaum, A. S., Akin, C., Wu, Y., Rottem, M., Goff, J. P., Beaven, M. A., Rao, V. K. and Metcalfe, D. D. (2003) Characterization of novel stem cell factor responsive human mast cell lines LAD 1 and 2 established from a patient with mast cell sarcoma/leukemia; activation following aggregation of FcepsilonRI or FcgammaRI. *Leuk Res* **27**, 677–682.
19. Ko, J., Yun, C. Y., Lee, J. S., Kim, D. H., Yuk, J. E. and Kim, I. S. (2006) Differential regulation of CC chemokine receptors by 9-cis retinoic acid in the human mast cell line, HMC-1. *Life Sci* (in press).
20. Demehri, S., Corbin, A., Loriaux, M., Druker, B. J. and Deininger, M. W. (2006) Establishment of a murine model of aggressive systemic mastocytosis/mast cell leukemia. *Exp Hematol* **34**, 284–288.

21. Razin, E., Cordon-Cardo, C. and Good, R. A. (1981) Growth of a pure population of mouse mast cells in vitro with conditioned medium derived from concanavalin A-stimulated splenocytes. *Proc Natl Acad Sci USA* **78**, 2559–2561.
22. Haig, D. M., McMenamin, C., Redmond, J., Brown, D., Young, I. G., Cohen, S. D. and Hapel, A. J. (1988) Rat IL-3 stimulates the growth of rat mucosal mast cells in culture. *Immunology* **65**, 205–211.
23. Saito, H., Ebisawa, M., Tachimoto, H., Shichijo, M., Fukagawa, K., Matsumoto, K., Iikura, Y., Awaji, T., Tsujimoto, G., Yanagida, M., Uzumaki, H., Takahashi, G., Tsuji, K. and Nakahata, T. (1996) Selective growth of human mast cells induced by Steel factor, IL-6, and prostaglandin E2 from cord blood mononuclear cells. *J Immunol* **157**, 343–350.
24. Abe, T. and Nawa, Y. (1987) Reconstitution of mucosal mast cells in W/WV mice by adoptive transfer of bone marrow-derived cultured mast cells and its effects on the protective capacity to Strongyloides ratti-infection. *Parasite Immunol* **9**, 31–38.

15

Murine Macrophages

A Technical Approach

Luisa Martinez-Pomares and Siamon Gordon

Summary

In this chapter, we describe current protocols used for the characterization of macrophages (Mϕ) in mouse tissues and in cell suspensions from spleen and lymph nodes. Also, we include a brief description of a complementary approach: culture of primary Mϕ. Although culture Mϕ are extremely useful for analysing the basic biology of Mϕ and their receptors, it should not be forgotten that the term Mϕ encompasses a wide range of different types of cells with phenotypic characteristics dependent on their activation state and tissue of origin. In our view, there is no perfect Mϕ marker and analysis of the expression profile of several markers, and functional studies are required to make an informed guess of the cellular characteristics and function of the Mϕ population of interest.

Key Words: Mouse; macrophages; surface markers; cell differentiation.

1. Introduction

The study of macrophages (Mϕ) represents a challenge for cellular immunologists. These cells are phenotypically diverse through adaptation to the unique conditions encountered in different tissues and alter their activation state in response to changes in their environment such as those taking place during inflammation. Advances in this area have been facilitated through the use of Mϕ differentiation markers. The usefulness of these reagents does not only reside in their restricted expression pattern; these molecules are helping to define in vitro distinct cellular activation programmes that can then be investigated in situ by determining the array of markers expressed by a specific cell population using immunolabelling of tissue sections or flow cytometry analysis of complex cell suspensions derived from tissues. Here, we focus on less widely known aspects of Mϕ characterization within mouse tissues. We also provide references to methods to isolate, culture and characterize standard Mϕ preparations in vitro, that are well described in the literature.

The knowledge provided by the identification of markers present in a particular cell population, placed in the context of anatomical localization, greatly facilitates the discernment of cellular function. This is particularly relevant in the area of myeloid cell differentiation; markers originally described as Mϕ-specific are being found in cells displaying a dendritic cell-like phenotype. This conspicuous overlap is probably indicative of a closer than anticipated relationship between Mϕ and myeloid dendritic cells and is a reason for caution in using stringent classification criteria.

2. Materials

2.1. Analysis of Mϕ In Situ by Immunolabelling

2.1.1. Tissue Collection

1. Plastic moulds (Raymond Lamb, Eastbourne, UK, E10.8/L15, E10.6/2001/33).
2. Tissue-Tek O.C.T compound (Bayer Diagnostics, Newbury, UK, 4583).
3. Iso-pentane (VWR, Lutterworth, UK, 29452 4E).
4. Dry ice.
5. Ice container.
6. Dissecting board.
7. Dissecting tools.

2.1.2. Tissue Processing

1. Fixative: methanol-free paraformaldehyde (16% solution Electron Microscope Sciences, 15710-S). To prepare 200 ml of fixative: 25 ml paraformaldehyde, 175 ml dH$_2$O, 1.8 g NaCl, 0.08 g CaCl$_2$, 1.5 ml 1 M HEPES Buffer, pH 7.4 (GIBCO™, Invitrogen Ltd, Paisley, UK).
2. Hydrophobic marker pen (Shandon, Thermo Scientific, Runcorn, UK).
3. Quenching solution: To prepare 50 ml of quenching solution, add 50μl 1 M sodium azide and 0.09 g glucosa to 50 ml of 0.1 M phosphate buffer. Warm to 37 °C and add 20 μl glucose oxidase (Sigma, Gillingham, UK G0543) just before use.
4. Immunostaining slide tray (Raymond Lamb, E103.2).
5. Avidin/Biotin Blocking Kit (Vector Laboratories, Orton Southgate, Peterborough, UK, SP-2001).
6. Blocking solution: 5% normal serum in phosphate-buffered saline (PBS); sources of normal sera: donkey (Sigma, D 9663), goat (Vector Laboratories, S-1000), mouse

(Sigma, M 5905), rabbit (Vector Laboratories, S-5000) or rat (Sigma, R 9759). Purified rat IgG (Sigma, I-4131).

2.1.3. Primary Antibodies

There is a wide range of antibodies suitable for the detection of antibodies in tissue samples [immunohistochemistry (IHC)] and in cell suspensions (flow cytometry). See **Table 1** for antibodies commonly used in the study of myeloid cells. Additional antibodies useful in assessing the activation state of the cells and localization within tissues have also been included. Information on Mϕ markers can be found in **ref. (1)**.

2.1.4. Secondary Reagents

2.1.4.1. BRIGHT FIELD

1. Biotinylated rabbit anti-rat (Vector Laboratories, UK, BA-4001).
2. Biotinylated goat anti-rat (Jackson Immunoresearch, Stratech Scientific Ltd. Newmarket, UK, 112-065-167).
3. Peroxidase-conjugated goat anti-rat (Chemicon, Europe, Ltd. Chandlers Ford, UK, AP183P).
4. Alkaline phosphatase-conjugated goat anti-rat (Chemicon, UK, AP183A).
5. ABC Peroxidase Vector Vectastain Elite ABC Kit (Vector laboratories, UK, PK6100).
6. ABC Alkaline Phosphatase Vector Vectastain Elite ABC Kit (Vector laboratories, UK, AK5000). For both ABC kits, add two drops of reagent A to 10 ml PBS and mix well followed by two drops of reagent B and mix well; incubate for 30 min at room temperature before using.
7. DAB (3,3´-diaminobenzidine tetrahydrochloride) (5 × 10 mg; Polysciences Europe Eppelheim, Germany 04008). To be used with peroxidase-conjugated secondary reagents (*see* **Note 1**). Equipment: 20-ml syringe, 22-μm syringe filter, two needles, 0.5–10 μl pipette and tip, two beakers, 10 mM imidazole solution (kept at 4°C) and hydrogen peroxide (H_2O_2). Preparation of developing solution: pour imidazole into beaker and draw 20 ml into syringe. Pour any excess back into the bottle. Push two needles into DAB vial and inject contents of syringe through. Mix. Pipette 6 μl H_2O_2 into beaker. Withdraw DAB solution with syringe and filter into beaker containing H_2O_2. Mix. The solution is now ready for use.
8. Substrate Kit IV BCIP/NBT (Vector laboratories, SK5400). To be used with alkaline phosphatase-conjugated secondary reagents. To 5 ml of 100 mM Tris–HCl pH 9.5 buffer, add one drop of levamisole (a reagent to quench endogenous alkaline phosphatase activity within tissues, Vector Laboratories, SP 5000) and mix. Add two drops of reagent 1 and mix well. Add two drops of reagent 2 and mix well. Add three drops of reagent 3 and mix well.
9. Methyl Green (Vector, H-3402).

Table 1
Antibodies Commonly Used for Analysis of Myeloid Cells

Antigens	Reference	Cellular expression	Clone	Isotype	Source	Use
F4/80 antigen Emr1	(1)	Mφ, Langerhans cells, eosinophils.	F4/80	IgG2b	Serotec, UK	IHC/F
CD68[a]	(5,6)	Mφ	FA11	IgG2a	Serotec, UK	IHC/F[b]
Sialoadhesin	(7)	Selected Mφ, B cell follicle- associated migratory dendritic cells.	3D6 MOMA-1	IgG2a	Serotec, UK	IHC/F
SignR-1	(8)	Selected Mφ	ER-TR9	IgM	DPC Biermann (Bad Nauheim, Germany)	IHC/F
CD301 (mMGL)	(9)	Connective tissue Mφ	ER-MP23	IgG2a	AbCam	IHC/F
Dectin-1/β-glucan receptor	(10,11)	Mφ, neutrophils Selected dendritic cells	2A11	IgG2b	Serotec, UK	IHC/F[c]
Mannose receptor (CD206)	(12)	Mφ, hepatic, splenic and LN lymphatic endothelia, kidney mesangial cells	5D3	IgG2a	Serotec, UK	IHC[d]/F
Marco	(13)	Selected Mφ Myeloid cells	ED31	IgG1	Serotec, UK	IHC/F
CD11b, CR3			5C6	Rat IgG2b	Serotec, UK	IHC/F
Scavenger Receptor A (SR-A)	(14)	Selected Mφ, hepatic endothelia Perivascular Mφ (CNS)	2F8	Rat IgG2b	Serotec, UK	IHC«F[e]
Dectin-2	(15)	Selected Mφ, and DC	11E4	Rat IgG2a	Serotec, UK	IHC/F
MHCII		Activated Mφ, constitutively DC and B cells	M5/114	Rat IgG2b		IHC/F

IgD	Naïve B cells	11-26c.2a	Rat IgG	BD biosciences	IHC
CD3	T cells	KT-3 209	Rat IgG	BD biosciences	IHC/F
FDC-M2. Activated Complement C4	Follicular dendritic cells		Rat IgG	(16)	IHC
CD21/CD35 Complement Receptor 1/2	Follicular dendritic cells, some B cells	8C12	Rat IgG	BD biosciences	IHC
CD11c (17)	Dendritic cells	N418/HL3	Hamster IgG	BD biosciences	IHC/F
CD40	Activation markerAntigen-presenting cell	3/23	Rat IgG2a	Serotec	F
CD80	Activation markerAntigen-presenting cell	16-10A1	Ham IgG	BD biosciences	F
CD86	Activation markerAntigen-presenting cell	PO3RMMP-2	Rat IgG2bRat IgG2a	BD biosciences Serotec, UK	F
CD205 (DEC-205) (18)	Dendritic cells and selected Mϕ e.g., intestine	NLDC-145	Rat IgG2a	Cedarlane Labs, USASerotec, UK	IHC/F

IHC, immunohistochemistry; F, flow cytometry.

[a] CD68 is a pan intracellular marker of Mϕ. It is expressed in most tissue Mϕ and its presence indicates a developed endolysosomal compartment. Expression in dendritic cells is restricted and punctate.

[b] For detection by flow cytometry, cells have to be fixed and permeabilized (see Note 10).

[c] mAb 2A11 is better suited to flow cytometry than immunohistochemistry.

[d] Labelling with 5D3 is fixation sensitive and should be performed on recently fixed tissue or cells.

[e] 2F8 does not detect SR-A in C57BL/6 mice because of a sequence polymorphism. This antibody is best suited for flow cytometry.

10. Histoclear (Raymond Lamb, C78-G).
11. DPX (VWR, 36125 4D).
12. AquaPerm (Shandon, Thermo).

2.1.4.2. FLUORESCENT LABELLING

1. Donkey anti-rat A488 labelled (Molecular probes, Invitrogen A-21208).
2. Cy5-Streptavidin (Jackson Immunoresearch, 016-170-084).
3. DAPI (4´,6-diamidino-2-phenylindole,dilactate; Sigma, D9564). Prepare concentrate stock in PBS (1 mg/ml), aliquot and store at −20 °C. Dilute in PBS at 400 ng/ml just before use.
4. Fluorescent mounting medium (DakoCytomation, Denmark A/S Produktionsuej 42 DK-2600 Glostrup Denmark, S3023).

2.2. Analysis of Mφ Populations by Flow Cytometry

2.2.1. Preparation of Cell Suspensions

1. Liberase blendzyme II (Roche Diagnostics Ltd. Burgers Hill, UK). Working solution 0.46 U/ml (9.2 U/ml stock in RPMI diluted 1 in 20 in RPMI).
2. Gey's Solution: Solution A − 35 g NH_4Cl, 1.85 g KCl, 1.5 g $Na_2HPO_4.12H_2O$, 0.12 g KH_2PO_4, 5 g glucose in 1 l H_2O (optional: add 50 mg of phenol red). Solution B − 0.42 g $MgCl_2.6H_2O$, 0.34 g $CaCl_2$, 0.14 g $MgSO_4.7H_2O$ in 100 ml H_2O. Solution C − 2.25 g $NaHCO_3$ in 100 ml H_2O. All solutions are filter sterilized and kept at 4°C. 1 × Gey's solution − solution A : solution B : solution C : H_2O in ratio of 20:5:5:70.

2.2.2. Staining

1. Fluorescence-activated cell sorting (FACS) blocking solution: 5% heat-inactivated (HI) rabbit serum; 0.5% bovine serum albumin (BSA); 2 mM NaN_3; 5 mM EDTA in PBS. It can be aliquoted and stored at −20 °C.
2. mAb 2.4G2 (IgG2b, BD Biosciences, Oxford, UK). 2.4G2 is an anti-FcγRII/III blocking antibody and helps to prevent non-specific background antibody staining.
3. FACS washing solution: 0.5% BSA; 2 mM NaN_3; 5 mM EDTA in PBS. Maintain at 4 °C.
4. Primary antibodies (see **Table 1**).
5. 2× fixation buffer: 2% formaldehyde in PBS.

2.3. Culture of Peritoneal Mφ

2.3.1. Harvesting of Peritoneal Cells

1. Specific pathogen-free mice.
2. 70% ethanol in water.
3. Scissors and forceps.
4. Sterile PBS, Ca^{2+} Mg^{2+} free (GIBCO™, Invitrogen Ltd).
5. 10-ml syringe (one per mouse).

6. 25-gauge needle (one per mouse).
7. 50-ml polypropylene tubes (BD Biosciences).

2.3.2. Stimuli Commonly Used to Elicit Inflammatory Peritoneal Mϕ

1. Bio-Gel P-100 polyacrylamide beads (fine, hydrated size 45–90 µm; Bio-Rad, Richmond, CA, USA). Wash 2 g of Bio-Gel beads twice in 20 ml endotoxin-free water or PBS. Pellet by centrifugation for 5 min at $400 \times g$ and resuspend in 100 ml PBS to give a 2% w/v solution. Autoclave at 15 lb/m^2 for 20 min before use.
2. Brewer's complete thioglycollate broth (Difco Laboratories, West Molesy, UK). Preparation: resuspend 15 g of dehydrated thioglycollate medium in 500 ml of distilled water in a one-litre Erlenmeyer flask. Heat over a flame until dissolved and remove from flame immediately after boiling. The solution should change from brown to red. Aliquot into 25- or 50-ml bottles and autoclave for 20 min, 121°C. Yield of inflammatory cells will increase if the solution is aged in a dark room at room temperature before use. This is thought to be due to an increase in glycation products.

2.3.3. Selection of Peritoneal Mϕ by Adhesion

1. Turk's white cell counting fluid: 1 ml glacial acetic acid, 1 ml gentian violet, 1% aqueous solution. Make up to 100 ml in distilled water.
2. Mϕ culture medium: RPMI 1640 supplemented with 10% HI fetal calf serum (FCS) containing 50 IU of penicillin, 50 µg streptomycin and 2 mM glutamine per ml (GIBCO™, Invitrogen Ltd).
3. Mϕ washing medium: Opti-MEM containing 50 IU of penicillin, 50 µg streptomycin and 2 mM glutamine per ml (GIBCO™, Invitrogen Ltd).
4. Sterile PBS (37 °C).
5. Tissue culture-treated six- or 24-well plates (BD biosciences).
6. Non-treated (bacteriologic plastic) six-well plates (BD Falcon).

3. Methods

3.1. Analysis of Mϕ In Situ by Immunolabelling

3.1.1. Peroxidase Labelling

1. Collect tissues and carefully place in mould containing OCT-Lab-Tek compound, and push towards the bottom to ensure they are all in the same plane (*see* **Note 2**).
2. Put moulds within a glass beaker containing iso-pentane previously placed on dry ice within an ice container with (*see* **Fig. 1**). Once the contents of the mould are frozen, they can be safely placed at −20 °C until cutting.
3. Perform cutting in a cryostat at −20 °C at 5–6 µm. Sections are collected onto gelatine-coated slides or charged slides, left for 1 h at room temperature and placed at −20 °C in sealed boxes containing silica gel in porous bags.
4. Thaw tissues at room temperature for 1 h, draw a circle around the tissues using a hydrophobic marker pen and place in ice-cold 2% paraformaldehyde solution for 10 min on ice in a Coplin jar (*see* **Note 3**).

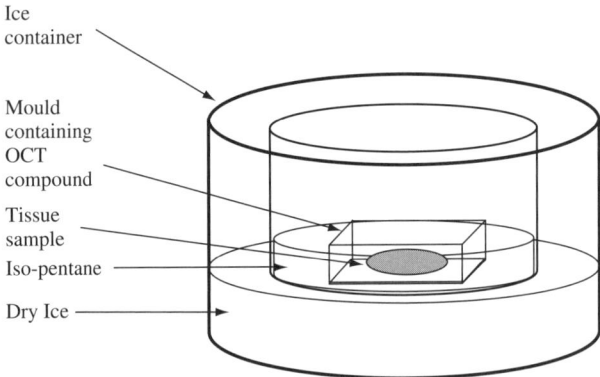

Fig. 1. Schematic representation of set up used for freezing tissue samples. Place dry ice at the bottom of ice container and position a basin containing iso-pentane on the dry ice. Leave iso-pentane to cool down. Fill labelled plastic mould with OCT-compound and immerse appropriate tissue into the OCT avoiding the formation of bubbles. Place mould into the iso-pentane and leave until OCT-compound solidifies. Store frozen tissues in sealed container at $-20°C$ until required.

5. Rinse three times in PBS.
6. Permeabilize the tissues in PBS + 0.1% Triton for 5 min at room temperature.
7. Place the slides in quenching solution for 15 min at 37 °C.
8. Rinse three times in PBS.
9. Slides can now be placed on the staining tray for the incubations with primary and secondary reagents (*see* **Table 2** and **Notes 4–7**).
10. Wash slides between steps by placing them in a Coplin jar containing PBS for 10 min at room temperature on a shaker at low speed. Repeat twice.
11. After final washes, drain slides and place them on racks in the tray.
12. Using a disposable pipette, cover tissue with DAB solution and monitor by eye or time for reaction, using control slide to assess specificity.
13. After desired reaction has taken place, remove slides from rack and place in a jar containing PBS; this will stop further colour development (*see* **Note 8**).
14. Wash again with new change of PBS, and transfer to distilled water for a quick rinse.
15. Place slides in staining jar and stain for 5 min using previously filtered Methyl Green stain. Tip off stain (it can be reused) and rinse slides in dH_2O.
16. Dehydrate tissues in 70, 80, 96 and 100% ethanol and then clear by immersing slides twice in Histoclear. Mount slides with DPX and leave to set at 37 °C.

For the use of peroxidase-based three-step detection systems, refer to **Table 3**.

Table 2
Two-Step Single Labelling

Bright field – peroxidase	Bright field – alkaline phosphatase
Blocking (30 min)[a]	Blocking (30 min)[a]
Primary antibody (60 min)[b]	Primary antibody (60 min)[b]
Secondary antibody – peroxidase-conjugated (30–45 min)[c]	Secondary antibody – alkaline-phosphatase-conjugated (30–45 min)[c]
Development – DAB (until colour develops)[d]	Development – BCIP/NBT (until colour develops)[e]
Counterstaining – Methyl Green[d]	Counterstaining – not recommended[e]
Mounting – DPX[d]	Mounting – AquaPerm[e]

[a] See **Note 5**.
[b] See **Note 6**.
[c] See **Note 7**.
[d] See **Subheading 3.1.1.** and **Note 8**.
[e] See **Subheading 3.1.2.** and **Notes 9** and **10**.

Table 3
Three-Step Single Labelling

Bright field – peroxidase	Bright field – alkaline phosphatase
Blocking (30 min)[a]	Blocking (30 min)[a]
Avidin/Biotin blocking – (15 min + 15 min)[b]	Avidin/Biotin blocking – (15 min + 15 min)[b]
Primary antibody (60 min)[c]	Primary antibody (60 min)[c]
Biotinylated secondary antibody (30–45 min)[d]	Biotinylated secondary antibody (30–45 min)[d]
ABC Peroxidase Vector Vectastain Elite ABC Kit (30 min)[e]	ABC Alkaline Phosphatase Vector Vectastain Elite ABC Kit (30 min)[e]
Development – DAB (until colour develops)[f]	Development – BCIP/NBT (until colour develops)[g]
Counterstaining – Methyl Green[f]	Counterstaining – not recommended[g]
Mounting – DPX[f]	Mounting – AquaPerm[g]

[a] See **Note 5**.
[b] See **Note 11**.
[c] See **Note 6**.
[d] See **Note 7**.
[e] See **Note 12**.
[f] See **Subheading 3.1.1.** and **Note 8**.
[g] See **Subheading 3.1.2.** and **Notes 9** and **10**.

3.1.2. Alkaline Phosphatase Labelling

1. Perform steps 1–6 (*see* **Subheading 3.1.1.**).
2. Place slides on the staining tray for the incubations with primary and secondary reagents (*see* **Table 2**).
3. Wash slides between steps by placing them in a Coplin jar containing PBS for 10 min at room temperature on a shaker at low speed. Repeat twice.
4. After final washes, drain slides and place them on racks in the tray.
5. Using a disposable pipette, cover tissue with BCIP/NBT solution (*see* **Note 9**) and monitor by eye or time for reaction, using control slide to assess specificity.
6. After desired reaction has taken place, remove slides from rack and place in a jar containing PBS; this will stop further colour development.
7. Wash again with new change of PBS, and transfer to distilled water for a quick rinse.
8. Cover tissue with a thin layer of AquaPerm (*see* **Note 10**).
9. Leave to solidify at room temperature.

For the use of alkaline phosphatase-based three-step detection systems, refer to **Table 3**.

For the detection of two antibodies using two rat IgG using peroxidase and alkaline phosphatase detection systems, refer to **Table 4**.

3.1.3. Double Immunofluorescence Labelling

1. Perform **steps 1–6** (*see* **Subheading 3.1.2.**).
2. Incubate slides as described in **Table 5**.
3. After final washes, drain slides and place them on Coplin jar containing DAPI solution (*see* **Note 17**).
4. Wash slides in PBS.
5. Mount slides with fluorescence-mounting medium and leave to set at 4 °C (*see* **Note 18**).

3.2. Analysis of Mϕ Populations by Flow Cytometry

3.2.1. Preparation of Cell Suspensions

3.2.1.1. COLLAGENASE DIGESTION OF SPLEENS

1. Inject 'liberase blendzyme II' into spleen and put at 37 °C for 10 min.
2. Cut spleen into small chunks and harvest loose cells.
3. Add more liberase and continue digestion for 10 min.
4. After 5 min, pipette up and down to loosen tissue (use a cut blue tip).
5. Harvest loose cells and repeat a third time (*see* **Note 19**).
6. Transfer harvested cells to 5 ml of 'Stop' medium, which is RPMI with 20% FCS (*see* **Note 20**).
7. Centrifuge at 350 × g for 10 min and remove supernatant.

Table 4
Double Immunolabelling Bright Field

Double labelling with two rat IgG (Bright field)
Blocking (30 min)^a
Avidin/Biotin blocking (15 min + 15 min)^b
Primary antibody (60 min)^c
Secondary antibody – peroxidase (30–45 min)^d
100 µg/ml rat IgG (30 min)^e
Biotinylated rat IgG^e
ABC Alkaline Phosphatase Vector Vectastain Elite ABC Kit^f
Development – BCIP/NBT (until colour develops)^g
Development – DAB (until colour develops)^h
Counterstaining – not recommended^i
Mounting – Aquaperm^i

^{a} See **Note 5**.
^{b} See **Note 11**.
^{c} See **Note 6**.
^{d} See **Note 7**.
^{e} See **Note 13**.
^{f} See **Note 12**.
^{g} See **Notes 9** and **14**.
^{h} See **Note 8**.
^{g} See **Note 10**.

3.2.1.2. COLLAGENASE DIGESTION OF LYMPH NODES

1. Place the organs in the collagenase solution and break capsule with two needles.
2. Process as described in **steps 1–7** (*see* **Subheading 3.2.1.1.**).

3.2.1.3. LYSIS OF RED BLOOD CELLS (SPLEEN ONLY)

1. Add 1 ml of cold Gey's hypotonic solution and resuspend cell pellet from one to two spleens.
2. Add a further 9 ml of Gey's solution and incubate at 4°C for 3–6 min.
3. Use a long glass Pasteur pipette to underlay solution with 2 ml of 100% FCS.
4. Centrifuge at 300 × *g* for 10 min (*see* **Note 21**).
5. Resuspend cells in FACS blocking solution and count them.

3.2.2. Labelling

1. Count cells and resuspend the desired number in FACS blocking solution with 2.4G2 at 4 µg/ml at 4°C (*see* **Note 22**).
2. Leave in FACS blocking solution for at least 30 min (*see* **Note 23**).

Table 5
Double Immunolabelling Fluorescence

Double labelling with two rat IgG (immunofluorescence)

Blocking (30 min)^a
Avidin/Biotin blocking (15 min + 15 min)^b
Primary antibody (60 min)^c
Secondary antibody – Alexa-488/FITC (30–45 min)^d
100 µg/ml rat IgG (30 min)^e
Biotinylated rat IgG (60 min)^c
Streptavidin Cy5 (30 min)^{f,g}
Counterstaining – DAPI (2 min)^h
Mounting – fluorescent mounting medium

^{a} See **Note 5**.
^{b} See **Note 11**.
^{c} See **Note 6**.
^{d} See **Note 7**.
^{e} See **Note 13**.
^{f} See **Note 15**.
^{g} See **Note 16**.
^{h} See **Note 17**.

3. Take 50-µl aliquots of cells into 96-well chilled V-bottom bacterial plastic plates on ice and add 50 µl of 2× antibody mix (antibodies diluted in FACS washing solution; see **Table 1**).
4. Leave at 4°C for 1 h.
5. Spin down at 350 × g to remove antibodies and wash 2× with FACS washing solution (see **Note 24**).
6. Wash three times as above, and then fix cells with 1% formaldehyde (resuspend cells in FACS wash and add to FACS tube containing equal volume of 2% formaldehyde to cells).
7. Store at 4 °C in the dark (see **Notes 25–29**).

3.3. Culture of Peritoneal Mɸ

3.3.1. Harvesting of Peritoneal Cells

1. For Bio Gel-elicited Mɸ, inject mice intraperitoneally with 1 ml of 2% Bio Gel suspension and harvest the peritoneal cells on day 4 or 5. For Brewer's complete thioglycollate broth-elicited Mɸ, inject mice intraperitoneally with 1 ml of medium and harvest the peritoneal cells on day 4.
2. To harvest the cells, kill the mice by CO_2 asphyxiation or cervical dislocation.
3. Sterilize abdomens with 70% ethanol.

4. Make a small off-centre skin incision over the caudal half of the abdomen with scissors and expose the underlying abdominal wall by retraction.
5. Inject 10 ml sterile PBS into the caudal half of the peritoneal cavity using a 25-gauge needle (bevelled side up).
6. Gently shake the entire body for 10 s.
7. Slowly withdraw saline containing resident peritoneal cells by inserting a 25-gauge needle, bevelled side down, into the cranial half of the peritoneal cavity (*see* **Note 30**).

3.3.2. Selection of Peritoneal Mϕ by Adhesion

1. Collect peritoneal cells by centrifugation at $250 \times g$ at $4\,°C$.
2. Resuspend cells in Mϕ culture medium and count them (*see* **Note 31**).
3. Plate the peritoneal cells at a suitable density (for instance 10^6 cells per well in a six-well plate provide a good monolayer) and incubate for 2 h at $37°C$ (*see* **Note 32**).
4. After the adherence step, wash thioglycollate-elicited cells three times in Opti-MEM, a defined serum-free medium (to provide a source of calcium in the case of bacteriologic plastic, PBS can be used if using tissue culture plastic), and place in culture medium, ready to use. Bio Gel-elicited Mϕ require more thorough washing as the beads can be difficult to remove (*see* **Notes 33** and **34**).
5. Culture Mϕ in a humidified incubator at $37\,°C$ containing 5% CO_2.

The bone marrow offers an additional source of primary Mϕ. Resident bone marrow Mϕ can be obtained through collagenase digestion of bone marrow plugs and express high levels of sialoadhesin *(2)*. Bone marrow-derived Mϕ are generated in culture by incubating bone marrow cells in the presence of the growth factor M-CSF. Further information on the culture of murine Mϕ can be found in **ref.** *(3)* (*see* **Notes 35 –37**).

4. Notes

1. DAB is a potential carcinogen and it should be used in a ventilated hood. It is most dangerous in powder form, and is not to be removed from the container in this state.
2. We routinely place several organs with similar consistency per block as this does not affect cutting.
3. Some antibodies are fixation sensitive. We have found that fixation with paraformaldehyde is suitable for most of the antibodies used. Alternative fixatives are acetone or methanol.
4. Several labelling procedures have been outlined in **Tables 2–5** to provide an idea of the different combinations available when doing immunolabelling. These will need to be altered if antibodies raised in different species (i.e., rabbit or goat) are used and if labelling for more than two antibodies within the same sample is required.

5. The species of origin of the normal serum to be used during the blocking step should be the same as the one used to raise the secondary antibody.
6. Primary antibodies are diluted in blocking solution. If using purified reagents, depending on the quality of the reagent, concentrations ranging from 1 to 10 µg/ml should be used. Preliminary titration experiments are recommended. Hybridoma supernatants can also be used.
7. Secondary antibodies are diluted in blocking solution. Most commercial reagents can be successfully used at a 1:100 dilution, but titration experiments are recommended.
8. The tray and all contents should be placed in the sink and filled with ~5% solution of bleach, making sure everything is submerged or filled with the solution. It should then be left overnight, after which it can be flushed away with copious amounts of water, and the contents disposed of appropriately to sharps box/bin.
9. Developing solution should appear clear prior to use. If deposits are observed it should be filtered.
10. The precipitate formed using BCIP/NBT can be easily extracted with most counterstaining procedures. AquaPerm is an aqueous-based mounting medium that offers a safe way to protect tissue reacted using alkaline phosphatase. Once applied, this reagent will form a solid thin film over the tissues after overnight incubation. If desired, tissue can be mounted using DPX once the AquaPerm layer has solidified.
11. For the Avidin/Biotin blocking step, solutions provided by manufacturer are used neat. Tissues are incubated 15 min with avidin, washed in PBS and then incubated 15 min with biotin.
12. Prepare reagent at least 30 min before needed.
13. Rat IgG is diluted in blocking solution. This blocking step ensures that the anti-rat IgG reagent used in the previous step will not recognize the biotinylated rat IgG used for the detection of a secondary antibody.
14. For double labelling, development of the alkaline phosphatase activity prior to the peroxidase activity is recommended.
15. The combination of fluorochromes to be used will depend on the markers of interest. Cy5 should be avoided for antibodies present at low amounts. We have successfully used biotinylated KT-3 (anti-CD3) and streptavidin Cy5 to determine location of T-cell areas in lymphoid tissues. This approach can also be used to determine co-localization of two markers within a particular population using two rat mAbs. We have used this procedure to analyse the presence of mannose receptor in $CD68^+$ cells.
16. Streptavidin should be diluted in PBS, not in blocking solution, as it will bind to the biotin present in serum.
17. DAPI counterstaining can be performed for 2–5 min depending on the tissue. Such labelling is useful in assessing the general morphology of the tissues under the fluorescence microscope and will greatly facilitate focusing of the images. Too much labelling will make this more difficult.

18. We have been successful in analysing double-labelled tissues counterstained with DAPI using a fluorescence microscope, but for more than two markers or for tissues with high levels of autofluorescence, confocal microscopy is recommended.
19. By this time, the blue tip will probably not need to be cut.
20. As an option, filter cells through a 70-μm nylon cell strainer to get a single cell suspension.
21. You should obtain a mostly white cell pellet.
22. Cell counts must be precise for mixed populations to allow accurate determination of cell numbers after FACS.
23. To detect intracellular antigens, such as CD68, or to quantify intracellular pools of antigen, fix cells at 4°C for 30 min using a solution of 2% paraformaldehyde in PBS and add 0.5% saponin to all buffers. This procedure requires vigorous washing to facilitate the elimination of unbound Ab within the cells and, therefore, reduction of non-specific labelling. The use of biotinylated reagents is not recommended in this case.
24. If using biotinylated primaries, resuspend cells after two washes in 100 μl of 1/100 streptavidin–fluorochrome [phycoerythrin (PE) gives the highest signal to noise and allophycocyanin (APC) is next best] for 30 min at 4°C. Always include one sample that has streptavidin conjugate, but no primary antibody (or preferably a subclass control). Streptavidin conjugates must be diluted in FACS washing solution and not FACS blocking solution.
25. Mφ are notorious for displaying non-specific binding to primary and secondary reagents used in flow cytometry. For that reason, efforts have to be made to minimize non-specific interactions during the labelling procedure. This can be facilitated by performing the blocking and labelling steps in the presence of purified mAb 2.4G2. When employing 2.4.G2, a rat IgG, directly conjugated primary antibody should be used to avoid recognition of 2.4G2 itself by the anti-rat reagents and problems with cross-reactivity between species. Antibodies directly conjugated to a fluorochrome will also facilitate labelling of multiple surface markers in the same cell population.
26. Direct labelling would be suitable for markers well represented at the cell surface. For other markers, such as receptors mostly restricted to the endocytic compartment, i.e., the mannose receptor, scavenger receptor class A or MARCO, amplification through the use of biotinylated primary antibody and an additional labelling step with fluorochrome-labelled streptavidin (BD biosciences) is recommended. A better signal will be obtained if a strong fluorochrome like PE or APC is used.
27. Although not perfect, the inclusion of isotype-matched antibodies as controls in the staining procedure will provide an indication of the specificity of the labelling.
28. The most common fluorochromes used are fluorescein isothiocyanate (FITC)/Alexa 488 (FL-1), PE (FL-2), CyChrome (PE-Cy5); peridinin-chlorophyll-protein (PerCP); PerCP-Cyanine5.5 (PerCP-Cy5.5) (FL-3) and APC (FL-4). Note

that Mφ have higher levels of autofluorescence than other cells and this should be taken into consideration when designing the labelling procedure. Autofluorescence is low in FL-4; therefore this channel might be preferred for the detection of markers expressed at low levels.

29. Number of cells to be used per staining will depend on the complexity of the population. For spleen cells, 10^6 cells per labelling are recommended to ensure detection of minor populations. In the case of cultured Mφ 1×10^5 to 3×10^5 cells, per labelling is recommended.
30. Material harvested should appear pale, and presence of red blood cells will indicate blood contamination. These samples should be discarded. Assess levels of blood contamination before pooling samples from several animals.
31. Trypan blue exclusion can be used to determine cellular viability. The use of Turk's fluid will allow the morphological identification of Mφ as these cells display a unique 'fried egg'-like appearance under these conditions.
32. The choice of surface to be used will depend on the intended use for the cells. In studies involving preparation of cell lysates and analysis of secreted products, tissue culture plastic is recommended as both Bio Gel- and Thioglycollate-elicited Mφ would readily bind to this surface. For studies requiring the recovery of cells for flow cytometry analysis upon treatment or for the analysis of scavenger receptor class A function, bacteriologic plastic is recommended as cells can be easily recovered without damage using 10 mM EDTA and 4 mg/ml of lidocaine in PBS, and scavenger receptor class A function, such as bacterial uptake, is not compromised by its involvement to adhesion to serum coated tissue culture plastic *(4)*. Bio Gel-elicited Mφ do not adhere very well to bacterial plastic and a longer adhesion step might be required.
33. Use of plastic Pasteur pipettes for the washes in order to resuspend the beads during each wash facilitates this procedure, and most beads will be removed after three washes. Once this is achieved, culture medium can be added. Alternatively, harvested cell suspensions kept in ice-cold PBS can be filtered through a 70 μM cell strainer (BD Falcon) to eliminate the beads prior to plating.
34. Plating of Mφ on acid-treated cover slips or Lab-Tek chamber slides (Nunc) is recommended for the phenotypic analysis of Mφ in situ. Both surfaces provide an appropriate support for immunolabelling as described in **Subheading 3.1**.
35. Bio Gel beads cannot be phagocytosed or digested by Mφ; therefore Bio Gel-elicited Mφ do not contain intracellular debris.
36. Thioglycollate-elicited Mφ ingest large amounts of agar but retain active endocytic and phagocytic functions on isolation. In addition, even sterile thioglycollate broth often contains low levels of lipopolysaccharide, or other microbial contaminants that could affect Mφ behaviour.
37. For bulk culture, bacteriologic plastic vessels facilitate cell detachment.

Acknowledgements

This work has been supported by the Medical Research Council, UK and The University of Nottingham. The authors thank Philip R. Taylor, Elizabeth Darley and Richard Stillion for providing useful information on the different techniques described.

References

1. Taylor, P.R., L. Martinez-Pomares, M. Stacey, H.H. Lin, G.D. Brown, and S. Gordon. 2005. Macrophage receptors and immune recognition. *Annu Rev Immunol* 23:901–944.
2. Crocker, P.R., and S. Gordon. 1986. Properties and distribution of a lectin-like hemagglutinin differentially expressed by murine stromal tissue macrophages. *J Exp Med* 164:1862–1875.
3. Davies, J.Q., and S. Gordon. 2004. Isolation and culture of murine macrophages. In *Basic Cell Culture Protocols*. C.D. Helgason and C.L. Miller, editors. Humana Press Inc. Totowa, N.J., 91–100.
4. Peiser, L., P.J. Gough, T. Kodama, and S. Gordon. 2000. Macrophage class A scavenger receptor-mediated phagocytosis of Escherichia coli: role of cell heterogeneity, microbial strain, and culture conditions in vitro. *Infect Immun* 68: 1953–1963.
5. Holness, C.L., R.P. da Silva, J. Fawcett, S. Gordon, and D.L. Simmons. 1993. Macrosialin, a mouse macrophage-restricted glycoprotein, is a member of the lamp/lgp family. *J Biol Chem* 268:9661–9666.
6. Rabinowitz, S., H. Horstmann, S. Gordon, and G. Griffiths. 1992. Immunocytochemical characterization of the endocytic and phagolysosomal compartments in peritoneal macrophages. *J Cell Biol* 116:95–112.
7. Crocker, P.R., S. Mucklow, V. Bouckson, A. McWilliam, A.C. Willis, S. Gordon, G. Milon, S. Kelm, and P. Bradfield. 1994. Sialoadhesin, a macrophage sialic acid binding receptor for haemopoietic cells with 17 immunoglobulin-like domains. *EMBO J* 13:4490–4503.
8. Geijtenbeek, T.B., P.C. Groot, M.A. Nolte, S.J. van Vliet, S.T. Gangaram-Panday, G.C. van Duijnhoven, G. Kraal, A.J. van Oosterhout, and Y. van Kooyk. 2002. Marginal zone macrophages express a murine homologue of DC-SIGN that captures blood-borne antigens in vivo. *Blood* 100:2908–2916.
9. Raes, G., L. Brys, B.K. Dahal, J. Brandt, J. Grooten, F. Brombacher, G. Vanham, W. Noel, P. Bogaert, T. Boonefaes, A. Kindt, R. Van den Bergh, P.J. Leenen, P. De Baetselier, and G.H. Ghassabeh. 2005. Macrophage galactose-type C-type lectins as novel markers for alternatively activated macrophages elicited by parasitic infections and allergic airway inflammation. *J Leukoc Biol* 77:321–327.
10. Reid, D.M., M. Montoya, P.R. Taylor, P. Borrow, S. Gordon, G.D. Brown, and S.Y. Wong. 2004. Expression of the beta-glucan receptor, Dectin-1, on murine

leukocytes in situ correlates with its function in pathogen recognition and reveals potential roles in leukocyte interactions. *J Leukoc Biol* 76:86–94.
11. Willment, J.A., H.H. Lin, D.M. Reid, P.R. Taylor, D.L. Williams, S.Y. Wong, S. Gordon, and G.D. Brown. 2003. Dectin-1 expression and function are enhanced on alternatively activated and GM-CSF-treated macrophages and are negatively regulated by IL-10, dexamethasone, and lipopolysaccharide. *J Immunol* 171: 4569–4573.
12. Martinez-Pomares, L., D.M. Reid, G.D. Brown, P.R. Taylor, R.J. Stillion, S.A. Linehan, S. Zamze, S. Gordon, and S.Y. Wong. 2003. Analysis of mannose receptor regulation by IL-4, IL-10, and proteolytic processing using novel monoclonal antibodies. *J Leukoc Biol* 73:604–613.
13. Kraal, G., L.J. van der Laan, O. Elomaa, and K. Tryggvason. 2000. The macrophage receptor MARCO. *Microbes Infect* 2:313–316.
14. Hughes, D.A., I.P. Fraser, and S. Gordon. 1995. Murine macrophage scavenger receptor: in vivo expression and function as receptor for macrophage adhesion in lymphoid and non-lymphoid organs. *Eur J Immunol* 25:466–473.
15. Taylor, P.R., D.M. Reid, S.E. Heinsbroek, G.D. Brown, S. Gordon, and S.Y. Wong. 2005. Dectin-2 is predominantly myeloid restricted and exhibits unique activation-dependent expression on maturing inflammatory monocytes elicited in vivo. *Eur J Immunol* 35:2163–2174.
16. Taylor, P.R., M.C. Pickering, M.H. Kosco-Vilbois, M.J. Walport, M. Botto, S. Gordon, and L. Martinez-Pomares. 2002. The follicular dendritic cell restricted epitope, FDC-M2, is complement C4; localization of immune complexes in mouse tissues. *Eur J Immunol* 32:1888–1896.
17. Metlay, J.P., M.D. Witmer-Pack, R. Agger, M.T. Crowley, D. Lawless, and R.M. Steinman. 1990. The distinct leukocyte integrins of mouse spleen dendritic cells as identified with new hamster monoclonal antibodies. *J Exp Med* 171: 1753–1771.
18. Kraal, G., M. Breel, M. Janse, and G. Bruin. 1986. Langerhans' cells, veiled cells, and interdigitating cells in the mouse recognized by a monoclonal antibody. *J Exp Med* 163:981–997.

16

Clinical Analysis of Dendritic Cell Subsets
The Dendritogram

Anne Hosmalin, Miriam Lichtner, and Stéphanie Louis

Summary

Dendritic cells (DCs) are crucial in adaptive immunity because they are the only antigen-presenting cells that can present antigens to naive T lymphocytes. Plasmacytoid DCs (pDC) are also the main producers of type I Interferons in response to infection. We have shown that circulating myeloid DC (mDC) and pDC numbers are reduced in chronic as well as primary HIV infection. Data from different laboratories indicate that pDC counts, obtained by flow cytometry and rare event analysis, correlate inversely with the viral load, may be an early marker of recovery after antiretroviral treatment, and may predict better immune control of HIV replication. PDC counts may also be predictive of severe illness in dengue virus infection or of successful treatment against *Mycobacterium tuberculosis*. DC counts, or the "dendritogram", may therefore become useful in the clinical assessment of different infectious diseases.

Key Words: Plasmacytoid dendritic cells; myeloid dendritic cells; monocytes; flow cytometry; rare event analysis.

1. Introduction

Dendritic cells (DCs) are considered as "professional" antigen-presenting cells. In the periphery, DCs circulate in an immature state in which they exhibit a high capacity for antigen uptake and processing. After encountering infectious agents, they become activated and migrate to the draining lymphoid organs where they are competent to activate T cells. In humans, two main subsets have been identified in the peripheral blood: myeloid and plasmacytoid DC *(1, 2)*. Plasmacytoid DC (pDC), also called "Natural Interferon Producing Cells," represent 0.2–0.8% of peripheral blood cells in both humans and mice and have also been found in the spleen, bone marrow, tonsil, and lymph nodes *(3–5)*. They participate in innate immunity with a specialized production of type I interferons (IFNα, β, ω) *(6)*, which have a strong antiviral activity and enhance the cytotoxicity of NK cells and $CD8^+$ T cells *(7)*. The pDCs also contribute to adaptive immunity by promoting IFNγ production by $CD8^+$ and $CD4^+$ T cells. When driven to mature by virus infection such as influenza, they strongly stimulate allogeneic mixed reactions and specific $CD4^+$ and $CD 8^+$ T cell responses; thus, pDC represent a important link between innate and adaptative immune responses *(8)*. Myeloid DC (mDC) display a high plasticity in their function depending on the type of maturation stimuli and the kinetics of activation *(9, 10)*. Present in all the tissues as interstitial DC and in blood, they initiate adaptive immune responses. In response to pathogens, they produce IL12, a cytokine required for IFNγ production by NK and T cells and for cytotoxic T-cell activity.

A decrease in the number of both DC subsets has been observed in viral infections such as HIV or HCV *(11–15)*, with an inverse correlation between pDC counts and viral load and a correlation between pDC counts and $CD4^+$ T-cell counts being found in several studies. Moreover, a partial recovery of pDC counts was observed after highly active antiretroviral therapy (HAART) during chronic infection *(16)* and probably during primary infection *(17)*. Recovery of pDC counts during HAART may be predictive of immune control of viral replication after HAART interruption *(18)*. Correlations were found more rarely between mDC counts and viral loads or $CD4^+$ T-cell counts *(19)*. In most cases, pDC counts correlate with in vitro type I IFN production, which was shown to recover earlier than $CD4^+$ T-cell counts during HAART *(20)*. Thus, DC counts, or a "dendritogram," may provide relevant information for the prediction of immune control of HIV replication during HAART or after HAART interruption. The "dendritogram" might be used in addition to the traditional parameters that are measured to try and assess HIV infection prognosis: baseline level $CD4^+$ T-cell counts and plasma HIV RNA, the nadir $CD4^+$ T-cell count, and parameters at the time of treatment interruption such as $CD4^+$ T-cell counts, proviral DNA load in PBMC, and anti-p24 lymphoproliferative response *(21)*. The "dendritogram" may also be predictive in other infections and other conditions. For instance, in dengue virus infection, children who developed hemorrhagic dengue fever had an early decrease in circulating pDC levels that was not found in those who had more benign forms of the disease *(22)*. Also, during *Mycobacterium tuberculosis* infection, mDC and pDC counts are reduced, but pDC numbers return to normal after successful antibiotic treatment, which is otherwise often difficult to evaluate *(23)*.

Different four-color labelings are available in commercial kits from Becton-Dickinson *(24, 25)*, Beckman-Coulter *(26)* or Miltenyi Biotec *(27)*. Plasmacytoid DCs are relatively well characterized by the lack of certain lineage-specific markers, the high expression of human leukocyte antigen (HLA)-DR and either the expression of the interleukin 3 (IL3) receptor (CD123) or the expression of BDCA2 or BDCA4 *(27)*. In contrast, the distinction of myeloid DC from monocytes, from which they originate, is more difficult, as they share the expression of many molecules such as CD11c, HLA-DR, CD14, CD33, or ILT3. Moreover, the current commercial kits for DC quantification do not allow to count the "intermediate" population ($CD14^+CD16^+$) between myeloid DC and monocytes. The quantification of "intermediate" monocytes may be interesting because they are expanded in different pathological situations such as HIV infection or autoimmune diseases *(28, 29)*. Recently, the emergence of multicolor cytometers allows the study of the two DC subsets and monocytes in a single test.

Therefore, the following methods are described for different uses:

1. Four-color labeling using the whole blood single platform TruCOUNT assay, most accurate for clinical DC counts.
2. Four-color labeling using a dual platform assay, available widely, on whole blood or on fresh peripheral blood mononuclear cells (PBMC), for clinical DC counts and/or characterization.
3. Five-color labeling allowing DC and monocyte counts and determination of other parameters on these populations, on fresh PBMC.

In any case, DC quantification requires the acquisition of large numbers of events and a rare event analysis. The protocols described here will certainly evolve in time to keep up with the standardization that is in progress and that is required for clinical use.

2. Materials

2.1. Single Platform TruCOUNT Assay and Dual Platform Assay with Whole Blood

1. Peripheral blood collected into glass whole blood tubes with K_3EDTA (ethylenediamine tetraacetic acid) (Ref: 366450, Becton Dickinson, Le Pont de Claix, France). It must be kept at room temperature until use. Blood counts, including total leukocyte, lymphocyte, and monocyte counts, must be obtained from another platform, that is, an automated hematology blood analyzer to determine absolute cell numbers if the single platform TruCOUNT assay is not used (*see* **Note 1**).
2. TruCOUNT tubes (Becton Dickinson).
3. 5-ml polystyrene round-bottom tubes (Becton Dickinson).

4. FACS lysing solution 10× (Becton Dickinson) diluted to 1× in distilled water. It must be kept at room temperature.
5. Phosphate-buffered saline (PBS) 1×: PBS 10× (Gibco, Invitrogen, Cergy Pontoise, France) diluted at 1:10 in sterile, distilled water, can be kept at 4°C.
6. Human serum from HIV-negative, AB blood group, male donors (ABS), previously decomplemented at 56°C for 30 min (Biowest) and kept in aliquots at −20°C.
7. Fixation buffer: PBS with 1% paraformaldehyde (PFA) (Sigma, St Quentin Fallauier, France) (*see* **Note 2**). A large amount of PFA dissolved in PBS can be prepared, aliquoted and stored at −20°C. Owing to its instability, an aliquot must be thawed before each use.

2.2. Isolation of Peripheral Mononuclear Cells

1. Peripheral blood collected into glass whole blood tube with K_3EDTA (Ref: 366450, Becton Dickinson) or BD Vacutainer® plasma tubes with 17UI/ml of lithium heparin (Ref: 366480, Becton Dickinson). It must be kept at room temperature until use. Blood counts, including total leukocyte, lymphocyte, and monocyte counts, must be obtained from another platform, that is, an automated hematology blood analyzer to determine absolute cell numbers if the single platform TruCOUNT assay is not used (*see* **Note 1**).
2. Ficoll-Paque Plus (Amersham Biosciences, Burkinghamshire, United Kingdom), kept at room temperature.
3. 0.9% sodium chloride solution (NaCl) (Baxter, Maurepas, France), kept at room temperature.
4. 50-ml polypropylene conical tubes (Becton Dickinson).

2.3. Immunolabelings in Flow Cytometry

1. 5-ml polystyrene round-bottom tubes (Becton Dickinson).
2. EDTA 0.5 M pH 8.0 (Gibco), kept at room temperature.
3. Fetal bovine serum (FBS) previously decomplemented at 56°C for 30 min (Biowest, Nuaillé, France) and kept in aliquots at −20°C.
4. Staining buffer: PBS 1× with 5 mM EDTA and 2% FBS. It can be kept at 4°C (*see* **Note 3**).
5. Saturation buffer: PBS with 5% ABS. It must be prepared extemporaneously (*see* **Note 4**).
6. Fixation buffer: PBS with 1% PFA.

2.4. Antibodies (see Table 1)

1. Lineage Cocktail 1 (CD3, CD14, CD16, CD56, CD19, CD20) labeled with fluorescein isothiocyanate (FITC) (Becton Dickinson).
2. Anti-HLA-DR labeled with peridinin chlorophyll protein (PerCP) or with allophycocyanin (APC), clone L243 (Becton Dickinson).

3. Anti-CD11c labeled with phycoerythrin (PE) or APC, clone B-ly 6 (Becton Dickinson).
4. Anti-CD123 labeled with PE, clone 9F5 (Becton Dickinson) or labeled with APC, clone AC145 (Miltenyi Biotec, Paris, France).
5. Anti-CD14 labeled with PE-cyanin 7 (PE-Cy7), clone M5E2 (Becton Dickinson).
6. Anti-CD16 labeled with APC-cyanin 7 (APC-Cy7), clone 3G8 (Becton Dickinson).
7. Anti-ILT3 labeled with PE-cyanin 5 (PE-Cy5), clone ZM3.8 (Beckman Coulter, Villepinte, France).
8. Anti-CD45 labeled with PerCP, clone 2D1 (Becton Dickinson).
9. Isotype controls corresponding to all the previous antibodies (*see* **Note 5**).

2.5. Acquisition and Analysis Software and Hardware

1. Cellquest Software version 3.3 (Becton Dickinson) (*see* **Note 6**).
2. DIVA Software version 1.0 (Becton Dickinson) (*see* **Note 6**).
3. USB key minimal size 126 Mo or rewritable disks for transfer of data (*see* **Note 7**).

3. Methods

3.1. Whole Blood Single-Platform TruCOUNT Assay (see Note 8)

3.1.1. Immunolabeling Based on the BD Commercial Lineage (see Note 9)

1. Check integrity of the TruCOUNT bead pellet and its location within the metal retainer. If not correct, discard.
2. Prepare the two mixes of antibodies diluted in PBS: (1) lin FITC (1:10) HLA-DR APC (1:40) CD45 PerCP (1:20) CD123 PE (1:10), (2) lin FITC (1:10) HLA-DR APC (1:40) CD45 PerCP (1:20) CD11c PE (1:20).
3. Pipette 20 µl of mixed antibodies above the steel retainer (*see* **Note 10**).
4. Add 100 µl of well-mixed whole blood onto the side of each tube, just above the retainer (*see* **Note 11**).
5. Cap tubes and vortex gently.
6. Incubate 20 min in the dark at room temperature.
7. Add 450 µl of FACS Lysing Solution 1×. Cap tubes and vortex gently.
8. Incubate 10 min maximum in the dark at room temperature.
9. Acquire events within 3 h. The number of acquired events is crucial to determine a significant number of DC (*see* **Note 12**). All data must be collected using identical instrument setting. The threshold is set on PerCP fluorescence.

3.1.2. Analysis and Quantification of pDC and mDC by Rare Event Analysis (see Note 12)

1. Set the threshold on PerCP fluorescence. Create a first dot plot with SSC/CD45 and draw a gate on live PBMC (R1) excluding fragments and dead cells (low CD45) and polymorphonuclear cells (high SSC, CD45+) (see **Fig. 1A** and **Note 13**).

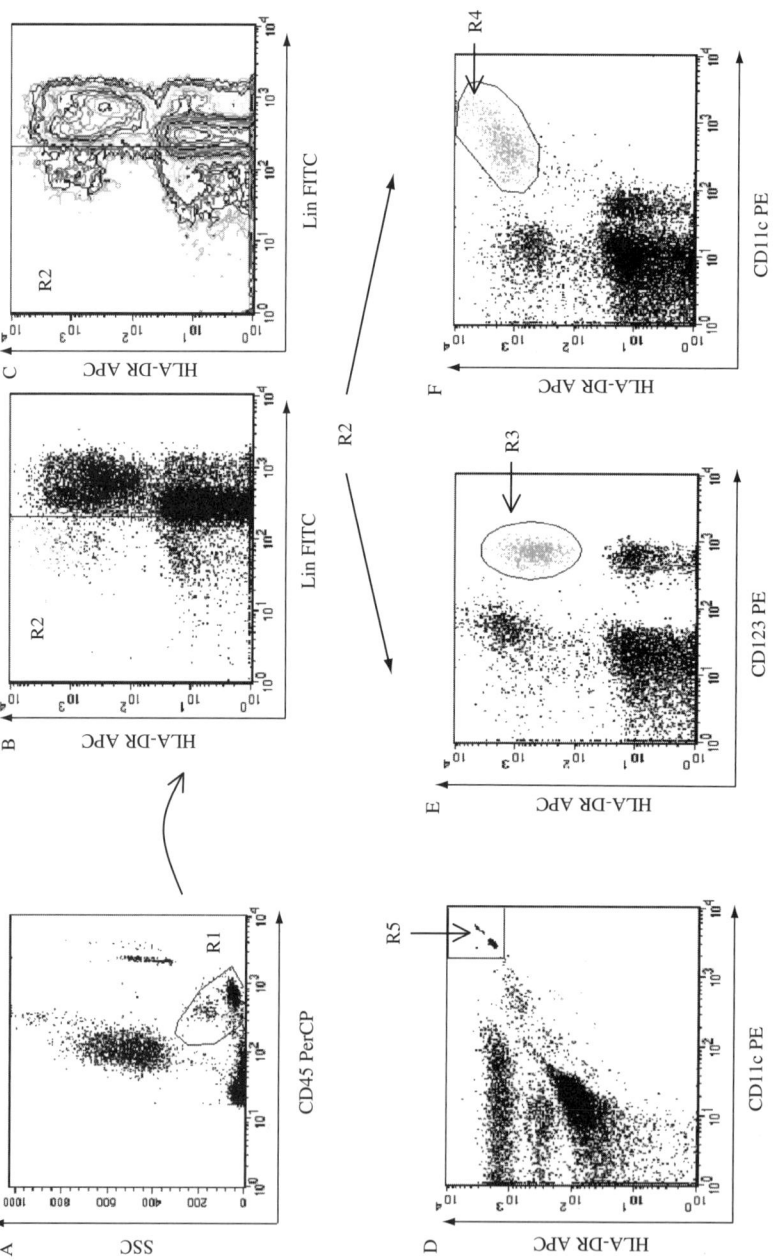

Fig. 1. "Dendritogram" on whole blood with the single platform TruCOUNT assay. After gating on live PBMC (R1; **A**), events are distributed in a dot plot (**B**) or a contour plot (**C**) of HLA-DR versus lin. Events included in a gate R2, drawn on the lin⁻ population, are shown on a dot plot of HLA-DR versus CD123 (**E**) or versus CD11c (**F**). Thus, pDCs correspond to the HLA-DR^{++} CD123$^+$ (R3) and mDCs to the HLA-DR^{++} CD11c^{++} (R4). To count DC subsets, a gate R5 was drawn on beads, which are fluorescent in all the channels (**D**). This analysis was performed with the Cellquest software.

2. Display events from gate R1 in a dot plot (see **Fig. 1B**) or a contour plot (see **Fig. 1C**) of lin FITC versus HLA-DR APC. Draw a new gate (R2) on the lin$^{-/low}$ populations, without excluding lineage low cells too drastically because some mDC express CD14; contour plot can be used to draw the lineage limit where the lowest HLA-DR expression is found between the lin$^{-/low}$ and the lin^{high} events (see **Fig. 1B and C**). The lymphocytes (T, B, NK) and most of the monocytes are excluded from R2 (see **Fig. 1B and C**).
3. For the first tube, display events from gate R2 in a dot plot of HLA-DR APC versus CD123 PE. Draw a gate (R3) on the HLA-DR^{++}CD123^{+} population, corresponding to pDC (see **Fig. 1E**).
4. For the second tube, display events from gate R2 in a dot plot of HLA-DR APC versus CD11c PE. Draw a gate (R4) on the HLA-DR^{++}CD11c^{++} population, corresponding to mDC (see **Fig. 1F**).
5. To determine bead count, in a dot plot of HLA-DR APC versus CD11c PE or CD123 PE (all the events), draw a gate (R5) on APC^{++} PE^{++} events (see **Fig. 1D** and **Note 14**).
6. Calculate the percentage of pDC (% pDC) and mDC (% mDC) among PBMC gate (R1).
7. To obtain the absolute values of pDC and mDC, use the following calculation:

pDC per ml of blood = (cell number in gate R3/bead number in R5) ×

(bead number in TruCOUNT tube/blood volume in ml)

m DC per ml of blood = (cell number in gate R4/bead number in R5) ×

(bead number in TruCOUNT tube/blood volume in ml)

3.2. Whole Blood Dual Platform Assay

3.2.1. Immunolabeling Based on the BD Commercial Lineage

1. Prepare, in the polystyrene tubes, the mix of antibodies: lin FITC (1:10) HLA-DR PerCP (1:20) CD123 PE (1:10) CD11c APC (1:20).
2. Add 100 µl of well-mixed whole blood with 5% ABS. Vortex gently.
3. Incubate 30 min in the dark at room temperature.
4. Lyse red cells with 1 ml of FACS lysing solution 1×, for 10 min, in the dark at room temperature.
5. Centrifuge at 150 g (850 rpm in a centrifuge with a radius of 12 cm) for 5 min at room temperature.
6. Remove the supernatant and wash with 2 ml of PBS 1×.
7. Centrifuge at 150 g for 5 min at room temperature.
8. Resuspend the pellet in 200 µl of fixation buffer.
9. Acquire events within 3 h. Cells must be kept at +4°C until acquisition.

3.2.2. Analysis and Quantification of pDC and mDC

1. Create a first dot plot with SSC/FSC and draw a gate on live PBMC (R1) excluding fragments and dead cells (low FSC) and polymorphonuclear cells (high FSC, SSC) (*see* **Fig. 2A** and **Note 13**).
2. Display events from gate R1 in a dot plot of lin FITC versus HLA-DR PerCP. Draw a new gate (R2) on the lin$^{-/\text{low}}$ populations, as described in the **Subheading 3.1.2** (*see* **Fig. 2B**).

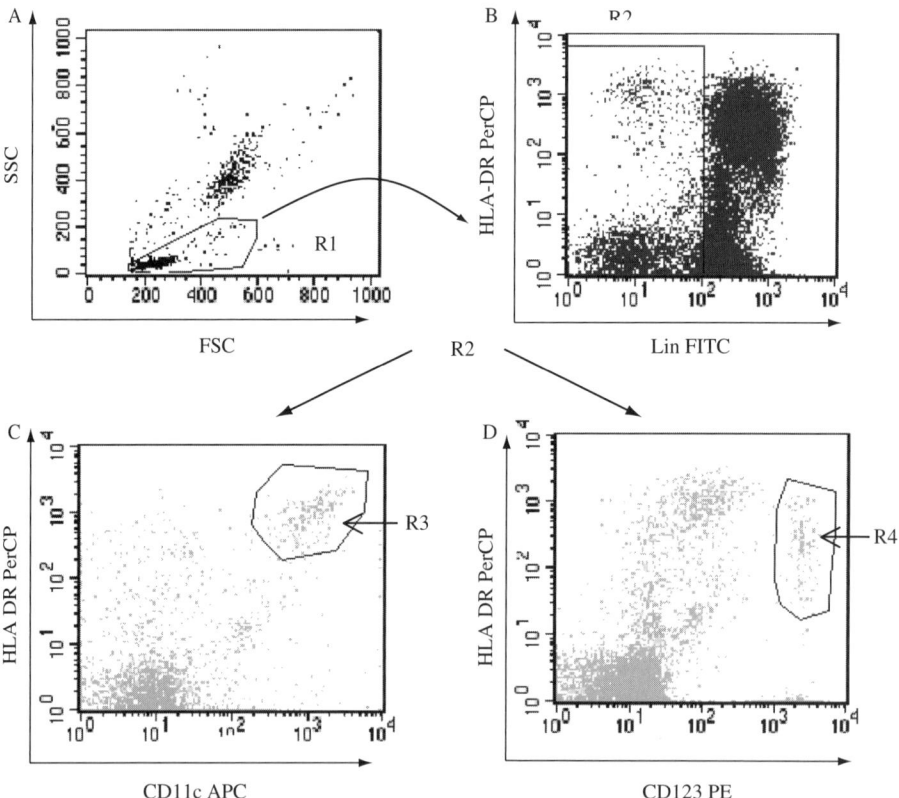

Fig. 2. "Dendritogram" on whole blood with a dual platform assay. After gating on live PBMC (R1; **A**), events are distributed in a dot plot of HLA-DR versus lin (**B**). Events included in a gate R2 (**B**), drawn on lin$^-$ events, are shown on a dot plot of HLA-DR versus CD11c (**C**) or versus CD123 (**D**). Thus, mDCs correspond to the HLA-DR^{++} CD11c$^+$ (R3), and pDCs correspond to the HLA-DR^{++} CD123$^+$ events (R4). This analysis was performed with the Cellquest software.

3. Display events from gate R2 in a dot plot of CD11c APC versus HLA-DR PerCP. Draw a gate (R3) on the HLA-DR^{++}CD11c^{++} population, corresponding to mDC (*see* **Fig. 2C**).
4. Display events from gate R2 in a dot plot of CD123 PE versus HLA-DR PerCP. Draw a gate (R4) on the HLA-DR^{++}CD123^{+} population, corresponding to pDC (*see* **Fig. 2D**).
5. Calculate the percentage of pDC (% pDC) and mDC (% mDC) among the PBMC gate (R1).
6. To obtain the absolute values of pDC and mDC, use the following calculation:

$$\text{pDC per ml of blood} = [\% \text{ pDC} \times (\text{lymphocyte count} + \text{monocyte count})] \times 10.$$

$$\text{mDC per ml of blood} = [\% \text{ mDC} \times (\text{lymphocyte count} + \text{monocyte count})] \times 10.$$

3.3. Isolation of PBMC (see Note 15)

1. Fresh blood samples are diluted in half with 0.9% NaCl.
2. A volume of Ficoll equivalent to that of undiluted blood is added into the 50-ml tubes.
3. Diluted blood (volume = 2/3) is carefully layered onto Ficoll (volume = 1/3) to minimize mixing of blood with Ficoll.
4. Centrifuge at 850 *g* (2000 rpm in a centrifuge with a radius of 12 cm) for 20 min at 18–20°C without break.
5. Draw off the upper layer containing plasma and platelets using a pipette, leaving the layer of mononuclear cells undisturbed at the interface. The upper layer of half-diluted plasma, which is essentially cell-free, may be saved for later use.
6. Collect the layer of mononuclear cells at the interface into new tubes.
7. Fill the tubes with 0.9% NaCl (*see* **Note 16**).
8. Centrifuge at 450 *g* (1500 rpm in a centrifuge with a radius of 12 cm) for 15 min at 18–20°C. Remove the supernatant and resuspend the pellets in NaCl.
9. Centrifuge at 400 *g* (1400 rpm in a centrifuge with a radius of 12 cm) for 10 min at 18–20°C. Remove the supernatant and resuspend the pellets in PBS 1×.
10. Determine cell number (*see* **Notes 17** and **18**).

3.4. DC Quantification Using Four-Color Labeling on PBMC (see Notes 19)

3.4.1. Immunolabeling Based on the BD Commercial Lineage

1. Distribute 10^6 cells in each polystyrene tube.
2. Centrifuge at 400 *g* for 5 min at +4°C. Remove the supernatant.
3. Resuspend the pellet in 200 µl of saturation buffer and incubate at +4°C for at least 15 min.

4. Prepare the mix of antibodies diluted in staining buffer: lin FITC (1:10) HLA-DR PerCP (1:20) CD123 PE (1:10) CD11c APC (1:20).
5. Centrifuge the cells at 400 g for 5 min at +4°C. Remove the supernatant.
6. Resuspend the pellet in 100 µl of antibody mix and incubate at +4°C for 30 min.
7. Add 3 ml of staining buffer and centrifuge at 400 g for 5 min at +4°C.
8. Remove the supernatant and repeat the previous step.
9. Resuspend the pellet in 100 µl of fixation buffer.
10. Acquire events within 4 h. Preferably, labeling must be measured immediately on the cytometer (*see* **Note 20**).

3.4.2. Analysis and Quantification of pDC and mDC

1. Create a first dot plot with SSC/FSC and draw a gate on live PBMC (P1) excluding fragments and dead cells (low FSC) and remaining polymorphonuclear cells (high FSC, SSC) (*see* **Fig. 3A** and **Note 13**).
2. Display events from gate P1 in a dot plot of lin FITC versus HLA-DR PerCP (*see* **Fig. 3B**). Draw a new gate (P2) on the lin$^{-/\text{low}}$ events, as described in the **Subheading 3.1.2**.
3. Display events from gate P2 in a dot plot of CD11c APC versus HLA-DR PerCP. Draw a gate (P3) on the HLA-DR^{++}CD11c^{++} population, corresponding to mDC (*see* **Fig. 3C**).
4. Display events from gate P2 in a dot plot of CD123 PE versus HLA-DR PerCP. Draw a gate (P4) on the HLA-DR^{++}CD123^{+} population, corresponding to pDC (*see* **Fig. 3C**).
5. Calculate the percentage of pDC (% pDC) and mDC (% mDC) among the PBMC gate (P1).
6. To obtain the absolute values of pDC and mDC, use the following calculation:

pDC per ml of blood $= [\% \, \text{pDC} \times (\text{lymphocyte count} + \text{monocyte count})] \times 10$.

mDC per ml of blood $= [\% \, \text{mDC} \times (\text{lymphocyte count} + \text{monocyte count})] \times 10$.

3.5. DC Quantification Using Five-Color Labeling on PBMC (see Note 21)

3.5.1. Immunolabeling Based on ILT3 Expression

1. Perform **steps 1–3** as described in **Subheading 3.2.1**.
2. Prepare the mix of antibodies diluted in staining buffer: CD14 PE-Cy7 (1:50) CD16 APC-Cy7 (1:50) ILT3 PE-Cy5 (1:50) CD123 PE (1:20) CD11c APC (1:20).
3. Perform **steps 5–10** as described in **Subheading 3.2.1**.
4. Acquire events within 4 h. Preferably, labeling is measured immediately on the cytometer (*see* **Note 20**).

Clinical Analysis of DC Subsets

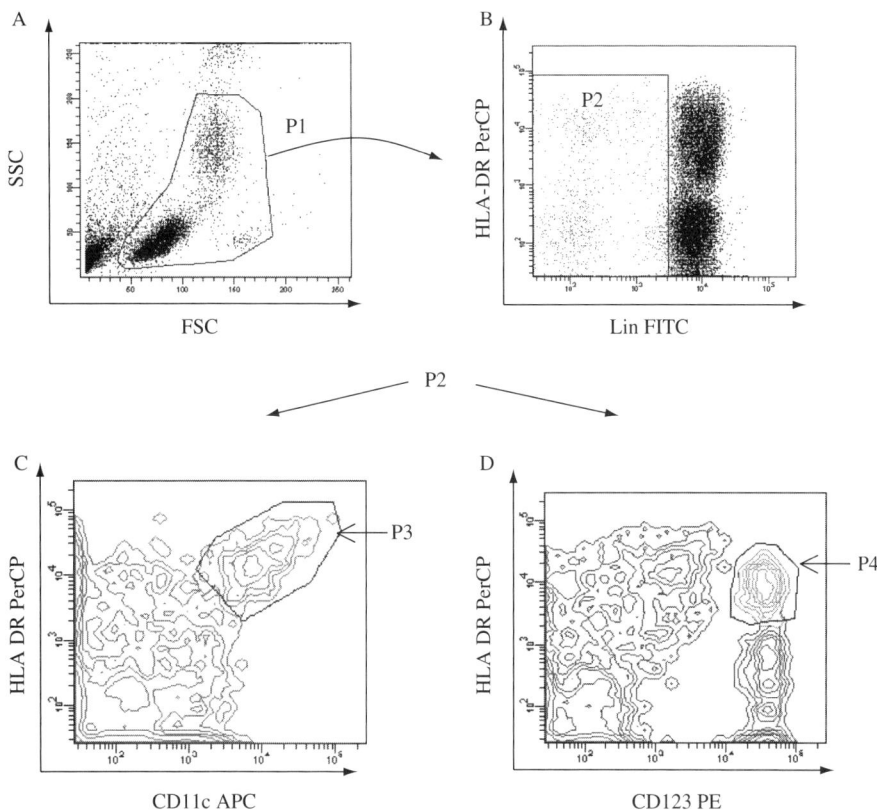

Fig. 3. "Dendritogram" on PBMC using four-color labeling. After gating on live PBMC (P1; **A**), events are distributed in a dot plot of HLA-DR versus lin (**B**). Events included in a gate P2, drawn on lin⁻ events, are shown on a dot plot of HLA-DR versus CD11c (**C**) or versus CD123 (**D**). Thus, the mDCs correspond to the HLA-DR^{++} CD11c^{++} (P3) and the pDCs to HLA-DR^{++} CD123^{+} (P4). This analysis was performed with the Diva software.

3.5.2. Analysis and Quantification of pDC and mDC, "Classical" and "Intermediate" Monocytes

1. Create a first dot plot with SSC/FSC and draw a gate on live PBMC (P1) excluding fragments and dead cells (low FSC) and resting polymorphonuclear cells (high FSC, SSC) (*see* **Fig. 4A** and **Note 13**).
2. Display events from gate P1 in a dot plot of SSC versus ILT3 PE-Cy5 (*see* **Fig. 4B**). Draw a new gate (P2) on ILT3^{+} events (*see* **Note 22**).

3. Display events from gate P2 in a dot plot of CD14 PE-Cy7 versus CD16 APC-Cy7. Three populations can be distinguished (*see* **Fig. 4C**). The first population (CD14^{++}CD16$^-$) corresponds to the "classical" monocytes (gate P3). The second population (CD14$^+$CD16$^+$) represents the "intermediate" monocytes between monocytes and myeloid DCs (gate P4).
4. Draw a gate (P5) on the CD14$^{-/\text{low}}$CD16$^-$ population (*see* **Fig. 4C**).

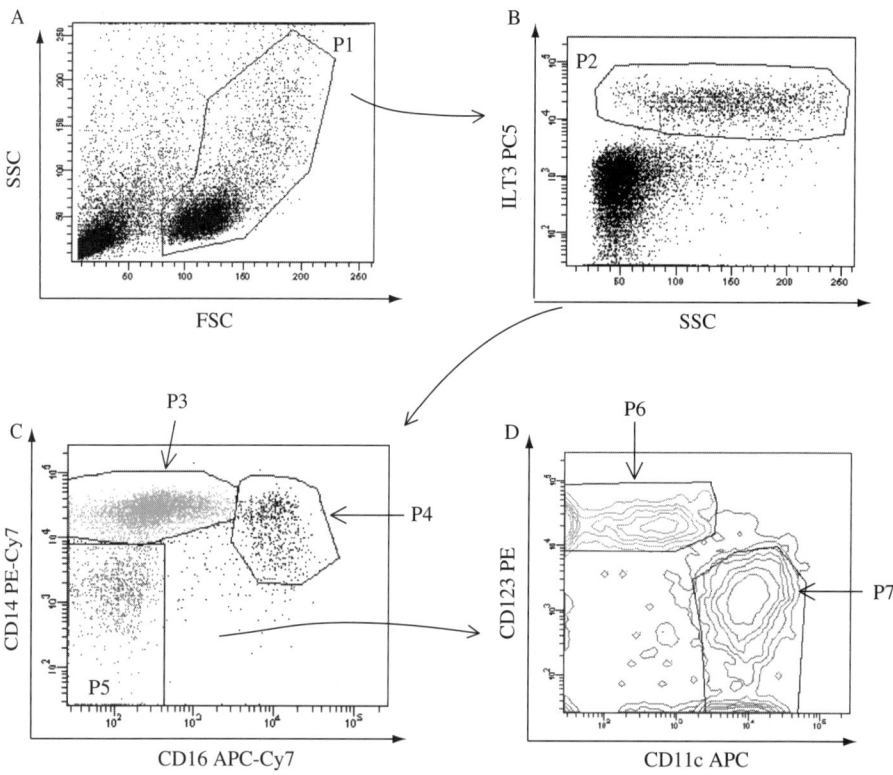

Fig. 4. "Dendritogram" on PBMC using five-color labeling. After gating on live PBMC (P1; A), events are distributed in a dot plot of ILT3 versus SSC (**B**). Events included in a gate P2, drawn on ILT3$^+$ events, are shown on a dot plot of CD14 versus CD16 (**C**). Three populations are defined: CD14^{++}CD16$^-$ population corresponding to the "classical" monocytes (P3), CD14$^+$CD16$^+$ to the "intermediate" monocytes between monocytes and myeloid dendritic cells (DCs) (P4), and CD14$^{-/\text{low}}$CD16$^-$ to DCs (P5). Events of P5 are distributed in a contour plot CD123 versus CD11c (**D**). Thus, the mDCs correspond to the CD123$^-$CD11c^{++} (P7) and the pDC to CD123^{++}CD11c$^-$ (P6). This analysis was performed with the Diva software.

5. Display events from gate P5 in a dot plot of CD123 PE versus CD11c APC. Draw two gates on the CD123^{low}CD11c^{++} (P6) and CD123^{+}CD11c^{-} (P7) populations, corresponding respectively to mDC and pDC (see **Fig. 4D**).
6. Calculate the percentage of « classical » monocytes (%CD14^{++}CD16^{-}), "intermediate" monocytes (%CD14^{+}CD16^{+}), pDC (% pDC) and mDC (% mDC) among PBMC (P1).
7. To obtain the absolute values of these populations, use the following calculation:

$$\text{pDC per ml of blood} = [\% \text{ pDC} \times (\text{lymphocyte count} + \text{monocyte count})] \times 10.$$

$$\text{mDC per ml of blood} = [\% \text{ mDC} \times (\text{lymphocyte count} + \text{monocyte count})] \times 10.$$

$$\text{«classical» monocytes per ml of blood} = [\%CD14^{++}CD16^{-} \times (lymphocyte\ count + monocyte\ count)] \times 10$$

$$\text{« intermediate» monocytes per ml of blood} = (\% \ CD14^{+}CD16^{+} \times (lymphocyte\ count + monocyte\ count)] \times 10$$

3.6. Important Troubleshooting and Final Notes (see Notes 23–25)

4. Notes

1. To determine the absolute values of DC subsets and monocytes, one needs the blood counts of the donor, mainly the lymphocyte and monocyte counts, corresponding to PBMC (lymphocyte and monocyte counts are given per mm^3, i.e., per μl of blood). This method displays some variations. For a more accurate determination of DC counts, you can use the TruCOUNT technology (Becton Dickinson) (see **Note 8**). Results obtained with fresh blood and careful pipetting are however close to those obtained with TruCOUNT.
2. Fixation with PFA (for at least 30 min) is required to inactivate human pathogens. PFA must be dissolved in PBS at 60°C with agitation under a chemical safety aspiration hood (toxic fumes). Then, the pH must be adjusted to 7.2–7.4.
3. EDTA is essential to avoid cell aggregates.
4. Saturation with human immunoglobulins, which are contained in human serum, is needed to block non-specific binding and binding to the Fc receptors, which are expressed highly on DC.
5. To adjust cytometer parameters, use the control isotype corresponding to each antibody. This control isotype must be from the same species, be of the same class have the same fluorochrome and be provided by the same supplier as the specific antibody (see **Table 1**). They also must be used at the same concentration as the specific antibody.

Table 1
Characteristics of Antibodies Used for the Different Methods of "Dendritogram"

	Antibodies	Fluorochrome	Clone	Ig class	Dilution	Supplier
Lineage Cocktail	CD3	FITC	SK7	IgG1		
	CD14	FITC	MφP9	IgG2b		
	CD16	FITC	3G8	IgG2b	1.10	Becton Dickinson
	CD56	FITC	NCAM16.2	IgG1		
	CD19	FITC	SJ25C1	IgG1		
	CD20	FITC	L27	IgG1		
	HLA-DR	PerCP or APC	L243	IgG2a	1:20 and 1:40	Becton Dickinson
	CD11c	PE or APC	B-ly 6	IgG1	1:20 and 1:20	Becton Dickinson
	Control isotype	APC	MOCP-21	IgG1	1:80	Becton Dickinson
	CD123	PE	9F5	IgG1	1:10	Becton Dickinson
	Control isotype	PE	MOCP-21	IgG1	1:40	Becton Dickinson
	CD123	APC	AC145	IgG2a	1:20	Miltenyi Biotec
	CD14	PE-Cy7	M5E2	IgG2a	1:50	Becton Dickinson
	CD16	APC-Cy7	3G8	IgG1	1:50	Beckman Coulter
	ILT3	PE-Cy5	ZM3.8	IgG1	1:50	Becton Dickinson
	CD45	PerCP	2D1	IgG1	1:20	Becton Dickinson

For each antibody (all of which are murine), labeled fluorochrome, clone, immunoglobulin class, dilution, and supplier are indicated. These characteristics are needed to choose the correct control isotype. Two examples of control isotypes were given for CD11c APC and for CD123 PE.

6. Analyses were performed either with the Cellquest software or with the DIVA software from Becton Dickinson. Nevertheless, a similar reasoning may be applied with other analysis softwares (FlowJo, CXP). Caution: DIVA software version 1.0 does not allow overlays.
7. Always backup source data and leave them untouched.
8. This method *(23,25)*, when done correctly, gives the most accurate DC counts. TruCOUNT tubes are however very expensive. TruCOUNT tubes must be used when the indicator inside the box is not blue.
9. This four-color staining has the great advantage of being readable on most cytometers (Dako, Becton Dickinson, Beckman Coulter).
10. Caution: do not touch the pellet.
11. Pipetting into the TruCOUNT tubes requires putting the drop of liquid on the wall of the tube. To sample exactly the volume of whole blood, reverse pipettes need to be calibrated carefully and regularly by weighing 100 µl of water (100 mg) with a precision scale.
12. For the analysis of cells such as DC that represent a small proportion (<5%) of the population, it is important to ensure that sufficient numbers of events are recorded to provide reasonable precision. For example, for a precision of 5%, 400 positive events must be recorded.
13. Importantly, optimal viability (>90%) is recommended to obtain correct DC counts. If not, DC counts will be overestimated.
14. Beads are fluorescent in all the channels. The number of beads that are in the tubes, as indicated in the batch specifications, and the number of events in the dot plot must be close. Approximately 26,000 beads must be acquired.
15. To avoid a decrease in cell output, all the buffers and centrifugations for PBMC isolation must be at room temperature. Blood should be transported, stored, and manipulated between 15 and 32°C. Separation of mononuclear cells using Ficoll is indicated here because it is our standard procedure to study the expression of different molecules on DC. If the aim of the technique is DC counts (the "dendritogram"), counting should rather be done on whole blood.
16. During washing, the volume of NaCl must be adapted to the number of cells (a small volume of NaCl for a low number of cells).
17. The estimation of cells may be performed with either a precision cell count glass cell (Malassez cell type or Burker chamber) or, to avoid using glass devices with potential infectious hazard, Kova Glasstic Slides with grids (Hycor Biomedical).
18. The following labeling can be also performed with thawed cells. To freeze PBMC, resuspend them after Ficoll isolation in cold FBS at a concentration of 10^7 cells/ml. Distribute 500 µl of cell suspension in each cold cryotube (Cryo.S sterile non-pyrogenic DNase and RNase free, Greiner Bio-One). Prepare, extemporaneously FBS with 20% dimethyl sulfoxide (DMSO) (Sigma) and cool it in the refrigerator. Add 500 µl of cold FBS 20% DMSO to each cryotube. Because of DMSO toxicity,

19. Centrifugations and incubations for the immunolabelings must be performed at +4°C, that is, any labeling lasting more than 3 min must be performed on crushed ice. Incubations are done in the refrigerator. All the buffers must be kept in the refrigerator and taken out just before use.

   all these steps must be performed around 4°C, on crushed ice or ideally in a freezing container (Nalgene). Keep the cryotubes at -80°C overnight; then, transfer them to liquid nitrogen.

20. Immediate reading on the cytometer avoids a decrease in fluorescence intensity of fluorochromes and a split of tandem fluorochromes (PE-Cy7, PE-Cy5, APC-Cy7). Nevertheless, for surface (not for intracellular) immunolabelings, cells can be kept at +4°C and read within 48 h.
21. The five-color labeling allows to use a free channel, FL1 (FITC), to investigate the expression of other parameters such as activation markers (CD80, CD86, CD40) on the DC subsets.
22. ILT3 is exclusively expressed on monocytes and DCs (26).
23. All labeling and gating methods must be tested in the laboratory for contaminating cell types within the gates of interest before starting a research or clinical project. An antibody directed against each cell lineage, if possible using another clone than that already used in the labeling, or else with a higher concentration, must be tested in comparison with its isotype. The following antibodies must be tested: anti-CD19 or 20 for B lymphocytes, anti-CD3 or T-cell receptor for T lymphocytes, anti-CD56 or CD57 for NK cells, anti-CD13 for monocytes, anti-CD15 for polymorphonuclear cells, anti-CD34 for stem cells.
24. For all the different labelings, CD123 may be replaced by BDCA2, which was shown to be more specific for pDC.
25. The present methods do not take into account a more minor mDC population labeled by BDCA3 that was shown to be decreased in HIV infection (*19,27*).

Acknowledgments

We thank Isabelle Kamga, Sandrine Kahi, Concepción Marañón, Ludovic Fery, and Michelina Nascimbeni, for participating in method setup, Nadine Fievet for reading the manuscript and Axel Kahn for suggestions. This work was supported by the ANRS and Sidaction.

References

1. Banchereau, J. and Steinman, R. M. (1998) Dendritic cells and the control of immunity. *Nature* **392**, 245–252.
2. Shortman, K. and Liu, Y. J. (2002) Mouse and human dendritic cell subtypes. *Nat Rev Immunol* **2**, 151–161.
3. Grouard, G. and Clark, E. A. (1997) Role of dendritic and follicular dendritic cells in HIV infection and pathogenesis. *Curr Opin Immunol* **9**, 563–567.

4. Siegal, F., Kadowaki, N., Shodell, M., Fitzgerald-Bocarsly, P., Shah, K., Ho, S., et al. (1999) The nature of the principal type 1 interferon-producing cells in human blood. *Science* **284**, 1835–1837.
5. Cella, M., Jarrossay, D., Facchetti, F., Alebardi, O., Nakajima, H., Lanzavecchia, A., et al. (1999) Plasmacytoid monocytes migrate to inflamed lymph nodes and produce large amounts of type I interferon. *Nat Med* **5**, 919–923.
6. Colonna, M., Trinchieri, G. and Liu, Y. J. (2004) Plasmacytoid dendritic cells in immunity. *Nat Immunol* **5**, 1219–1226.
7. Biron, C. A. (2001) Interferons alpha and beta as immune regulators–a new look. *Immunity* **14**, 661–664.
8. McKenna, K., Beignon, A. S. and Bhardwaj, N. (2005) Plasmacytoid dendritic cells: linking innate and adaptive immunity. *J Virol* **79**, 17–27.
9. Banchereau, J., Briere, F., Caux, C., Davoust, J., Lebecque, S., Liu, Y. J., et al. (2000) Immunobiology of dendritic cells. *Annu Rev Immunol* **18**, 767–811.
10. Lanzavecchia, A. and Sallusto, F. (2001) The instructive role of dendritic cells on T cell responses: lineages, plasticity and kinetics. *Curr Opin Immunol* **13**, 291–298.
11. Grassi, F. R., Hosmalin, A., McIlroy, D., Calvez, V., Debré, P. and Autran, B. (1999) CD11c-positive dendritic cells are depleted in the blood of HIV-infected patients. *AIDS* **13**, 759–766.
12. Pacanowski, J., Kahi, S., Baillet, M., Lebon, P., Deveau, C., Goujard, C., et al. (2001) Reduced blood CD123+ and CD11c+ dendritic cell numbers in primary HIV-1 infection. *Blood* **9**, 3016–3021.
13. Donaghy, H., Gazzard, B., Gotch, F. and Patterson, S. (2003) Dysfunction and infection of freshly isolated blood myeloid and plasmacytoid dendritic cells in patients infected with HIV-1. *Blood* **101**, 4505–4511.
14. Hosmalin, A., Lebon, P. (2006) Type I interferon production in HIV-infected patients. *J Leuko Biol* **80**, 984–93.
15. Longman, R. S., Talal, A. H., Jacobson, I. M., Rice, C. M. and Albert, M. L. (2005) Normal functional capacity in circulating myeloid and plasmacytoid dendritic cells in patients with chronic hepatitis C. *J Infect Dis* **192**, 497–503.
16. Finke, J. S., Shodell, M., Shah, K., Siegal, F. P. and Steinman, R. M. (2004) Dendritic cell numbers in the blood of HIV-1 infected patients before and after changes in antiretroviral therapy. *J Clin Immunol* **24**, 647–652.
17. Kamga, I., Develioglu, L., Lichtner, M., Kahi, S., Marañón, C., Deveau, C. et al. (2005) Type I interferon production is profoundly impaired in primary HIV-1 infection. *J Infect Dis* **192**, 303–310.
18. Pacanowski, J., Develioglu, L., Kamga, I., Sinet, M., Desvarieux, M., Girard, P. M. et al. (2004) Early plasmacytoid dendritic cell changes predict plasma HIV rebound in primary infection. *J Infect Dis* **190**, 1889–1892.
19. Chehimi, J., Campbell, D. E., Azzoni, L., Bacheller, D., Papasavvas, E., Jerandi, G., et al. (2002) Persistent decreases in blood plasmacytoid dendritic cell number and function despite effective highly active antiretroviral therapy and increased blood myeloid dendritic cells in HIV-infected individuals. *J Immunol* **168**, 4796–4801.

20. Siegal, F. P., Fitzgerald-Bocarsly, P., Holland, B. K. and Shodell, M. (2001) Interferon-alpha generation and immune reconstitution during antiretroviral therapy for human immunodeficiency virus infection. *AIDS* **15**, 1603–1612.
21. Thiebaut, R., Pellegrin, I., Chene, G., Viallard, J. F., Fleury, H., Moreau, J. F., et al. (2005) Immunological markers after long-term treatment interruption in chronically HIV-1 infected patients with CD4 cell count above 400 x 10(6) cells/l. *AIDS* **19**, 53–61.
22. Pichyangkul, S., Endy, T. P., Kalayanarooj, S., Nisalak, A., Yongvanitchit, K., Green, S., et al. (2003) A blunted blood plasmacytoid dendritic cell response to an acute systemic viral infection is associated with increased disease severity. *J Immunol* **171**, 5571–5578.
23. Lichtner, M., Rossi, R., Mengoni, F., Vignoli, S., Colacchia, B., Massetti, A. P., et al. (2006) Circulating dendritic cells and interferon-alpha production in patients with tuberculosis: correlation with clinical outcome and treatment response. *Clin Exp Immunol* **143**, 329–337.
24. Olweus, J., BitMansour, A., Warnke, R., Thompson, P. A., Carballido, J., Picker, L. J., et al. (1997) Dendritic cell ontogeny: a human dendritic cell lineage of myeloid ontogeny. *Proc Natl Acad Sci USA* **94**, 12551–12556.
25. Vuckovic, S., Gardiner, D., Field, K., Chapman, G. V., Khalil, D., Gill, D., et al. (2004) Monitoring dendritic cells in clinical practice using a new whole blood single-platform TruCOUNT assay. *J Immunol Methods* **284**, 73–87.
26. Cella, M., Dohring, C., Samaridis, J., Dessing, M., Brockhaus, M., Lanzavecchia, A., et al. (1997) A novel inhibitory receptor (ILT3) expressed on monocytes, macrophages, and dendritic cells involved in antigen processing. *J Exp Med* **185**, 1743–1751.
27. Dzionek, A., Fuchs, A., Schmidt, P., Cremer, S., Zysk, M., Miltenyi, S., et al. (2000) BDCA-2, BDCA-3, and BDCA-4: three markers for distinct subsets of dendritic cells in human peripheral blood. *J Immunol* **165**, 6037–6046.
28. Thieblemont, N., Weiss, L., Sadeghi, H. M., Estcourt, C. and Haeffner-Cavaillon, N. (1995) CD14lowCD16high: a cytokine-producing monocyte subset which expands during human immunodeficiency virus infection. *Eur J Immunol* **25**, 3418–3424.
29. Kawanaka, N., Yamamura, M., Aita, T., Morita, Y., Okamoto, A., Kawashima, M., et al. (2002) CD14+,CD16+ blood monocytes and joint inflammation in rheumatoid arthritis. *Arthritis Rheum* **46**, 2578–2586.

17

Clinical Analysis of Human Natural Killer Cells

Pascale André and Nicolas Anfossi

Summary

Natural killer (NK) cells are major actors of innate immune responses against viruses, bacteria, parasites, and other mediators of pathology such as malignant transformation. These cells are also directly implicated in the link between innate and adaptive immunity, shaping T-cell responses. It is now obvious that manipulation of this lymphocyte subset could be the basis of new therapeutic approaches for cancer and/or pathogen-driven pathology. Hence, techniques enabling the phenotypic and functional analysis of patient NK cells are of major importance. In this chapter, we present an extensive immunophenotyping of patient NK cells as well as a recently described method to assess NK cell functional activity at the single-cell level.

Key Words: NK cell; CD107; LAMP; IFN-γ; immunophenotyping.

1. Introduction

Natural killer (NK) cells are large granular lymphocytes involved in innate immunity against viruses, bacteria, parasites, and other mediators of pathology such as malignant transformation *(1, 2)*. Peripheral NK cells represent 5–15% of circulating lymphocytes and exert two major effector functions: (i) cytolytic activity toward a spectrum of stressed cells as well as toward antibody-coated cells [antibody-dependent cell cytotoxicity (ADCC)] and (ii) production of various chemokines and cytokines. This latter characteristic enables NK cells to shape immune responses, making a link between innate and adaptive immunity *(3)*. Hence, dissecting NK cell biology has represented an important challenge for the past three decades as these cells play crucial roles in immunity and could be used to develop new therapeutics for cancer and/or pathogen-driven diseases.

From: *Methods in Molecular Biology, vol. 415: Innate Immunity*
Edited by: J. Ewbank and E. Vivier © Humana Press Inc., Totowa, NJ

In this section, we present a simple immunophenotyping method using a small volume of whole-blood sample to analyze the distribution of cell-surface and intracellular markers of peripheral NK cells. These markers are all implicated in the regulation of NK cell functions. We also describe a recent method to assess functional activity of these cells. Indeed, although a lot of methods are available to study chemokine and cytokine secretion, the commonly used method to study cytotoxic activity is the ^{51}chromium release assay. As powerful it is, a major limitation of this test is that the readout of killer cell activity is indirect (i.e., target cell lysis), and as a consequence, no information is available at a single effector cell level. This limitation prompted researchers to find new tools to assess cytotoxic activity at this single-cell level. The recently described CD107 cell-surface mobilization assay represents the most promising answer to this problem *(4–6)*. CD107/LAMP molecules (CD107a, CD107b) are expressed on the inner membrane of cytolytic granules. Upon activation, these granules fuse with the cell membrane to release the cytotoxic molecular machinery (i.e., perforin and granzymes) at the effector cell/target cell interface. At this stage, CD107 molecules are transiently exposed at the cell surface. Adding anti-CD107 antibodies during the stimulation allows staining of these transiently exposed molecules. Using fluorochrome-labeled anti-CD107 antibodies, we are now able to detect cells that have effectively undergone granule release by flow cytometry at a single-cell level. One advantage of this method is that it can be coupled to method of intracellular cytokine detection.

2. Materials

2.1. Whole-Blood Immunophenotyping

1. Whole blood collected on EDTA (3 mL for this staining; this includes sample test and controls).
2. Phosphate-buffered saline (PBS, Gibco, Invitrogen, Carlsbad, CA, USA).
3. Fluorescence-activated cell sorting (FACS) staining buffer: 100% PBS, 0.2% bovine serum albumin, 0.02% sodium azide NaN_3.
4. Cytometer and acquisition/analysis software.
5. Polypropylene 5-mL tubes.
6. Optilyse C [Beckman Coulter (BC), Villepinke, France].
7. Intraprep reagent (Beckman Coulter).
8. Antibodies: PerCP-Cy5.5 anti-CD3 [SP34.2, Becton Dickinson, San Jose, CA, USA (BD)], Allophycocyanin (APC) anti-CD56 (N901, BC), fluorescein isothiocyanate (FITC)-labeled anti-CD45 (J33, BC), and phycoerythrin (PE)-labeled antibodies as follows—IgG1 control (BC), PE IgG2a control (BC), anti-CD159a (Z199, BC), anti-NKG2C (134591, R&D systems, Minneapolis, USA), anti-NKG2D (ON72, BC), anti-CD158a, h (EB6, BC), anti-CD158b1, b2 (GL183, BC), anti-CD158e1, e2 (Z27,

BC), anti-CD158i (FES172, BC), anti-NKp30 (Z25, BC), anti-NKp44 (Z231, BC), anti-NKp46 (Bab280, BC), anti-CD94 (HP-3B1, BC), anti-CD16 (IgG1, 3G8, BC), anti-CD25 (M-A251, BD), anti-CD69 (FN50, BC), anti-CD85j (HP-F1, BC), anti-CD244 (C1.7.1, BC), anti-CD161 (191B8, BC), anti-CD226 (DX11, BD), anti-perforin (δG9, BD), and anti-granzyme B (GB11, eBioscience, ebioscience : San Diego, USA).

2.2. Functional Analysis

1. Peripheral blood mononuclear cells (PBMCs): PBMCs are isolated by Ficoll-Hypaque density gradient centrifugation (Amersham Pharmacia Biotech, Amersham bioscience, piscataway, USA).
2. U-shape 96-well plates.
3. Monensin (Sigma): stock solution in ethanol at 2 mM.
4. Complete medium (CM): RPMI 1640 (Gibco) supplemented with 10% fetal calf serum, penicillin 50 U/mL, streptomycin 50 μL/mL, and L-glutamine 2 mM (Gibco).
5. EDTA (100 mM solution, Gibco).
6. PBS.
7. Target cells: K562 and P815.
8. Intraprep reagent (BC).
9. FACS staining buffer.
10. Antibodies: PerCP-Cy5.5 anti-CD3 (BD), APC-labeled anti-CD56 (BC), FITC-labeled anti-CD107a (BD), FITC-labeled anti-CD107b (BD), PE-labeled anti-interferon (IFN)-γ (BD), purified anti-CD16 (3G8, BD), and purified anti-lymphoma (Accurate Chemicals, Aumate chemicals: westbury, NY, USA).
11. Flow cytometer and acquisition/analysis software.

3. Methods

3.1. Whole-Blood Immunophenotyping

3.1.1. Cell-Surface Marker Detection

1. Prepare a mix containing anti-CD45, anti-CD3, and anti-CD56 (5 μL/mAb/sample). The total volume of this mix depends on sample number.
2. In polypropylene 5-mL tubes, aliquot 15 μL of antibody mix.
3. Add 10 μL of PE antibody against cell-surface marker (*see* **Table 1** and **Note 1**).
4. Add 100 μL whole blood to each tube.
5. Mix gently.
6. Incubate at room temperature (RT) in the dark for 20 min.
7. Wash once with 4 mL PBS.
8. Discard supernatant by aspiration.
9. Add 500 μL/tube Optilyse and vortex immediately for 5 s.

Table 1
Whole-Blood Four-Color Immunophenotyping of Human Natural Killer (NK) Cells

Sample	Function	FITC	PE	PerCP-Cy5.5	APC
Cell-surface staining					
1	Isotype control	CD45	IgG1	Anti-CD3	Anti-CD56
2	Isotype control	CD45	IgG2a	Anti-CD3	Anti-CD56
3	Isotype control	CD45	IgG2b	Anti-CD3	Anti-CD56
7	NKR	CD45	Anti-CD158a,h	Anti-CD3	Anti-CD56
8	NKR	CD45	Anti-CD158b1, b2, j	Anti-CD3	Anti-CD56
9	NKR	CD45	Anti-CD158e1, e2	Anti-CD3	Anti-CD56
10	NKR	CD45	Anti-CD158i	Anti-CD3	Anti-CD56
11	NKR	CD45	Anti-CD94	Anti-CD3	Anti-CD56
4	NKR (inhibitory)	CD45	Anti-CD159a (NKG2A)	Anti-CD3	Anti-CD56
5	NKR (activating)	CD45	Anti-NKG2C	Anti-CD3	Anti-CD56
18	NKR (inhibitory)	CD45	Anti-CD85j (ILT-2)	Anti-CD3	Anti-CD56
12	Activating	CD45	Anti-CD16	Anti-CD3	Anti-CD56
13	NCR	CD45	Anti-NKp30	Anti-CD3	Anti-CD56
14	NCR	CD45	Anti-NKp44	Anti-CD3	Anti-CD56
15	NCR	CD45	Anti-NKp46	Anti-CD3	Anti-CD56
16	Activation	CD45	Anti-CD25	Anti-CD3	Anti-CD56
17	Activation	CD45	Anti-CD69	Anti-CD3	Anti-CD56
6	Co-activating	CD45	Anti-NKG2D	Anti-CD3	Anti-CD56
19	Co-activating	CD45	Anti-CD161 (NKRP1-A)	Anti-CD3	Anti-CD56
21	Co-activating	CD45	Anti-CD226 (DNAM-1)	Anti-CD3	Anti-CD56
20	Co-activating	CD45	Anti-CD244 (2B4)	Anti-CD3	Anti-CD56
Intracellular staining					
22	Isotype control	CD45	IgG1 (intra)	Anti-CD3	Anti-CD56
23	Cytotoxic molecule	CD45	Anti-granzyme B (intra)	Anti-CD3	Anti-CD56
24	Isotype control	CD45	IgG2b (intra)	Anti-CD3	Anti-CD56
25	Cytotoxic molecule	CD45	Anti-perforin (intra)	Anti-CD3	Anti-CD56

NKR: Natural Killer Cell Receptor
NCR: Natural Cytotoxicity Receptor

10. Incubate at RT for 10–15 min in the dark.
11. Add 500 µL/tube PBS.
12. Incubate at RT for 5–10 min in the dark.
13. Wash once with 4 mL PBS.
14. Discard supernatant by aspiration.
15. Resuspend pellet in 300 µL FACS staining buffer.
16. Analyze samples on flow cytometer (*see* **Note 2**).

3.1.2. Intracellular Marker Detection

1. Follow **steps 1–6** (*see* **Subheading 3.1.1.**).
2. Add 100 µL of Intraprep reagent 1 (formaldehyde-containing fixative solution).
3. Incubate at RT for 15 min in the dark.
4. Wash once with 4 mL PBS.
5. Discard supernatant by aspiration.
6. Add 50 µL Intraprep reagent 2 (saponin-containing permeabilizing solution) to pellet without vortexing.
7. Incubate at RT for 5 min in the dark.
8. Gently shake by hand.
9. Dilute antibody directed against intracellular marker in Intraprep reagent 2 (*see* **Table 1**).
10. Add 50 µL/tube of the latter solution.
11. Gently vortex.
12. Incubate at RT for 20 min in the dark.
13. Wash once with 4 mL PBS.
14. Discard supernatant by aspiration.
15. Resuspend pellet in 300 µL FACS staining buffer.
16. Analyze samples on flow cytometer.

3.2. CD107 Cell-Surface Mobilization and IFN-γ Production on PBMCs (see Fig. 1)

1. Prepare in CM a solution containing 10 µM monensin.
2. To each 50 µL of this solution, add 5 µL of FITC-labeled anti-CD107a and 5 µL of FITC-labeled anti-CD107b monoclonal antibody (*see* **Note 1**).
3. Pipet 50 µL of this solution per well of a U-shape 96-well plate.
4. Add 50 µL per well of a solution containing the desired number (usually 0.5 to 1×10^6 cells) of PBMCs (effector cells) in CM (*see* **Note 3**).
5. Add 50 µL per well of a solution of target cells in CM to have an effector/target cell ratio = 10 (*see* **Table 2**).
6. Add 50 µL per well of a solution containing antibodies used for redirected killing in CM (*see* **Note 4**).
7. Centrifuge at $300 \times g$ for 1 min.

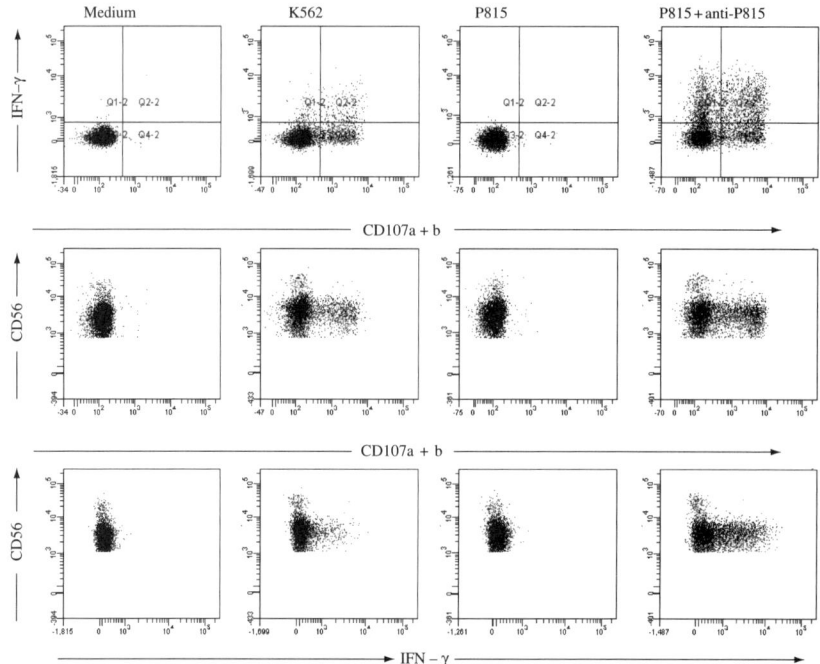

Fig. 1. CD107 cell-surface mobilization and interferon (IFN)-γ production. Peripheral blood mononuclear cells (PBMCs) were stimulated for 4 h by the indicated stimuli in the presence of anti-CD107a and anti-CD107b antibodies. After incubation, cells were stained with anti-CD3 and anti-CD56 antibodies, then fixed and permeabilized. Finally, intracellular IFN-γ production was measured using an anti-IFN-γ antibody. Dot plot are gated on CD3⁻CD56⁺ NK cells.

8. Incubate in incubator at 37°C for 4 h.
9. Centrifuge at 300 × g for 3 min.
10. Flick the plate to eliminate the supernatant.
11. After this step, if using target cells as stimulus, wash wells with 150 μL of a PBS solution containing 2 mM EDTA to disrupt cell conjugates.
12. Centrifuge at 300 × g for 3 min.
13. Flick the plate to eliminate the supernatant.
14. Then wash twice with FACS staining buffer.
15. Flick the plate to eliminate the supernatant.
16. Prepare a mix containing anti-CD3 and anti-CD56 antibodies in FACS staining buffer (*see* **Note 5**).
17. Add 50 μL/well of this mix.
18. Incubate in the dark for 20 min at 4°C.

Table 2
Stimuli Used for Assessing Human Natural Killer Cell Functional Activity

1	No stimulus
2	P815
3	P815 + anti-lymphoma
4	P815 + anti-CD16
5	P815 + anti-receptor of interest
6	K562

19. Wash twice with FACS staining buffer.
20. Flick the plate to discard the supernatant.
21. Add 100 μL of Intraprep reagent 1 (formaldehyde-containing fixative solution).
22. Incubate at RT in the dark for 15 min.
23. Wash once with FACS staining buffer.
24. Centrifuge at 300 × g for 3 min (see **Note 6**).
25. Flick the plate to eliminate the supernatant and vortex gently.
26. Repeat the last three steps.
27. Add 50 μL of Intraprep reagent 2 (saponin-containing permeabilizing solution).
28. Incubate at RT in the dark for 5 min.
29. Prepare in Intraprep reagent 2 a solution containing a saturating concentration of directly labeled anti-IFN-γ (see **Notes 5** and **7**).
30. Add 50 μL/well of this antibody solution.
31. Incubate at RT in the dark for 20 min.
32. Wash once with FACS staining buffer.
33. Centrifuge at 300 × g for 3 min.
34. Flick the plate to eliminate the supernatant and vortex gently.
35. Repeat the last three steps.
36. Distribute 150 μL of FACS staining buffer in each well.
37. Transfer the suspension of each well into cytometer tubes already containing 150 μL of FACS staining buffer.

Cells are ready for analysis on a flow cytometer (see **Notes 8–14**).

4. Notes

1. The monoclonal antibody combination can be adapted accordingly to your flow cytometer. The example presented here is for acquisition on a BD flow cytometer such as FacsCalibur. If using a BC flow cytometer such as FC500, the fluorochrome combination could be FITC, PE, PE-Cy5, PE-Cy7, for example.

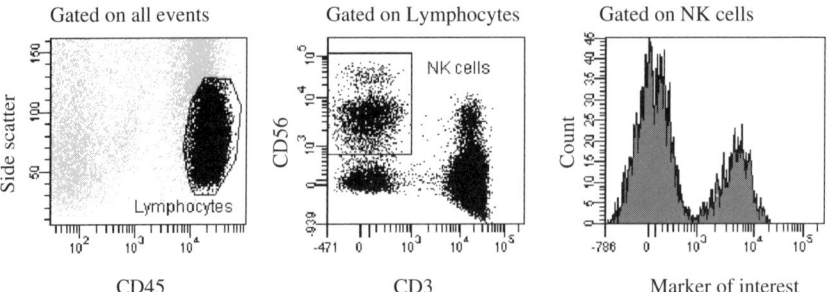

Fig. 2. Whole-blood phenotypic analysis of human natural killer (NK) cells. Whole blood was stained with anti-CD45, anti-CD3, anti-CD56, and antibodies of interest. Red blood cells were then lysed by the optilyse treatment. Using these settings, it is also possible to gate on $CD56^{dim}$ and $CD56^{bright}$ NK cell subsets.

2. In whole blood, the NK cell subset is characterized by a lymphocyte forward and side scatter and a $CD45^{high}CD3^-CD56^+$ phenotype (see **Fig. 2**).
3. Freshly isolated PBMCs or frozen PBMCs can be used for this assay. In the later case, cells should be thawed and left at least 2 h in CM at 37°C before starting the assay.
4. For redirected killing, it is not mandatory to coat target cells with mAb before incubation with effector cells. All three partners (effector, target cells, and antibody) can be mixed together. For ADCC, we advise to first stain P815 with anti-lymphoma, wash the cells, and then mix with the effector cells.
5. If a five- or six-color flow cytometer is available, other antibodies directed against cell-surface markers or intracellular molecules could be added to the anti-CD3/anti-CD56 combination (see **step 15**) or to anti-IFN-γ (see **step 28**).
6. Note that after fixation, the appearance of the cell pellet will be changed and it is usually more difficult to detect.
7. Other cytokines may be detected instead of IFN-γ [i.e., tumor necrosis factor TNF-α, IL-6, IL-8, etc]. In such a case, replace anti-IFN-γ by an antibody directed against the cytokine of interest.
8. If you are unable to run your samples following the staining, you have to keep cells in 1% formaldehyde containing PBS or an equivalent fixative solution, and store them in the dark at 2–8°C until analysis.
9. We advise whenever possible to also perform a chromium release assay as well as the CD107 mobilization test. Indeed, the two methods are complementary and give different information. Briefly, ^{51}Cr [100 µCi (3.7 MBq)/1 × 10^6 cells]-labeled target cells(3×10^3) are mixed with PBMCs at different effector–target (E/T) cell ratios (from 1 to 100) and incubated at 37°C. Reconstitution of cell lysis is performed in the presence of the indicated mAb (see **Table 2**). After 4 h of

incubation, 50 μL supernatant is removed, and the ^{51}Cr release is measured. All samples are analyzed in duplicate or triplicate, and the percentage of specific lysis is determined as follows: 100 × (mean cpm experimental release − mean cpm spontaneous release)/(mean cpm total release − mean cpm spontaneous release).

10. In PBMCs, the NK cell subset is characterized by a lymphocyte forward and side scatter and a $CD3^-CD56^+$ phenotype.
11. K562, a human major histocompatibility complex (MHC) class I negative erythroleukemic cell line, is a classical target cell line for spontaneous cytotoxicity (positive control). P815 is a murine mastocytoma that can be used to test ADCC by adding the Ig fraction of rabbit anti-mouse anti-sera. As this cell line also expresses Fc receptors, it is also used for redirected killing assays in the presence of monoclonal antibodies such as anti-CD16 or directed against other NK cell surface markers.
12. A different CD107 cell-surface mobilization assay has been described where anti-CD107 mAbs are added at the end of the stimulation instead of during the culture *(7)*. Using this assay, CD107 molecules that have been internalized will be not stained by the anti-CD107 mAb.
13. The E/T ratios used in the CD107 cell-surface mobilization assay are usually lower than those used in chromium release assay. We usually used 10/1 for PBMCs versus target cells in the CD107 assay.
14. In such a CD107 test, the percentage of $CD107^+$ NK cell is usually between 5 and 30%, depending on the stimulation.

Acknowledgment

N.A. is partially funded by EEC grant Allostem (number 503319).

References

1. Moretta, A., C. Bottino, M. C. Mingari, R. Biassoni, and L. Moretta. (2002) What is a natural killer cell? *Nat. Immunol.* 3: 6–8.
2. Moretta, L., and A. Moretta. (2004) Unravelling natural killer cell function: triggering and inhibitory human NK receptors. *EMBO J.* 23: 255–9.
3. Martin-Fontecha, A., L. L. Thomsen, S. Brett, C. Gerard, M. Lipp, A. Lanzavecchia, and F. Sallusto. (2004) Induced recruitment of NK cells to lymph nodes provides IFN-gamma for T(H)1 priming. *Nat. Immunol.* 5: 1260–5.
4. Alter, G., J. M. Malenfant, and M. Altfeld. (2004) CD107a as a functional marker for the identification of natural killer cell activity. *J. Immunol. Methods* 294: 15–22.
5. Betts, M. R., J. M. Brenchley, D. A. Price, S. C. De Rosa, D. C. Douek, M. Roederer, and R. A. Koup. (2003) Sensitive and viable identification of antigen-specific CD8+ T cells by a flow cytometric assay for degranulation. *J. Immunol. Methods* 281: 65–78.

6. Rubio, V., T. B. Stuge, N. Singh, M. R. Betts, J. S. Weber, M. Roederer, and P. P. Lee. (2003) Ex vivo identification, isolation and analysis of tumor-cytolytic T cells. *Nat. Med.* 9: 1377–82.
7. Bryceson, Y. T., M. E. March, D. F. Barber, H. G. Ljunggren, and E. O. Long. (2005) Cytolytic granule polarization and degranulation controlled by different receptors in resting NK cells. *J. Exp. Med.* 202: 1001–12.

18

Lentiviral Transduction of Immune Cells

Louise Swainson, Cedric Mongellaz, Oumeya Adjali, Rita Vicente and Naomi Taylor

Summary

Gene transfer into mammalian cells has been of crucial importance for studies determining the role of specific genes in the differentiation and cell fate of various hematopoietic lineages. Until recently, the majority of these studies were performed in transformed cell lines due to difficulties in achieving levels of transfection of greater than 1–3% in primary hematopoietic cells. Vectors based on retrovirus and lentivirus backbones have revolutionized our ability to transfer genes into primary hematopoietic cells. These vectors have allowed extensive ex vivo and in vivo studies following introduction of a gene of interest and have been used clinically in individuals suffering from cancers, infections, and genetic diseases. Ex vivo lentiviral gene transfer can result in efficient transduction of progenitor cells (>80%) that can then be further differentiated into immune lineage cells including T, B, dendritic, or natural killer cells. Alternatively, differentiated immune cells can themselves be transduced ex vivo with lentiviral vectors. Here, we discuss optimization of technologies for human immunodeficiency virus (HIV)-based gene transfer into murine and human progenitor and immune cell lineages.

Key Words: Lentiviral vectors; HIV; gene transfer; transduction; progenitors; CD34; T cells; B cells; dendritic cells.

1. Introduction

Human immunodeficiency virus (HIV)-based lentiviral vectors have been used for the last 10 years to achieve gene transfer in immune cells *(1)* (and reviewed in **refs.** *(2)* and *(3)*. During this decade, various improvements have been made to these vectors, and notably, the efficiency of gene transfer has been significantly augmented by the introduction of the HIV-1 central polypurine tract and central termination sequence *(4–8)*. This sequence creates a plus strand overlap, the central DNA flap, which is a *cis*-determinant of nuclear import *(4)*. Other improvements have been based on safety considerations and have included the deletion of enhancer sequences in the HIV-1 long terminal repeat (LTR) *(9)* (*see* **Fig. 1**). More recent efforts have focused on the introduction of insulator sequences in order to reduce the influence of chromosomal

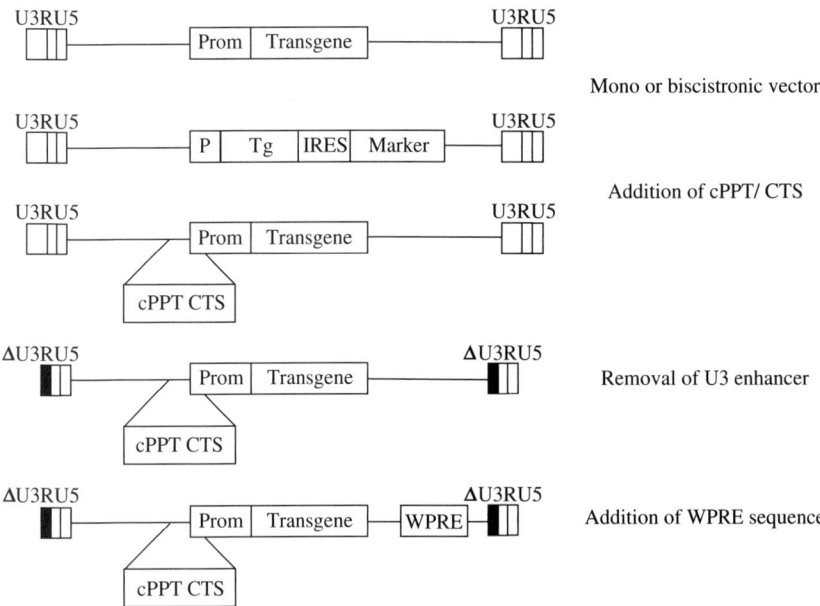

Fig. 1. Evolution of lentiviral vectors. First-generation lentiviral vectors were designed such that transgenes are expressed from an internal promoter (P) in the absence or presence of a marker gene. Marker genes are generally expressed downstream of an internal ribosome entry site (IRES) resulting in a bicistronic vector. Insertion of the human immunodeficiency virus (HIV) central polypurine tract (cPPT) and central termination sequence (CTS) improves nuclear import and integration *(4,5)*. Deletion of 400 nucleotides from the U3 enhancer at the 3′-end of the lentiviral vector leads to the subsequent removal of U3 in both long terminal repeats (LTRs) of the integrated vector. This deletion, which includes the TATA box of the LTR, abrogates promoter activity, and as such, the resulting self-inactivating vector (SIN) has enhanced biosafety *(9)*. Addition of the Woodchuck hepatitis virus post-transcriptional regulatory element (WPRE) stimulates nuclear RNA export and probably affects RNA processing, thereby resulting in increased transgene expression *(46,47)*.

sequences flanking the site of vector integration on expression of the lentiviral transgene *(10,11)*.

Two major packaging plasmids have been used to generate virions harboring HIV-based lentiviral virions. One of the first packaging plasmids, pCMV•R8.2, expresses all the HIV proteins with the exception of *env (11)*, whereas newer generation plasmids such as pCMV•R8.91 have removed the virulence genes *vif*, *vpu*, *vpr*, and *nef (12)*. Although there is one report of superior transduction with virions generated in the presence of these accessory proteins *(13)*, many studies have shown equivalent levels of transduction, and as such, most scientists use the pCMV•R8.9 vector for virion production.

The level of transgene expression will be influenced by the site of vector integration. In addition, the average expression can be significantly modulated by the use of different promoters. Although the use of tissue-specific promoters in retroviral vectors has met with multiple difficulties, this approach has been successfully used in lentiviral vectors *(14–16)*. Notably though, the level of transgene expression is generally significantly higher under conditions wherein the internal lentiviral promoter is derived from a murine leukemia virus (MLV) as compared with a ubiquitous or tissue-specific cellular promoter. This appears to be the case in multiple hematopoietic cell types as demonstrated in **Fig. 2**, and as such, the choice of promoter should be based on the expression level that is desired.

HIV-based virions are generally pseudotyped with the vesicular stomatitis G (VSV-G) protein allowing efficient concentration of viral particles. Nevertheless, it is important to note that pseudotyping virions with envelopes that allow activation of specific cell types can target gene transfer to those cells that are activated *(17,18)*. Moreover, envelopes based on the extracellular and transmembrane domains of endogenous retroviral envelopes such as that of the gibbon ape leukemia virus (GALV), feline endogenous virus (RD114), or amphotropic MLV result in enhanced transduction of primary blood lymphocytes and $CD34^+$ cells *(19,20)*. Similarly, lentiviral vectors pseudotyped with the rabies-associated Mokola virus can efficiently transduce murine CNS cells. Finally, pseudotyping of lentiviral vectors with the ecotropic MLV envelope *(20,21)*, an envelope that allows transduction of murine but not human cells, can diminish safety risks and even improve gene transfer into certain murine hematopoietic cells. The methods described in this chapter can be employed with virions pseudotyped with VSV-G or other envelopes as described below (*see* **Subheading 3.6**).

The vast majority of wild type HIV and simian immunodeficiency virus (SIV) isolates are not able to efficiently infect mononuclear cells that are in

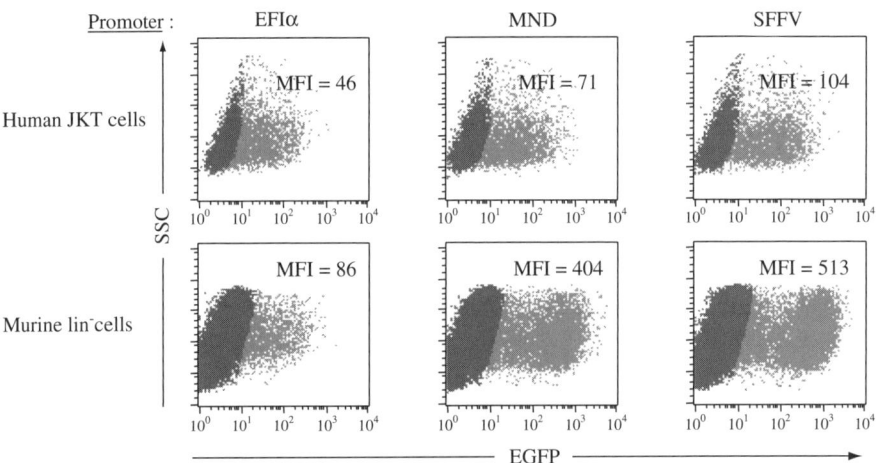

Fig. 2. Expression of a lentivirally expressed transgene is modulated by the internal promoter. Lentiviral vectors differing in the promoter used to drive expression of the EGFP transgene were used to transduce the human Jurkat T-cell line as well as murine lineage-negative bone marrow progenitor cells. EGFP was expressed from a ubiquitous promoter, EF1α, or from the retroviral long terminal repeats (LTRs): MND (Myeloproliferative sarcoma virus LTR-Negative control region Deleted) or SFFV (spleen-focus forming virus). The mean fluorescence intensity (MFI) of EGFP expression is shown 2 days post transduction.

a quiescent state (22). Accordingly, the ability of lentiviral vectors to achieve efficient gene transfer in lymphocytes (6,23), monocytes (24), and CD34 cells (25) that are in the G0 phase of the cell cycle is severely limited. Interestingly, this does not appear to be the case of quiescent neurons, a cell type that can be efficiently transduced with lentiviral vectors (1,12). The limitations in non-stimulated hematopoietic cells might be overcome by the use of lentiviral vectors derived from virus strains capable of infecting non-stimulated peripheral blood mononuclear cells (PBMCs). Indeed, it has recently been reported that vectors derived from the Pbj strain of SIV, a virus capable of replicating in unstimulated PBMCs, can transduce quiescent human monocytes (26). It is nonetheless important to note that in marked contrast to MLV-based vectors, lentiviral vectors are able to transduce cells that are in early phases of the cell cycle and do not require mitosis for efficient nuclear import.

In addition to HIV-based vectors, SIV- as well as FIV-based virions have been successfully used to achieve gene transfer in the majority of cells discussed here *(19, 27–32)*. This methods section will focus on transduction of various hematopoietic cell types with HIV-based lentiviral vectors following ex vivo stimulation of the cells of interest.

2. Materials

2.1. Plasmid Preparation

1. HIV-derived lentiviral vectors can be obtained from various commercial sources including Invitrogen or, alternatively, are generously provided by Didier Trono's laboratory following completion of a material transfer agreement that can be downloaded from the website http://tronolab.epfl.ch/page58114.html. Other academic sources include the laboratories of Pierre Charneau and Luigi Naldini. SIV-based lentiviral vectors can be obtained from the laboratory of FL Cosset. Other core facilities that can provide aid include, in the USA, the Penn vector core (http://www.uphs.upenn.edu/penngen/gtp/vcore_lv.html) and National gene vector laboratory (http://www.ngvl.org), in France, Genethon (http://www.genethon.fr), in Sweden, Lund University (http://www.rvec.lu.se/services.html), and the National gene vector laboratory of Ireland (http://www.nuigalway.ie/remedi/index.php).
2. Bacterial strains used for transformation should lack the recombinase gene (*rec*) to prevent recombination of plasmid and genomic DNA. Electrocompetent *Escherichia coli* DH5-α or Top10 is commonly used (purchased from ATCC, Manassas, VA or Eurogentec, Liege, Belgium).
3. Bacteria are grown in 2 × Luria-Bertani (LB) media (20 g Bacto Tryptone, 10 g Bacto Yeast Extract, 10 g NaCl, H_2O to 1 liter, pH from 7.2 to 7.5).
4. Ampicillin is dissolved in milliQ water at a concentration of 100 mg/ml, stored at −20°C and used at a final concentration of 50 μg/ml.
5. Plasmid amplification kits are available from commercial sources, e.g., Nucleobond PC500, Macherey-Nagel, Easton, PA.

2.2. Cell Culture and Transfections

1. 293T cells can be obtained from the American Tissue Type Culture Collection, ref. CRL-11268.
2. Dulbecco's Modified Eagle's Medium (DMEM) supplemented with 10% fetal calf serum (FCS) and 1% antibiotics (penicillin/streptomycin).
3. Dulbecco's Phosphate Buffer Saline (D-PBS) without calcium and magnesium.
4. 2.5% Trypsin and 0.5 M ethylenediaminetetraacetic acid (EDTA).

2.3. Virus Production and Titering

1. 1 × hepes buffered saline (HBS) solution for transfection can be made in 500 ml batches by adding 4.1 g NaCl, 0.186 g $Na_2HPO_4 \cdot 2H_2O$, 0.6 g dextrose, 3 g HEPES, qsp deionized H_2O. Buffer the solution to four different pHs ranging from 7.01 to 7.10 and pass through a 0.2-μm filter (*see* **Note 1**).

2. CaCl$_2$ is dissolved in deionized water to a final concentration of 2 M and also passed through a 0.2-μm filter.
3. Serum-free media for virion production can include CellGro or Optimem (Invitrogen, Carlsbad, CA).
4. 0.45-μm pore filters (Sartorius, Goettingen, Germany).
5. 20- or 50-ml syringes.
6. Ultracentrifugation is performed on a SW28 rotor with swinging bucket with adapted ultraclear centrifuge tubes (Beckman Coulter, ref. 344058, Fullerton, CA).
7. Ultrafiltration can be performed using columns such as Centricon Plus-70 100,000 MWCO (Millipore, ref. UFC710008, Billenca, MA).
8. Sucrose is dissolved in Tris-NaCl-EDTA (TNE) buffer: 10 mM Tris–HCl pH 7.5, 100 mM NaCl, 1 mM EDTA.
9. p24 enzyme-linked-immunosorbent assay (ELISA) kits are commercialized by multiple companies including Beckman Coulter.
10. Primers can be obtained from multiple companies including Sigma-Genosys, Poale, UK.
11. Genomic DNA is extracted using a commercial DNA extraction kit.
12. Real-time amplifications are performed using the QuantiTect SybrGreen kit (Qiagen, Valencia, CA) on a LightCycler system (Roche, Meylan, France).

2.4. Immune Cell Transductions

1. Human immune cells can be isolated from whole blood, available from blood banks.
2. Human progenitor cells can be isolated from umbilical cord blood, bone marrow, or granulocyte colony-stimulating factor (G-CSF)-mobilized peripheral blood. These samples require collaboration with hospital services or alternatively can be purchased from Cambrex BioSciences (Rockville, MD).
3. Mice immune cells can be isolated from lymph nodes or spleen and progenitor cells from bone marrow. To increase the relative percentage of progenitor cells, one can treat mice with 5-fluorouracil (6.25 mg per mouse via IP administration) 4 days prior to bone marrow harvest.
4. The cell type of interest can be purified using commercial antibody selection kits by either negative or positive depletion methods.
5. Cytokines are purchased from Peprotech (Paris, France) and murine as well as human T cell expander beads are available from Dynal (Invitrogen). Anti-CD3 and anti-CD28 antibodies are purchased from Becton Dickinson BioSciences, San Jose, CA. Concanavalin A and bacterial lipopolysaccharide (LPS) from *E. coli* O55:B5 can be obtained from Sigma. Anti-CD40 antibody can be obtained from Becton Dickinson.
6. Recombinant fibronectin is available from Takara (Otsu, Shiga, Japan).
7. For the culture of progenitor cells, cells are resuspended in a progenitor cell media [i.e., Stemspan medium (Stemcell Technologies, Vancouver, Canada) or X-Vivo-20 (Cambrex BioSciences)] supplemented with stem cell factor (SCF) (100 ng/ml),

interleukin (IL)-3 (10 ng/ml), and IL-6 (100 ng/ml). T cells are grown in RPMI supplemented with 10% FCS. Culture of murine cells requires the addition of β-mercaptoethanol (50 μM final concentration).

3. Methods

3.1. Preparation of Plasmids

1. Add 200 ng of plasmid DNA to 50 μl of electrocompetent *E. coli* (DH5-α) bacteria on ice under sterile conditions.
2. Mix and transfer to a chilled electroporation cuvette and electroporate at 2.5 kV, 200 ohms, 25μF. The time constant reading should be >4.5.
3. Immediately add 500 μl of ice-cold 2 × LB.
4. Spread dilutions of transformed bacteria on LB agar–ampicillin plates and incubate at 37°C overnight.
5. Pick three colonies and use each to inoculate 2 ml of 2 × LB–ampicillin medium (containing 50 μg/ml ampicillin). Incubate for 6 h at 37°C with agitation (at least $30 \times g$).
6. Use each cultured colony to inoculate 500 ml 2 × LB–ampicillin medium in a conical flask of at least 2-l capacity and incubate overnight at 37°C with agitation.
7. Transfer 2 ml of the 500 ml culture volume to an Eppendorf tube for miniprep verification of plasmid.
8. Centrifuge the remaining 500 ml of bacterial culture at $4500 \times g$ for 10 min at 4°C.
9. Discard the supernatant. Pellets may be stocked at −20°C until maxipreps are made.
10. Perform a miniprep with a commercial kit. Verify the amplified plasmid with restriction enzyme digestion and agarose gel electrophoresis.
11. Large-scale preparation of plasmid DNA from the frozen bacterial pellet can then be carried out (*see* **Note 2**).
12. Verify the amplified plasmid with restriction enzyme digestion and store at −20°C at a concentration of 1 mg/ml.

3.2. Culture and Transfection of HEK 293T Cells

1. Human Embryonic Kidney (HEK) 293T is the most commonly used packaging cell line for the production of VSV-G-pseudotyped lentiviral vectors. 293T are adherent cells cultured in DMEM High Glucose (4.5 mg/l) supplemented with 10% FCS and 1% antibiotics (penicillin/streptomycin).
2. Recover cells from plates using a 0.4% trypsin–PBS solution. Incubate for 5 min at 37°C (*see* **Note 3**).
3. Ensure that the cells are visibly detached and homogenize thoroughly by pipetting to dissociate cellular aggregates.
4. Add 5 ml of complete medium. The presence of FCS at this step is necessary to stop the trypsin enzymatic reaction. Resuspend vigorously with a 5-ml pipette to again dissociate cellular aggregates and transfer to a centrifuge tube.

5. Centrifuge cells at 370 × g for 5 min, and then discard the supernatant and resuspend thoroughly in complete medium. Seed 3×10^6 293T cells on a 10-cm plate in 7 ml of complete DMEM.
6. Following overnight culture, cells should be 60–70% confluent. This level of confluence is essential for optimal transfection efficiency.
7. Mix the three vectors, lentiviral vector, Gag/Pol, and Env-expressing plasmids, in a 1.5-ml tube at a 3:3:1 ratio. Do not exceed 20 μg of total plasmid DNA per reaction to minimize toxicity. Add 500 μl of 1× HBS solution to the plasmid mix, followed by 33 μl of 2 M $CaCl_2$, and vortex quickly for 10 s (*see* **Note 4**).
8. Gently add the transfection mix to the plate dropwise, taking care not to detach the producer cells. Incubate overnight at 37°C (*see* **Notes 5 and 6**).

3.3. Production of Lentiviral Virions

1. Following overnight culture after transfection, replace the medium by gently adding 6 ml of serum-free pre-warmed medium (i.e., CellGro or Optimem) to the side of the plate. Incubate for 30–48 h at 37°C.
2. Harvest the supernatant in a 50-ml tube, centrifuge at 400 × g for 5 min to pellet cellular debris, and then pass the viral supernatant through a 0.45-μm pore filters (*see* **Note 7**).
3. Concentrate VSV-G-pseudotyped virions from the supernatant by ultracentrifugation (*see* **Note 8**). At this stage, it is also possible to centrifuge over a 20% sucrose cushion or a double 20–50% sucrose cushion to obtain a supernatant of significantly higher purity (*see* **Fig. 3**).
4. Transfer the supernatant to an ultracentrifuge tube (Ultraclear 38.5 ml Beckman Coulter). Use serum-free medium to balance tubes to within ±0.01 g, and then load in an SW28 rotor. Centrifuge at 13,000 × g at 4°C for 2 h. The brake must be inactivated to prevent disturbance of the viral pellet during deceleration.
5. Discard the supernatant by inversion and leave the tube inverted for 2 min on absorbent tissue paper.
6. Add the desired volume of resuspension medium (PBS or serum-free medium) and incubate for 1 h on ice to allow the pellet to soften.
7. Resuspend carefully by pipetting, then store at −80°C in small aliquots (*see* **Note 9**).

3.4. Virus Titering

1. The viral "titer" may relate to the concentration of either physical particles or transducing units. Determine the concentration of physical particles from the p24 content using a commercial ELISA kit. For determining the concentration of transducing units, use either adherent (e.g., 293T) or suspension (e.g., Jurkat) target cells (*see* **Note 10**).
2. Count and seed target cells. For Jurkat cells, seed 1×10^6 cells in 1 ml per well on a 24-well plate (*see* **Note 11**).

Lentiviral Transduction of Immune Cells

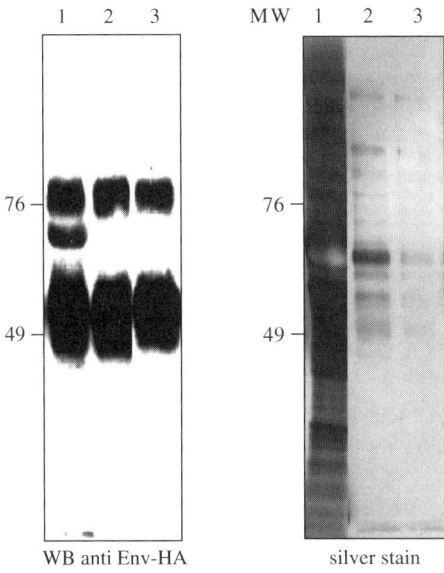

Fig. 3. Sucrose gradient centrifugation results in enhanced purity of virion preparations. To assess the purity of a lentiviral virus preparation pseudotyped with an fowl plague virus-hemagglutinin (FPV-HA) envelope, the viral supernatant was divided in three equivalent volumes and the following steps were performed: simple ultracentrifugation and resuspension of the viral pellet (lane 1); centrifugation through 1.5 ml of a 20% sucrose solution (lane 2); or centrifugation through a double sucrose cushion with 1.5 ml of a 20% sucrose solution and 1.0 ml of a 50% sucrose solution. The pellets from the first two conditions and the interface from the third condition were electrophoresed, and the level of incorporated envelope was assessed with an anti-HA antibody. As can be noted in this anti-HA western blot, wherein the top band represents immature envelope precursors and the bottom band is mature envelope, the level of incorporated envelope was similar in all three lanes. Silver staining of these three conditions, however, demonstrates the loss of 90–95% of non-virion proteins following centrifugation over a 20% sucrose cushion and a purification of 99% by centrifugation through a double 20%/50% sucrose gradient. Note that the middle band, which is lost between lanes 2 and 3, is not envelope protein. (This previously unpublished figure generously provided by Dr. Virginie Sandrin and Dr. Francois-Loic Cosset, ENS, Lyon, France.) MW: molecular weight (kDa).

3. Thaw an aliquot of virions on ice, and then dilute 1 µl in 100 µl of culture medium. Mix serial dilutions of the diluted virions with target cells and incubate for 72 h.
4. Analyze the percentage of cells expressing the transgene marker by fluorescence-activated cell sorting (FACS) (see **Fig. 4**). Calculate the titer from a dilution generating no more than 30% of transgene-expressing cells according to the formula

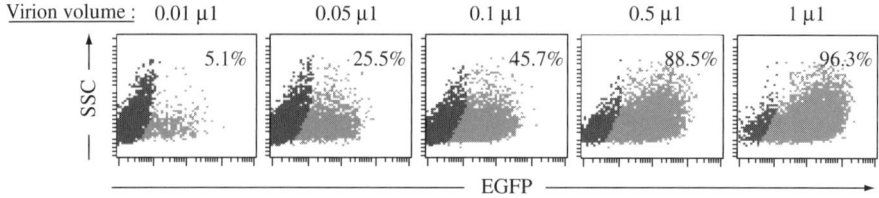

Fig. 4. Determining titers of lentiviral vectors. Limiting dilutions of virions are added to cells on which the virions are titered. In the example shown here, various volumes of EGFP-expressing lentiviral virions were added to 1×10^6 Jurkat cells cultured in a volume of 1 ml. Two days following transduction, the percentage of EGFP$^+$ cells was determined by flow cytometry. As vector copy number is generally greater than one at percentage transductions of >30% *(42)*, it is important to determine titers on virion dilutions resulting in transductions of <30%. In the example presented here, titers can therefore be determined from the dilutions of 0.01 µl or 0.05 µl of virions. Starting with a volume of 0.01 µl of virions, there was transduction efficiency of 5.1%, indicating that the titer is: $(5.1\%/100 \times 10^6)/0.0001$ ml = 5.1×10^9 transduction units/ml.

indicated below, wherein the number of cells refers to the number of cells present in the well at the time point when the virions are added to the culture:

$$\text{Titer (transduction units/ml)} = \frac{(\% \text{ transgene}^+ \text{ cells}/100) \times \text{number of cells per well}}{\text{total volume of virions (ml)}}$$

5. Alternatively, calculate the transducing units titer by quantitative polymerase chain reaction (q-PCR) analysis of the number of viral DNA copies in each dilution. Use a cell line expressing one viral copy per cell or plasmid dilutions for the standard control. Extract DNA from the transduced cells using a commercial DNA extraction kit and perform q-PCR using primers to amplify an exogenous transgene such as EGFP, or the 5´-LTR-Gag product. Commonly used primers for the 5´-LTR-Gag are M667 5´-GGCTAACTAGGGAACCCACTG-3´ and M661 5´-CCTGCGTCGAGAGAGCTCCTCTGG-3´.

3.5. Transduction of Human Hematopoietic Progenitor Cells

1. CD34$^+$ cells are cultured at 1×10^6 cells/ml in serum-free media harboring cytokines. Common cytokine cocktails are as follows: SCF, IL-3, and IL-6 (50 ng/ml); or SCF, thrombopoietin (TPO), Flt3-L (100 ng/ml), and IL-3 (10 ng/ml) for 24 h (*see* **Note 12**).
2. Following 24 h of cytokine stimulation, virions are added directly to the wells and culture is continued for an additional 24 h under the desired cytokine conditions.

A multiplicity of infection (MOI) of at least 100 is generally used to obtain transduction efficiencies of >50%. Reports using MOIs of 1000 are common, but this may result in decreased engraftment if the cells are transferred in vivo. Unlike other hematopoietic cell types, efficient transduction of human CD34$^+$ cells does not require the addition of polybrene and is not augmented by the presence of retronectin.
3. Cells are then washed and cultured as per the conditions required for further differentiation of the cells (*see* **Note 13**).
4. Transgene expression can be assessed within 48 h (*see* **Note 14**).

3.6. Transduction of Murine Hematopoietic Progenitor Cells

1. Seed lineage-negative (lin$^-$) progenitor cells at 1×10^6 cells/well in 500 µl of medium on tissue-culture-treated 24-well plates. Commonly used cytokine cocktails are SCF (100 ng/ml), IL-3 (10 ng/ml), and IL-6 (100 ng/ml); or SCF and TPO (50 ng/ml). Culture overnight at 37°C (*see* **Note 15**).
2. Prepare retronectin-coated plates. Reconstitute lyophilized retronectin in sterilized water at a concentration of 1 mg/ml. This stock solution should be stored at −20 °C. Dilute the stock to 32 µg/ml in PBS and add 500 µl of this working solution per well on non-tissue-culture-treated 24-well plates. Incubate at 37°C for 2 h. Remove the retronectin and store at 4°C; this working solution can be reused up to three times. Rinse wells with PBS, then saturate with PBS containing 5% FCS for 20 min at 37°C.
3. The stromal cells will have adhered to the overnight culture plate. Recover the non-stromal suspension lin$^-$ cells by thorough pipetting, and rinse the well twice to collect any remaining cells. Wash cells with PBS 5% fetal calf serum (SVF) and resuspend in fresh complete medium at 2×10^6/ml.
4. Remove PBS 5% FCS from the retronectin plate and seed the cell suspension at 500 µl/well.
5. Add the virions and incubate for 30 min at 37°C. MOIs of 100 are typically used but can be modulated depending on the level of transduction that is desired.
6. Spinoculate the cells by centrifugation at $1000 \times g$, 37°C for 2 h, and then place the transduction plate in an incubator at 37°C.
7. Gene transfer efficiencies may be augmented by performing a second round of transduction 12 h later by simply adding fresh viral supernatant. The addition of 5% FCS can also augment transduction but is often not used as it may decrease engraftment of the transduced cells.
8. The following day, lin$^-$ cells will be attached to the retronectin. Collect the cells by thorough pipetting and wash the well out twice with PBS to collect remaining cells. Detach remaining adherent cells with 300 µl of non-enzymatic cell dissociation buffer for 5 min. Rinse well several times with PBS and verify by microscopy that all cells have been collected.

9. Centrifuge the cells at 600 × g for 7 min and resuspend in complete medium for cell culture, or wash with PBS again if transduced cells are to be used for in vivo studies. Verify transduction efficiency by monitoring transgene expression 48 h later.

3.7. Lentiviral Transduction of Human T Lymphocytes

1. Human T cells can be activated through the T cell receptor (TCR) using CD3/CD28 T-cell expander beads at a concentration of 1 bead per cell or, alternatively, by culture on αCD3-/αCD28-coated plates (*see* **Note 16**). For the latter, add 1 ml of PBS containing 1 μg/ml each of stimulating αCD3 and αCD28 mAbs per well of a 24-well non-tissue-culture-treated plate and incubate for 2 h at 37°C. Rinse wells with PBS before seeding T cells at a density of 1×10^6 cells/ml in RPMI. T-cell stimulation may also be achieved by exposure to cytokines such as recombinant (r)IL-7 (10 ng/ml).
2. Different forms of T-cell activation will generate distinct windows of cell permissivity to lentiviral transduction and thus alter the necessary kinetics of virion exposure. Following TCR engagement, transduction is optimal 1–2 days later, whereas following IL-7 stimulation, expose T cells to virions 4–6 days later.
3. Add virions directly to the T cells in culture, based on the MOI that is desired (*see* **Note 17**). Maintain T cells at a density of 1×10^6 cells/ml in the appropriate activation medium.
4. Incubate for 24 h at 37°C.
5. Wash the cells with PBS to remove virions and reseed T cells in culture medium containing IL-2 (100 U/ml) (if cells have been activated through the TCR) or with other cytokines such as IL-7 (10 ng/ml) depending on the initial activation.
6. Analyze transduction efficiency 48 h later by monitoring transgene expression.

3.8. Lentiviral Transduction of Murine T Lymphocytes

1. Murine T cells can be activated through the TCR using murine expander beads or alternatively through concanavalin A (2 μg/ml) supplemented with IL-2 (100 U/ml) at a density of 1×10^6 to 2×10^6 cells/ml (*see* **Note 18**).
2. Following 24 h of activation, cells should be seeded on 24-well retronectin-coated plates (*see* **Subheading 3.6., step 2**) at a density of 1×10^5 to 5×10^5 cells/ml in the presence of IL-2.
3. Virions are added (MOI 10–100) and plates are spinoculated by centrifugation at 1200 × g for 1 h at 30°C (*see* **Note 19**).
4. Cells are incubated at 37°C for 48–72 h prior to assessing transgene expression.
5. If cells are to be maintained for longer time periods, it is optimal to culture them at densities of 10^6 cells/ml in the presence of IL-2 (100 U/ml) or IL-7 (10 ng/ml).

3.9. Lentiviral Transduction of Human B Cells

1. Efficient lentiviral transduction of human B cells has proven to be difficult *(23,33)* and requires optimal B-cell stimulation. This can be achieved by coculture on fibroblasts transfected with CD40L and rIL-4 (5 ng/ml) *(34)* or by coculture on the murine EL-4 B5 thymoma cell line *(35)*. To avoid coculture, more recent work has shown efficient transduction following stimulation of B cells with the 21 bp CG-rich oligonucleotide C274 *(36,37)*. Alternatively, CD34 progenitor cells can be transduced (*see* **Subheading 3.5.**) and the cells can then be differentiated into B cells.
2. For transduction of mature peripheral B cells, plate 5×10^4 cells/ml B cells in 200 µl of RPMI media supplemented with 10% FCS and HEPES buffer with 2.5 µg/ml CpG DNA, goat F(ab´2) anti-human IgA/IgG/IgM (H+L), rIL-2 (50 ng/ml), and rIL-10 (10 ng/ml) *(36)*.
3. Add virions to cells, at a MOI of 10–20 (higher MOIs are toxic) at day 2 or 3 post stimulation. Expression from an internal CMV promoter has been shown to be approximately threefold higher than from an EF1α promoter *(36)*.
4. Assess transduction 2 days later. Note that in contrast to T cells, exposure of B cells to lentiviral vectors is toxic and can result in apoptosis.

3.10. Lentiviral Transduction of Murine B Cells

1. Lentiviral transduction of murine B cells is superior when virions are pseudotyped with ecotropic MLV envelope as compared with VSV-G *(34)*, and as such, the former virions may provide more optimal results.
2. Culture B cells at 1×10^6 cells/ml in the presence of either LPS (50 µg/ml) or anti-CD40 antibodies (2 µg/ml) and mouse (m)IL-4 (10 ng/ml) for 24–48 h.
3. Add virions at an MOI of 5 (*see* **Note 20**). Virions may be centrifuged onto B cells on 24-well plates for 1 h at $700 \times g$ at 20°C, but under some conditions, this can result in increased cell death. Alternatively, polybrene can be added at 6 µg/ml.
4. Assess transduction at 24–48 h.

3.11. Lentiviral Transduction of Human Monocyte-Derived Dendritic Cells (see Note 21)

1. PBMCs are obtained from normal donors and isolated by Ficoll density centrifugation. Cells can be further purified to obtain the $CD14^+$ monocyte fraction, or monocytes can be enriched by adhesion on tissue-culture-treated plates.
2. Activate cells in RPMI media containing 10% FCS and supplemented with granulocyte–macrophage (GM)-CSF (800 µg/ml) and rIL-4 (100 U/ml). Replace with fresh media containing GM-CSF and rIL-4 every 2 days.
3. Transduce with lentiviral virions at an MOI of 20 prior to inducing maturation of dendritic cells, at time points ranging from 24 h to 7 days of culture. Transductions

should be performed in the presence of polybrene (8 µg/ml) *(38,39)*, but efficiencies may be improved by substituting polybrene with protamine sulfate (5 µg/ml) *(40)*.
4. Remove virions after 16 h, wash cells by centrifugation, and continue to culture in fresh media under the same conditions.

4. Notes

1. Transfection efficiencies are extremely sensitive to the pH of the HBS–Ca solution. It is thus recommended that four batches of HBS be made with pHs of 7.01, 7.03, 7.07, and 7.10. The efficiency of transfection with the four batches can then be tested and one batch maintained. HBS can then be aliquoted and stored at –20°C for long-term use.
2. Plasmids can easily be purified using the column kits sold by a large number of different biotech companies. Nevertheless, the purity of the plasmid preparations is often higher following purification by cesium chloride gradient.
3. 293T are not overtly adherent cells and may be detached from the plate using less toxic 0.5 mM EDTA instead of trypsin.
4. Safety specifications for culture of 293T cells transfected with plasmids resulting in the production of virions whose envelope allows transduction of human cells vary from country to country. In France, for example, these experiments must be performed in a BL3 facility if the transgene encoded in the lentiviral vector is an oncogene. Virions can be handled in a BL2 facility only following testing to ascertain that there are no replication-competent virus recombinants.
5. Low transfection efficiencies can be due to mycoplasma infection. Importantly, >30% of cells in most cell culture facilities are estimated to be mycoplasma-positive. Routine testing, using either commercially available PCR kits or Hoescht staining to assess whether there is DNA in the cytoplasm, is highly encouraged. In the event of a positive result, the two options are to re-purchase the 293T cell line from a cell culture collection and treat the cells with an anti-mycoplasma agent such as Baytril (Bayer Puteaux, France) *(41)*.
6. Even if your lentiviral vector does not encode a marker gene, it is important to assure that your levels of transfection are high (>80%) before continuing to virion production. This is most easily done by transfecting an EGFP-harboring vector and assessing fluorescence of the tissue-culture plate by microscopy.
7. For pilot experiments wherein titers of $>10^6$ are not necessary, supernatants harboring virions may be used immediately without any further concentration. In this case, it is advantageous to change the media on the 293T cells carefully 24 h post transfection to the media that is used for culture of the target cells that will be transduced.
8. Because of the secondary structure of the envelope, virions harboring VSV-G envelopes are easily concentrated through ultracentrifugation. Although titers of virions harboring retroviral envelopes with a surface-transmembrane (SU-TM) structure will also be increased by ultracentrifugation, the overall yield may be

reduced as some virions will be rendered non-infectious because of shedding of the SU.
9. Because of the stability of the VSV-G envelope, titers of VSV-G-pseudotyped virions decrease by approximately 10% upon freeze-thawing. By contrast, the titers of virions pseudotyped with retroviral envelopes with an SU/TM structure tend to decrease by approximately 1-log upon freeze-thawing. It is therefore optimal to freeze virion preps in small aliquots.
10. The titers of virions can be assessed in various manners. Titers based on assessing expression of a transgene are generally simplest, especially if the transgene can be monitored by flow cytometry. To avoid problems due to multiple integrations, it is best to determine titers at dilutions wherein the percentage of cells expressing the transgene is <30%. At these transduction efficiencies, the number of integrated copies is generally one *(42)*.

   Importantly though, virions can infect cells without resulting in transgene expression, and as such, it might be more accurate to monitor reverse transcription of the lentiviral vector. Finally, reverse transcription can also be modulated by the cellular environment and quantification of physical particles can be determined by monitoring levels of the p24 protein incorporated in the virions. This last test, however, will not provide any information regarding the ability of the virions to transduce a target cell.
11. Titers, or transducing units, are very difficult to "translate" from one laboratory to another and are biased by the target cells used to test transduction. Thus, the titer that will be obtained on NIH3T3, 293T, HeLa, or Jurkat cells can differ by up to 1-log. Therefore, it is important to standardize within your own laboratory and use the same cell type plated at the same density for titering.
12. Transductions of $CD34^+$ cells can be performed without cytokine stimulation, but the relative transduction is extremely low *(5,25)*.
13. The choice of cytokines for culture of the $CD34^+$ cells is dictated by the further use of the cells. For example, IL-7 is added for further lymphoid differentiation.
14. The timing at which transgene expression is assessed can be important. Although dependent on the cell type, transgene expression generally requires 24–48 h. Nevertheless, if transgene expression is assessed at these early time points, it is important to validate the data at a later time as early expression can be due to pseudotransduction. This refers to transgene protein from the virion producer cell which is incorporated into the virion and then transferred to the target cell rather than new transgene transcription in the cell from an integrated copy of the vector.
15. Lineage-negative cells include stromal precursors that are difficult to deplete with the use of antibodies. Selected cells are therefore cultured on tissue-culture-treated plates overnight to allow the adhesion of stromal cells. This step decreases, although does not eliminate completely, the stromal cell population. As discussed for human progenitor cells, the choice of cytokines is modulated by the further use of the transduced cells. For example, it has recently been shown that the

combination of SCF and TPO appears to be optimal for engraftment of murine progenitor cells in vivo *(43)*.
16. T cells can be activated through the TCR using either anti-CD3/ IL-2 or a combination of anti-CD3/anti-CD28 antibodies. The latter is preferable under situations where it is important to maintain a polyclonal repertoire as it has been shown that anti-CD3/IL-2 stimulation results in skewing of the TCRBV repertoire *(44)*.
17. At MOIs of 1 and 10, you can expect a level of human T-cell transduction ranging from 5 to 15% and 35 to 60%, respectively, on TCR-stimulated T cells. For IL-7-stimulated T cells, it is recommended to use an MOI >20 to achieve transductions of >20%.
18. Activation and transduction of murine immune cells (including T and progenitor cells) will differ depending on the strain of mice used, and thus, comparisons between different experiments will be facilitated by consistently using the same strain.
19. As indicated above, the efficiency of transduction of murine T cells with lentiviral vectors is significantly lower than that obtained with conventional MLV vectors. It is therefore recommended that lentiviral vectors be pseudotyped with ecotropic envelope *(21)* for transduction of these cells.
20. Use of higher MOI may result in higher transduction efficiencies, but they are also likely to be associated with higher cell mortality.
21. Lentiviral transduction of murine dendritic cells is generally performed on bone marrow-derived progenitor cells that are cultured ex vivo in the presence of GM-CSF (1000 U/ml) and IL-4 (1000 U/ml) and then transduced as indicated in this section *(45)*.

Acknowledgments

We are very grateful to all members of our laboratory, past and present, for their precious input into improving gene transfer technology. A special thanks to Dr. Els Verhoeyen, Dr. F.L. Cosset, and Dr. M. Sitbon for their generosity in sharing theoretical and practical guidance on lentiviral gene transfer technology. We are very grateful to Dr. Virginie Sandrin and Dr. F.L. Cosset for generously allowing us to use their unpublished data in Fig. 3 of this chapter. This work has been supported by the AFM and by the European Community (contract LSHC-CT-2005-018914 "ATTACK").

References

1. Naldini, L., U. Blomer, P. Gallay, D. Ory, R. Mulligan, F.H. Gage, I.M. Verma and D. Trono. 1996. In vivo gene delivery and stable transduction of nondividing cells by a lentiviral vector. *Science* **272**:263–267.
2. Delenda, C. 2004. Lentiviral vectors: optimization of packaging, transduction and gene expression. *J Gene Med* **6 Suppl 1**:S125–S138.

3. Sinn, P.L., S.L. Sauter and P.B. McCray, Jr. 2005. Gene therapy progress and prospects: development of improved lentiviral and retroviral vectors – design, biosafety, and production. *Gene Ther* **12**:1089–1098.
4. Zennou, V., C. Petit, D. Guetard, U. Nerhbass, L. Montagnier and P. Charneau. 2000. HIV-1 genome nuclear import is mediated by a central DNA flap. *Cell* **101**:173–185.
5. Follenzi, A., L.E. Ailles, S. Bakovic, M. Geuna and L. Naldini. 2000. Gene transfer by lentiviral vectors is limited by nuclear translocation and rescued by HIV-1 pol sequences. *Nat Genet* **25**:217–222.
6. Dardalhon, V., B. Herpers, N. Noraz, F. Pflumio, D. Guetard, C. Leveau, A. Dubart-Kupperschmitt, P. Charneau, et al. 2001. Lentivirus-mediated gene transfer in primary T cells is enhanced by a central DNA flap. *Gene Ther* **8**:190–198.
7. Sirven, A., F. Pflumio, V. Zennou, M. Titeux, W. Vainchenker, L. Coulombel, A. Dubart-Kupperschmitt and P. Charneau. 2000. The human immunodeficiency virus type-1 central DNA flap is a crucial determinant for lentiviral vector nuclear import and gene transduction of human hematopoietic stem cells. *Blood* **96**: 4103–4110.
8. Manganini, M., M. Serafini, F. Bambacioni, C. Casati, E. Erba, A. Follenzi, L. Naldini, S. Bernasconi, et al. 2002. A human immunodeficiency virus type 1 pol gene-derived sequence (cPPT/CTS) increases the efficiency of transduction of human nondividing monocytes and T lymphocytes by lentiviral vectors. *Hum Gene Ther* **13**:1793–1807.
9. Zufferey, R., T. Dull, R.J. Mandel, A. Bukovsky, D. Quiroz, L. Naldini, D. Trono and D. Nagy. 1998. Self-inactivating lentivirus vector for safe and efficient in vivo gene delivery. *J Virol* **72**:9873–9880.
10. Ramezani, A., T.S. Hawley and R.G. Hawley. 2003. Performance- and safety-enhanced lentiviral vectors containing the human interferon-beta scaffold attachment region and the chicken beta-globin insulator. *Blood* **101**:4717–4724.
11. Rivella, S., J.A. Callegari, C. May, C.W. Tan and M. Sadelain. 2000. The cHS4 insulator increases the probability of retroviral expression at random chromosomal integration sites. *J Virol* **74**:4679–4687.
12. Zufferey, R., D. Nagy, R.J. Mandel, L. Naldini and D. Trono. 1997. Multiply attenuated lentiviral vector achieves efficient gene delivery in vivo. *Nat Biotechnol* **15**:871–875.
13. Chinnasamy, D., N. Chinnasamy, M.J. Enriquez, M. Otsu, R.A. Morgan and F. Candotti. 2000. Lentiviral-mediated gene transfer into human lymphocytes: role of HIV-1 accessory proteins. *Blood* **96**:1309–1316.
14. Cui, Y., J. Golob, E. Kelleher, Z. Ye, D. Pardoll and L. Cheng. 2002. Targeting transgene expression to antigen-presenting cells derived from lentivirus-transduced engrafting human hematopoietic stem/progenitor cells. *Blood* **99**:399–408.
15. May, C., S. Rivella, J. Callegari, G. Heller, K.M. Gaensler, L. Luzzatto and M. Sadelain. 2000. Therapeutic haemoglobin synthesis in beta-thalassaemic mice expressing lentivirus-encoded human beta-globin. *Nature* **406**:82–86.

16. Adjali, O., G. Marodon, M. Steinberg, C. Mongellaz, V. Thomas-Vaslin, C. Jacquet, N. Taylor and D. Klatzmann. 2005. In vivo correction of ZAP-70 immunodeficiency by intrathymic gene transfer. *J Clin Invest* **115**:2287–2295.
17. Verhoeyen, E. and F.L. Cosset. 2004. Surface-engineering of lentiviral vectors. *J Gene Med* **6 Suppl 1**:S83–S94.
18. Verhoeyen, E., V. Dardalhon, O. Ducrey-Rundquist, D. Trono, N. Taylor and F.L. Cosset. 2003. IL-7 surface-engineered lentiviral vectors promote survival and efficient gene transfer in resting primary T lymphocytes. *Blood* **101**:2167–2174.
19. Sandrin, V., B. Boson, P. Salmon, W. Gay, D. Negre, R. Le Grand, D. Trono and F.L. Cosset. 2002. Lentiviral vectors pseudotyped with a modified RD114 envelope glycoprotein show increased stability in sera and augmented transduction of primary lymphocytes and CD34+ cells derived from human and nonhuman primates. *Blood* **100**:823–832.
20. Hanawa, H., P.F. Kelly, A.C. Nathwani, D.A. Persons, J.A. Vandergriff, P. Hargrove, E.F. Vanin and A.W. Nienhuis. 2002. Comparison of various envelope proteins for their ability to pseudotype lentiviral vectors and transduce primitive hematopoietic cells from human blood. *Mol Ther* **5**:242–251.
21. Schambach, A., M. Galla, U. Modlich, E. Will, S. Chandra, L. Reeves, M. Colbert, D.A. Williams, et al. 2006. Lentiviral vectors pseudotyped with murine ecotropic envelope: increased biosafety and convenience in preclinical research. *Exp Hematol* **34**:588–592.
22. Zack, J.A., S.J. Arrigo, S.R. Weitsman, A.S. Go, A. Haislip and I.S. Chen. 1990. HIV-1 entry into quiescent primary lymphocytes: molecular analysis reveals a labile, latent viral structure. *Cell* **61**:213–222.
23. Unutmaz, D., V.N. KewalRamani, S. Marmon and D.R. Littman. 1999. Cytokine signals are sufficient for HIV-1 infection of resting human T lymphocytes. *J Exp Med* **189**:1735–1746.
24. Neil, S., F. Martin, Y. Ikeda and M. Collins. 2001. Postentry restriction to human immunodeficiency virus-based vector transduction in human monocytes. *J Virol* **75**:5448–5456.
25. Sirven, A., E. Ravet, P. Charneau, V. Zennou, L. Coulombel, D. Guetard, F. Pflumio and A. Dubart-Kupperschmitt. 2001. Enhanced transgene expression in cord blood CD34(+)-derived hematopoietic cells, including developing T cells and NOD/SCID mouse repopulating cells, following transduction with modified trip lentiviral vectors. *Mol Ther* **3**:438–448.
26. Muhlebach, M.D., N. Wolfrum, S. Schule, U. Tschulena, R. Sanzenbacher, E. Flory, K. Cichutek and M. Schweizer. 2005. Stable transduction of primary human monocytes by simian lentiviral vector PBj. *Mol Ther* **12**:1206–1216.
27. Mangeot, P.E., D. Negre, B. Dubois, A.J. Winter, P. Leissner, M. Mehtali, D. Kaiserlian, F.L. Cosset, et al. 2000. Development of minimal lentivirus vectors derived from simian immunodeficiency virus (SIVmac251) and their use for gene transfer into human dendritic cells. *J Virol* **74**:8307–8315.

28. Negre, D., P.E. Mangeot, G. Duisit, S. Blanchard, P.O. Vidalain, P. Leissner, A.J. Winter, C. Rabourdin-Combe, et al. 2000. Characterization of novel safe lentiviral vectors derived from simian immunodeficiency virus (SIVmac251) that efficiently transduce mature human dendritic cells. *Gene Ther* **7**:1613–1623.
29. Schnell, T., P. Foley, M. Wirth, J. Munch and K. Uberla. 2000. Development of a self-inactivating, minimal lentivirus vector based on simian immunodeficiency virus. *Hum Gene Ther* **11**:439–447.
30. Hanawa, H., P. Hematti, K. Keyvanfar, M.E. Metzger, A. Krouse, R.E. Donahue, S. Kepes, J. Gray, et al. 2004. Efficient gene transfer into rhesus repopulating hematopoietic stem cells using a simian immunodeficiency virus-based lentiviral vector system. *Blood* **103**:4062–4069.
31. Dupuy, F.P., E. Mouly, M. Mesel-Lemoine, C. Morel, J. Abriol, M. Cherai, C. Baillou, D. Negre, et al. 2005. Lentiviral transduction of human hematopoietic cells by HIV-1- and SIV-based vectors containing a bicistronic cassette driven by various internal promoters. *J Gene Med* **7**:1158–1171.
32. Poeschla, E.M., F. Wong-Staal and D.J. Looney. 1998. Efficient transduction of nondividing human cells by feline immunodeficiency virus lentiviral vectors. *Nat Med* **4**:354–357.
33. Serafini, M., L. Naldini and M. Introna. 2004. Molecular evidence of inefficient transduction of proliferating human B lymphocytes by VSV-pseudotyped HIV-1-derived lentivectors. *Virology* **325**:413–424.
34. Janssens, W., M.K. Chuah, L. Naldini, A. Follenzi, D. Collen, J.M. Saint-Remy and T. VandenDriessche. 2003. Efficiency of onco-retroviral and lentiviral gene transfer into primary mouse and human B-lymphocytes is pseudotype dependent. *Hum Gene Ther* **14**:263–276.
35. Bovia, F., P. Salmon, T. Matthes, K. Kvell, T.H. Nguyen, C. Werner-Favre, M. Barnet, M. Nagy, et al. 2003. Efficient transduction of primary human B lymphocytes and nondividing myeloma B cells with HIV-1-derived lentiviral vectors. *Blood* **101**:1727–1733.
36. Kvell, K., T.H. Nguyen, P. Salmon, F. Glauser, C. Werner-Favre, M. Barnet, P. Schneider, D. Trono, et al. 2005. Transduction of CpG DNA-stimulated primary human B cells with bicistronic lentivectors. *Mol Ther* **12**:892–899.
37. Marshall, J.D., K. Fearon, C. Abbate, S. Subramanian, P. Yee, J. Gregorio, R.L. Coffman and G. Van Nest. 2003. Identification of a novel CpG DNA class and motif that optimally stimulate B cell and plasmacytoid dendritic cell functions. *J Leukoc Biol* **73**:781–792.
38. Dyall, J., J.B. Latouche, S. Schnell and M. Sadelain. 2001. Lentivirus-transduced human monocyte-derived dendritic cells efficiently stimulate antigen-specific cytotoxic T lymphocytes. *Blood* **97**:114–121.
39. Rouas, R., R. Uch, Y. Cleuter, F. Jordier, C. Bagnis, P. Mannoni, P. Lewalle, P. Martiat, et al. 2002. Lentiviral-mediated gene delivery in human

monocyte-derived dendritic cells: optimized design and procedures for highly efficient transduction compatible with clinical constraints. *Cancer Gene Ther* **9**:715–724.
40. Lizee, G., M.I. Gonzales and S.L. Topalian. 2004. Lentivirus vector-mediated expression of tumor-associated epitopes by human antigen presenting cells. *Hum Gene Ther* **15**:393–404.
41. Uphoff, C.C., C. Meyer and H.G. Drexler. 2002. Elimination of mycoplasma from leukemia-lymphoma cell lines using antibiotics. *Leukemia* **16**:284–288.
42. Kustikova, O.S., A. Wahlers, K. Kuhlcke, B. Stahle, A.R. Zander, C. Baum and B. Fehse. 2003. Dose finding with retroviral vectors: correlation of retroviral vector copy numbers in single cells with gene transfer efficiency in a cell population. *Blood* **102**:3934–3937.
43. Mostoslavsky, G., D.N. Kotton, A.J. Fabian, J.T. Gray, J.S. Lee and R.C. Mulligan. 2005. Efficiency of transduction of highly purified murine hematopoietic stem cells by lentiviral and oncoretroviral vectors under conditions of minimal in vitro manipulation. *Mol Ther* **11**:932–940.
44. Ferrand, C., E. Robinet, E. Contassot, J.M. Certoux, A. Lim, P. Herve and P. Tiberghien. 2000. Retrovirus-mediated gene transfer in primary T lymphocytes: influence of the transduction/selection process and of ex vivo expansion on the T cell receptor beta chain hypervariable region repertoire. *Hum Gene Ther* **11**:1151–1164.
45. He, Y., J. Zhang, Z. Mi, P. Robbins and L.D. Falo, Jr. 2005. Immunization with lentiviral vector-transduced dendritic cells induces strong and long-lasting T cell responses and therapeutic immunity. *J Immunol* **174**:3808–3817.
46. Schambach, A., H. Wodrich, M. Hildinger, J. Bohne, H.G. Krausslich and C. Baum. 2000. Context dependence of different modules for posttranscriptional enhancement of gene expression from retroviral vectors. *Mol Ther* **2**:435–445.
47. Donello, J.E., J.E. Loeb and T.J. Hope. 1998. Woodchuck hepatitis virus contains a tripartite posttranscriptional regulatory element. *J Virol* **72**:5085–5092.

19

Axenic Mice Model

Antoine Giraud

Summary

The interaction of an organism with surrounding microbial communities has a profound impact on its anatomy, physiology, and behavior. The innate immune system plays a major role in the crosstalk between bacteria and the hosts they colonize. Axenic mice provide a powerful *in vivo* controlled model to decipher the particular interactions between the host and one defined strain, or group of strains, isolated from the "noise" of the rest of the flora. This chapter briefly reviews the specificities of axenic mice and describes the main guidelines to derive, house, and work with axenic mice.

Key Words: Axenic; mice; isolator; intestinal flora; gnotoxenic; germ free.

1. Introduction

We are living in a microbial world. On Earth, the total biomass of the 5×10^{30} Bacteria and Archea is larger than the biomass of all other life forms together *(1)*. Microbes are ubiquitous, colonizing every environment from deep ocean to mesosphere, and have even been detected in rocks. All exposed surfaces and particularly epithelial surfaces are colonized. The density of microorganisms living in the gastrointestinal tract is the highest known. In humans, the gut is the residence of up to 10^{14} bacteria belonging to hundreds of species *(2)*. Sterile at birth, the intestine is colonized immediately after delivery by a few pioneer species. Initial colonization by facultative anaerobes lowers the redox potential of the intestinal lumen, a prerequisite for subsequent colonization by the anaerobes. The complexity of the flora increases with age and the diversification of the alimentation during the first year of life. Anaerobic bacteria belonging to the genera *Bacteroides, Bifidobacterium, Eubacterium, Clostridium, Peptococcus, Peptostreptococcus,* and *Ruminococcus* are predominant, whereas aerobes (*Escherichia, Enterobacter, Enterococcus, Klebsiella, Lactobacillus, and Proteus*) are less abundant. The density and composition of the flora vary between the mouth, the stomach, upper small bowel, lower small bowel, and rectum. For example, the facultative anaerobes represent 25% of the total bacteria in the initial part of the right colon versus 1% in the distal colon *(3)*.

The composition of the gut flora is influenced by different factors, mainly nutritional habit and aging *(4)*. Obviously, medical treatments, such as antibiotic therapy, strongly affect the flora in a more or less transient way *(5)*. Furthermore, the colonization ability of particular strains, within the same bacterial species, can differ dramatically due to genetic characteristics.

The composition of the flora is also influenced by the host's genotype. Indeed, the fecal floras of monozygotic twin siblings were found to be much more alike than those of dizygotic twin siblings *(6)*. In addition to undefined gene loci, the major histocompatibility complex (MHC) alone has a pronounced effect, because mice with different MHC in the same genetic background have significantly different fecal flora *(7)*. Thus, the composition of the flora varies from one individual to another *(8)*, and if the population size of a defined genus seems to remain stable over time within the same individual, its composition in terms of strains is subject to important fluctuations *(9)*.

The intimate association of this huge microbial community with the host makes determining how the animal anatomy, physiology, or even behavior is dependent on the presence of the colonizing microorganisms complex. Such questions have haunted the scientific community for a long time. In 1885, Pasteur expressed interest in breeding animals from birth isolated from microorganisms to determine whether any microbes are useful or even necessary for life. He even proposed a model and an approach to experimentally address this question *(10)*. How an observable characteristic of an individual results from the interaction with or reaction against the resident microbial flora can only be gained by comparing the same trait in animals that have no flora (axenic), all other factors being equal between the animals. Only in the absence of any difference can one attribute the feature to an animal's own characteristic. The difficulties in setting up a reliable experimental model delayed for decades the exploration of the impact of the microflora on the host. In more recent years, it has been shown that many features of the biochemistry, physiology, and immunology of the animal host are, in fact, microflora-associated characteristics.

The flora has a critical role in the maturation of the gastrointestinal tract. Bacteria participate in the development and maintenance of gut sensory and motor functions, including the promotion of intestinal propulsive activity *(11)*. The flora has an impact on epithelial villus length, increases the renewal time of epithelial cells *(12)*, increases the number of the mucin-producing Goblet cells *(13)*, changes the composition of secreted mucins *(14)*, and induces the development of villus capillary networks *(15)*. All these modifications enhance the protection of the epithelial layer and also increase the intestinal absorbance capacity. The latter is further enhanced by the induction upon colonization of host genes involved in the processing and absorption of nutrients upon colonization *(16)*.

Axenic mice have an excess of food consumption compared with conventionally raised animals. Indeed, it is now well established that the bacterial population participates in the digestion of the diet. Intestinal bacteria metabolize the mucus produced by epithelial cells and indigestible dietary compounds *(17)*. Fermentation of dietary carbohydrate, which produces short-chain fatty acids absorbed by the host, results in energy and nutrient salvage *(18)*, such as butyrate (principal energy source of colonic epithelial cells), propionate (substrate for gluconeogenesis in the liver), and acetate (substrate of hepatic lipogenesis). Besides energy uptake, the intestinal bacteria are potent inducers of storage, increasing body fat deposition *(19)*. The resident microflora participates in the production of amino acids absorbed by the host *(20)*. Bacteria are also a source of vitamin K and some of group B vitamins for the host *(21)*. The dramatic cecal enlargement observed in axenic mice may be the result of inefficient digestion and absorption.

In addition, the resident flora is involved in the establishment of the host mucosal immune defenses. In axenic mice, the spleen and lymph nodes are immature *(22)*, the Payer's patches are poorly developed *(23)*, the lamina propria of the ileum and the cecum contains few lymphoid cells, and the number of plasmocytes secreting type A immunoglobulins (IgA) is low *(24)*. All of these characteristics are reverted upon colonization. Stimulation of the mucosal immune system by the resident flora induces a response that, on the one hand, limits the translocation of bacteria in the body, mainly through the secretion of IgA *(25)*, and on the other hand, leads to the development of a local tolerance to microbial epitopes to avoid an inflammatory response deleterious for both the host and the flora *(26)*. It is necessary however to maintain a potent response against pathogenic bacteria. As commensal and pathogenic bacteria share determinants recognized by the Toll-like receptors (TLRs) involved in the initiation of innate immune responses, discrimination of pathogens can be achieved through the modulation and/or the compartmentalization of the expression of

TLR by the host *(27)* or down-regulation of immunostimulant antigens by the bacteria, as observed for flagellin by an *Escherichia coli* commensal strain.

An additional feature of the flora is the protection of the host against pathogenic invasion. The presence of a particular bacterial strain or species or association of strains or species in the digestive tract has been reported to abrogate efficiently the colonization ability of another bacteria *(28)*. This has been extensively documented for pathogenic bacteria, but the same conclusions may be drawn for commensal strains. These antagonist activities of the resident flora against invading strains may be due to the occupancy of attachment sites and competition for essential nutriments. It has often been argued that secreted antimicrobial molecules are largely responsible for this exclusion of competing strains, but solid *in vivo* evidence is lacking. The stimulation of the host innate immune response by some bacterial species may participate in the clearance of local competitors *(29)*.

Axenic mice colonized with one bacterial strain (monoxenic) allow us to study the impact of this strain on its host. When analyzing results, it is important to differentiate the strain's specific effect from the above-mentioned microflora-associated characteristics that have been detected by axenic/conventional comparisons. Moreover, in monoxenic mice, the location and the size of the bacterial population may be different than that in conventional conditions. Thus, it is sometimes necessary to have two groups of mice inoculated with a complex set of strains with and without the strain of interest. Monoassociated mice remain, however, a powerful tool in the understanding of host–bacterial interactions.

When following the mouse response to a particular flora, it may be relevant to monitor in parallel the evolution of the bacterial population(s) size and activity by classical microbiology or molecular biology techniques *(2)*. As the clarity of the results of an experiment often depends on the quality of the controls, it is sometimes necessary to breed conventional mice in an isolator under the exact same conditions as axenic mice.

Lastly, axenic mice are above all mice; thus, protocols developed for non-axenic mice remain valid.

2. Materials

2.1. Working with Isolators

2.1.1. The Isolator

The isolator is the actual barrier that separates the controlled environment from the external atmosphere. It can be composed of flexible plastic film, rigid plastic, or stainless steel allied with glass. The latter provides optimum resistance. Rigid isolators are suitable to work under negative pressure, but a rigid structure supporting a flexible isolator (semi-rigid isolator) is sufficient for such an application. The flexible isolators are probably the least expensive and easiest to repair. The isolator requires different devices for the following purposes.

2.1.1.1. TRANSFER

Airlocks: These are solid plastic cylinders sealed on the isolator. They can vary in size depending on purpose. Large airlocks can allow connection to a transfer device. The smallest are closed with caps; the others with flexible plastic caps are fixed with rubber bands. The small size airlock allows the transfer of tubes containing inoculums into the isolator and removal of small size samples (blood, feces, etc.). When introducing material into the isolator, it is necessary to wait approximately half an hour to sterilize the airlock with peracetic acid (see 3.1.2.1).

One convenient feature to reduce the transfer time is the double-door transfer port. This system allows the rapid transfer of components or products between a pre-sterilized container and the isolator without breaking its integrity. It is made up of two parts: a fixed component mounted on the wall of the isolator and a mobile component on the container. When fixed to each other, the two parts hermetically associate and can be opened from the inside of the isolator. This system is particularly recommended when working with pathogenic strains as it allows removal of infected material or samples out of the isolator into a confined container without exposing experimenters.

2.1.1.2. MANIPULATION

Glove sleeves allow the operator to work in the protected area. The sleeve is fixed to a sealed attachment on the wall of the isolator and the glove fixed to the sleeve. The number of glove sleeves depends on the size of the isolator; it should allow the access of the entire isolator area. The isolator can possess two pairs of glove sleeves facing each other on opposite walls to allow the simultaneous work on the same animal by two experimenters.

2.1.1.3. AERATION

The internal volume of the isolator is maintained at positive or negative pressure by ventilation through high-efficiency particular air (HEPA) filters to provide the required quality of air. A HEPA filter must retain all particles as small as 0.3 µm in size with an efficiency rating of 99.97%. A pre-filter grid is recommended to protect the filters from the dust of the isolator.

2.1.2. Container

Containers are devices that can be hermetically closed, sterilized, and then connected to an isolator through an airlock or a double-port transfer system. Containers can be sterilized by heat, autoclave; others that are equipped with HEPA filters can be sterilized by gas as for isolators.

2.1.3. Preparation of the Isolator and Sterilization

1. Washing solution: soap in water.
2. Decontamination solution: aqueous solution of 0.5% peracetic acid, stable at 4°C up to 24 h. Prepare under a hood and wear gloves, protection glasses, and a lab coat.
3. Sterilization solution: aqueous solution of 5% peracetic acid, stable at 4°C up to 2 h. Prepare under a hood and wear gloves, protection glasses, and a lab coat.
4. Material for the isolator: cages and their covers (it is convenient to have one extra cage to put the mice when cleaning a housing cage), mice bottles and their covers, stock of aliment, stock of water, stock of cage bedding, a plastic bag for the waste, box of hermetically closable tubes for samples, forceps to handle the mice, scissors if needed, marker pen to identify sample tubes or mark mice tail, and syringes and intragastric tube for inoculation. Equipment and supplies that enter the isolator should be either acid-resistant (plastic, glass, and stainless steel) or pre-sterilized wrapped in a hermetically closed acid-resistant device. All hermetically closed storage devices should contain sterilized material (autoclaved bottle of water, bag/box of irradiated diet, autoclaved bag/box of tubes for sample, autoclaved box of cage bedding, etc.).

2.2. Validation of the Sterilization Procedure

1. Bacterial strain: *Geobacillus stearothermophilus* (ATCC 12980) spores are used to test steam, peracetic acid, and gas plasma sterilization. Spore solutions and spores fixed on strips are commercially available.
2. Nutrient broth: autoclaved solution of 5 g/l peptone and 3 g/l meat extract in distilled water.
3. Nutrient broth: autoclaved solution of 15 g/l agar, 5 g/l peptone, and 3 g/l meat extract in distilled water.
4. Sporulation plate: nutrient agar supplemented with 1 mg of Mn^{2+}/l.

2.3. Assessment of Isolator Integrity

1. Concentrated ammonia solution.
2. Blotting paper wet with Thymol blue solution in water (yellow at pH = 7). Ammoniac detectors are commercially available.

2.4. Axenic Mice Production by Aseptic Hysterectomy

1. Heat lamp or exam lamp placed close to the foster mother isolator.
2. A beaker and a sterile tube hermetically closable containing 10 ml of a non-toxic skin disinfectant such as iodophors.
3. Sterile scissors and forceps.
4. A thin plastic film disinfected with the decontamination solution.
5. Sterile scissors, forceps, and tissues or paper are placed into the isolator housing the foster mother several days in advance in order to minimize perturbation to the foster mother the day of the cesarian.

2.5. Assessment of the Microbiologic Sterility

1. LB: autoclaved solution of tryptone 10 g/l, yeast extract 5 g/l, and NaCl 10 g/l in distilled water.
2. Sabouraud: autoclaved solution of peptone 10 g/l and glucose 20 g/l in distilled water.
3. BHI agar: autoclaved solution of brain–heart infusion 37 g/l, yeast extract 5 g/l, and agar 15 g/l in distilled water.
4. MRS agar: autoclaved solution of proteose peptone 10 g/l, beef extract 10 g/l, yeast extract 5 g/l, dextrose 20 g/l, sodium acetate 5 g/l, potassium phosphate 2 g/l, ammonium citrate 2 g/l, polysorbate 80 1 g/l, magnesium sulfate 0.1 g/l, manganese sulfate 0.05 g/l, and agar 15 g/l in distilled water, pH 5.4.
5. ML agar: autoclaved solution of meat-liver base 30 g/l, glucose 20 g/l, and agar 15 g/l in distilled water.
6. Anaerobic growth tube: These are long (400 mm) tubes of small diameter (8 mm) that allow the growth of anaerobic strains at the bottom of the tube. The size of the tubes can vary depending on the supplier; however, it is important that the diameter of the tube is small compared to the length.

3. Methods

3.1. Working with the Isolator

When working with the isolator, before every manipulation, it is necessary to mentally visualize where the contaminant microorganisms lie. One can imagine the isolator as a vacuum and the external environment as a gas colored with bacteria.

3.1.1. Preparation of the Isolator and Sterilization

Peracetic acid is a suitable agent for isolator sterilization. It is sporicidal, tuberculocidal, bacteriocidal, fungicidal, and virucidal, while remaining safe for users and the environment. It decomposes into oxygen and acetic acid. Alternatively, hydrogen peroxide can be used, but the desorption time is longer as it migrates into plastics, and because it is odorless it makes leaks more difficult to detect.

1. Empty the isolator.
2. Disconnect all the removable parts: shelves, doors, gloves, sleeves, caps, filters, and joints.
3. Wash all removable parts with the washing solution.
4. Rinse with autoclaved sterile water and dry with clean absorbent paper.
5. Wash the contact surfaces between removable parts and the isolator with the decontamination solution (*see* **Note 1**).
6. Re-attach the removable parts on the isolator.
7. Wash internal surfaces of the isolator with the washing solution (*see* **Note 2**).
8. Rinse the washed surfaces with autoclaved sterile water and dry with clean absorbent paper.
9. Wash internal surfaces of the isolator with the decontamination solution (*see* **Note 3**).
10. Fill the isolator with the material needed for the experiment (*see* **Note 4**). All hermetically closed storage devices must have been beforehand sterilized.
11. Wash the plastic and glass material surfaces with the decontamination solution.
12. Dry all washed surfaces with clean absorbent paper.
13. Open all inside doors of airlocks. If the isolator has a double-door transfer port, connect a container and open the door.
14. Display the material on one half of the isolator with minimal contact surface to each other and to the isolator's floor or walls (*see* **Note 5**).
15. Connect the isolator's air inlet to the sterilization apparatus filled with sterilization solution and outlet to a gas scrubber or to the building extraction system.
16. Start sterilization and adjust the flow rate (*see* **Note 6**).
17. Every 20 min, turn the material upside down and transfer to the other half of the isolator to expose all surfaces to the sterilizing agent (*see* **Note 5**).
18. Flush the isolator for about 12 h and dry condensate vapors before introduction of animals.

3.1.2. Bringing Material in and out of an Isolator

3.1.2.1. THROUGH AIRLOCKS

This procedure describes the movement of a tube containing a bacterial suspension. The main steps of the process are, however, valid for any other material.

1. Place the inoculation solution into a sterile tube and close the tube hermetically (*see* **Note 7**).
2. Wash the surface of the tube with the decontamination solution.
3. Check that the internal cap of the airlock is closed and open the external cap of the airlock.
4. Spray the sterilization solution into the tubule with an atomizer and remove excess solution (*see* **Note 8**).

Axenic Mice Model

5. Close the airlock hermetically.
6. Wait 20 min until the sterilization is complete.
7. Open the airlock from the inside of the isolator and remove the tube and close the airlock (*see* **Note 9**).
8. Inoculate the mice (*see* **Note 10**).
9. Replace the tube into the airlock and close the cap, and harvest the tube from the outside.
10. Spray the sterilization solution into the tubule with an atomizer and close the airlock.

3.1.2.2. Through Double-Door Transfer

1. Spray the disinfection solution on the isolator and sterile container door.
2. Cover the entire surface of the doors and on the joints with the solution.
3. Connect the container door to the isolator door.
4. Open the joint doors from the inside of the isolator.
5. Transfer the material.
6. Close the door.
7. Disconnect the container from the isolator.
8. Take the material out of the container and sterilize the container before reuse.

3.2. Validation of the Sterilization Procedure

3.2.1. Preparation of Spore Suspensions

1. Grow *G. stearothermophilus* strain overnight in 10 ml nutrient broth at 55°C.
2. Plate the whole culture on 10 sporulation plates, and incubate at 55°C for 2 weeks (*see* **Note 11**).
3. Pour 5 ml of distilled water on each plate, scrape the plate surface, and collect the resulting solution.
4. Centrifuge for 10 min at $8000 \times g$, decant supernatant, and suspend the pellet in an equal amount of distilled water. Repeat this step twice to wash the spore suspension. In the final wash, suspend the pellet in 5 ml of water.
5. Plate dilutions of the spore suspension on nutrient agar plates pre-incubated at 55°C for 24 h and incubate 1 week at 55°C to determine concentration of viable spores.
6. Dilute the spore suspension to a concentration of 1×10^7 colony-forming units/ml, and store aliquots at 4°C for up to 8 months.

3.2.2. Evaluation of the Sterilization Procedure

1. Put a 100μl drop of the spore suspension containing 1×10^6 spores on empty Petri dishes and allow to dry at room temperature under a hood.
2. Place all but one Petri dish face open in various locations of the isolator (*see* **Note 12**).

3. Start sterilization (see 3.1.1).
4. Retrieve Petri dishes.
5. Pour 5 ml of distilled water on each plate, scrape the plate surface to suspend spores, and collect the resulting solution.
6. Plate dilutions of the spore suspension on nutrient agar plates pre-incubated at 55°C for 24 h and incubate for 1 week at 55°C to determine the concentration of viable spores.

3.3. Assessment of Isolator Integrity

Before starting detection for ammoniac leaks, one can assess the sealing of an isolator under positive pressure by passing a hand over its surface. An air leak gives a cold sensation. A drop of water with soap can help to localize a leak.

1. Introduce a bottle of ammonia, a beaker, and one paper wet with the indicator solution into the isolator (*see* **Note 13**).
2. Close all the isolator airlocks, doors, and filters.
3. Open the bottle and pour the ammonia into the beaker. The indicator should rapidly turn from yellow to blue.
4. Wait half an hour to saturate the isolator with the ammonia vapors.
5. Place the indicator for a few seconds on the isolator and check for color change. Move the indicator to another place until the whole surface of the isolator has been explored.
6. Pour the ammonia from the beaker back into the bottle.
7. Connect the isolator's air outlet to a gas scrubber or to the building extraction system, flush the isolator with air for about 12 h, and rinse with water before any other manipulation (*see* **Note 14**).

3.4. Axenic Mice Production by Aseptic Hysterectomy

The axenization of a mouse strain is performed by scarifying a gravid mother just before delivery and transferring the uterus into an isolator. The pups must be hand fed or foster nursed by an existing axenic mother, the latter being easiest. The production of axenic mice can also be performed by an embryo transfer technique *(30,31)*.

The first step is to mate the mouse strain to be derived axenic and the already axenic foster mother simultaneously. To synchronize estrus of females, they are isolated from the odor of males for several weeks. Following exposure to male urine, they will enter estrus after 3 days and be receptive to mating for about 4 days. The cesarian is started when the first foster mother has started delivering a few pups.

Axenic Mice Model

It is very important that the operation last for the shortest time possible. Thus, check carefully that all requested material is available and practice.

3.4.1. Cesarian

1. Kill the gravid mother by cervical dislocation (*see* **Note 15**).
2. Dip the mother into disinfectant bath and dry with sterile absorbent paper.
3. Lay the mother on the operating area inside a hood.
4. Cut the skin over the abdomen with scissors and peel the skin away.
5. Wear new sterile gloves and wrap the body with a disinfected transparent flexible thin plastic film.
6. With new forceps and scissors, lift the skin together with the film, and incise longitudinally from the abdomen to the thorax.
7. Remove the muscular abdominal wall (*see* **Note 16**).
8. Gently press the abdomen with finger to drive out the gravid uterus (*see* **Note 17**).
9. Isolate the uterus from other tissues.
10. Place the uterus into a sterile tube filled with warmed skin disinfectant and close hermetically (*see* **Note 18**).
11. Wash the external surface of the tube with the decontamination solution and place it into a sterile container (*see* **Note 19**).
12. Transfer the tube into the sterile isolator and wipe the decontamination solution on the surfaces.
13. Remove the uterus from the tube, wipe away excess skin disinfectant, and put it on a pre-warmed tissue or paper nest under the heat lamp (*see* **Note 20**).
14. Open the uterus with scissors from the ends of the uterine horns, and brush aside the uterine wall to expose the young mice (*see* **Note 21**).
15. Wipe amniotic fluid carefully away from the muzzle to free it of any obstruction and gently rub the thorax to stimulate breathing (*see* **Note 22**).
16. When breathing has started, clamp the umbilical cord, and separate the pups from their placenta.
17. Transfer the mice to the axenic foster mother.

3.4.2. Fostering

1. Remove the foster mother from the cage, collect feces and urine drops, and place her in a separate cage in the isolator where she cannot see her offspring (*see* **Note 23**).
2. Transfer one new pup into the foster mother nest, roll it around the foster mother natural litter, and put aside. Repeat for every pup (*see* **Note 24**).
3. Remove the same number of pups from the foster mother litter as the number of pups added (*see* **Note 25**).
4. Wipe feces and urine from the foster mother on the backs of the pups.
5. Cover the pups with the foster nest bedding.
6. Bring the foster mother back to her nest and avoid further disturbance.

3.5. Assessment of the Microbiologic Sterility

1. Melt agar medium. Heat BHI, MRS, and ML agar melted in a boiling water bath for at least 30 min and preserve at 50°C to keep melted (*see* **Note 26**). Prepare for each sample on aliquot of 5.5 ml of LB and one aliquot of 10 ml of all media.
2. Take a food, water, and litter sample from the stock in different tubes.
3. Catch a mouse by the tail with forceps, put it on the cover of the cage, and collect passing feces in a tube (*see* **Note 27**).
4. Take the samples out of the isolator through an airlock.
5. Under microbiologically sterile condition, transfer the samples in a tube containing 5.5 ml of LB and crush with a pipette.
6. Inoculate 1 ml of the resulting solution to the 10 ml of aliquoted melted agar medium (BHI, MRS, and ML) and pour into the anaerobic growth tube. Cool in a cold-water bath to solidify the medium. Incubate for 72 h at 37°C (*see* **Note 28**).
7. Inoculate 1 ml of the sample suspension to 10 ml of Sabouraud medium. Incubate for 3 weeks at 22°C.
8. Inoculate 1 ml of the sample suspension to 10 ml of LB medium. Incubate for 72 h at 37°C.
9. Examine the sample solution directly under the microscope.

Acknowledgments

I thank Pascal Guillaume, Jean Dabard, and Michel Fons from the National Institute for Agronomic Research (INRA), Unit on Ecology and Physiology of the Digestive Tract (UEPSD), 78352 Jouy-en-Josas Cedex, France, for teaching me how to handle and work with axenic mice in isolators.

4. Notes

1. Always wear gloves, lab coat, and protection glasses when working with the decontamination and sterilization solutions.
2. It is essential to wash carefully until all visible traces of dust are eliminated as the efficiency of the decontamination decreases in biofilms.
3. In flexible isolators, this step can be performed only if the isolator is under pressure. Thus, it is necessary to fill the isolator and to close the airflow routes to maintain the inflation of the isolator.
4. To limit the risks of contamination, it is necessary to minimize transfer of material in or out of the isolator during the experiment. Thus, it is convenient to fill the isolator with all the material required for the experiment. The space in the isolator is however restricted and sometime limiting. Space for the manipulations required for the experiment must be saved.
5. During the sterilization process, all surfaces of the isolator and of the material it contains have to be in contact with the vaporized sterilization solution (*see* **step 17**).

6. The automated sterilizer is connected to a source of pressurized air that bubbles into a heated peracetic acid solution to saturate the air with acid vapors that flow into the isolator. Adjust the flow rate to inflate the glove sleeves. Gas generators should not be assumed to be equivalent to each other, and the volume of peracetic acid required for sterilization depends on the model. The sterilization time required depends on both the sterilizer and the volume of the device to be sterilized. In any case, the efficiency of the sterilization procedure has to be validated on its action on *G. stearothermophilus* spores (see 3.2).
7. A leak in the tube will kill the inoculums during sterilization. The incoming solution should also be prepared using sterile technique to eliminate the risk of possible contaminants. If the strain(s) to be inoculated are sensitive to oxygen, the inoculum must be prepared in an anaerobic hood. The volume introduced into the isolator should be larger than necessary to inoculate all animals that occupy the isolator.
8. It is important to wet all surfaces of the airlock with the disinfection solution. As the airlock will be open from the inside of the isolator, for the animal's comfort it is, however, necessary to limit the entry of residual sterilization solution.
9. It is better to dry the tube of remaining sterilization solution before opening it.
10. The inoculation route depends on the experiment. For gut colonization, it can by performed per os or by adding the bacteria into the drinking water. Axenic mice can also be inoculated by placing a "contaminated" mouse into the cage.
11. Place the plates in a plastic bag to limit dehydration.
12. As you do not expect to see any growth, it is essential to have a growth control plate inoculated from the spores doted on a plate that had not been exposed to sterilization.
13. The beaker allows better evaporation of the ammoniac solution.
14. The mixture of peracetic acid with ammoniac is explosive.
15. Drugs may affect the pups.
16. At any step of the dissection, disrupting the gastrointestinal tract will instantaneously contaminate the working area and then greatly compromise the success of the manipulation. Ablation of the abdominal wall facilitates the extraction of the uterus.
17. Do not pull with tweezers that could impair the uterus.
18. **Steps 10–12** should be performed very rapidly; the uterus should remain in the skin disinfectant solution for only 1 min. The tube containing the disinfectant is preheated in a water bath at 37°C.
19. The container should be opened under the hood and stay open for the shortest time necessary for transfer. This procedure supposes that the isolator and the container are equipped with the double-door port system. If not, a flexible plastic tube with one end fixed to an isolator out-port airlock and the other end dipping into a bath of disinfection solution can be used. This airlock should be decontaminated as other airlocks before the operation (see 3.1.2).

20. It is crucial to keep the pups warm until breathing has started. It is possible to enter into the isolator a bottle of hot water to warm the nest.
21. As the uterine wall is thin, be careful not to injure the pups.
22. Pouring a drop of cold water (room temperature) on the pup can stimulate breathing, but it is necessary to dry immediately. The pups should be moved and turned upside down by catching them carefully at the neck with forceps, as the mother would.
23. To minimize the stress of the foster mother, it is better that she familiarized to such manipulation. If she does not excrete instantaneously, do not insist. Fresh feces can be retrieved from the foster mother cage (stick to the forceps when touching).
24. **Steps 2, 4,** and **5** should maximize the olfactory recognition of the fostered litter by the foster mother.
25. If the fostered litter is smaller than the natural one, the remaining natural pups must be identifiable by their fur or eye color or other mark to be removed before maturity.
26. Boiling the medium lowers its redox potential and favors the cultivation of anaerobic bacteria.
27. Take feces from at least one mouse per cage in the isolator. It is possible to pool all the feces collected in one isolator.
28. If the tubes are empty of detectable bacterial colonies, incubate for another 72 h. Bacteria growing into the medium make tiny colonies. To check that the spots are colonies rather than remains of non-digested food, break the tube, collect the suspect spot, and make a direct observation under the microscope. Gas bubbles in the medium are an indicator of bacterial fermentation.

References

1. Whitman, W. B., Coleman, D. C., and Wiebe, W. J. (1998) Prokaryotes: the unseen majority. *Proc Natl Acad Sci USA* **95,** 6578–83.
2. Zoetendal, E. G., Vaughan, E. E., and de Vos, W. M. (2006) A microbial world within us. *Mol Microbiol* **59,** 1639–50.
3. Marteau, P., Pochart, P., Dore, J., Bera-Maillet, C., Bernalier, A., and Corthier, G. (2001) Comparative study of bacterial groups within the human cecal and fecal microbiota. *Appl Environ Microbiol* **67,** 4939–42.
4. Mueller, S., Saunier, K., Hanisch, C., Norin, E., Alm, L., Midtvedt, T., Cresci, A., Silvi, S., Orpianesi, C., Verdenelli, M. C., Clavel, T., Koebnick, C., Zunft, H. J., Dore, J., and Blaut, M. (2006) Differences in fecal microbiota in different European study populations in relation to age, gender, and country: a cross-sectional study. *Appl Environ Microbiol* **72,** 1027–33.
5. De La Cochetiere, M. F., Durand, T., Lepage, P., Bourreille, A., Galmiche, J. P., and Dore, J. (2005) Resilience of the dominant human fecal microbiota upon short-course antibiotic challenge. *J Clin Microbiol* **43,** 5588–92.

6. Stewart, J. A., Chadwick, V. S., and Murray, A. (2005) Investigations into the influence of host genetics on the predominant eubacteria in the faecal microflora of children. *J Med Microbiol* **54,** 1239–42.
7. Toivanen, P., Vaahtovuo, J., and Eerola, E. (2001) Influence of major histocompatibility complex on bacterial composition of fecal flora. *Infect Immun* **69,** 2372–7.
8. Franks, A. H., Harmsen, H. J., Raangs, G. C., Jansen, G. J., Schut, F., and Welling, G. W. (1998) Variations of bacterial populations in human feces measured by fluorescent in situ hybridization with group-specific 16S rRNA-targeted oligonucleotide probes. *Appl Environ Microbiol* **64,** 3336–45.
9. Kimura, K., McCartney, A. L., McConnell, M. A., and Tannock, G. W. (1997) Analysis of fecal populations of bifidobacteria and lactobacilli and investigation of the immunological responses of their human hosts to the predominant strains. *Appl Environ Microbiol* **63,** 3394–8.
10. Pasteur, L. (1885) Observation relative à la note présentée par M. Duclaux. *CR Acad Sci* **100,** 68–69.
11. Barbara, G., Stanghellini, V., Brandi, G., Cremon, C., Di Nardo, G., De Giorgio, R., and Corinaldesi, R. (2005) Interactions between commensal bacteria and gut sensorimotor function in health and disease. *Am J Gastroenterol* **100,** 2560–8.
12. Savage, D. C., Siegel, J. E., Snellen, J. E., and Whitt, D. D. (1981) Transit time of epithelial cells in the small intestines of germfree mice and ex-germfree mice associated with indigenous microorganisms. *Appl Environ Microbiol* **42,** 996–1001.
13. Kandori, H., Hirayama, K., Takeda, M., and Doi, K. (1996) Histochemical, lectin-histochemical and morphometrical characteristics of intestinal goblet cells of germfree and conventional mice. *Exp Anim* **45,** 155–60.
14. Bry, L., Falk, P. G., Midtvedt, T., and Gordon, J. I. (1996) A model of host-microbial interactions in an open mammalian ecosystem. *Science* **273,** 1380–3.
15. Stappenbeck, T. S., Hooper, L. V., and Gordon, J. I. (2002) Developmental regulation of intestinal angiogenesis by indigenous microbes via Paneth cells. *Proc Natl Acad Sci USA* **99,** 15451–5.
16. Hooper, L. V., Wong, M. H., Thelin, A., Hansson, L., Falk, P. G., and Gordon, J. I. (2001) Molecular analysis of commensal host-microbial relationships in the intestine. *Science* **291,** 881–4.
17. Hooper, L. V., Midtvedt, T., and Gordon, J. I. (2002) How host-microbial interactions shape the nutrient environment of the mammalian intestine. *Annu Rev Nutr* **22,** 283–307.
18. Wong, J. M., de Souza, R., Kendall, C. W., Emam, A., and Jenkins, D. J. (2006) Colonic health: fermentation and short chain fatty acids. *J Clin Gastroenterol* **40,** 235–43.
19. Backhed, F., Ding, H., Wang, T., Hooper, L. V., Koh, G. Y., Nagy, A., Semenkovich, C. F., and Gordon, J. I. (2004) The gut microbiota as an environmental factor that regulates fat storage. *Proc Natl Acad Sci USA* **101,** 15718–23.

20. Metges, C. C. (2000) Contribution of microbial amino acids to amino acid homeostasis of the host. *J Nutr* **130,** 1857S–64S.
21. Mills, S., Benjarattanaporn, P., Bennett, A., Pattalung, R. N., Sundhagul, D., Trongsawad, P., Gregorich, S. E., Hearst, N., and Mandel, J. S. (1997) HIV risk behavioral surveillance in Bangkok, Thailand: sexual behavior trends among eight population groups. *AIDS* **11 Suppl 1,** S43–51.
22. Bauer, H., Horowitz, R. E., Levenson, S. M., and Popper, H. (1963) The response of the lymphatic tissue to the microbial flora. Studies on germfree mice. *Am J Pathol* **42,** 471–83.
23. Yamanaka, T., Helgeland, L., Farstad, I. N., Fukushima, H., Midtvedt, T., and Brandtzaeg, P. (2003) Microbial colonization drives lymphocyte accumulation and differentiation in the follicle-associated epithelium of Peyer's patches. *J Immunol* **170,** 816–22.
24. Moreau, M. C., Thomasson, M., Ducluzeau, R., and Raibaud, P. (1986) Cinétique d'établissement de la microflore digestive chez le nouveau-né humain en fonction de la nature du lait. *Reprod Nutr Dev* **26,** 745–53.
25. Macpherson, A. J., and Uhr, T. (2004) Induction of protective IgA by intestinal dendritic cells carrying commensal bacteria. *Science* **303,** 1662–5.
26. Mueller, C., and Macpherson, A. J. (2006) Layers of mutualism with commensal bacteria protect us from intestinal inflammation. *Gut* **55,** 276–84.
27. Harris, G., KuoLee, R., and Chen, W. (2006) Role of Toll-like receptors in health and diseases of gastrointestinal tract. *World J Gastroenterol* **12,** 2149–60.
28. Lievin-Le Moal, V., and Servin, A. L. (2006) The front line of enteric host defense against unwelcome intrusion of harmful microorganisms: mucins, antimicrobial peptides, and microbiota. *Clin Microbiol Rev* **19,** 315–37.
29. Lysenko, E. S., Ratner, A. J., Nelson, A. L., and Weiser, J. N. (2005) The role of innate immune responses in the outcome of interspecies competition for colonization of mucosal surfaces. *PLoS Pathog* **1,** e1.
30. Inzunza, J., Midtvedt, T., Fartoo, M., Norin, E., Osterlund, E., Persson, A. K., and Ahrlund-Richter, L. (2005) Germfree status of mice obtained by embryo transfer in an isolator environment. *Lab Anim* **39,** 421–7.
31. Okamoto, M., and Matsumoto, T. (1999) Production of germfree mice by embryo transfer. *Exp Anim* **48,** 59–62.

20

In Vivo Analysis of Zebrafish Innate Immunity

Jean-Pierre Levraud, Emma Colucci-Guyon, Michael J. Redd,
Georges Lutfalla, and Philippe Herbomel

Summary

Among vertebrate model species, the zebrafish embryo combines at an unprecedented level optical accessibility with easy genetic manipulation. As such, it is gaining recognition as a powerful model to study innate immunity. In this chapter, we provide a protocol for the generation of zebrafish embryos deficient in a protein of interest for innate immune signaling using antisense morpholino oligonucleotides, the systemic or local infection of these embryos with bacteria, and the assessment of various aspects of the following immune response with emphasis on microscopic observation. This example can be easily adapted to study the role of other genes, either knocked down or overexpressed, and in response to any other challenge, from purified microbial compounds to pathogenic viruses. This protocol is aimed at people not necessarily familiar with zebrafish biology and handling.

Key Words: Zebrafish; innate immunity; macrophage; granulocyte; morpholino; experimental infection; chemotaxis; DIC microscopy.

1. Introduction

The zebrafish is being increasingly viewed as a powerful tool for the study of innate immunity *(1,2)*. As a vertebrate, it is capable of adaptive immune reactions, but its main experimental advantages over other model species imply its early life stages, when only innate immunity is involved in antimicrobial defenses. First, external fertilization and small, transparent embryos allow for easy and non-invasive live microscopic observation up to the subcellular level.

Second, transient genetic manipulations are readily performed by microinjection at the single-cell stage: gene expression can be knocked down by injection of antisense morpholino oligonucleotides *(3)*; ubiquitous or mosaic gene overexpression can be obtained by injection of mRNA or plasmid DNA, respectively. Furthermore, its genome is now almost entirely sequenced, and an ever-increasing number of stable mutant and transgenic lines are available.

We provide in this chapter protocols for the analysis of innate immune responses (phagocytosis, chemotaxis, gene induction, and microbe clearing) following experimental infection of zebrafish embryos with fluorescent bacteria. To assess the role of a gene of interest (here, Myd88) in these responses, the protocol begins with the generation of Myd88-deficient embryos using morpholinos. This is intended as an example and can be easily adapted to study the role of other genes, in response to other challenges, e.g., viruses, bacteria, eukaryotic parasites, or purified compounds.

2. Materials

1. Zebrafish facility and adult breeding fishes: If only wild-type fish are used, hobbyist-type aquariums can suffice. Large dedicated facilities, such as AHAB systems (Aquatic Eco-Systems, Apopka, FL), are necessary to handle various fish strains and transgenic or mutant lines.

   For breeding, we recommend about 10 males and 10 females per 10-l tank, 6–18 months old. We fit plastic boxes containing marbles inside the tanks or use specialized breeding tanks.

   Many wild-type and mutant lines are available from the Zebrafish International Resource Center (ZIRC, Eugene, OR). For the experiments described in this chapter, we use the following strains: wild-type AB (*see* **Note 1**), *Nacre* mutants (*see* **Note 2**), and *Fli1::gfp* transgenics (*see* **Note 3**).

2. Embryo water: bottled spring water (we use Volvic water) supplemented with 0.3 µg/ml of methylene blue.

3. Non-culture-treated petri dishes.

4. 6-, 12-, and 24-well culture plates.

5. Two ventilated heating/cooling incubators, one constantly set at 28°C and another one that can be set at any temperature from 22 to 34°C (*see* **Note 4**).

6. Morpholinos (Gene Tools, Philomath, OR): Morpholinos are antisense oligonucleotides with a modified backbone chemistry that, if injected before the four-cell stage, can induce specific gene knockdown in zebrafish embryos for several days *(3)*. They can be used to block translation of mRNAs or to block splicing (*see* **Note 5**). To obtain morpholinos specific for a gene of interest, one should simply provide Gene Tools with the sequence around the translation start site or a mandatory splice site.

a. The Myd88-specific morpholino used here is derived from Van der Sar et al. (2006) *(4)*. Myd88-UTR morpholino: 5´-TAGCAAAACCTCTGTTATCCAGCGA-3´.
b. 5-mismatch control morpholino: 5´-TACCAAAAGCTCTCTTATCGAG GGA-3´.
c. Morpholino buffer: KCl 120 mM, HEPES 20 mM, pH 7.2. The initial morpholino stock is dissolved in sterile morpholino buffer at a concentration of 1–2 mM and aliquoted as 10 µl in tubes kept at −20°C.
d. Red morpholino buffer: morpholino buffer containing 0.1% phenol red.
e. Dextran-10000 coupled with rhodamine (Cat. number D-1816) or oregon green (Cat. number D-7170) (Molecular Probes, Eugene, OR). Make a stock solution at a 3-mM concentration and 300-µM working aliquots in 200 mM KCl; keep at −20°C in the dark.

7. Microinjection apparatus: We recommend to install it in a room with a temperature around 22°C, low enough to slightly slow down egg development without being deleterious.

   a. Stereomicroscope with wide-field eyepieces including a reticle, long-distance working objective, and internal zoom. The combined magnification range should cover at least ×10 to × 50, e.g., Stemi 2000-C (Zeiss, Oberkochen, Germany) with ×10 eyepieces and ×1 objective.
   b. Microinjector, e.g., Picospritzer III (Parker Hannifin; Fairfield, NJ). This model requires a separate compressed air source.
   c. Mechanical xyz micromanipulator arm, e.g., M-152 (Narishige, Tokyo, Japan).
   d. Borosilicate glass capillaries, e.g., Cat. number GC100F-15 for capillaries containing a filament, GC100-15 without filament (Harvard apparatus, Edenbridge, UK).
   e. Pipette puller; e.g., model 720 (David Kopf Instruments, Tujunga, CA) (*see* **Note 6**).
   f. Microloader pipette tips, Cat. number 5242956.003 (Eppendorf, Hamburg, Germany).
   g. Embryo-positioning chamber for single-cell embryos: a microscope slide glued with cyanoacrylate on a 10-cm petri dish lid with edges partially removed (*see* **Fig. 1A**).
   h. Embryo-positioning chamber for 24 hpf (hours post fertilization) or older embryos: a petri-dish containing an 1.5% agarose gel in embryo water with V-shaped channels (*see* **Fig. 2B and C**). A description of the plastic mold to create the channels can be found in The Zebrafish Book *(5)*, page 5.3, or online at http://zfin.org/zf_info/zfbook/chapt5/5.1.html.

8. Tools for embryo manipulation:

   a. Finest possible paintbrushes (purchased in your favorite art shop).
   b. Disposable wide-bore plastic pipettes, e.g., 3-ml pastettes, Cat. number LW4111 (Alpha Laboratories, Eastleigh, UK).

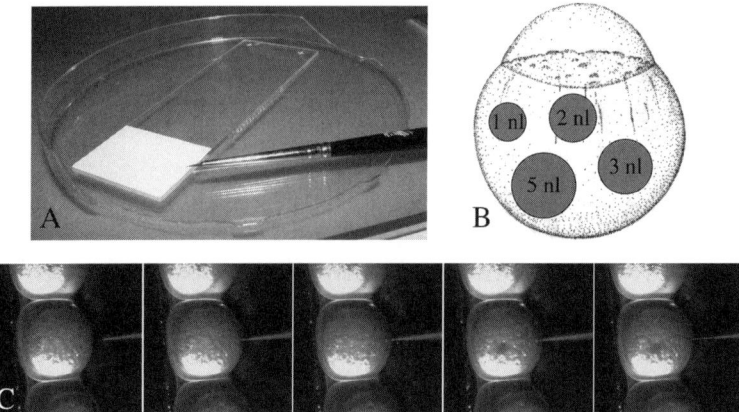

Fig. 1. Morpholino injections. (**A**) Injection stage for egg injection, with a row of eggs set in place for injection with the help of a fine paintbrush. (**B**) A visual help to estimate the volume of the injection. (**C**) Morpholino microinjection process; from left to right: approaching the chorion, entering chorion, entering yolk, injection, and withdrawal.

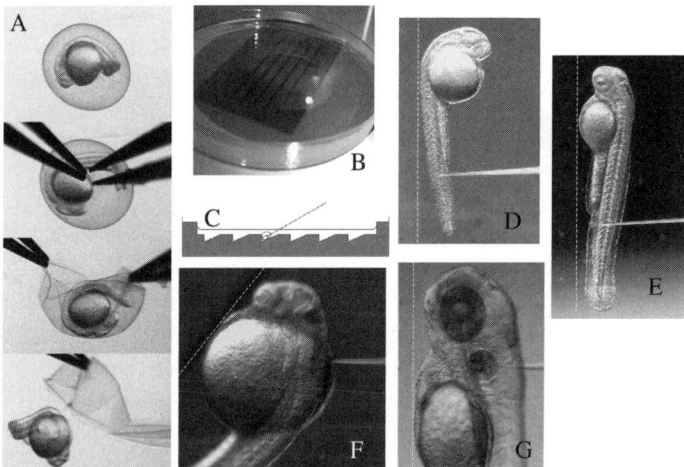

Fig. 2. Embryo experimental infection. (**A**) Dechorionation of a 25 hpf embryo. (**B**) The injection plate. (**C**) Schematic section of the injection plate. (**D**) Intravenous injection in a 28 hpf embryo. (**E**) Intravenous injection in a 50 hpf embryo. (**F**) Injection in the hindbrain ventricle of a 25 hpf embryo. (**G**) Injection in the left ear of a 72 hpf embryo. In (**D**) to (**G**), the dotted line delineates the vertical wall of the channel.

c. Disposable fine-tip plastic pipettes, e.g., 1-ml pastettes, Cat. number LW4635 (Alpha Laboratories).
d. Very fine forceps, e.g., Dumont number 5 (Moria, Antony, France).

9. Phenylthiourea (PTU), Cat. number P7629 (Sigma, St-Louis, MO). A 10 mM (50×) stock solution is prepared by dissolving powder in a beaker of sterile water, heated to 60°C for several hours under agitation. Store at 4°C. Warning: PTU is toxic and the powder must be handled under a chemical hood.

10. Tricaine, Cat. number A5040 (Sigma). To prepare a 25× solution (4 mg/ml), dissolve 1.2 g in 300 ml of water and adjust pH to 7.2 with a few drops of Tris-base 1 M; aliquot in 15-ml tubes and freeze. Once thawed, a tube may be kept at 4°C for at least 1 week.

11. Fluorescence stereomicroscope with green fluorescent protein (GFP) and DsRed filters, e.g., fluoIII MZ-16 (Leica, Solms, Germany), and a fluorescence camera, e.g., DS-5Mc (Nikon, Tokyo, Japan).

12. Fluorescent bacteria.

   We use *Escherichia coli* as a standard non-pathogenic gram-negative bacterium.

   a. Very bright, red fluorescent *E. coli* were obtained by transfecting XL-10 cells with the pRZT3 vector *(6)*.
   b. Tetracyclin, 5 mg/ml stock solution in methanol (store at -20°C).
   c. Isopropyl-β-O-A-thiogalactopyranoside (IPTG), 200 mg/ml stock solution in water (store at -20°C).
   d. Phosphate-buffered saline (PBS) 1×.
   e. Red PBS: PBS 1× containing 0.1% phenol red.

13. Equipment to mount embryos for observation under the compound microscope:

   a. Microscope depression slides. Try different suppliers to find slides deep enough to accommodate embryos. Fisher Cat. number N01851 slides (Fisher Bioblock Scientific, Illkirch, France) are fine but not available in the USA.
   b. Standard 20 × 20 mm number 1 (i.e., 0.17 mm thick) coverslips.
   c. Possible alternatives: coverslip-bottom chambers—Lab-tek slides Cat. number 155411 (Nalge Nunc, Naperville, IL) or MatTek dishes Cat. number P35G-1.5-10-C (MatTek Corp., Ashland, MA).
   d. Low-melting agarose.
   e. Methylcellulose: Prepare a 3% solution in embryo water (dissolving methylcellulose at such a high concentration requires overnight stirring with mild heating) and keep at –20°C.

14. Compound microscope equipped with differential interference contrast (DIC) optics (also known as Nomarski optics) and fluorescence, e.g., Eclipse 80i (Nikon), and camera suitable for fluorescence, e.g., DS-5Mc (Nikon). In addition, for DIC videomicroscopy, we recommend true (analogical) video three-CCD cameras, such as HV-C20, HV-D25, or HV-C20 (Hitachi, Tokyo, Japan), as described in *(7,8)*.

15. Lysis buffer for bacterial plating: PBS 1× + 1% Triton X-100.
16. Bacterial culture plates containing tetracyclin.
17. Equipment for quantitative PCR:

    a. Lysis buffer for RNA extraction: 4.5 M Guanidine-Cl, 30% Triton-X100, 50 mM Tris–Cl, 0.14 M 2-mercaptoethanol, pH 6.6.
    b. Glass fiber fleece kits for total RNA purification, Cat. number 1 828 665 (Roche, Zürich, Switzerland) or Cat. number 740 955 (Macherey-Nagel, Düren, Germany).
    c. Reverse transcription primer Spl2XhoT18: CAAGGATGGATGCGTGGT-GCTCGAGTTTTTTTTTTTTTTTTTT.
    d. RNaseBlock, Cat, number 300151-51 (Stratagene, La Jolla, CA).
    e. Moloney murine leukemia virus (M-MLV) reverse transcriptase (RT), Cat. number 28025-013 (Invitrogen, Carlsbad, CA).
    f. DNA fragment purification column kit, Cat. number 740-609 (Macherey-Nagel).
    g. PCR primers:
    zGAPDH5 ACTTTGTCATCGTTGAAGGT
    zGAPDH3 TGTCAGATCCACAACAGAGA
    zTNFa.5 GTGATAGTGTCCAGGAGGAA
    zTNFa.3 CATCTCTCCAGTCTAAGGTCT
    h. Light Cycler instrument (Roche) with SYBR Green format.

18. Neutral red (NR), Cat. number N-4638 (Sigma). Prepare a small amount of 0.5% solution in water and store at 4°C.
19. Reagents for Sudan Black B (SB) staining:

    a. Methanol-free 10% formaldehyde solution (Polysciences, Warrington, PA). Fixative buffer: four parts formaldehyde solution, one part PBS 10×, and five parts water; keep at 4°C and use within 24 h.
    b. PBT (post-fixation buffer): PBS 1× + 0.1% Tween-20.
    c. SB ready-to-use staining solution, Cat. number 3801(Sigma).
    d. 70% ethanol in water.
    e. 30% ethanol in PBT.
    f. 30% glycerol: three parts glycerol and seven parts PBT.
    g. 60% glycerol: six parts glycerol, two parts PBT, and two parts water.
    h. 90% glycerol: nine parts glycerol and one part water.
    i. Horizontal rotatory shaker.

3. Methods

3.1. Typical Work Plan

We describe in this chapter several in vivo assays of innate immune responses to microbes in zebrafish embryos: phagocytosis, evolution of microbial burden, survival, induction of inflammatory genes, and leukocyte chemotaxis toward an infected site. The dependence of these biological responses on a given protein is tested by generating knockdown embryos with morpholinos. These assays require different timings, even though all begin with a common step of morpholino injection. Some may be easily combined in the same experiment. Rather than providing detailed schedules for each assay, typical timings are summarized in **Fig. 3**.

We also provide in **Fig. 3** some guidelines regarding the number of embryos that are required to obtain statistically significant results. When compared with similar tests performed in mice, assays requiring experimental infection in zebrafish embryos are subject to two additional sources of variability: genetic heterogeneity, as fishes are not truly inbred, and increased variation in the injection volume, due to the microinjection process. Therefore, the recommended number of infected animals is higher than usual standards for mice.

3.2. Generation of Myd-88-Deficient Embryos

Methods for raising zebrafish are described in detail in *The Zebrafish Book (5)*, but see also **Note 7**.

3.2.1. Obtaining Eggs

1. The afternoon before the planned experiment, set up for spawning three or four tanks containing adults of the desired strain by putting at the bottom of each tank a low plastic box filled with two layers of marbles.
2. On the day of injection, do not disturb the fishes for the first 30 min after switching on the light to allow mating onset.
3. About 45 min after light onset, collect the boxes (carefully keep as much water as possible to avoid losing eggs, but do not embark adults) and check whether they contain eggs. If not, replace the box immediately in the tank and do not check again until at least 30 min.
4. If eggs are present, pour the contents of the box over a wide-bore sieve positioned over a suitable container. The marbles will be retained in the sieve whereas the eggs are flushed into the container. After a brief rinse, put the marbles back into the box and the box back into the tank; more eggs (in total several hundred) may be collected regularly during the following hour. Pour eggs in a fine-bore strainer and gently rinse them with embryo water. Transfer the eggs to a petri dish filled with embryo water.
5. Briefly sort the eggs under the dissecting scope to remove feces, scales, and, occasionally, older embryos from eggs that may have been laid the previous evening.

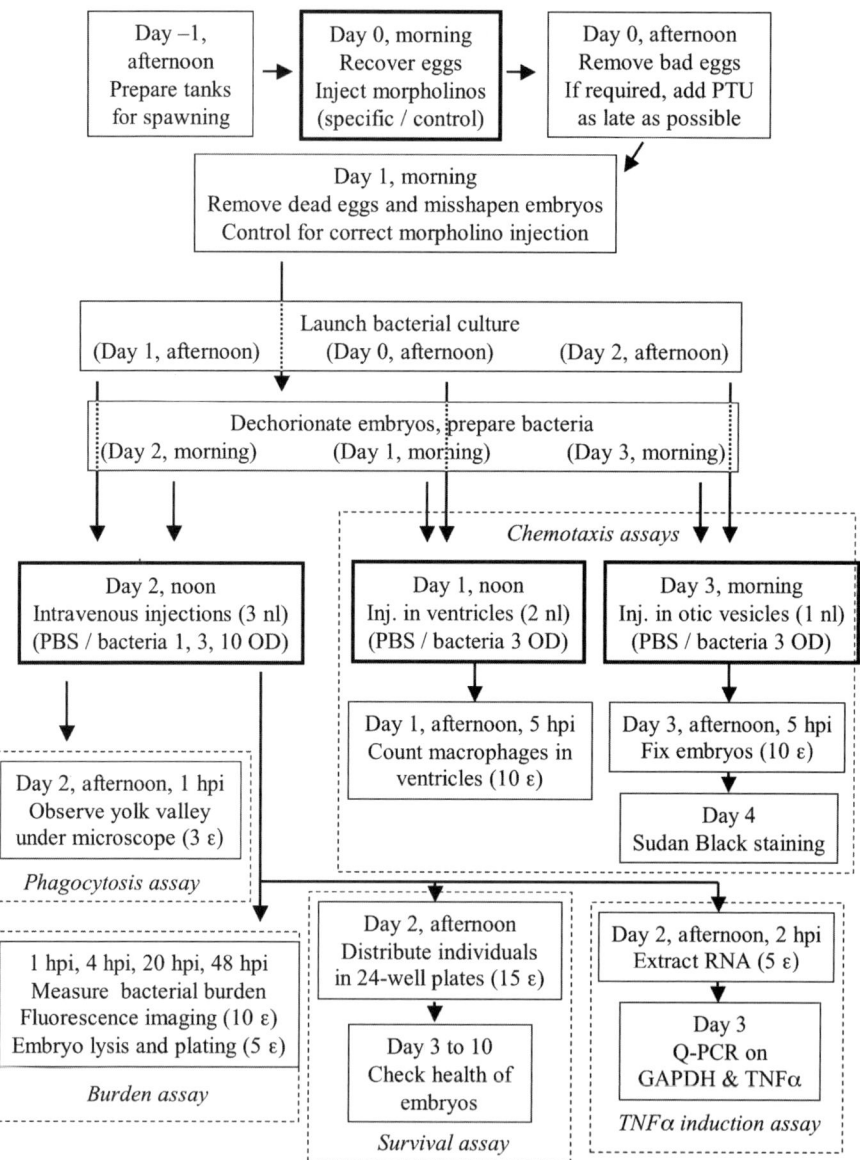

Fig. 3. Typical work plan for different assays. hpi: hours post injection; ε: embryo; OD: optical density units at 600 nm. The number of embryos is the recommended number for each subgroup.

3.2.2. Preparing the Injection

As there is little time (~1 h) to inject eggs once they have been fertilized, this preparation step is better carried out while fish are starting to spawn (of course, part of it will have to be done at a later time when the capillary containing the first morpholino is to be replaced by one containing the second morpholino).

1. Thaw one tube of the Myd88-specific morpholino 1 mM stock solution and one of the control mismatch morpholino.
2. Heat them for 10 min at 65°C, spin briefly, and put on ice (this step is recommended by the manufacturer as morpholinos tend to precipitate with time).
3. Prepare the injection solutions under sterile conditions:
   a. 2 µl of 1 mM morpholino stock solution.
   b. (if fishes are not GFP transgenics) 1µl of 300 µM dextran 10000–oregon green solution.
   c. Adjust to 10 µl with sterile red morpholino buffer; mix gently and then keep on ice.
4. Position the micromanipulator roughly, and switch the microinjector and compressed air source on. Make a first approximate setting of the injector (with the Picospritzer III, we use 50-ms injection time and 20-psi pressure as a starting point).
5. Load 3 µl of the red injection solution into the back end of a capillary with filament, using a microloader tip. Because the capillary contains a filament, some injection solution should quickly accumulate at the opposite, elongated end.
6. Mount the loaded capillary onto the micromanipulator, position its tip under the dissecting scope at the center of the field, zoom to maximum magnification onto the tip, and break it close to its end, where its external width is about 20 µm. To break the tip, one may use either fine forceps, or a scalpel and a petri dish edge like a hammer and anvil.
7. Operate the microinjector a few times to chase bubbles and ensure that the tip is not clogged. Then set finely the injection time to obtain the desired volume. As the diameter of the opening at the broken tip will never be exactly the same for each capillary, this needs to be done for each new capillary. To measure the injected volume, lift the tip into the air and focus on it at the highest magnification, perform 10 pulses (as the measure is more precise with big drops), and immediately measure the diameter of the drop with the help of the reticle in the eyepiece, before the drop starts to evaporate. The volume of a spherical drop in nanoliters is ($167 p \times D^3$), where D is the diameter of the sphere expressed in millimeters. Divide the volume by 10 (to account for the 10 pulses) and adjust the injection time.

   For example, with the Myd88-specific morpholino, the desired dose is 2 ng, as Van der Sar et al. have shown that the efficiency range of this morpholino is from 1.7 to 5 ng *(4)*. With a 200 µM injection solution, and a molecular weight of ~8500, 2 ng of morpholinos are delivered in a volume of 1.2 nl. Suppose that, after 10 50-ms

pulses, you obtain a drop with an apparent diameter of 12 mm with a ×1 objective, ×5 zoom, and ×10 eyepieces. This translates into a real diameter of 12/50 = 0.24 mm and a volume of 7.3 nl, i.e., 0.73 nl per pulse. To get 1.2 nl per pulse, you have to increase the injection time to 50 × 1.2/0.73 = 82 ms.
8. When finished, rinse the tip in a drop of water and then lift it in the air, well above the injection stage in order not to damage the tip by accidentally touching it while you are installing the eggs.

3.2.3. Morpholino Microinjection

1. Set up a dry egg-positioning chamber on the scope stage, slide edge perpendicular to the capillary.
2. Pipet about 20 eggs near the edge of the glass slide, and then remove almost all the water with a fine-tip disposable pasteur pipette.
3. Align the eggs along the side of the slide, using a very fine paintbrush (*see* **Fig. 1A**). It is best to orient them with animal pole toward you.
   Check that their stage is not too advanced, and discard eggs advanced beyond the two-cell stage. This is because the morpholino will distribute evenly in the embryos injected only up to the four-cell stage, and although the first cleavage occurs 45 min after fertilization, the next nine cleavages then occur every 15 min.
4. Center the first egg in the row in the field of view and then lower the capillary and, at maximum magnification, position its tip perpendicular to the chorion surface.
5. Gently enter the chorion. Bring the tip perpendicular to the egg surface, targeting the interface between yolk and blastodisc.
   Enter the egg halfway, move very slightly back, and inject the red solution. Check by the size of the red spot that the injected volume is the correct one; as a rule of thumb, a 1-nl drop has a diameter about one-sixth that of the yolk cell (124 µm vs. 700 µm) (*see* **Fig. 1B**). Withdraw smoothly the capillary from egg and chorion. **Figure 1C** illustrates the complete injection process.
6. Move the injection plate to present the next egg, and proceed to inject it, and so on until the end of the row.
7. Once the row of eggs is injected, flush them into a clean petri dish containing embryo water (it is advisable not to pipet the eggs for the next hour) and transfer them to 28°C.
8. Wipe the injection tray dry, and proceed to the next series.

An experienced user may inject up to three hundred eggs in 2 h.

In addition to the two groups of injected eggs (one with the specific morpholino, one with the control morpholino), keep some non-injected eggs from the same batch.

3.2.4. Egg Incubation and Quality Control

1. One to two hours after morpholino injection, check your injected eggs and remove all dead or leaky eggs, which would favor contamination. One may incubate up to 100 embryos in a 10-cm-diameter petri dish containing 40 ml of embryo water.
2. About 6 h after the injection, check the eggs again and remove the non-fertilized eggs, which by now are clearly visible under the dissecting scope as no cleavage occurred. The percentage of non-fertilized eggs should be similar in injected and non-injected eggs of the same clutch.
3. If your experiment requires microscopic examination and embryos are from a pigmented strain, add PTU to the dish (800 μl of the 50× stock solution to a petri dish containing 40 ml of embryo water) just before you leave the laboratory. As PTU is toxic, label the dish as containing PTU, manipulate with gloves, and treat waste accordingly.
4. The following morning, you will probably find a significant amount of dead eggs and more or less severely misshapen embryos, due to bad injection and/or to the morpholino itself. A 30% loss rate is common. Remove these embryos; if the water appears turbid, transfer the remaining embryos to a new dish.
5. If the morpholino was co-injected with fluorescent dextran, observe the embryos under the fluorescence stereomicroscope. The distribution of fluorescent dextran is used as an indicator for the distribution of the morpholino, which has a similar molecular weight. The embryo body should appear as homogeneously fluorescent, while the yolk should not, except for a weak autofluorescence. Embryos that are less fluorescent than the others should be discarded; generally, this is caused by immediate leakage of the injected material. These embryos can often be recognized under transmitted light, as the space between embryo and chorion is cloudy.

3.3. Infection with Bacteria

3.3.1. Dechorionation

Embryos must be removed from their protective chorion before blocking them in proper position for precise injections. The more advanced the embryos are in their development, the easier dechorionation is, and the less they tend to stick to plastic. Dechorionation is accomplished under the dissecting scope, delicately tearing the chorions open with very fine Dumont number 5 forceps (*see* **Fig. 2A**). It is not necessary to anaesthetize the fish. To process many embryos, one may use pronase treatment as an alternative (*5*), but we prefer to avoid this method as it damages the embryo periderm.

3.3.2. Bacteria Preparation

Escherichia coli transfected with pRZT3 have the strongest DsRed expression after a few hours of stationary-phase culture in presence of IPTG.

1. In the afternoon before the day of injection, launch a culture tube containing:
    a. 2 ml of LB medium.
    b. 4 µl of tetracyclin stock solution (i.e., 10 µg/ml).
    c. 4 µl of IPTG stock solution (i.e., 400 µg/ml).
       Seed with a red colony of *E. coli*-pRZT3 kept on a tetracyclin LB plate.
2. Incubate overnight at 37°C with shaking.
3. One hour before starting the injection, transfer 200 µl of the bacterial culture in a sterile 1.5-ml tube, centrifuge, and wash twice with 1× PBS. The pellet should appear bright pink.
4. Resuspend in 100 µl of 1× PBS and measure the optical density (OD) at 600 nm.
5. Adjust the bacterial suspension at the desired concentration with red PBS.

The desired bacterial concentration depends on the experiment to be performed. An OD of 1 at 600 nm of *E. coli* roughly corresponds to 10^9 bacteria/ml; thus, an inoculum of 1 nl of bacteria at 1 OD contains about 10^3 bacteria.

We typically perform intravenous injections with bacteria at 1 OD for phagocytosis assays, and 1, 3, and 10 OD for survival assays. Concentrations over 10 OD result in capillary clogging.

Local injections in the brain ventricle or inner ear are usually performed with bacteria concentrated at 3 OD.

3.3.3. Preparing the Injection of Bacteria

1. About 30 min before starting the injections, the agar injection plate (*see* **Fig. 2B**) is taken out of the cold room to allow it to warm to room temperature and soaked with 20 ml of embryo water containing tricaine at 1.5× its usual concentration (i.e., 0.24 mg/ml).
2. In addition to the plastic pipettes and paintbrush to manipulate embryos, the following pieces of equipment should be at hand:
    a. An ice bucket to store the laden capillary between injections.
    b. A small petri dish used as anaesthesia chamber.
    c. Petri dishes (or six-well plates) containing embryo water to keep the injected embryos.
    d. A beaker for trash (e.g., poorly injected embryos).
    e. A security needle disposal box.
3. The capillary is loaded with bacteria and broken as described in **Subheading 3.2.2.**, except that a capillary without filament is used. A long pulse (~1 s) is needed to push the bacterial suspension down the capillary after the tip is broken. Determine injection volume as in **Subheading 3.2.2**.
4. The tip is rinsed and the capillary is delicately put on ice until the last moment.

3.3.4. Anaesthesis

Fill a 3.5-cm-diameter petri dish with 3 ml of embryo water containing tricaine at 1.5× concentration (0.24 mg/ml). Pipet the embryos into the dish, and wait for at least 2 min. It will be necessary to replace the anaesthesis medium from time to time because it gets diluted with each new embryo added (*see* **Note 8**).

3.3.5. Intravenous Injections

1. Orient the injection plate filled with tricaine-containing water so that the channels are perpendicular to the capillary, with the vertical side away from it (*see* **Fig. 2C**).
2. Pipet the anaesthetized embryos from the tricaine dish to the injection tray, and then position them laterally in the channels using the paintbrush. Depending on their stage of development, embryos are positioned with either ventral (24–48 hpf) or dorsal (>48 hpf) side facing the capillary (*see* **Fig. 2D and E**, respectively). This is the position in which they are the less prone to roll during the injections.
3. Unless the microinjector offers the option of a constant basal pressure, some water will always tends to enter through the tip opening by capillary force. As a rule, just before pricking the embryo, flush the capillary with one or more pulses made in the water, until some red solution comes out.
4. Pierce the skin with the capillary tip near the uro-genital opening, aiming at the caudal vein (CV) which widens there. If the caudal artery located just dorsal to it is hit instead, it is still OK.
5. When the capillary tip is at the correct location, perform one pulse to inject the bacteria. The phenol red contained in the injection medium allows one to check whether the injection is made at the right place, but the injected volume cannot easily be gauged. Occasionally, injections are extravascular and the embryo should be discarded; a sudden change of the refringence of muscle fibers near the capillary tip indicates an intramuscular injection; a pink bubble on the ventral side or a deformation of the yolk tube indicates a subcutaneous injection.
6. Quickly withdraw the capillary after the injection, otherwise some blood will be sucked back by capillary force.
7. When a series of embryos has been injected, put the capillary on ice, and then pipet the embryos back into embryo water without tricaine.
8. Install and orient the next group of embryos, put the capillary back in the micromanipulator, and inject the next series.

3.3.6. Local Infections in Closed Body Cavities

Except for the embryo orientation and site of injection, the procedure is basically the same as for intravenous injections in **Subheading 3.3.5**.

With young embryos (up to 48 hpf), the most conspicuous cavity is the hindbrain ventricle. Embryos are positioned in the injection plate, ventral side to the vertical side of the channels. Embryos are injected through the ventricle roof. As the ventricle is located more or less anteriorly depending on the developmental stage, the plate channels are oriented at a suitable angle with the capillary (*see* **Fig. 2F**). Care should be taken not to push the capillary tip too far, in order not to damage the underlying neuroepithelium. Once in position, the ventricle is filled with the red medium.

When embryos get older than 48 hpf, the ventricle shrinks and becomes more difficult to inject. By contrast, the inner ear (otic vesicle) increases in size with development and is thus chosen for older embryos. For ear injections, embryos are positioned laterally, ventral side toward the vertical wall of the channel (*see* **Fig. 2G**). The injection has to be very smooth, with a small volume, otherwise the ear may rupture.

3.4. Visualization of Leukocyte–Bacteria Interactions

We describe here mounting techniques specific to live zebrafish embryos. For a description of imaging techniques, see **ref. (8)**.

3.4.1. Observations Under the Dissecting Scope

If bacteria used for infection are strongly fluorescent, a general view of the degree and distribution of infection can be obtained quickly using a fluorescence stereomicroscope. The embryos may be imaged as often as needed.

1. Prepare a 3.5-cm petri dish containing embryo water with 1× tricaine (0.16 mg/ml).
2. Pipet the embryo into the dish and wait 1 min for it to get anaesthetized.
3. Using a paintbrush, position it in a suitable orientation—generally in lateral view—and image it.
   It may be difficult to position an embryo laterally in a stable enough position to allow for fluorescence imaging, which often require long exposure times. Several tricks may be used: embryos may be positioned in an injection tray or immobilized with the help of a viscous solution of methylcellulose. Depression slides may also be used without a coverslip; their slopes help the embryo to stay in position.
   Mounting embryos under a coverslip as described in **Subheading 3.4.2** is also possible, but more time-consuming and slightly risky for embryos observed repeatedly.
4. After observation and photography, pipet the embryo back into normal embryo water.

3.4.2. Mounting Embryos for Observation Under the Compound Microscope

For high-power microscopic observation, the embryo has to be perfectly immobile under a 0.17 mm thick coverslip. There are two possibilities: either in a depression slide or in a "Lab-Tek" type chamber. The depression slide fits all microscopes whereas the Lab-Tek is suitable for inverted microscopes only. Lab-tek chambers may be better for long observations, and allow one to change conditions during the course of observation, as the medium can be renewed and is aerated. Embryos are, however, more difficult to orient precisely in Lab-Tek chambers and may not be recovered after observation. In addition, older embryos/larvae (72 hpf and more) are harmed by agarose entering the mouth and gills. MatTek glass-bottom dishes combine the advantages of Lab-Tek chambers with easy orientation but are more expensive.

3.4.2.1. To Mount an Embryo in a Depression Slide

1. Install a depression slide of appropriate depth under a dissection scope.
2. Put the embryo to sleep in 1× tricaine-containing embryo water, and pipet it into the depression with a few drops. Orient it roughly with a paintbrush.
3. Install a coverslip on the opposite side of the depression, and slide it slowly toward the embryo.
   When it reaches the embryo, gently tuck it under the coverslip edge with the paintbrush, and orient it carefully.
   Slowly slide the coverslip fully over the depression; ensure that the embryo has not turned and that the depression is completely filled with water.
4. Draw off excess fluid using a paper towel, to "seal" the coverslip. The embryo should be immobilized but blood should still be flowing over the yolk.
5. When the observation is over, the embryo can be set free by unsealing the coverslip with a drop of water on the edges. Delicately push the coverslip aside, add fresh embryo water to the depression, and pipet the embryo back to a petri dish.

3.4.2.2. To Mount an Embryo in a Lab-Tek Chamber (or a MatTek Dish)

1. About 1 h before mounting the embryo, prepare a 1% solution of low-melting agarose in embryo water. After boiling, set it to cool in a 33°C water bath. Just before mounting the embryo, check that the agarose is still liquid and add tricaine to 1× final concentration.
2. Put the embryo to sleep in × tricaine-containing water and transfer it in the chamber, leaving as little water as possible.
3. Add two drops of tricaine-containing agarose, and orient the embryo with a paintbrush.
4. Do not move the chamber for a few minutes until gelling has occurred. Once the gel is well set, cover it with a few millimeters of tricaine-containing water.
5. Keep the lid on whenever possible so as not to let water evaporate too much.

3.4.3. Sites of Observation in Live Embryos

In principle, all regions of the embryos may be imaged with Nomarski (DIC) optics or confocal fluorescence microscopy to image single planes. In practice, the quality of DIC imaging is the better when there is less refractile or polarizing tissue between the focal plane of interest and the observer. A proper orientation of the embryo is crucial to achieve this purpose.

Some guidelines will be given for specific regions of interest.

3.4.3.1. THE YOLK SAC CIRCULATION VALLEY

In the zebrafish embryo up to 60 hpf, the venous blood flow, as it arrives in the yolk sac, is no longer enclosed in a vessel but flows directly on the yolk surface, in a depression that we call the "yolk sac circulation valley" *(7)*. The flow is wider there than anywhere else in the embryo and occurs just under the skin, making it an ideal site for observing leukocyte/microbe interactions by DIC videomicroscopy (*see* **Fig. 4A**). Embryos are mounted laterally. Because the bottom of the valley is still close to the coverslip, ×63 or even ×100 objectives have a sufficient working distance for in vivo observation there; a ×40 objective however usually offers a better contrast and can still reveal some subcellular details.

Typically, the kinetics of phagocytosis in vivo can be determined at this site.

3.4.3.2. CAUDAL HEMATOPOIETIC TISSUE

As intravenous injections are performed at the anterior end of this area, many bacteria will usually be trapped in the vicinity, not only because some were not properly injected in the targeted vessel, but also because the multi-branched CV plexus is an efficient trap for circulating particulate material. In addition, larval hematopoiesis develops around the CV plexus from about 40 hpf onward *(9)*, so that many granulocytes born there, and some primitive macrophages that are part of the hematopoietic stroma, all reside in this area. Therefore, although many phagocytosis events will occur there, the complexity of the tissue often makes it difficult to determine whether these events occur intra- or extra-vascularly, or which cells are involved.

To image this caudal area, as the fish is slimmer there, to orient it properly, one may take advantage of the slope of a depression slide, with the head toward the center of the depression and the tail toward the edge, aiming at a lateral orientation. Adding methylcellulose can also help.

3.4.3.3. HINDBRAIN VENTRICLE

The hindbrain ventricle normally contains no leukocytes up to ∼35 hpf, and then only two to four macrophages *(10)*, but in case of local infection, some will promptly come and clear the bacteria *(7)*. Although imaging quality is not homogeneous throughout the ventricle due to its depth and curvature, it is easy to count the incoming cells. Even more conveniently, these cells can be visualized as GFP-positive over a dark background in *Fli1::GFP* embryos aged no more than 36 hpf (*see* **Fig. 4B**).

Embryos are viewed dorsolaterally.

3.4.3.4. EAR

Otic vesicles normally do not contain any leukocyte. Upon local infection, many come in the vicinity and some enter the otic cavity.

A fully dorsal view allows imaging of both ears (the injected one and the control one) at once in a comparable fashion, but at relatively low magnification. For a closer look, e.g., with a ×40 objective, a lateral orientation is optimal.

3.5. Bacterial Growth Assay

Imaging may provide a rough estimate of the level of infection. To obtain precise quantitative data, one should perform a classical lysis/plating experiment using the whole embryo.

In addition to the experimentally infected embryos, a few non-infected embryos should be included as controls.

3.5.1. Embryo Wash

This step is meant to remove most free bacteria from the pharyngeal cavity and the water surrounding the embryo. Should one desire to eliminate bacteria attached to the surfaces, an antibiotic treatment step should be considered, after checking that the antibiotic does not cross the embryonic skin (for instance, erythromycin does not).

1. For each group of embryo that is to be washed separately, prepare 10 ml of embryo water containing 1× tricaine. In a 24-well culture plate, distribute 2 ml of tricaine water in the four wells of a row.
2. Transfer embryos in the first well, wait for 5 min; then transfer them from well to well every 2 min, gently shaking the plate from time to time.

3.5.2. Embryo Lysis

1. Pipet the washed, anaesthetized embryos individually to sterile 1.5-ml microcentrifuge tubes, and withdraw the water with a micropipette.
2. Add 200 µl of lysis buffer (PBS 1× + 1% Triton X-100) and disrupt the embryo by thoroughly pipetting up and down with a micropipette or a fine-gauge syringe. The older the embryos, the more resistant they are and the longer this step needs to be.
3. When disruption is completed, incubate the tubes for 5 min at room temperature, and then vortex them at maximum speed for 30 s.

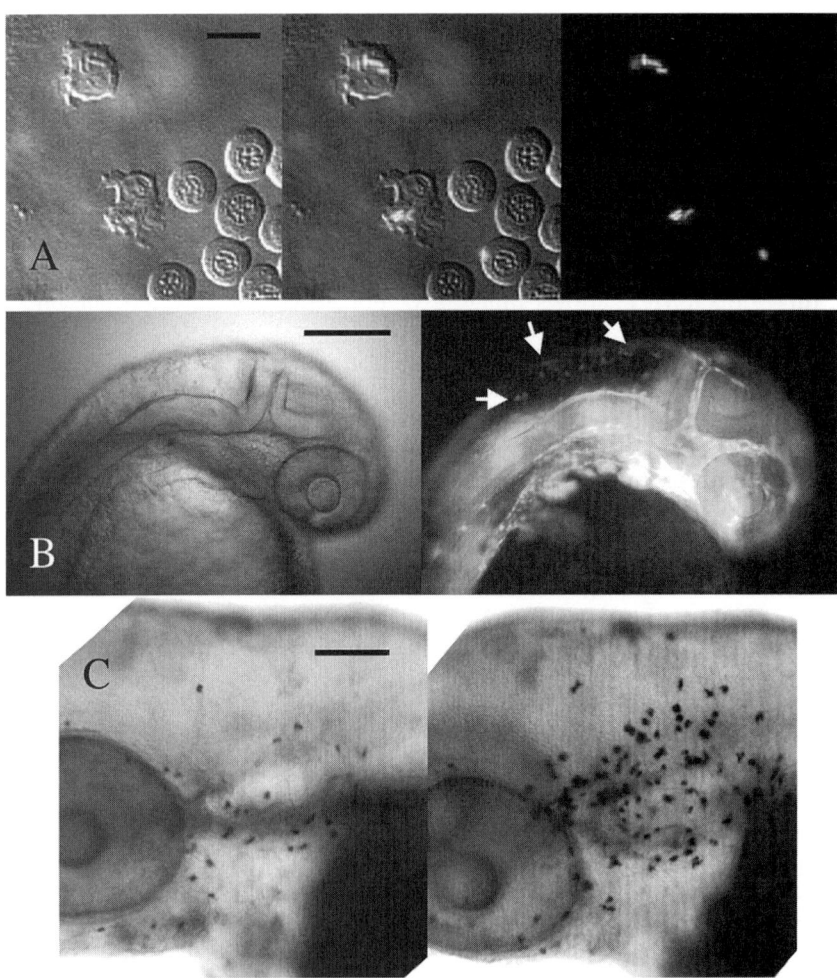

Fig. 4. Examples of infection imaging. (A) Phagocytosis assay: view in the yolk circulation valley of a living Myd88 morpholino-injected, phenylthiourea (PTU)-treated 29 hpf AB embryo, 45 min after intravenous injection of $\sim 3 \times 10^3$ DsRed-expressing *Escherichia coli* bacteria. Left: differential interference contrast (DIC) image; right: wide-field fluorescence with DsRed filter; middle: merging of the two images. Images were extracted and cropped from a video recording. Notice two young macrophages that have engulfed three to four bacteria each. The fluorescent dot on the bottom left corresponds to an out-of-focus bacterium loosely stuck on the surface of an erythrocyte. Scale bar, 10 μm.(B) Hindbrain ventricle macrophage chemotaxis assay: head region of a living, PTU-treated 30 hpf *Fli1::gfp* embryo, 4 h after injection of $\sim 6 \times 10^3$ DsRed-expressing *E. coli* in the hindbrain ventricle.

3.5.3. Lysate Plating

1. Plate the lysate from each embryo on three antibiotic-containing bacterial culture plates. Plates without antibiotics are suitable also, but the background noise measured in non-infected embryos is higher; thus, the sensitivity of the test is lower. One plate receives 1/100th of the lysate, another receives 1/10th, and the third one receives the remaining of the lysate.
2. Count colonies after an overnight growth step in a 37°C incubator.

3.6. Survival Assay

The duration of the assay is an important parameter. Embryos are usually infected when they are 1–3 days old. They may start feeding at 5 days post fertilization (dpf), but if they are not fed, they survive on their yolk supply up to 12–14 dpf (at 28°C). The need for survival past the food-independent stage considerably increases individual variability (and work, to feed several times a day dozens of individual wells, and clean the resulting waste). Therefore, experiments that measure relatively short-term survival differences, within the week following injection and without the addition of any food, are much more reliable and convenient to perform.

3.6.1. Setting up the Assay

It is necessary to isolate embryos from each other to evaluate survival, as one dying embryo will cause a bacterial bloom in the well.

1. Prepare 24-well plates by filling each well with 1 ml of embryo water.
2. Pipet individual infected embryos in individual wells; carefully label the lid.
3. Place plates inside clear boxes to avoid mixing infected embryos with regular embryos and reduce evaporation, and put in the incubator.

Fig. 4. (*Continued*) Images were taken with a fluorescence stereomicroscope. Left: transmitted light (oblique illumination); right: fluorescence with green fluorescent protein (GFP) filters. Notice the ∼15 GFP-positive macrophages that have entered the ventricle; three are indicated by white arrows. Scale bar, 200 μm. (**C**) Inner ear granulocyte chemotaxis assay: lateral views (anterior to left, dorsal to top) of heads of two Sudan Black B (SB)-stained, PTU-treated 77 hpf AB embryos, fixed 5 h after injection of ∼1 nl in the left ear. Images taken with a dissection scope. Left: embryo that received red PBS only; right: embryo that received ∼3 × 10^3 DsRed-expressing *E. coli*. The massive influx of granulocytes in response to local infection is obvious. Scale bar, 100 μm.

3.6.2. Health Monitoring

Infected embryos should be observed under the dissecting scope at least once a day (if possible more often during the critical stages).

Tease embryos with a soft tool (paintbrush or shortened microloader tip) to bring them in the center of the well—optical quality is too poor close to the well walls. This may take several trials with a healthy embryo, but it will tire after a few escape dashes.

Check for heartbeat (one may want to record cardiac rhythm), blood flow, absence of necrotic spots, absence of edema, and reaction to touch. At 4–5 dpf, healthy embryos should acquire a swim bladder.

An embryo is considered to be dead when it does not react to touch and no heartbeat at all is visible (though it may visibly have lost some other body parts already!).

3.7. Measurement of TNFα Induction

Considering the small size of zebrafish embryos, quantitative PCR is the method of choice to measure the induction of inflammatory genes such as TNFα.

We describe here methodological details to generate good-quality cDNA from a small number of zebrafish embryos; for a description of the quantitative PCR method itself, see **ref. *(11)*.**

Single embryos older than 24 hpf can yield enough RNA to perform the analysis. With 24 hpf embryos, however, we usually lyse 10 embryos so that the expression of other genes may be quantified. For 72 hpf embryos, yields are good starting from three embryos.

1. Anaesthetize the embryos by transferring them to embryo water containing 1× tricaine.
2. Transfer them in RNase-free microcentrifuge tubes, and withdraw the maximum amount of water.
3. Pipet in 350 µl of lysis buffer and lyse the embryos immediately by pipetting up and down thoroughly with a fine-gauge syringe.
4. If embryos are aged <20 hpf, perform a chloroform extraction step to get rid of the fat from the yolk.
5. Extract total RNAs on glass fiber fleece according to the Roche or Macherey-Nagel kits, including the DNase treatment.
6. Incubate up to 5 µg of total RNA for 90 min at 37°C in a final volume of 50 µl of reverse transcription mix (50 mM Tris–Cl, 75 mM KCl, 3 mM $MgCl_2$, 10 mM dithiothreitol, 500 µM each dNTP, 1 mM Spl2XhoT18, RNaseBlock 40 u/ml, and M-MLV RT 4000 U/ml).

7. Purify cDNAs using a DNA fragment purification column kit, after which they are directly used for quantitative PCR (Q-PCR) analysis.
   Elution volume depends on the input of RNA and the number of genes to be quantified by Q-PCR. Typically, using the RNA from three 72 hpf embryos, perform elution in 50 μl, which will allow the expression of eight genes to be quantified.
8. Perform Q-PCR on a light cycler, amplifying TNFα (with primers zTNFa.5 and zTNFA.3) and GAPDH (with primers zGAPDH5 and zGAPDH3) transcripts. Briefly, the amounts of GAPDH and TNFα cDNAs are measured in each reverse transcription product using standards, and the TNFα/GAPDH ratio can be compared from sample to sample.

3.8. In Situ Staining of Leukocyte Subsets

We describe in this paragraph two simple staining procedures used to label macrophages in live embryos and granulocytes in fixed embryos. Protocols for the standard methods of histological examination of zebrafish embryos, such as whole-mount immunohistochemistry (IHC) and whole-mount in situ hybridization (ISH) with RNA antisense probes, are available in *The Zebrafish Book (5)* (Chaps. 8 and 9; see online zfin.org/zf_info/zfbook) and in the literature *(12)*. As combining various analysis methods may be very useful, we provide in **Note 9** the information we have regarding the compatibility of these methods with the stainings described below.

3.8.1. NR Staining

NR is a vital dye that is able to cross the embryo skin and accumulates in the lysosomes of endocytic cells. It will therefore label a sizable subset of macrophages, which become easy to spot under the dissecting scope. After their transition to the microglial phenotype at 55 hpf, macrophages in the brain, retina, and a few other sites become even more strongly stained with NR *(10)*.

Embryo viability (and macrophage mobility) is diminished by NR staining. Thus, after imaging NR-stained embryos, they should not be kept to follow the outcome of the infection. They may, however, be fixed for further analysis, but fixation will cause the red staining to vanish quickly.

1. For each group of embryos to stain, prepare 5 ml of staining solution by adding 2.5–5 μl of 0.5% NR stock solution to 5 ml of embryo water, and mix well.
2. Fill wells of a six-well plate with 5 ml of staining solution, pipet the embryos into the wells (they do not have to be anaesthetized), and incubate them at 28°C for 2 h in the dark.
3. After the incubation step, rinse the embryos by transferring them into another six-well plate containing embryo water.

4. Anaesthetize and mount the embryos as described in **Subheading 3.4.**, and image them under either the dissecting scope at high magnification or the compound microscope.

3.8.2. SB Staining

SB stains lipids and more intensely and irreversibly the granules of granulocytes. Thus, SB staining of fixed embryos followed by 70% ethanol washes leaves only the granulocytes stained *(13)*.

3.8.2.1. Embryo Fixation

1. Prepare the 4% formaldehyde fixation solution under a chemical hood and keep on ice. One milliliter of solution is suitable to fix up to 20 embryos.
2. Over-anaesthetize the embryos with embryo water containing tricaine at 4× its usual concentration (i.e., 0.64 mg/ml).
3. Transfer them to a 1.5-ml tube and drain the remaining water with a micropipette. Add 1 ml of fixation solution and mix gently by inverting the tube a few times, ensuring that no embryo remains stuck to the walls of the tube.
4. Incubate at 4°C overnight (2 h fixation at room temperature with gentle agitation is also suitable).
5. After the fixation step, wash embryos once with cold PBT (fixed embryos stick to plastic surfaces if PBS is used).

It is best to stain directly the embryos after fixation. They may, however, be kept in PBT (not methanol—*see* **Note 9**) at 4°C for no more than 1 week.

3.8.2.2. Staining

The whole staining procedure is carried out at room temperature, and all incubations are made with gentle agitation on a rotatory shaker.

1. Pipet the fixed embryos in the staining container, which can be either a 12-well plate or in individual 1.5-ml tubes, up to 10 embryos per well/tube. Remove remaining PBT.
2. Cover the embryos with 1 ml of the ready-to-use SB staining solution and incubate them for 15 min. The SB solution contains phenol and therefore should be handled under a chemical hood.
3. Pipet out the SB solution from the well/tube. Be careful at this step, as the solution is totally opaque and an embryo is easily carried along! To avoid losing embryos, instead of directly eliminating the removed fluid, we pour it first on a tilted petri dish: an inadvertently caught embryo will stop at mid-slope, where it can be recovered.
4. Make two quick rinses with 1 ml of 70% ethanol; carefully eliminate any SB "grains."

In Vivo Analysis of Zebrafish Immunity

5. Wash at least twice with 1 ml of 70% ethanol for more than 30 min each time. At the end, the rinsing medium should not be visibly stained.
6. If embryos were thus far in tubes, transfer them now to small petri dishes or 12-well plates.
   The embryo body will still appear black at first sight but should be mostly transparent with intensely stained granulocytes when viewed under the dissecting scope under incident (not transmitted) light, except for the yolk that usually retains some diffuse stain.
7. Replace the 70% ethanol with 30% ethanol in PBT.
8. If embryos retain some melanin-based pigment pigmentation, it can be cleared at this step (*see* **Note 10**).
9. Drain the ethanol and replace it with 1 ml of 30% glycerol solution, and incubate for at least 1 h (alternatively, overnight at 4°C).
10. Repeat the step, moving to 60% glycerol.
11. Repeat the step gain, moving to the 90% glycerol solution. The embryos may now be conserved in glycerol for several months at 4°C.

For optimal imaging of the embryos under the dissecting scope, provide illumination from above with a reflecting white background. Granulocytes are easy to visualize, image, or count (*see* **Fig. 4C**). They can also be viewed under the compound microscope with or without DIC optics; individual granules can then be discerned.

4. Notes

1. Unlike for mice, there is no standard, stable, pure inbred zebrafish line. The AB laboratory strain is a partially inbred strain, which has been purged many generations ago from recessive lethal embryonic mutations. There is, however, a significant genetic diversity in the strain *(14)*, which is deliberately kept to ensure vigor and efficient breeding. The easiest way to avoid genetic drift is to reorder each year or so a batch of AB embryos from the ZIRC, and raise them to adulthood. These fishes or their F1 descendants may be used for breeding purposes.
   We recommend not to rely on fish purchased from local pet shops. Although cheap and convenient, this practice results in unpredictable outcomes regarding the health of the clutches.
2. *Nacre* embryos lack melanophores *(15)* which otherwise hinder microscopic observation. Instead of relying on such mutant fish, one may prevent melanin synthesis by adding PTU to the water, a routine practice in zebrafish labs. In addition to being toxic to humans, however, we have repeatedly observed that PTU-treated fishes were slightly more susceptible than their untreated siblings in several viral or bacterial survival assays. Therefore, whenever possible, we prefer to use non-pigmented mutants such as *nacre*. *Albino* mutants are another option; a minor

practical advantage of *nacre* over *albino* embryos is that because the retina remains pigmented, it is easier to see the embryos to count or pipet them.

3. *Fli1::gfp* fishes express enhanced green fluorescent protein (EGFP) under control of the Fli1 promoter *(16)*. The transgene is expressed in all endothelial cells during the lifespan of the fish. This provides very useful spatial information, especially for confocal microscopy. Additionally, neural crest cells and primitive macrophages also contain GFP up to about 36 hpf, which is useful for tracking purposes. Other GFP transgenic fish lines may be useful, such as the *H2AF/Z-GFP* fishes that express a histone–GFP fusion protein in all nuclei *(17)*. A list of existing mutant and transgenic zebrafish lines is available on the zebrafish information network (ZFIN) website (http://zfin.org), along with a myriad of other useful data.

4. Zebrafish are cold-blooded animals of subtropical origin; their optimal temperature range is from 24 to 29°C. The reference temperature for developmental staging is 28.5°C *(18)*. They tolerate temperatures a few degrees above or below these limits; one should, however, carefully control for potential developmental defects or immune suppression caused by too high or too low temperature maintained for too long. Varying temperatures is useful to accelerate or delay development, or to test interactions with cold- or heat-preferring microbes. It is also worth noting that cold-blooded animals offer the possibility of using temperature-sensitive mutants.

5. Not all morpholinos work, so controlling for their efficiency and lack of toxicity is a previous, obligatory step. The toxicity and efficiency range has to be determined for each morpholino. Because AUG-specific morpholinos block translation, the appropriate way to check their efficiency is to measure the amount of specific protein in the treated embryos with a specific antibody. Unfortunately, such antibodies are often unavailable. An elegant way to test for blocking efficiency is described in **ref. *(4)***: create a fusion mRNA with the targeted sequence in 5′ (including the 5′-UTR) and with GFP-coding sequence in frame at the 3′-end. When co-injected with this mRNA in single-cell embryos, the morpholino will inhibit GFP production if it effectively blocks translation of the target protein mRNA.

Splice-blocking morpholinos are easier to control, as their action will result in incorrectly spliced mRNAs. By RT-PCR with primers located in distant exons upstream and downstream of the splice, one should obtain bands of an incorrect size or, quite often, no detectable RNA.

Morpholino injection often has some side effects *(19)*, and the parallel use of a control morpholino, such as a 5-base mismatch, is required to demonstrate a specific effect. Finally, true confidence in the specificity of the observed phenotype usually entails the demonstration that the same effects are observed with a second morpholino specific to the same gene, but directed at a different, non-overlapping region of the mRNA.

6. The shape and resistance of capillary tips critically affect the ease of injections. As pipette pullers offer many parameters to play with, capillary elongation will

involve a good dose of trial and error at the beginning. With the Kopf vertical pipette puller we use, we obtain the best results with a U-shaped solenoid and the following parameters: heater = 11, solenoid = 3. For isolated experiments, one may also consider the purchase of pre-elongated, pre-opened capillaries (e.g., Eppendorf CustomTips), which are standardized but expensive.

7. Because zebrafish eggs must be injected in the first hour following spawning, it is mandatory to have a zebrafish facility at close range. The facility may be quite small: one single tank containing 10 males and 10 females of about 1 year of age may be sufficient. In order to increase reliability of egg production, however, we prefer to launch three or four such spawning tanks when we want to inject eggs. It is also possible to use fewer fish, even single pairs, but this increases the risk of not getting eggs on the desired day. Let us also emphasize that high-quality water and a diversified diet, including live sources such as newly hatched *Artemia* nauplii, are necessary for optimal egg output.

   Typically, zebrafish spawn during the 2 h following "sunrise." It is important to leave the fishes on a regular light/dark schedule (usually 14/10 h), with an automated switch. They also require some irregular bottom substrate, such as a grid or pebbles, to get to breed. Eggs are not sticky and simply fall on the bottom of the tank. Once breeding is over, the substrate also provides protection for the eggs, which are quite tasty for adult zebrafish, including their own parents.

8. Unlike adults, which generally do not recover from an anaesthesis lasting more than a few minutes, embryos survive for several hours in tricaine solution. We try, however, not to leave them in contact with tricaine for more than 20 min if possible.

9. Compatibility of NR and SB stainings with ISH and IHC.

   a. ISH and IHC may be performed following an NR staining, but the red staining will be lost following fixation, so it should be imaged before fixation to combine images.

   b. IHC can be performed after a SB staining, while keeping the black stain. Possibly some epitopes may suffer from the SB staining procedure, although we have not yet observed this. After the last 70% ethanol rinsing step, transfer the embryos to PBT instead of glycerol and proceed to the immunostaining protocol. We have only tried fluorescent detection of the bound antibodies; note that the potential fluorescence of the granulocytes will appear only outside of their black-stained granules.

   c. We were not able to make ISH work after a SB staining. Reciprocally, SB staining will not work following an ISH.

   d. Fixed embryos that have been stored in methanol at –20°C (the usual practice when ISH is to be performed later) are not suitable any more for SB staining.

10. Some melanin may remain in pigmented strains due to lack of or insufficient PTU treatment during their development. This residual pigmentation can be completely cleared by H_2O_2 treatment without affecting the SB staining. This is often useful to

uncover the granulocytes spread along the dorsal side of the embryo, where pigment cells are numerous. The procedure is as follows:

a. Rehydrate the embryos progressively from ethanol to PBT.
b. Replace PBT by a freshly made 1% H_2O_2, 1% KOH solution in water, and incubate under gentle horizontal agitation for 10–15 min, checking from time to time under the dissecting scope.
c. When the melanin pigments have disappeared, stop the reaction by replacing the H_2O_2 solution by PBT.

Acknowledgments

We are grateful to Karima Kissa-Marin, Dorothée Le Guyader, and Muriel Tauzin for tips regarding staining procedures, and to Valérie Briolat for excellent fish care. We are grateful to Dr. Wilbert Bitter (Amsterdam, Netherlands) for the pRZT3 plasmid and to Dr. Bruce Draper (Davis, CA), who originally designed the injection help pictured in **Fig. 1B**.

This work was supported with the help of an Inserm Avenir grant and of a Genanimal grant from the Ministère de la Recherche.

References

1. Traver, D., Herbomel, P., Patton, E., Murphey, R., Yoder, J., Litman, G., Catic, A., Amemiya, C., Zon, L., and Trede, N. (2003) The zebrafish as a model organism to study development of the immune system. *Adv. Immunol.* **81**, 253–330.
2. Trede, N., Langenau, D., Traver, D., Look, A., and Zon, L. (2004) The use of zebrafish to understand immunity. *Immunity* **20**, 367–379.
3. Nasevicius, A., and Ekker, S. (2000) Effective targeted "knockdown" in zebrafish. *Nat. Genet.* **26**, 216–220.
4. van der Sar, A., Stockhammer, O., van der Laan, C., Spaink, H., Bitter, W., and Meijer, A. (2006) Myd88 innate immune function in a zebrafish embryo infection model. *Infect. Immun.* **74**, 2436–2441.
5. Westerfield, M. (1993) *The Zebrafish Book. A Guide for the Laboratory Use of Zebrafish (Brachydanio rerio)*, The University of Oregon Press, Eugene.
6. van der Sar, A., Musters, R., van Eeden, F., Appelmelk, B., Vandenbroucke-Grauls, C., and Bitter, W. (2003) Zebrafish embryos as a model host for the real time analysis of *Salmonella typhimurium* infections. *Cell. Microbiol.* **5**, 601–611.
7. Herbomel, P., Thisse, B., and Thisse, C. (1999) Ontogeny and behaviour of early macrophages in the zebrafish embryo. *Development* **126**, 3735–45.
8. Herbomel, P., and Levraud, J.P. (2005) Imaging early macrophage differentiation, migration, and behaviors in live zebrafish embryos. *Methods Mol. Med.* **105**, 199–214.

9. Murayama, E., Kissa, K., Zapata, A., Mordelet, E., Briolat, V., Lin, H.F., Handin, R.I., and Herbomel, P. (2000) Tracking hematopridic precursor migration to successive hematopritic organs during zebrafish development. Immunity. 25: 963–975.
10. Herbomel, P., Thisse, B., and Thisse, C. (2001) Zebrafish early macrophages colonize cephalic mesenchyme and developing brain, retina, and epidermis through a M-CSF receptor-dependent invasive process. *Dev. Biol.* **238**, 274–288.
11. Lutfalla, G., and Uzé, G. (2006) Performing quantitative RT-PCR experiments, in *DNA Microarrays* (Kimmel, A., and Oliver, B., Eds.), Vol. 410, Elsevier Academic Press, San Diego, 383–397.
12. Novak, A., and Ribera, A. (2003) Immunocytochemistry as a tool for zebrafish developmental neurobiology. *Methods Cell Sci.* **25**, 79–83.
13. Sheehan, H., and Storey, G. (1947) An improved method of staining leukocyte granules with Sudan black B. *J. Pathol. Bacteriol.* **59**, 336.
14. Guryev, V., Koudijs, M., Berezikov, E., Johnson, S., Plasterk, R., van Eeden, F., and Cuppen, E. (2006) Genetic variation in the zebrafish. *Genome Res.* **16**, 497–497.
15. Lister, J., Robertson, C., Lepage, T., Johnson, S., and Raible, D. (1999) nacre encodes a zebrafish microphthalmia-related protein that regulates neural-crest-derived pigment cell fate. *Development* **126**, 3757–3767.
16. Lawson, N., and Weinstein, B. (2002) *In vivo* imaging of embryonic vascular development using transgenic zebrafish. *Dev. Biol.* **248**, 307–318.
17. Pauls, S., Geldmacher-Voss, B., and Campos-Ortega, J. (2001) A zebrafish histone variant H2A.F/Z and a transgenic H2A.F/Z:GFP fusion protein for in vivo studies of embryonic development. *Dev. Genes Evol.* **211**, 603–610.
18. Kimmel, C., Ballard, W., Kimmel, S., Ullmann, B., and Schilling, T. (1995) Stages of embryonic development of the zebrafish. *Dev. Dyn.* **203**, 253–310.
19. Heasman, J. (2002) Morpholino oligos: making sense of antisense? *Dev. Biol.* **243**, 209–214.

21

Reverse Genetics Analysis of Antiparasitic Responses in the Malaria Vector, *Anopheles gambiae*

Stephanie A. Blandin and Elena A. Levashina

Summary

Anopheles mosquitoes are the major vectors of human malaria parasites. Mosquito–parasite interactions are critical for disease transmission and therefore represent a potential target for malaria control strategies. Mosquitoes mount potent antiparasitic responses, and identification of mosquito factors that limit parasite development is one of the major objectives in the field. To address this question, we have developed a convenient reverse genetics approach by injection of double-stranded RNA (dsRNA) in adult mosquitoes, to evaluate the function of candidate genes in mosquito antiparasitic responses.

Key Words: Malaria; mosquito immunity; RNAi; fluorescence and confocal microscopy; immunofluorescence assays.

1. Introduction

Malaria persists today as one of the most widespread and devastating diseases in the world. It is caused by a protozoan parasite of the genus *Plasmodium* and is transmitted to humans when an infected *Anopheles* mosquito takes a blood meal. Currently, 60 anopheline species are bona fide vectors of malaria, whereas the majority of mosquitoes do not support parasite development. Moreover, large variations exist between individuals within a vector species. It has been observed in wild populations that only 2–10% of individuals are susceptible to infection and therefore contribute to transmission of malaria. What determines the capacity of *Anopheles* to support development of *Plasmodium* is

From: *Methods in Molecular Biology, vol. 415: Innate Immunity*
Edited by: J. Ewbank and E. Vivier © Humana Press Inc., Totowa, NJ

an important and intriguing question in fundamental host–pathogen biology. Previous studies have investigated this area using predominantly biochemical, molecular biology and microscopy approaches (reviewed in **refs.** *(1)* and *(2–4)*). Here, we describe the application of reverse genetics based on RNA interference (RNAi) to study the infection of *Anopheles gambiae* with the rodent malaria parasite, *Plasmodium berghei*. The *P. berghei* model presents a number of advantages that facilitate experimental procedures: *(1)* it is easy to propagate asexual parasite stages in mice; *(2)* *P. berghei* is infectious to *A. gambiae* and mice but not to humans and, therefore, can be used under normal safety conditions in the laboratory; *(3)* *P. berghei* is easily amenable to genetic manipulation *(5)*.

During their development in the mosquito vector, *Plasmodium* parasites undergo massive losses, which are caused, at least in part, by mosquito factors that act to limit parasite numbers. With the simple technique of gene silencing in adult mosquitoes described here, we are now in a position to identify genes whose invalidation alters mosquito responses to infection, i.e., reduces or increases parasite numbers *(6)*. Infections with a transgenic green fluorescent protein (*GFP*)-expressing *P. berghei* strain *(7)* are easily visualized and scored by microscopy. Our method for gene functional analysis can also be used to examine infections with other parasite species, including the more demanding experimental infections with the human parasite *Plasmodium falciparum*. This will allow the comparison of infections by *Plasmodium* species that are adapted to *A. gambiae* with infections by non-adapted species such as *P. berghei* or *Plasmodium yoelii*. The gained knowledge could have implications for the design of novel strategies for disease control.

2. Materials

2.1. Double-Stranded RNA Synthesis

1. pLL10 *(8)*: a pBluescript-based plasmid with two T7 promoter sequences flanking the polylinker region in opposite direction (*see* **Fig. 1A**).
2. Proteinase K stock solution: 20 mg/ml in sterile 20 mM Tris–HCl (pH 8), 1.5 mM $CaCl_2$, 50% glycerol. Aliquots can be stored at –20°C. Proteinase K buffer: 10 mM Tris–HCl pH8, 10 mM EDTA (pH 8), 5 mM NaCl, 2 mM $CaCl_2$. Proteinase K final solution: Add 1 µl of proteinase K (20 mg/ml) to 150 µl of buffer. Aliquot by 50 µl (one tube for two reactions) and store at –20°C.
3. Linearized plasmid and RNA purification: RNase-free diethyl pyrocarbonate (DEPC) treated water, phenol/chloroform/isoamyl alcohol (25/24/1), chloroform, isopropanol, and 70% ethanol. All reagents should be RNase-free.
4. Synthesis and purification of single stranded RNAs: T7 MEGAscript kit (Ambion, Austin, USA).

Functional Analysis of Mosquito Antiparasitic Responses

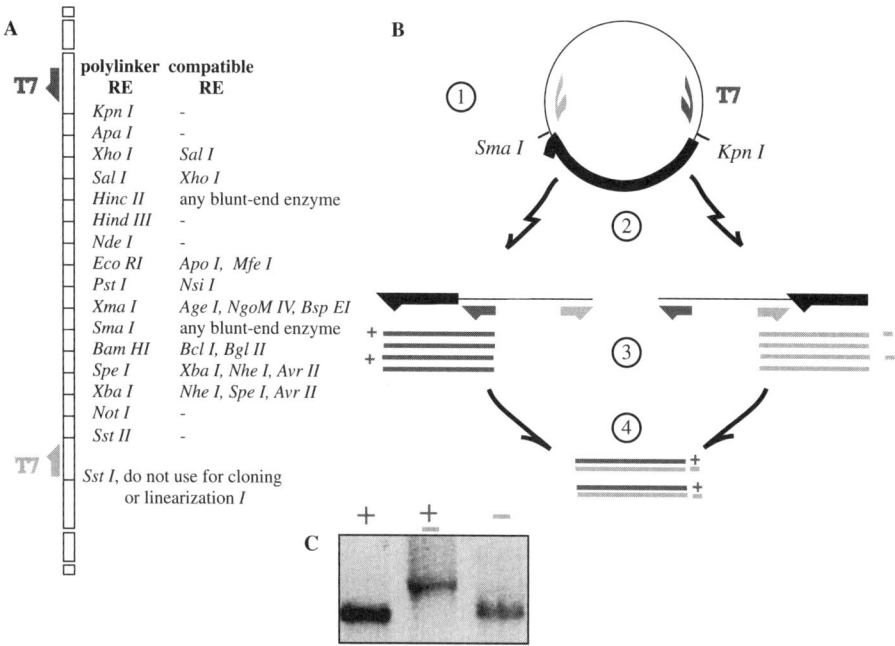

Fig. 1. Synthesis of double-stranded RNA (dsRNA). (**A**) Map of pLL10 polylinker. Restriction enzymes (RE) with compatible ends are listed in the "compatible RE" column. Position of the T7 promoters is indicated on the map. (**B**) The main steps in dsRNA synthesis are the following: *(1)* cloning of gene of interest in pLL10 between the two T7 promoters using polylinker enzymes; *(2)* linearization of plasmid on both sides of the insert; *(3)* synthesis of sense (+) and antisense (−) ssRNAs independently using the MEGAscript kit; *(4)* annealing of ssRNAs to form dsRNA (±). (**C**) Quality of dsRNA should be evaluated by electrophoresis: dsRNA migrates slower than corresponding ssRNAs.

2.2. Injection of Double-Stranded RNA into Adult Mosquitoes

1. Strains of *A. gambiae*: e.g., G3 and L3-5 (*see* **Note 1**). Anopheline strains can be obtained from the MR4 (*see* **Note 2**). Mosquitoes are bred at 27°C and 70% humidity, with a day/night period of 12 h/12 h.
2. Waxed paper cartons (e.g., ice-cream or drink containers), filter paper circles of a matching diameter to fit in the bottom of the cartons (#1, Whatman, Brentford, UK), fine nylon or cotton netting, tape, and small elastic bands.
3. Mosquito aspirator (pooter).
4. Absorbent cotton wool, 10% sugar solution in water, and small-sized Petri dish bottoms.

5. CO_2 bottle and CO_2 distributor and pad (InjectMatic, Geneva, Switzerland).
6. Injector Nanoject II (Drummond, Broomall, USA), capillaries (FT330B), a syringe and needle filled with mineral oil, a brush, and forceps.
7. Micropipette puller (P-97, Sutter Instrument Company, Novato, USA).

2.3. Infections with GFP-Expressing P. bergei Parasites

1. *Plasmodium berghei* clone *PbGFPcon* (*7*), where *GFP* is constitutively expressed under the control of the *EF1* promoter. This and other *P. berghei* clones can be requested from Dr. C.J. Janse (http://www.lumc.nl/1040/research/malaria/model.html).
2. Mice: CD1 strain.
3. Heparin sodium salt grade IA, 5000 U/ml (Sigma, St. Louis, USA). Prepare working solution at 300 U/ml in phosphate-buffered saline (PBS) (1 part heparin stock : 16 parts PBS) and store at 4°C.
4. Fluorescence-activated cell sorting (FACS) analyzer; running buffer for FACS analysis: BD FACS Flow™ Sheath Fluid (BD Biosciences, Franklin Lakes, USA).
5. Staining of blood cells from infected mice: methanol; rapid detection hematology staining (DADE Boehringer, Mannheim, Marburg, Germany).
6. Anesthetic: 5% solution of Rompun® and 10% solution of Ketamine® in PBS (100 µl for 10 g of mouse body weight).
7. Incubator at 21°C, 70% humidity with a day/night period of 12 h/12 h.

2.4. Analysis of Antiparasitic Responses by Fluorescence and Confocal Microscopy

1. 24-well plates; 1.5-ml Eppendorf tubes and a fine mesh (Nitex dominique Dutscher, Brumath, France, 100 µm). PBS: Prepare a 10× stock (130 mM NaCl, 7 mM Na_2HPO_4, 3 mM Na H_2PO_4) and autoclave. For working solution, dilute one part with nine parts of water. Adjust to pH 7.2 if necessary. *See* **Note 3**.
2. Paraformaldehyde (PFA), 16% stock solution (EM grade, EMS). Working 4% solution in PBS should be prepared fresh for each experiment with 1:3 PFA : PBS (*see* **Note 4**). PFA should be manipulated carefully under a safety hood; avoid inhaling and contact with skin.
3. Permeabilization solution: 0.2% (v/v) Triton X-100 and 1% bovine serum albumin (BSA) (w/v) in PBS. The solution should be prepared fresh for each experiment (25 ml per sample).
4. Primary antibody solution: To perform multiple stainings, combine primary antibodies produced in different animal species (e.g., rabbit and mouse, or rat), at relevant dilution in 500 µl of permeabilization buffer per sample.
5. Secondary antibody solution: Use the recommended dilution of anti-rabbit, anti-mouse, or anti-rat IgG antibodies conjugated with fluorophores emitting at wavelengths that should be distinct from each other and from those of GFP, the parasite marker (e.g., Cy3 or Alexa-546 and Cy5 or Alexa-647, *see* **Note 5**).

6. Microscope slides and cover slips (22 × 22 mm).
7. VectaShield mounting medium (Vector Laboratories, Burtingame, USA), with or without 4′, 6-diamidino-2-phenylindole (DAPI).

3. Methods

Silencing of gene expression by RNAi is a powerful approach to study antiparasitic responses in mosquitoes. This method includes several steps: *(1)* a candidate gene(s) is/are selected; *(2)* the optimal target sequence is cloned into an appropriate vector for double-stranded RNA (dsRNA) synthesis; *(3)* dsRNA is injected into adult mosquitoes and the efficiency of gene silencing is examined 1–4 days after injection; and *(4)* the effect of the gene knockdown is analyzed by counting the number of surviving parasites 7–10 days after infection, which serves as a readout for the early stages of parasite development within the mosquito, and/or by analysis of the kinetics of parasite killing within the mosquito midgut, 18–48 h after infection. This method can be extended to different mosquito species and strains to analyze responses to infections with different pathogens (bacteria, fungi, and viruses) *(8)*, or different *Plasmodium* species (*P. berghei, P. yoelii, P. falciparum,* etc).

3.1. dsRNA Synthesis

1. Clone a fragment of a gene of interest in pLL10 (*see* **Fig. 1A**). The optimal size of the fragment is determined by the efficiency of the RNA synthesis reaction and is between 500–1000 bp for the T7 MEGAscript kit. The choice of targeted region depends on the application: closely related genes could be co-silenced by designing the dsRNA against a highly conserved region, whereas the specificity of silencing of an individual gene is achieved through the selection of a variable region including 5′- or 3′-untranslated regions.
2. Prepare a mini prep of the constructed plasmid (*see* **Note 6**).
3. For the synthesis of sense and antisense ssRNAs, linearize 2 × 10 μg of plasmid separately with two different restriction enzymes, one on each side of the insert, in 50 μl (*see* **Fig. 1B**). Confirm that digest is complete by loading 1 μl of the reaction on 1% agarose gel (*see* **Note 7**).
4. Add 25 μl of the proteinase K final solution and 4 μl of 10% sodium dodecyl sulfate (SDS) to each digest. Incubate at 50°C for 30 min.
5. Add 80 μl phenol/chloroform/isoamyl alcohol. Vortex, and incubate 2 min at room temperature (RT). Centrifuge for 5 min at maximal speed and collect aqueous phase in a fresh tube.
6. Add 80 μl of chloroform. Vortex and incubate 2 min at RT. Centrifuge at maximal speed and collect aqueous phase in a fresh tube.

7. Add 56 µl of isopropanol, mix by inverting tube, and incubate in cold room for 15 min. Centrifuge 15 min at 4°C at maximal speed and discard supernatant.
8. Wash with 100 µl of 70% ethanol, centrifuge 5 min at maximal speed, discard supernatant, and allow the DNA pellet to air-dry on the bench.
9. Dissolve pellet in 20 µl of water. Check DNA quality and concentration (should be around 0.5 µg/µl) on a 1% agarose gel. Linearized plasmids can be stored at −20°C.
10. Thaw reagents of the T7 MEGAscript kit (Ambion). Keep ribonucleotides on ice, and transcription buffer at RT.
11. For each linearized plasmid, add in order: (8–x) µl of water, 2 µl of each NTP, 2 µl buffer, x µl linearized plasmid (1 µg), and 2 µl enzyme mix. If several reactions are assembled in parallel, prepare a master mix with all reagents except linearized plasmids, aliquot calculated volume in each tube, and then add linearized plasmids. Incubate overnight at 37°C (8–14 h). *See* **Notes 8** and **9**.
12. Digest DNA template by adding 1 µl of DNase I to each reaction. Incubate 15 min at 37°C.
13. Purification of ssRNAs: To the incubation mix, add 115 µl of water and 15 µl of ammonium acetate stop solution and mix thoroughly. Extract RNA with 150 µl of phenol/chloroform/isoamyl alcohol, followed by 150 µl of chloroform (*see* **steps 5** and **6**). Recover aqueous phase and transfer it to a fresh tube.
14. Precipitate RNA by adding 150 µl of isopropanol, mix well by inverting tube, and incubate 15 min at −20°C. Centrifuge 15 min at 4°C at maximal speed and discard supernatant. Air-dry pellet for a few minutes on the bench, and dissolve it in 20 µl of water.
15. While samples are drying, boil water in a 1–3 l beaker covered with aluminium foil.
16. Measure concentration of ssRNAs using a UV spectrophotometer. For this, dilute 1 µl of each ssRNA in 9 µl of 10 mM Tris buffer (pH 8). Take 1 µl of this dilution into 100 µl of the same 10 mM Tris buffer for measurement, and keep the remaining 9 µl at −20°C. [ssRNA] (µg/µl) = OD_{260}*dilution factor*40/1000 (dilution factor is 1000 in the given example).
17. Adjust concentration of sense and antisense ssRNAs to 3 µg/µl each and mix equal volumes of both (the rest of the ssRNAs is stored at −20°C or −80°C). Close tubes tightly and boil samples for 5 min in the beaker. Allow samples to slowly cool down to RT on the bench.
18. Check dsRNA quality on a 1.8% agarose gel. For this, denaturate the ssRNA dilutions from **step 16** at 95°C for 3 min and immediately cool on ice. Spin briefly. Mix 1 µl of these and 1 µl of dsRNA from **step 17** with 5 µl of DNA-loading buffer each, and load on gel. A clear shift should be observed in the migration patterns of ssRNAs and dsRNA: dsRNA migrates slower than the corresponding ssRNAs (*see* **Fig. 1C**). dsRNA is quite resistant to multiple freezing/thawing cycles and can be stored at −20°C or −80°C.

3.2. dsRNA Injection into Adult Mosquitoes

1. Mosquito breeding: see "Anopheles Culture" by M.Q. Benedict, CDC Atlanta, USA at the MR4 website (http://www2.ncid.cdc.gov/vector/vector.html).
2. Needles for injection: Pull glass capillaries using the needle puller. Each capillary gives two needles, the tips of which are sealed.
3. Prepare mosquito pots. Cut a cross-shaped opening in the side of a pot and seal it with tape. Place a filter paper circle at the bottom of the pot; it will blot excess sugar or mosquito droppings. Stretch netting over the top of the pot and secure with an elastic band.
4. Place around 60 1–2 day-old mosquito females into the pot using the pooter. Fill a small-sized Petri dish bottom with a cotton ball soaked in 10% sugar solution and place it on the netting.
5. Break the tip of the glass capillary with forceps, so that the tip is rigid enough, but the opening not too big. Fill the capillary with a mineral oil using a syringe and a needle. Assemble the Nanoject injector and fill the capillary with the dsRNA solution. Verify settings on the control block of the Nanoject and set the desired injection volume and speed. To analyze one gene, we usually inject each mosquito with 69 nl of dsRNA at the highest speed.
6. Immobilize mosquitoes in the pot by CO_2 treatment and align them with dorsal side up on the CO_2 pad. Using the injector and a brush, carefully inject dsRNA solution into the dorsal plate of the mosquito thorax (*see* **Notes 10** and **11**). During injections, limit exposure of mosquitoes to CO_2 using the pedal-controlled distributor (*see* **Note 12**).
7. After all mosquitoes are injected, gently place them back in paper pot using a brush, stretch the netting, fix it with an elastic band, and place a sugar cotton on the top. Keep mosquitoes until infectious blood-feeding in the 27°C incubator (usually for 4 days) (*see* **Note 13**).

3.3. Infections with GFP-Expressing P. bergei Parasites

1. Parasite maintenance: Infect CD1 mice by intraperitoneal injection of infected erythrocytes containing asexual blood stages. To assure infectivity of gametocytes to mosquitoes, avoid passing the parasite cultures through a mouse more than 10 times. Please consult the website of the Malaria group of Leiden University Medical Center (LUMC) (http://www.lumc.nl/1040/research/malaria/model04.html).
2. In the morning of the day of infection, remove sugar cotton balls from mosquito pots and place injected mosquitoes in the incubator at 21°C.
3. To check the parasitemia and gametocytemia of the infected mouse, we routinely use FACS-based counting of fluorescent parasites (*see* **Fig. 2A**). Collect a blood drop from the tail vein of a mouse and dilute it in 5 μl of heparin solution. Dilute the sample 1000 times in a FACS running buffer. Adjust the program according to your appliance. Blood cells infected with fluorescent parasites correspond to the

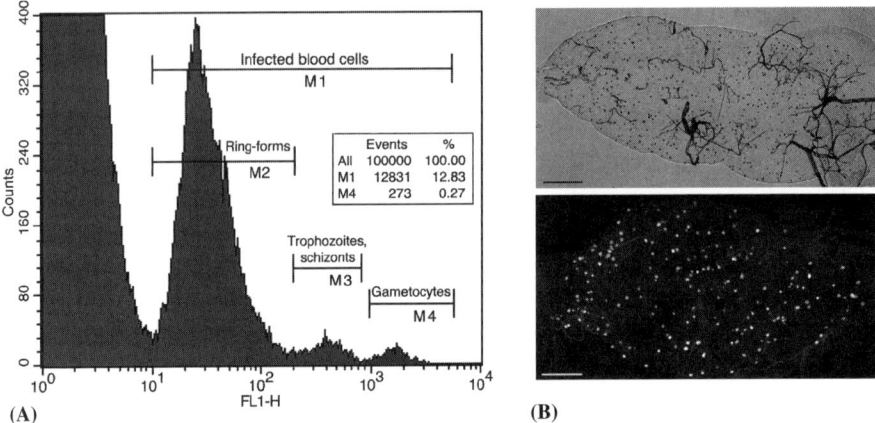

Fig. 2. Infections with PbGFPcon. (**A**) Counting parasitemia (percentage of infected blood cells) and gametocytemia (percentage of blood cells containing gametocytes) of a mouse infected with *Plasmodium berghei* parasites expressing green fluorescent protein (GFP) using—flow cytometry. Infected erythrocytes (M1) are easily distinguished from uninfected cells by the intensity of GFP fluorescence. The different parasite life stages are clearly separated: the M2 fraction represents ring-forms, and M3 fraction includes trophozoites and schizonts and M4, gametocytes. In this example, parasitemia (M1) is 12.8%, and gametocytemia (M4) is 0.27%; the mouse can be used to infect mosquitoes. (**B**) Upper panel, bright field picture of the midgut of a refractory mosquito with melanized parasites. Lower panel, fluorescence picture of the midgut of a susceptible mosquito with developing parasites. Midguts were dissected 7 days post infection and pictures were taken with a ×5 objective. Scale bar, 200 µm.

right-shifted peaks (M1) and represent three populations: (M2) blood cells infected with ring-forms; (M3) blood cells infected with trophozoites and schizonts; and (M4) female and male gametocytes. Mice with parasitemia around 10–12% and sufficient number of gametocytes (0.2–0.5%) are used for infections.

4. In the case of non-fluorescent parasite strains, we prepare blood smears stained with the rapid detection hematology staining kit. A blood drop from the tail vein of a mouse is smeared on a slide and fix rapidly in methanol. Allow slide to air-dry and incubate in staining solution I (red) for 10 s and immediately after in solution II (blue) for 2 min. Remove excess of staining with running water and allow slide to air-dry. Place a drop of immersion oil directly on smear and observe blood cells under the microscope using a ×100 objective. Please refer to the web page of the malaria group in LUMC for images and description of parasite stages http://www.lumc.nl/1040/research/malaria/model05.html. For infections, we use mice with parasitemia around 10–12% and detectable gametocytes.

5. Anesthetize infected mouse by an intraperitoneal injection of anesthetic (100 μl for 10 g of mouse body weight).
6. Protect mouse eyes with conventional eye drops and place the anesthetized mouse on the top of the mosquito pot at 21°C. Allow mosquitoes to feed for approximately 20 min. Remove the mouse, and place the sugar cotton ball on the top of the pot. Mosquitoes are kept in the 21°C incubator until the end of the experiment.
7. Discard unfed mosquitoes 7–24 h after infection. For this, mosquitoes are quickly put to sleep with CO_2 and spread on the CO_2 pad. Blood is easily visible by eye in the abdomen of fed females.

3.4. Analysis of Antiparasitic Responses by Fluorescence and Confocal Microscopy

1. Prepare baskets for dissections: Cut the lid and conical part of a 1.5-ml Eppendorf tube. Flame the bottom part of the tube that was cut with a Bunsen burner and place immediately on the fine mesh to glue the mesh on the tube. Make sure there is no hole between tube and mesh, and cut the mesh around the tube. The resulting basket is a convenient tool to transfer dissected tissues from one well to another. *see* **Note 14**.
2. Place dissection baskets in wells of the first row (one for each sample). Prepare 2 ml of fixation solution per well and aliquot in wells containing the dissection baskets. Distribute PBS (2 ml/well) in the three wells below each fixation well. Keep the 24-well plate at RT.
3. Put mosquitoes to sleep at 4°C.

   a. To analyze infection levels at oocyst stage using fluorescence microscopy, please go to **steps 4–10**. Mosquitoes are dissected 7–10 days post infection.
   b. To analyze interactions between a given protein and parasites or another mosquito protein, please go to **steps 11–26**. Mosquitoes are dissected at different time points after infection (usually 18/24/32/48 h post injection).

4. Sample preparation for classical fluorescence microscopy: Perform dissections on ice, on a glass plate, in a drop of PBS. With forceps, tear the cuticle at the level of the last two segments of the abdomen. Pull away the tip of the abdomen to extract the midgut (and Malpighian tubules). Proceed similarly with other mosquitoes. Place tissues in the dissection basket in fixation solution. Incubate in fixation solution for 15–20 min at RT.
5. After fixation, transfer baskets with midguts from fixation wells to the wells below, containing PBS, and incubate at RT for 10 min. Repeat washing twice (longer washing times could be used).
6. Label fresh slides with the name of samples using a pencil.
7. Mount the samples at RT. Transfer the dissected tissues to a glass plate in a drop of mounting medium (without DAPI). It is convenient to have a dark background below the glass plate (tissues are white). With forceps, remove cuticle pieces that

are still attached to midguts (*see* **Note 15**). Transfer midguts in a small drop of fresh mounting medium on the labeled slide and align them in columns to facilitate screening with microscope.
8. Carefully place a coverslip on samples avoiding bubbles. Place a folded Kimwipe on the top of the slide and gently press to remove excess of mounting medium. Once all samples are mounted, seal coverslips on slides with nail polish. Slides can be stored in a dark box at 4°C for several months.
9. Take pictures of each midgut with a fluorescence microscope, with a ×5 objective (higher magnifications might not accommodate the whole midgut in one picture), in the GFP channel and in bright field.
10. Count oocysts and melanized parasites manually on each picture (*see* **Fig. 2B**). Analyze results using Prism software (GraphPad). For statistical significance, it is recommended to have more than 20 midguts per sample, and experiment should be repeated at least three times.
11. Sample preparation for confocal microscopy: Perform dissections on ice, on a glass plate, in a drop of PBS. Using forceps, tear the abdomen at the limit between abdomen and thorax below the level of first abdominal segment and pull away abdomen from thorax and head.
12. Cut abdomen cuticle longitudinally using micro-scissors. It will open itself and float on the surface of the PBS, with the hydrophobic surface up and the midgut facing the PBS.
13. Transfer abdomen and midgut to fixation solution for 22–25 s, not more (otherwise it will be difficult to clean the blood from the midgut) (*see* **Note 16**).
14. Transfer them back to the drop of PBS on ice. Cut the midgut longitudinally with micro-scissors. Remove blood by scrapping gently with micro-scissors and by flushing with a glass Pasteur pipette filled with PBS: hold the abdomen with forceps and pipette up and down nearby the midgut. Midgut can be aspirated in the pipette, but be gentle or it will detach from abdomen. Transfer clean abdomen and midgut to fixation well.
15. Proceed similarly with dissection of the next mosquito (place dissected tissues in the same well as in previous step). A minimum of seven midguts should be dissected per sample. Once this is done, leave dissected tissues all in the same well in fixation solution for 45 min.
16. After fixation, transfer baskets with samples from fixation wells to the wells below, containing PBS, and incubate at RT for 10 min. Repeat washing twice (longer washing times could be used).
17. While washing, prepare permeabilization solution (PS) and incubate at RT on a rocking platform to dissolve BSA. Aliquot 2 ml of PS in a new 24-well plate (one well per sample; *see* **Note 14**), and transfer the baskets from last washing to PS. Incubate for 1–2 h at RT.
18. Meanwhile, prepare primary antibody solution (0.5 ml per well) and aliquot solution in wells of the row below that containing PS. Transfer baskets to primary

Functional Analysis of Mosquito Antiparasitic Responses 375

     antibody solution (make sure all samples are in solution), and incubate overnight at 4°C (*see* **Note 17**).

19. Aliquot 2 ml of PS in wells of the next row, below primary antibody solution, and transfer baskets to these for washing. Incubate at RT for 10 min, and then pipette solution out with a Pasteur pipette and replace with clean PS. Repeat this step twice.
20. While washing, prepare secondary antibody solution (0.5 ml per well) and aliquot in wells of the row below that containing the baskets washing in PS. Transfer baskets to secondary antibody wells (make sure all samples are in solution), and incubate at RT for 1 h. From this step on, samples should be protected from light to preserve the intensity of the fluorophores that are conjugated to secondary antibodies.
21. Repeat washing as described in the **step 19**.
22. Label fresh slides with the name of samples and the date using a pencil.
23. Mount the samples at RT. Transfer dissected tissues to a glass plate in a drop of mounting medium. It is convenient to have a dark background below the glass plate. With forceps, remove ovaries that are still attached to abdomen and separate midgut and Malpighian tubules from abdominal cuticle. Place two small drops of mounting medium on the labeled slide and transfer all cuticles in one drop and all midguts in the other (*see* **Note 18**). Cuticles should be spread with the interior part up. Align cuticles and midguts in columns to facilitate screening by microscopy.
24. Carefully place a coverslip on samples; get rid of bubbles. Place a folded Kimwipe on the top of the slide and gently press to remove excess of mounting medium. Once all samples are mounted, seal coverslips on slides with nail polish. Slides can be stored in a dark box at 4°C for several months.
25. Samples are examined using fluorescence microscope (×40, ×63, and ×100 objectives) equipped with an Apotome module (Zeiss, Oberkochen, Germany) or by confocal microscopy to detect the presence of a protein of interest on parasite surface, co-localization of two proteins, where the protein is produced, etc.

4. Notes

1. G3 mosquitoes are susceptible to parasite development. The refractory L3-5 strain completely blocks parasite development and melanizes dead ookinetes.
2. MR4 (Malaria Research and Reference Reagent Resource) is a center that stores and provides reagents to the malaria research community (http://www.malaria.mr4.org/).
3. Solutions are stored at RT, unless otherwise stated.
4. PFA is conditioned in 10-ml sealed glass ampoules. Once opened, the rest can be aliquoted in 1.5-ml Eppendorf tubes and stored at −20°C for further use. Thaw aliquots in a 37°C heating block and vortex to dissolve any precipitate.

5. Alexa dyes are usually stronger than Cy dyes. Note that A488 and Cy2 cannot be used as these dyes have the same spectra as that of GFP, which is used to mark parasites.
6. Sequencing of an insert cloned in pLL10 can be done using universal M13 forward and reverse primers.
7. It is important to make sure that all plasmid is digested, as RNA polymerases are very processive and generate long heterogeneous transcripts from circular plasmids.
8. When transcription is optimal, the reaction mixture at the end will be rather viscous.
9. Sense and antisense ssRNAs can be produced in a single reaction, e.g., when amplified from a PCR fragment that was amplified with primers bearing T7 promoters. However, this does not ensure that both strands are synthesized with the same efficiency. We therefore prefer to prepare sense and antisense ssRNAs separately and measure their respective concentrations before annealing to make sure equal quantities of both strands are mixed and to obtain reproducible quantities of dsRNA.
10. The maximum volume that can be injected at once is 69 nl. Should you need to inject larger volumes, press the injection button twice or three times. In our hands, mosquitoes tolerate up to 300 nl per injection, which roughly corresponds to four repeated injections.
11. With one filled needle, about 60 mosquitoes can be injected with 69 nl.
12. Excess of CO_2 treatment is toxic to mosquitoes.
13. As the efficiency of silencing can vary between genes, we usually verify that expression of the targeted gene is indeed affected by the dsRNA treatment. We recommend performing this analysis at the protein level when it is possible (e.g., by immunoblotting if antibodies are available, or by mass spectrometry), or otherwise at the mRNA level (e.g., by quantitative real-time PCR).
14. The 24-well plates and dissection baskets can be carefully washed with distilled water and reused for the next experiment.
15. It is important to remove cuticle pieces as they are thick and increase space between slide and coverslip, which makes it more difficult to see all parasites in the midgut in the same focal plane. Malpighian tubules can also be removed although it is not necessary and it is more time consuming.
16. A short time of fixation before opening midgut to remove blood prevents the midgut muscles from contracting during the dissection. Midgut tissues are therefore flat and are easy to analyze in the same focal plane.
17. To save precious antibodies, primary antibody solutions can be kept at 4°C and reused for other experiments within 1–2 weeks.
18. It is practical to keep the abdominal cuticle for immunofluorescence assays as it can be further used for analysis of immune responses in the fat body cells and hemocytes.

Acknowledgments

The authors first developed this methodology while working at the EMBL in the group of Professor Fotis C. Kafatos. They further acknowledge the continuous support and interest of Professor Jules A. Hoffmann and members of the Strasbourg laboratory for constructive discussions. This work was supported by grants from CNRS, INSERM, and Schlumberger Foundation For Education and Research (FSER) and by the 6th European Commission Programme "Networks of Excellence" BioMalPar. S.A.B. is a fellow of the European Molecular Biology Organization. E.A.L. is an International Scholar of the Howard Hughes Medical Institute.

References

1. Richman, A. M., Dimopoulos, G., Seeley, D., and Kafatos, F. C. (1997) Plasmodium activates the innate immune response of *Anopheles gambiae* mosquitoes. *EMBO J.* **16**, 6114–6119.
2. Vizioli, J., Bulet, P., Charlet, M., Lowenberger, C., Blass, C., Muller, H. M., Dimopoulos, G., Hoffmann, J., Kafatos, F. C., and Richman, A. (2000) Cloning and analysis of a cecropin gene from the malaria vector mosquito, *Anopheles gambiae*. *Insect Mol. Biol.* **9**, 75–84.
3. Dimopoulos, G., Christophides, G. K., Meister, S., Schultz, J., White, K. P., Barillas-Mury, C., and Kafatos, F. C. (2002) Genome expression analysis of *Anopheles gambiae*: responses to injury, bacterial challenge, and malaria infection. *Proc. Natl. Acad. Sci. USA* **99**, 8814–8819.
4. Han, Y. S., and Barillas-Mury, C. (2002) Implications of Time Bomb model of ookinete invasion of midgut cells. *Insect Biochem. Mol. Biol.* **32**, 1311–1316.
5. Janse, C. J., Franke-Fayard, B., Mair, G. R., Ramesar, J., Thiel, C., Engelmann, S., Matuschewski, K., van Gemert, G. J., Sauerwein, R. W., and Waters, A. P. (2006) High efficiency transfection of *Plasmodium berghei* facilitates novel selection procedures. *Mol. Biochem. Parasitol.* **145**, 60–70.
6. Blandin, S., Shiao, S. H., Moita, L. F., Janse, C. J., Waters, A. P., Kafatos, F. C., and Levashina, E. A. (2004) Complement-like protein TEP1 is a determinant of vectorial capacity in the malaria vector *Anopheles gambiae*. *Cell* **116**, 661–670.
7. Franke-Fayard, B., Trueman, H., Ramesar, J., Mendoza, J., van der Keur, M., van der Linden, R., Sinden, R. E., Waters, A. P., and Janse, C. J. (2004) A *Plasmodium berghei* reference line that constitutively expresses GFP at a high level throughout the complete life cycle. *Mol. Biochem. Parasitol.* **137**, 23–33.
8. Blandin, S., Moita, L. F., Kocher, T., Wilm, M., Kafatos, F. C., and Levashina, E. A. (2002) Reverse genetics in the mosquito *Anopheles gambiae*: targeted disruption of the *Defensin* gene. *EMBO Rep.* **3**, 852–856.

22

Drosophila Immunity

Methods for Monitoring the Activity of Toll and Imd Signaling Pathways

Yves Romeo and Bruno Lemaitre

Summary

Invertebrates lack an adaptive immune system and rely on innate immunity to resist pathogens. The response of *Drosophila melanogaster* to bacterial and fungal infections involves two signaling pathways, Toll and Imd, both of which activate members of the nuclear factor (NF)-κB family of transcription factors, leading to antimicrobial peptide (AMP) gene expression. In this chapter, we present the current methods used in our laboratory to monitor the activity of both signaling pathways.

Key Words: *Drosophila* immunity; Toll and Imd signaling pathways; antimicrobial peptides; reporter genes; RT-qPCR; survival analyses.

1. Introduction

Innate immunity serves as a first-line defense against microbial invaders and is common to all metazoans. *Drosophila melanogaster*, the fruit fly, shows a potent host defense when challenged by various microorganisms. Studies on the fly immune response have provided evidence for similarities between mammalian and *Drosophila* innate immunity. Hence, this organism appears to be a suitable model system for studying the innate immune defense.

The *Drosophila* immune response consists of multiple cellular and humoral response mechanisms including activation of phagocytosis by specialized blood cells (hemocytes and plasmatocytes), melanization, coagulation, and synthesis of antimicrobial peptides (AMPs) by the fat body (functional equivalent of the mammalian liver). Seven classes of inducible AMPs with activity spectra directed against bacteria or fungi have been identified in *Drosophila*. Their expression is controlled by members of the nuclear factor (NF)-κB family transcription factors, which are activated by two evolutionary conserved signaling pathways, Toll and Imd *(1,2)*. The Toll pathway is activated mainly by Gram-positive bacteria and fungi, whereas the Imd pathway mostly responds to Gram-negative bacterial infection *(3,4)*.

In this chapter, we describe the techniques currently used in our laboratory to monitor the Toll and Imd pathway activities by measuring AMP gene expression after infection. We also include methods used to perform survival analyses, because inactivation of either of these pathways can lead to increased susceptibility to microbial infection.

2. Materials

2.1. Fly Strains

1. Wild-type flies: Canton^S and Oregon^R.
2. Toll pathway mutant strain: spz^{rm7}/TM6C.

    spz^{rm7} is a null mutation generated by ethyl methane sulfonate (EMS) mutagenesis in *spätzle* that encodes the ligand of Toll *(5)*. Several markers of the original stock (M317, Tübingen stock center) including *ebony* were removed by recombination *(6)*. spz^{rm7} homozygous flies are viable, but females are sterile and lay dorsalized embryos.

    Other mutations affecting the intracellular components of the Toll pathway (*Tl, pll, tub, dif, MyD88*) can be used instead of spz^{rm7}.
3. Imd pathway mutant strain: Rel^{E20}, e^+ *(3)*.

    Rel^{E20} is a deletion of *relish* (that also affects a nearby gene), which encodes the NF-κB targeted by the Imd pathway *(7)*.

    Other mutations affecting the Imd pathway (*imd, dredd*...) may be used as positive controls.
4. Strains carrying reporter genes are described in **Table 1**.

2.2. Bacterial Strains

1. Gram-negative bacteria: *Escherichia coli* strain 1106, *Erwinia carotovora carotovora* 15 rif^R (*Ecc15*) and *Pseudomonas entomophila* rif^R. After oral infection, *Ecc15* induces a systemic immune response in larvae but not in adults *(8)*, whereas *P. entomophila* is an entomopathogenic bacterium that can trigger a systemic immune response in both larvae and adults *(9)*.

Table 1
Fly Strains Carrying Reporter Genes

Name	Genotype	Reporter genes	Chromosomal location	References
Dpt-lacZ	$P[Dpt - lac, ry^+]$; $ry506$	Dpt-lacZ	I	(21)
DD1	$y, w, P[Dpt - lacZ, ry^+]$, $P[Drs - GFP, w^+]$	Dpt-lacZ Drs-GFP	I	(19)
Drs-lacZ	$w, P[Drs - lac, w^+]$	Drs-lacZ	I	(19)
DIG	$w; P[Dpt - GFP, w^+]D3 - 2$, $P[Dpt - GFP, w^+]D3 - 4$	Dpt-GFP	III	(9)

Drs, Drosomycin; *Dpt, Diptericin*. Flies carrying each of these reporter genes in combination with Toll and Imd mutations are used as positive controls. The expression of the *Drs-lacZ* reporter gene is strong and shows a significant basal activity in larvae *(19)*.

2. Gram-positive bacteria: *Enterococcus faecalis* and *Micrococcus luteus*. *E. faecalis*, but not *M. luteus*, kills flies after injection *(4)*.
3. Fungi: *Beauveria bassiana* and *Candida albicans*. *B. bassiana*, but not *C. albicans*, kills flies after infection.

2.3. Buffers, Chemicals, and Primers

1. Luria Bertani (LB) and YPGA (5 g of yeast extract, 5 g of Bacto Peptone, 10 g of glucose, and 15 g of agar per liter) media.
2. Corn-meal fly medium, per liter: 8 g agar, 80 g polenta, 40 g yeast, 40 g sucrose, and 53.6 ml moldex.
3. Malt-agar medium, per liter: 1 g peptone, 20 g glucose, 20 g malt extract, and 15 g agar.
4. Apple juice medium, per liter: 17.5 g agar, 25 g sucrose, and 250 ml apple juice.
5. Phosphate-buffered saline (PBS) 10×: 80 g NaCl, 11.5 g Na_2HPO_4, 2 g KH_2PO_4, 2 g KCl, water qsp 100 ml (adjust pH to 7.3/7.5 with 10 M HCl). Autoclave.
6. Glutaraldehyde 25% solution.
7. X-gal: 5-bromo-4-chloro-3-indolyl β-D-galactoside, 5% in dimethylformamide.
8. Staining buffer: 10 mM NaH_2PO_4/Na_2HPO_4, 150 mM NaCl, 1 mM $MgCl_2$, 3.5 mM K_3FeCN_6, 3.5 mM K_4FeCN_6 (adjust to pH 7.2). Add 30 µl of X-gal per ml of solution (*see* **Note 1**).
9. Z buffer: 60 mM Na_2HPO_4, 60 mM NaH_2PO_4, 10 mM KCl, 1 mM $MgSO_4$, 50 mM β-mercaptoethanol (adjust pH to 8 with NaOH).
10. ONPG (*o*-nitrophenol-β-D-galactoside).
11. TRIZOL® Reagent (Invitrogen Cergy Pontoise, France).
12. Chloroform, isopropanol, and ethanol.
13. Random hexamer primers, dNTP.
14. SuperScript™ II Reverse Transcriptase (SSII-RT), RNaseOUT™ Recombinant Ribonuclease Inhibitor (RNaseOUT™ RRI, Invitrogen).

Table 2
Sequence of Oligonucleotide Primers Used for Quantitative PCR (26)

Primer name	Gene	5´–3´ Sequence	PCR efficiency
DiptericinF	Dpt	GCTGCGCAATCGCTTCTACT	1.9864
DiptericinR	Dpt	TGGTGGAGTGGGCTTCATG	
DrosomycinF	Drs	CGTGAGAACCTTTTCCAATATGATG	1.9908
DrosomycinR	Drs	TCCCAGGACCACCAGCAT	
Ribosomal protein 49F	rp49	GACGCTTCAAGGGACAGTATCTG	1.891
Ribosomal protein 49R	rp49	AAACGCGGTTCTGCATGAG	

PCR efficiency has been set up for each couple of primers using the LightCycler® 2.0 System.

15. LightCycler® (LC) 2.0 System, LC FastStart DNA Master SYBR Green I, LC Capillaries, LC Carousel Centrifuge 2.0 (Roche Neuilly sur Seine, France).
16. Oligonucleotide primers (*see* **Table 2**).

3. Methods

The patterns of AMP expression can be classified into three categories *(10)*. (1) Systemic response: Injection of microbes into the body cavity or natural infection by specific microbes (*B. bassiana*, *Ecc15*, *P. entomophila*) induces a strong expression of AMP genes in the fat body and a low expression in a fraction of hemocytes. (2) Inducible local response: Studies have shown that many epithelia can express a subset of AMP genes and that the expression can be enhanced upon natural bacterial infection via the Imd pathway *(8,9,11)*. (3) Constitutive local expression: Several tissues constitutively express AMP genes [e.g., *Drosomycin* (*Drs*) is constitutively expressed in the spermathecae of female flies]. This response is not regulated by the Toll or the Imd pathway *(12)*.

Monitoring AMP gene expression is one of the easiest ways to determine Toll and Imd pathway induction. The *Diptericin* (*Dpt*) gene encodes an antibacterial peptide secreted by the fat body in response to Gram-negative bacterial infection. This gene is tightly regulated by the Imd pathway, and its expression profile provides an accurate readout of Imd pathway activity. *Drs* encodes an antifungal peptide. In contrast to *Dpt*, which is exclusively regulated by the Imd pathway, *Drs* is largely regulated by the Toll pathway but is also partially induced by the Imd pathway after Gram-negative bacterial infection *(3)*. Both genes are generally used as reporter genes of Toll and Imd pathway activation.

Analysis of *Dpt* and *Drs* transcripts by real-time quantitative PCR (RT-qPCR) together with the use of reporter genes is commonly performed to monitor the pattern of AMP gene expression. The methods described below outline *(1)* the preparation of microbes, *(2)* the infection of larvae and adults flies, *(3)* the use of reporter genes, *(4)* the β-galactosidase titration, and *(5)* the RT-qPCR assays to check AMP gene expression.

Besides, mutants deficient in the Toll and/or Imd pathway succumb to microbial infection. Monitoring fly survival after infection is thus another way to estimate the activity of both pathways. Corresponding protocols are described in **Subheading 3.4**.

3.1. Preparation of Microorganisms

3.1.1. Bacterial Pellets

1. Inoculate 50 ml of LB medium or YPGA with the microorganism and let the culture grow overnight at 37°C, except for *P. entomophila*, *Ecc15*, and *C. albicans*, which grow at 30°C.
2. Concentrate microorganisms to $OD_{600nm} = 200$, except for *E. faecalis*, $OD_{600nm} = 30$.
3. Store pellets at 4°C for 1 week.

3.1.2. Fungal Spores

1. Spread *B. bassiana* spores on Petri dishes with malt-agar and incubate at 25–29°C. Use glass beads to obtain well-covered plates. The fungal hyphae will germinate after 3–5 days at 25°C.
2. After 10–30 days, check for the presence of dust-like spores. Store well-sporulated plates at 4°C for infection experiments and keep for 3 months.

To store *B. bassiana* spores, proceed as following:

1. Collect spores into a 50-ml vial by washing the Petri dishes with 10 ml of sterile water and separating spores from hyphal bodies through a funnel lined with glass fibers.
2. Centrifuge spores at $2600 \times g$ for 15 min at 4°C.
3. Quantify the number of spores per ml using a hemacytometer (generally around 10^9 to 10^{12} spores per ml).
4. Store spores in 20% glycerol for several months at 4°C or several years at –70°C.

3.2. Infection of Drosophila Larvae and Adults by Microorganisms

Infection of flies is performed either by introducing microbes directly into the body cavity or by natural infection without injury. Additional information on infection procedures can also be found in **ref. *13***.

3.2.1. Septic Injury

3.2.1.1. ADULTS

Infection of adult flies by septic injury requires the use of a thin metal needle (0.5 mm diameter) mounted on a small handle. Sterile injury of flies is used as an internal control to ensure that pricking has no effect on the studied mutants.

1. Dip the needle into a microbe solution.
2. Prick the lateral side of the thorax of a CO_2-anesthetized fly.
3. Separate the fly from the needle with a brush and put them in a clean vial containing corn-meal fly medium.

3.2.1.2. LARVAE

For larval challenge, a tungsten wire, previously sharpened in a 0.1 M NaOH solution by electrolysis, is used.

1. Wash third instar wandering larvae in water and place them in a small drop of water on a black rubber block.
2. Prick larvae on their posterior lateral side.
3. Transfer larvae to a Petri dish containing apple juice medium and seal the plate carefully.

Following pricking, a dark spot appears at the site of injury corresponding to activation of the melanization cascade, which is a good way to ensure that the flies/larvae have been infected. Larvae are more sensitive to septic injury than adults. In some cases, septic injury can be replaced by injection (*see* **Note 2**).

3.2.2. Natural Infection

Natural infection is a way to infect *Drosophila* in the absence of artificial injury. It relies on oral ingestion of bacteria or on covering flies with spores of invasive fungi.

3.2.2.1. BACTERIAL NATURAL INFECTION

Natural infection is performed with the bacterial strains *Ecc15* or *P. entomophila*. *Ecc15* is not lethal for *Drosophila* and induces a strong systemic immune response only in larvae, whereas in adults it stimulates local expression of AMPs in several epithelial tissues. Conversely, *P. entomophila* is able to induce a local and systemic immune response in both adults and larvae but in addition leads to death of wild-type larvae and flies.

3.2.2.1.1. Adults

1. Dehydrate adults for 2 h in a dry vial in the absence of food.
2. Put a filter paper into a vial with corn-meal fly medium.
3. Hydrate filter paper with a mix of 100 µl of a bacterial pellet ($OD_{600nm} = 200$) and 100 µl of a food solution containing a 5% sucrose solution.
4. Transfer flies into the prepared vial and incubate at 25°C.

It is possible to rehydrate the paper each day with additional food solution.

3.2.2.1.2. Larvae

1. Put 200 µl of a bacterial pellet ($OD_{600nm} = 200$) and 400 µl of crushed banana into a 2-ml microfuge tube.
2. Place approximately 200 third instar larvae in this tube.
3. Mix thoroughly by shaking the tube.
4. Insert a piece of foam into the tube to prevent larvae from wandering away from the bacterial mixture and to ensure air supply at the same time.
5. Place the tube at room temperature for 30 min.
6. Transfer the larvae together with the bacteria to a vial containing a standard corn-meal fly medium and incubate at 29°C.

3.2.2.2. FUNGAL NATURAL INFECTION

Fungal natural infection is usually performed with the entomopathogenic fungus *B. bassiana*, which has the ability to cross the cuticle of insects, through the secretion of proteases and lipases, and causes a significant mortality in wild-type adult flies *(14,15)*.

1. Anesthetize adult flies and place them on a Petri dish containing a sporulating fungus.
2. Hand-shake thoroughly for 30 s to cover the flies with spores.
3. Transfer infected flies to a clean corn-meal fly medium and incubate at 29°C. Vials should be changed every other day.

Larvae can also be rolled on sporulated plates.

Several parameters can influence the infection process (*see* **Note 3**). In cases where the pathogens are lethal for mutant flies, it is recommended to use non-living compounds instead of the entire microorganisms, such as peptidoglycans purified from bacteria *(16,17)*, which are good elicitors of the immune response (*see* **Note 4**).

3.3. Measuring Drosophila AMP Gene Expression

AMP gene expression is measured on samples collected at different time points, according to the expression peak of the monitored gene and the type of infection. Moreover, previously characterized mutants affecting different components of *Drosophila* signaling pathways, like spz^{rm7} and Rel^{E20}, are included in each experiment as internal controls. **Table 3** gives the time points at which flies or larvae should be collected for a time course study. Because the range of AMP gene expression can vary from almost undetectable expression to high levels of transcription, it is essential to include unchallenged and bacterial-challenged wild-type flies in each experiment to determine the expression range of the monitored AMP. It is also important to perform experiments applying the same conditions (sex, age of flies, and time of day; *see* **Note 5**). Combining two independent methods (e.g., reporter genes and RT-qPCR) is the best way to obtain a significant result.

3.3.1. Reporter Genes

The use of reporter genes is an informative method to analyze the expression pattern of *Drosophila* immune genes. Hence, they are the best tools to monitor local expression of AMP genes, which is largely controlled by the Toll and Imd pathways. It is important to ascertain, however, that the reporter gene precisely reproduces the pattern of endogenous gene expression.

3.3.1.1. LacZ Reporter Genes

Drosophila lines carrying a *P*-transgene wherein AMP gene promoter sequences are fused upstream of *lacZ* (*see* **Table 1**) allow the analysis of the expression patterns of the corresponding genes by X-gal staining and the quantification of the expression levels by titration of LacZ activity.

Table 3
Time Points to Collect Flies and Larvae During a Microbial Infection

		Septic injury		Natural infection	
	Monitored peptide gene	Bacteria	Fungi	Bacteria	Fungi
Adults	Drs	6–12–**24**–48 h	24–48 h	–	24–48–72 h
	Dpt	1.5–3–**6–12**–24 h	–	6–**12–24** h	–
Larvae	Drs	6–**12–24** h	6–12–24 h	–	24 h
	Dpt	1.5–3–**6–12**–24 h	–	24 h	–

Time points in bold letters correspond to the highest expression of each AMP gene when flies are infected by non-persisting microbes *(3,27)*.

3.3.1.1.1. X-Gal Staining. This method provides an easy way to study reporter gene expression in larval or adult tissues (*see* **Note 6**).

1. Dissect larvae (or adults) in PBS and quickly place the dissected tissues in 1 × PBS on ice.
2. Fix 10 min in 1 × PBS with 0.5% glutaraldehyde on ice.
3. Wash three times in 1 × PBS on ice.
4. Incubate at 37°C in staining buffer until color develops (from 10 min to overnight).

3.3.1.1.2. β-Galactosidase Titration.

1. Collect three sets of five larvae or adults.
2. Incubate tubes at –20°C to freeze the flies (samples can be stored at this step for assaying later).
3. Thaw samples on ice.
4. Add 250 µl of buffer Z to each tube and homogenize for 30 s using an electric pestle or the PrecellysTM24 automated lyser (Berlin Technologies, Saint-Quentin en Yvelines, France).
5. Add 250 µl of buffer Z to each tube and quickly vortex the samples.
6. Centrifuge at 6000 × *g* for 5 min.
7. Collect the supernatant and estimate the protein concentration with classical methods such as the Bradford assay using bovine serum albumin (BSA) as a protein standard (samples can be stored for few hours on ice).
8. Aliquot 30 µl (*Dpt*) or 10 µl (*Drs*) of the samples into 96-well plates for the readout.
9. Add 250 µl of Buffer Z + ONPG ([ONPG]$_{final}$ = 0.35 mg/ml) to each well rapidly and place plate at 37°C.
10. Monitor β-galactosidase at regular time intervals (2–30 min) by measuring the OD at 420 nm using a microtiter.

According to Miller (1972), the β-galactosidase activity equals $((\Delta OD/\Delta T_{min})_V/(\text{protein concentration in } v)/0.0045$ *(18)*.

3.3.1.2. GFP-REPORTER GENES

Lines carrying AMP gene promoters fused to GFP have been described *(10, 11)*. Thus, AMP gene expression can be monitored in living larvae and adults. GFP reporter genes can be useful to analyze gene expression in tissues that are less accessible with classical staining methods, such as the tracheae (*see* **Note 7**). GFP-expressing *Drosophila* are analyzed directly under a stereomicroscope (e.g., Leica MZFLIII) equipped with epifluorescent illumination (*see* **Note 8**). *GFP* and *lacZ* reporter genes are complementary tools. Lines carrying both a *Drs-GFP* and a *Dpt-lacZ* reporter gene on the X chromosome are currently used to monitor the pattern of expression of both AMP genes in the same animal *(19)*.

3.3.2. RT-qPCR

The northern blot technique has been extensively used to analyze the infection-induced AMP gene expression. A better alternative to study this process is the RT-qPCR method. It consists of a kinetic quantification of a PCR product by measuring a fluorescence signal, which is directly proportional to the amount of accumulated double-stranded DNA. Absolute quantification of the DNA concentration of a sample allows the determination of the corresponding gene expression level. Besides, this technique is safer because it does not require radioactivity.

3.3.2.1. Total RNA Extraction

All steps are performed wearing gloves and using RNase-free products in order to prevent contamination by RNases.

1. Collect 20 adults (or larvae) in a 1.5-ml Eppendorf tube (*see* **Note 9**).
2. Place tubes at –80°C to freeze flies (samples can be stored at this point for assaying later).
3. Thaw samples on ice.
4. Add 250 µl of TRIZOL® Reagent to each tube and homogenize for 30 s using an electric pestle/mortar.
5. Add 250 µl of TRIZOL® Reagent and quickly vortex the samples.
6. Incubate at room temperature for 5 min.
7. Add 100 µl of chloroform and vortex for 2 × 30 s.
8. Centrifuge at 12,000 × g for 15 min at 4°C.
9. Collect 250 µl of the supernatant (aqueous phase) in a clean tube.
10. Precipitate RNA by adding 250 µl of isopropanol stored at room temperature and mix gently but thoroughly by inverting the tube three times.
11. Centrifuge immediately at 12,000 × g for 15 min at 4°C.
12. Remove supernatant by inverting tube.
13. Wash RNA pellet with 400 µl of room-temperature 70% ethanol.
14. Centrifuge at 12,000 × g for 10 min at 4°C.
15. Carefully decant the supernatant without disturbing the pellet.
16. Air-dry the pellet for 10 min and redissolve the RNA in 50 µl of water.
17. Determine the RNA concentration using UV spectrophotometry.

3.3.2.2. cDNA Synthesis

Filter tips are used to prevent contamination of the samples by foreign amplicons. The reverse transcription is performed in a 20 µl final volume. For each experiment, include two control samples: one without RNA and the other without SSII-RT.

1. In a clean 0.2-ml PCR tube add:

    a. RNA (500 ng/µl)   2 µl
    b. Random hexamers (50 ng/µl) 2 µl
    c. dNTPs mix (10 mM)   1 µl
    d. H$_2$O   7.5 µl

2. Denature samples at 65°C for 10 min.
3. Quick-chill on ice and centrifuge briefly.
4. In each tube add:

    a. 5× SSII-RT buffer   4 µl
    b. Dithiothreitol (DTT) (0.1 M) 2 µl
    c. RNaseOUT™ RRI   0.5 µl
    d. SSII-RT   0.5 µl

5. Place the tubes in a thermal cycler (heated lid at 70°C) and run the following program:

    a. Incubate at 42°C for 50 min.
    b. Denature at 60°C for 15 min.

6. Dilute RT products to 1/50 in water.
7. Store diluted RT products at –20°C.

3.3.2.3. RT-QPCR

Expression levels of the *Drs* and *Dpt* genes are used as readouts for the Toll and Imd pathways, respectively, after being normalized to *rp49* (constitutively expressed in *Drosophila*) expression level (*see* **Note 10**). Experiments are performed in 20-µl glass capillaries using the LightCycler® 2.0 System (other systems exist; *see* **Note 11**). Include a control sample containing water instead of RT product.

1. Prepare the following PCR master mix including for each sample:

    a. DNA Master SYBR Green I mix 1 µl
    b. MgCl$_2$ 25 mM   1.2 µl
    c. Forward primer 10 µM   0.5 µl
    d. Reverse primer 10 µM   0.5 µl
    e. PCR-grade H$_2$O   1.8 µl

2. Add in each glass capillary:

    a. PCR master mix   5 µl
    b. Diluted RT product   5 µl

3. Run PCR using the LightCycler® 2.0 System with the following program:
   a. Denaturation: 95°C for 10 min.
   b. Cycling: 95°C for 5 s, 60°C for 5 s, 72°C for 15 s, repeated 40 times.
   c. Melting curve: 95°C for 5 s, 60°C for 5 s, 72°C for 15 s, 70°C→95°C with 0.1°C/s ramping.
4. Monitor the absolute quantification using the LightCycler® software and calculate the gene expression level (R) using the formula: $R_x = [E_{rp49} \exp(CP_{rp49})]/[E_x \exp(CP_x)]$, where E = PCR efficiency, CP = crossing point, x = *Drs* or *Dpt*. The PCR efficiency for each couple of corresponding primers is indicated in **Table 2**.

3.4. Survival Analysis

Survival analyses are performed using several classes of bacteria that exhibit different interactions with *Drosophila*. Highly pathogenic bacteria (such as *E. faecalis*) kill flies after injection of low doses, weakly pathogenic bacteria induce low lethality as in the case of *Ecc15*, and non-pathogenic strains such as *E. coli* cause low lethality. Survival experiments must be carried out in the same conditions (methods, needle, experimenter, time) for each sample of flies. As a prerequisite, the fly strain must also exhibit a good viability in the absence of an immune challenge. A classical survival analysis in adults can be carried out by using *Ecc15*, *E. faecalis*, *C. albicans*, and *B. bassiana* as pathogens and Rel^{E20} and spz^{rm7} as positive controls.

1. Collect twenty 2–4 day-old adults.
2. Infect flies using a needle previously dipped into a bacterial or fungal solution. For *B. bassiana*, proceed as described in **Subheading 3.2.2.2**.
3. Incubate flies at 29°C and transfer them to a fresh vial every 2–3 days to ensure healthy medium conditions.
4. Count dead flies twice a day during 1 week.

Flies that die within 2 h after infection are excluded from the analysis (<5% as a norm; *see* **Note 12**). Infected larvae may be transferred onto a Petri dish containing apple juice agar or fly medium *(20)* to facilitate the sorting of dead versus live animals. A basic survival count for larvae includes the number of pupae and adults that emerge.

4. Notes

1. Preincubation at 37°C and centrifugation of the staining solution can prevent the formation of undesirable crystals. It can be stored for several months at 4°C.
2. Septic injury does not allow for an accurate quantification of injected microbes or microbial components. When necessary, a Drummond's Nanoject™ (automatic) injector can be used to deliver a defined volume of microbial solution. The glass

capillary tips are pulled under high heat, backfilled with mineral oil, and then mounted onto the Nanoject device calibrated for the specified injection volume. Injections of 4–73 nl into an adult fly are possible. This apparatus can be used for the injection of microorganisms, chemicals, and purified bacterial compounds, where highly accurate conditions are required.

3. Parameters that can influence the infection process are:

   The infection procedure. For instance, challenge via septic injury with *B. bassiana* triggers an immune response with characteristics and kinetics different from those elicited by a natural infection with the same fungus, suggesting that a different set of recognition signals is switched on in response to various infection methods. In the case of injection, the needle size and the site of injection may influence the infection process.

   The nature of the microbes and their concentration. The use of various types of bacteria of the same microbial class is recommended to compare the pattern of AMP gene expression, as the latter differs according to the microorganisms used *(15)*.

   Temperature. Flies live well between 16 and 32°C. Temperature can influence both the growth of the microbes and the physiology of the insect. Many fungal species and some *Bacillus*, as well as *Erwinia* species, favor growth at 30°C.

   The rearing conditions. Crowded conditions may induce more trauma in flies. Reduced food amount and contaminated medium can also lead to immune-compromised larvae or adult flies. Thus, all lines to be tested should be taken from healthy stocks to minimize preexisting disadvantages.

4. Bacterial compounds may be diluted in Ringer solution *(20)* or water and injected into flies.

5. Parameters influencing AMP gene expression:

   Fly. Males usually express AMP at a higher level than females after systemic response. Unchallenged females constitutively express *Drs* at a low level *(10)*. This basal level of expression can influence the induction rate after microbial infection.

   Age of flies. We usually use 2–4 day-old flies, because younger flies still possess remainders of the larval fat body. *Dpt* expression is strongly dependent on the larval stage *(21,22)*.

   Time of day. Microarray studies suggest that AMP gene expression can fluctuate with the circadian rhythm *(23)*.

6. *Drosophila* expresses an endogenous galactosidase in the midgut and few other tissues, the optimal pH of which is 6.5. It is possible to use fly lines deficient for the Drosophila *gal* gene to prevent monitoring of this background.

7. Strong induction of *Drs*-GFP in trachea is obtained by naturally infecting larvae with *Ecc15* and keeping them at 18°C for 2 days.
8. The use of GFP reporters has two major drawbacks. First, in order to fluoresce, the protein requires cyclization, which results in a lag time, and thus GFP detection occurs long after that of β-galactosidase with the same promoter. Second, GFP activity is difficult to quantify although quantification of a *Drs-GFP* reporter expression has been used to screen for regulators of the immune response using a spectrophotometer *(24)*.
9. RNA extraction can be also performed from fly/larval tissues like gut or fat body. In this case, collect about 50 samples of the tissue.
10. RT-qPCR analysis allows the course of a PCR to be visualized as a curve similar to a population growth curve. The initial lag phase lasts until the fluorescence signal from the PCR product is greater than the background fluorescence of the probe system. The exponential log phase starts when sufficient product has been generated to be detected above background and ends when the reaction enters the plateau phase and the reaction efficiency falls. Determination of the PCR cycle (called "crossing point" or CP) for which the fluorescence of a sample rises above the background fluorescence allows to calculate the expression level of the studied gene.
11. Alternatively, 96- or 384-well plates can be used. Besides, other suppliers have developed RT-qPCR systems, which can be used. However, the assays must be set up each time a parameter (primer, system, kit assay, etc.) is changed. In this case, PCR efficiency (which depends on the hybridization temperature of the primers), dilution of RT products, and melting curve (which gives the number of amplified products) have to be determined to ensure that only the designated PCR product is amplified.
12. Previous observations showed that survival rates may depend on the genetic background. For example, we noted that homozygous *ebony* fly stocks exhibit a low viability after infection as reported by Flyg et al. *(25)*. In order to examine the survival due exclusively to the mutation under analysis, we choose mutated chromosomes carrying a minimal number of markers *(3)*.

References

1. Lemaitre, B., and Hoffmann, J. (2007) The host defense of Drosophila melanogaster. *Anna Rev Immunol* **25**, 697–743.
2. Hultmark, D. (2003) Drosophila immunity: paths and patterns. *Curr Opin Immunol* **15**, 12–9.
3. Leulier, F., Rodriguez, A., Khush, R. S., Abrams, J. M., and Lemaitre, B. (2000) The Drosophila caspase Dredd is required to resist gram-negative bacterial infection. *EMBO Rep* **1**, 353–8.
4. Rutschmann, S., Kilinc, A., and Ferrandon, D. (2002) Cutting edge: the toll pathway is required for resistance to gram- positive bacterial infections in Drosophila. *J Immunol* **168**, 1542–6.

5. Morisato, D., and Anderson, K. (1994) The spätzle gene encodes a component of the extracellular signaling pathway establishing the dorsal-ventral pattern of the Drosophila embryo. *Cell* **76**, 677–88.
6. Lemaitre, B., Nicolas, E., Michaut, L., Reichhart, J., and Hoffmann, J. (1996) The dorsoventral regulatory gene cassette spätzle/Toll/cactus controls the potent antifungal response in Drosophila adults. *Cell* **86**, 973–83.
7. Hedengren, M., Asling, B., Dushay, M. S., Ando, I., Ekengren, S., Wihlborg, M., and Hultmark, D. (1999) Relish, a central factor in the control of humoral but not cellular immunity in Drosophila. *Mol Cell* **4**, 827–37.
8. Basset, A., Khush, R., Braun, A., Gardan, L., Boccard, F., Hoffmann, J., and Lemaitre, B. (2000) The phytopathogenic bacteria, *Erwinia carotovora*, infects *Drosophila* and activates an immune response. *Proc Natl Acad Sci USA* **97**, 3376–81.
9. Vodovar, N., Vinals, M., Liehl, P., Basset, A., Degrouard, J., Spellman, P. T., Boccard, F., and Lemaitre, B. (2005) Drosophila host defense after oral infection by an entomopathogenic Pseudomonas species. *Proc Natl Acad Sci USA* **102**, 11414–19.
10. Ferrandon, D., Jung, A. C., Criqui, M., Lemaitre, B., Uttenweiler-Joseph, S., Michaut, L., Reichhart, J., and Hoffmann, J. A. (1998) A drosomycin-GFP reporter transgene reveals a local immune response in Drosophila that is not dependent on the Toll pathway. *EMBO J* **17**, 1217–27.
11. Tzou, P., Ohresser, S., Ferrandon, D., Capovilla, M., Reichhart, J. M., Lemaitre, B., Hoffmann, J. A., and Imler, J. L. (2000) Tissue-specific inducible expression of antimicrobial peptide genes in Drosophila surface epithelia. *Immunity* **13**, 737–48.
12. Ryu, J. H., Nam, K. B., Oh, C. T., Nam, H. J., Kim, S. H., Yoon, J. H., Seong, J. K., Yoo, M. A., Jang, I. H., Brey, P. T., and Lee, W. J. (2004) The homeobox gene Caudal regulates constitutive local expression of antimicrobial peptide genes in Drosophila epithelia. *Mol Cell Biol* **24**, 172–85.
13. Tzou, P., Meister, M., and Lemaitre, B. (2002) Methods for studying infection and immunity in Drosophila. *Methods Microbiol* **31**, 507–29.
14. Clarkson, J. M., and Charnley, A. K. (1996) New insights into the mechanisms of fungal pathogenesis in insects. *Trends Microbiol* **4**, 197–203.
15. Lemaitre, B., Reichhart, J., and Hoffmann, J. (1997) *Drosophila* host defense: differential induction of antimicrobial peptide genes after infection by various classes of microorganisms. *Proc Natl Acad Sci USA* **94**, 14614–19.
16. Leulier, F., Parquet, C., Pili-Floury, S., Ryu, J. H., Caroff, M., Lee, W. J., Mengin-Lecreulx, D., and Lemaitre, B. (2003) The Drosophila immune system detects bacteria through specific peptidoglycan recognition. *Nat Immunol* **4**, 478–84.
17. Kaneko, T., Goldman, W. E., Mellroth, P., Steiner, H., Fukase, K., Kusumoto, S., Harley, W., Fox, A., Golenbock, D., and Silverman, N. (2004) Monomeric and polymeric gram-negative peptidoglycan but not purified LPS stimulate the Drosophila IMD pathway. *Immunity* **20**, 637–49.

18. Miller, J. H. (1972) *Experiments in Molecular Genetics*, Cold Spring Harbor Publisher, Cold Spring Harbor.
19. Manfruelli, P., Reichhart, J. M., Steward, R., Hoffmann, J. A., and Lemaitre, B. (1999) A mosaic analysis in Drosophila fat body cells of the control of antimicrobial peptide genes by the Rel proteins Dorsal and DIF. *EMBO J* **18,** 3380–91.
20. Ashburner, M. (1989) *Drosophila, A Laboratory Manual*, Cold Spring Harbor laboratory Press, Cold Spring Harbor.
21. Reichhart, J., Meister, M., Dimarcq, J., Zachary, D., Hoffmann, D., Ruiz, C., Richards, G., and Hoffmann, J. (1992) Insect immunity: developmental and inducible activity of the Drosophila diptericin promoter. *EMBO J* **11,** 1469–77.
22. Meister, M., and Richards, G. (1996) Ecdysone and insect immunity: the maturation of the inducibility of the diptericin gene in Drosophila larvae. *Insect Biochem Mol Biol* **26,** 155–60.
23. McDonald, M. J., and Rosbash, M. (2001) Microarray analysis and organization of circadian gene expression in Drosophila. *Cell* **107,** 567–78.
24. Jung, A. C., Criqui, M. C., Rutschmann, S., Hoffmann, J. A., and Ferrandon, D. (2001) Microfluorometer assay to measure the expression of beta-galactosidase and green fluorescent protein reporter genes in single Drosophila flies. *Biotechniques* **30,** 594–8, 600–1.
25. Flyg, C., Kenne, K., and Boman, H. G. (1980) Insect pathogenic properties of Serratia marcescens: phage-resistant mutants with a decreased resistance to Cecropia immunity and a decreased virulence to Drosophila. *J Gen Microbiol* **120,** 173–81.
26. Irving, P., Troxler, L., Heuer, T. S., Belvin, M., Kopczynski, C., Reichhart, J. M., Hoffmann, J. A., and Hetru, C. (2001) A genome-wide analysis of immune responses in Drosophila. *Proc Natl Acad Sci USA* **98,** 15119–24.
27. Rutschmann, S., Jung, A. C., Hetru, C., Reichhart, J. M., Hoffmann, J. A., and Ferrandon, D. (2000) The Rel protein DIF mediates the antifungal but not the antibacterial host defense in Drosophila. *Immunity* **12,** 569–80.

23

Investigating the Involvement of Host Factors Involved in Intracellular Pathogen Infection by RNAi in *Drosophila* Cells

Hervé Agaisse

Summary

Intracellular pathogens represent a serious threat to human health. Although the biology and the virulence factors involved in intracellular bacterial infection are relatively well documented, little is known about the host factors involved in the infection process. This situation is mainly due to the difficulty of conducting extensive genetic analyses in the targeted host cells and points to the need for developing genetic systems to model the infection process. Here, we describe a method (amenable to high-throughput analysis) that allows for the study of host factors involved in intracellular pathogen infection by using dsRNA-mediated gene expression knockdown [RNA interference (RNAi)] in *Drosophila* cell lines.

Key Words: Host/pathogen interaction; intracellular pathogen; innate immunity; *Drosophila*; *Listeria*; RNAi; immunofluorescence microscopy; high-throughput screening.

1. Introduction

Drosophila S2 cells resemble embryonic hemocytes/macrophages and are able to mount an innate immune response. For example, addition of lipopolysaccharide (LPS) to S2 cells leads to rapid expression of antimicrobial peptides *(1)*. Genome-wide expression-profiling studies have provided a broad overview of the transcriptional response to LPS *(2,3)*. Recent analyses of *Listeria* infection in S2 cells have also revealed striking similarities to infection in mammalian macrophages *(4,5)*, suggesting that S2 cells may be a good system to model human infections with intracellular pathogens. An important discovery was made by Clemens et al. *(6)*, who reported that the simple addition of dsRNA to S2 cells in culture reduces or eliminates the expression of target genes by RNA interference (RNAi), thus efficiently phenocopying loss-of-function mutations. The ease of RNAi methodology in S2 cells set the stage for conducting extensive genetic analyses of any cell-based process.

Here, we present an application of the RNAi methodology in *Drosophila* S2 cells to investigate the host factors involved in infection by intracellular pathogens, such as *Listeria (7)*.

2. Materials

2.1. Bacterial and Drosophila Cell Culture

1. Brain–heart infusion medium (BHI, Difco, BD Biosciences, Franklin Lakes, NJ, USA).
2. Schneider medium (Gibco).
3. Fetal bovine serum (FBS, Gibco, Invitrogen, Carlsbad, CA, USA).
4. 75-cm^2 culture flask (Corning, NY, USA 430641).

2.2. dsRNA Synthesis

1. Primers (*see* **Table 1**).
2. Trizol (Invitrogen Carlsbad, CA, USA).
3. Superscript II reverse transcription (RT) reaction kit (Invitrogen).
4. DNase I (Invitrogen).
5. Takara Taq polymerase (Fisher Morris Plains, NJ, USA).
6. 5× T7 MEGAscript kit for in vitro transcription (Ambion, Austin, TX, USA, Cat. number B1334-5).
7. RNAeasy purification kit (Qiagen, Valencia, CA, USA, Cat. number 74104)

2.3. RNAi and Infection and Imaging

1. 384-well plates, black with clear bottom (Corning 3712).
2. Gentamycin (50 mg/ml stock solution) (Sigma-Aldrich, St. Louis, MO, USA, G1397).

Table 1
Specific Primers Used for Generating *lacZ* and *RAB1* Double-Stranded RNA

Gene	Forward primer	Reverse primer
lacZ	GCATTATCCGAACCATCCGCTG	CAGAACTGGCGATCGTTCGGC
RAB1	CACCATCACGTCTTCATATTAT	TGGTGTGGTCGACTACTTTC

Note that a T7 tail needs to be added for in vitro transcription (see **Subheading 3.3**).

3. Dulbecco's phosphate-buffered saline (PBS) buffer (Gibco).
4. 24-pin vacuum manifold (Drummond, Broomall, PA, USA).
5. Formaldehyde (methanol free) (Polysciences, Inc., Warrington, PA, USA, Cat. number 18814).
6. Hoechst DNA stain (Molecular Probes Invitrogton, Carlsbad, CA, USA).

3. Methods

The approach presented here relies on the efficient production of double-stranded RNA (dsRNA) in order to interfere with the expression of a given gene by RNAi *(8)*. The analysis of the infection process is conducted by fluorescence microscopy using green fluorescent protein (GFP)-expressing *Listeria* but can be easily conducted by immunostaining for any intracellular pathogen not amenable to genetic manipulation. *Listeria* has been studied for over four decades as a model for understanding mechanisms of intracellular pathogenesis and the associated aspects in the field of cell biology *(9)*. *Listeria* can readily infect professional phagocytes, such as mammalian macrophages and *Drosophila* S2 cells. In S2 cells, following internalization, *Listeria* escapes the phagocytic vacuole and gains access to the cytosol, where replication occurs *(4,5)*. After several hours of infection, the invaded cells are packed with bacteria

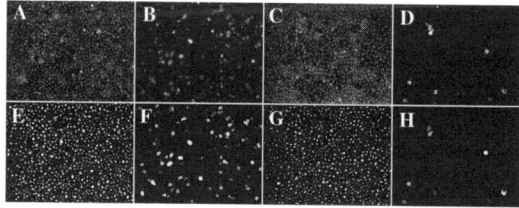

Fig. 1. Analysis of the RAB1 involvement in *Listeria* infection by RNA interference (RNAi) in *Drosophila* S2 cells. Cells were treated with double-stranded RNA (dsRNA) corresponding to *lacZ* (**A** and **B**) or *RAB1* (**C** and **D**) and infected with green fluorescent protein (GFP) *Listeria* as described in **Subheading 3.**, fixed, and treated with the Hoechst DNA stain. Nuclei were visualized in the DAPI channel (**A** and **C**) and bacteria were detected in the fluorescein isothiocyanate (FITC) channel (**B** and **D**). Images in the bottom panel (**E, F, G,** and **H**) show segmentation analysis of the corresponding images in the top panel (**A, B, C** and **D**, respectively). The MetaMorph software was used to count objects in **E, F, G,** and **H** and determine that **A** and **C** displays 1304 and 1165 nuclei, respectively, whereas **B** and **D** display 114 and 8 infected cells, respectively. We conclude that, in the control experiment, 9% of the cells are infected whereas this number drops to <0.7% in the RAB1 knockdown experiment, probably reflecting a strong entry defect.

and are easily identified by low-magnification immunofluorescence microscopy (*see* **Fig. 1A and B**). The following instructions describe an RNAi approach to show the role of the small GTPase RAB1 in this process.

3.1. RNA Extraction

1. Grow S2 cells in Schneider medium supplemented with 10% FBS.
2. When reaching 5×10^6 to 10×10^6 cells/ml, harvest floating cells and spin down.
3. Resuspend cell pellet in Trizol (1 ml for 5×10^6 cells).
4. Homogenize sample and add 200 µl of chloroform per ml of Trizol.
5. Shake for 30 s and centrifuge at $16,000 \times g$ for 15 min at 4°C.
6. Harvest the top phase (500 µl) and mix to 500 µl isopropanol.
7. Centrifuge at $16,000 \times g$ for 15 min at 4°C.
8. Remove isopropanol and wash the visible white pellet with 75% ethanol.
9. Air-dry pellet for 2 min and resuspend in 50 µl of diethylpyrocarbonate (DEPC) treated water.
10. Determine RNA concentration (UV spectrophotometer).

3.2. Reverse Transcription Reaction

1. Add 2 µl of DNase I to 10 µg of total RNA in 20 µl DEPC-treated water final.
2. Incubate at 37°C for 30 min and inactivate enzyme at 70°C for 10 min.
3. Add 1 µl of oligo dT and 9 µl water.
4. Incubate for 5 min, at 70°C, and cool down on ice.
5. Add 10 µl 5× RT buffer, 4 µl dNTP, 5 µl DTT, and 0.5 µl RNasin.
6. Add 2 µl of Super Script II RT and incubate at 42 °C for 2 h.

3.3. PCR Amplification

1. Design gene-specific primers with 5´ tails containing the T7 promoter sequence (TAATACGACTCACTATAGG) to amplify a ~500 bp target sequence.
2. Use 1 µl of the RT reaction as a template to set up a 50 µl PCR reaction using the Takara Taq polymerase and 20 pmol of the corresponding primers according to the recommendation of manufacturer.

The PCR program is as follows:

1. 94°C for 3 min.
2. 94°C for 30 s.
3. 60°C (–1°C/cyc) for 45 s.
4. 72°C for 1 min.
5. Repeat **steps 2–4** 10 times.
6. 94°C for 30 s.
7. 50°C for 45 s.
8. 72°C for 1 min.

9. Repeat **steps 6–8** 23 times.
10. 72°C for 10 min.
11. Hold at 4°C.

3.4. dsRNA Synthesis

1. Use 5 μl of the PCR reaction to perform the in vitro transcription reaction according to the recommendation of the manufacturer (Ambion), for 16 h to optimize yield.
2. Add 2 μl of DNase to the in vitro transcription product incubate at 37°C for 30 min and purify 100 μg dsRNA with Qiaquick RNAeasy kits according to the recommendation of manufacturer.
3. Determine dsRNA concentration and subject an aliquot of the reaction to quality control by gel electrophoresis analysis. dsRNA should run as a single ~ 500 bp band. A faint smear may be visible.

3.5. RNAi Treatment

1. Grow S2 cells at 25°C in Schneider medium supplemented with 10% FBS until reaching 5×10^6 to 10×10^6 cells/ml.
2. Centrifuge cells at $200 \times g$ for 5 min and resuspend at 2×10^6 cells/ml in serum-free Schneider medium.
3. Add 5 μl of the dsRNA stock solution (0.05 mg/ml) to the wells of a 384-well plate.
4. Add 10 μl of the cell suspension (20,000 cells/well), spin down at $200 \times g$, and incubate for 1 h at 25°C.
5. Add 30 μl Schneider medium supplemented with 10% FBS.
6. Incubate for 3 days at 25°C in humidified chamber.

3.6. Listeria Infection in 384-Well Plate Format

The day before the infection, set up an overnight *Listeria* culture in BHI medium at 30°C without shaking.

The day of the infection:

1. Add 50 μl of the overnight *Listeria* culture to 4 ml of Schneider medium supplemented with 10% FBS.
2. Add 10 μl of the diluted bacterial suspension to a given well.
3. Incubate for 1 h at 30°C to allow for bacterial invasion.
4. Add 50 μl of the Gentamycin stock solution to 4 ml of Schneider medium supplemented with 10% FBS.
5. Dispense 10 μl per well to kill extracellular bacteria.
6. Incubate at 30°C for 18 h in humidified chamber.

3.7. Preparing Samples for Imaging

1. At 18 h post-infection, centrifuge the 384 well plates at $200 \times g$ for 5 min.
2. Remove medium carefully using a 24-pin vacuum manifold.

3. Fix cells by adding 10 µl per well of PBS 4% formaldehyde for 20 min at room temperature or overnight at 4°C.
4. Wash wells with 30 µl of PBS.
5. Remove carefully PBS and stain with Hoechst stain in PBS 0.1% Triton for 20 min.
6. Remove Hoechst stain and add 10 µl of PBS with 0.1% sodium azide to prevent contamination.
7. Seal plates with aluminium foil and keep at 4°C until analysis.

3.8. Low- and High-Throughput Fluorescence Microscopy

The following instructions assume the use of a Nikon TE2000 microscope with a Plan fluor ×20/0.45 lens objective, a IEEE1394 CCD Camera C4742-80-12AG (Hamamatsu Corporation UAS, Bridgewater, USA), and MetaMorph acquisition software (Molecular Devices).

1. Use the images in the 4;6-diamidino-2phynylindole (DAPI) channel to determine the number of cells in a given field, thereby determining the potential effect of a given dsRNA on cell growth. Determine cell numbers using the count nuclei drop-in in the MetaMorph software (*see* **Fig. 1**).
2. Determine the extent of the infection by counting the number of "green" cells with the count nuclei drop-in preset with appropriate parameters with regard to the size of objects to be detected (8–15 µm in this case).
3. The treatment of large sets of samples, such as several 384-well plates, will be facilitated by the use of an automated imaging system controlling the position of the stage (Prior), the filter wheel and shutter (Ludl Electronic Products, Hawthorne, USA), and the focusing motor (Piezo Systems Inc., Cambridge, USA). Typically, perform automated focusing in the DAPI channel on Hoechst-stained DNA. Collect images from the stained nuclei and GFP-expressing bacteria at several sites in each well using preset exposure times. The system repeats the task across the entire plate. The collected data will be analyzed subsequently using journals designed to run the appropriate drop-ins for sequential analysis of the corresponding pictures.

4. Notes

1. RNA extraction with phenol-based reagent, such as Trizol, must be conducted in a fume hood. The cell pellet is easy to resuspend by pipetting up and down with a P1000. Upon recovery of the top phase after centrifugation, use a blue tip and a P1000 set to 500. This will avoid the recovery of any contaminant from the bottom phase and (white) interface.
2. The RNA pellet must be air-dried. An over-dried pellet will however be difficult, if not impossible, to resuspend in water. When a large amount of RNA is recovered, it may be necessary to heat the suspension at 75°C for 10–15 min to get RNA into solution and/or add more water.

3. The PCR amplification of the DNA fragment corresponding to the targeted genes can also be achieved by using genomic DNA as a template. We recommend, however, the use of an RT product generated from S2 cells RNA, as described in this protocol, as it ensures the actual expression of the targeted gene and the actual presence of the targeted sequence in the corresponding transcript.
4. PCR amplification may be difficult for some genes because (1) the expression of the targeted gene may be very low or the gene may not be expressed at all in S2 cells and (2) the prediction of the mRNA sequence may be incorrect.
5. It is important to confirm by BLAST analysis (http://flybase.net/blast/) that the targeted DNA regions do not display any identical 19 nucleotide stretches in the entire fly genome in order to avoid any off-target effect.
6. When possible, target a region harboring an intron, so it is easy to differentiate between the results of PCR amplification of (contaminant) genomic DNA and PCR amplification of an actual cDNA, based on difference in size of the PCR products.
7. It is highly recommended to clone and sequence the product of PCR amplification and confirm that it does correspond to the targeted sequence.
8. Starting the experiment with 20,000 cells will result in 100% confluency, but not overgrowth after 3 days of RNAi treatment. This is important because overgrown S2 cells stop being adherent and start floating in the medium. As a consequence, sucking up the medium upon the fixation procedure will result in the loss of the majority of the cells. To minimize this problem, it is important (1) to centrifuge the plates before sucking up the medium, (2) to set up the vacuum to the minimum required, and (3) to cover the edges of the manifold with several layers of rubber, so that the tips of the pins do not touch the bottom of the wells and some liquid remains in the well after aspiration.
9. If required (for instance, to conduct 4 day RNAi treatment), it is possible to seed fewer cells than 20,000 cells/well. We have been consistently successful in using as few as 15,000 cells/well. Depending on the physiological states of the cells, we observed, however, striking differences in cell growth from one experiment to the other. It is recommended to start from a flask of cells that just reached 5×10^6 to 10×10^6 cells/ml.
10. Cells in wells located on the edge of the 384-well plate tend to grow slower (edge effect). We observed that this effect may be attenuated by growing the cells in a "humidified chamber," a plastic box containing wet paper towels.
11. *Listeria* infections are performed at a multiplicity of infection of 5:1. These conditions were determine empirically as a nice balance between having enough infected cells per field of view and not having too many non-internalized bacteria making image analysis difficult. Note that procedures involving high MOI and clearing the non-internalized bacteria by multiple washing steps are not to be recommended: too many cells are washed off in the process.

Acknowledgments

This work was supported by the Mallinckrodt foundation and NIH Grant 1R21 NS056876-01.

References

1. Samakovlis, C., Asling, B., Boman, H. G., Gateff, E., and Hultmark, D. (1992) In vitro induction of cecropin genes–an immune response in a Drosophila blood cell line. *Biochem Biophys Res Commun* 188, 1169–75.
2. Boutros, M., Agaisse, H., and Perrimon, N. (2002) Sequential activation of signaling pathways during innate immune responses in Drosophila. *Dev Cell* 3, 711–22.
3. Silverman, N., Zhou, R., Erlich, R. L., Hunter, M., Bernstein, E., Schneider, D., and Maniatis, T. (2003) Immune activation of NF-kappaB and JNK requires Drosophila TAK1. *J Biol Chem* 278, 48928–34.
4. Cheng, L. W., Viala, J. P., Stuurman, N., Wiedemann, U., Vale, R. D., and Portnoy, D. A. (2005) Use of RNA interference in Drosophila S2 cells to identify host pathways controlling compartmentalization of an intracellular pathogen. *Proc Natl Acad Sci USA* 102, 13646–51.
5. Mansfield, B. E., Dionne, M. S., Schneider, D. S., and Freitag, N. E. (2003) Exploration of host-pathogen interactions using Listeria monocytogenes and Drosophila melanogaster. *Cell Microbiol* 5, 901–11.
6. Clemens, J. C., Worby, C. A., Simonson-Leff, N., Muda, M., Maehama, T., Hemmings, B. A., and Dixon, J. E. (2000) Use of double-stranded RNA interference in Drosophila cell lines to dissect signal transduction pathways. *Proc Natl Acad Sci USA* 97, 6499–503.
7. Agaisse, H., Burrack, L. S., Philips, J. A., Rubin, E. J., Perrimon, N., and Higgins, D. E. (2005) Genome-wide RNAi screen for host factors required for intracellular bacterial infection. *Science* 309, 1248–51.
8. Fire, A., Xu, S., Montgomery, M. K., Kostas, S. A., Driver, S. E., and Mello, C. C. (1998) Potent and specific genetic interference by double-stranded RNA in Caenorhabditis elegans. *Nature* 391, 806–11.
9. Portnoy, D. A., Auerbuch, V., and Glomski, I. J. (2002) The cell biology of Listeria monocytogenes infection: the intersection of bacterial pathogenesis and cell-mediated immunity. *J Cell Biol* 158, 409–14.

24

Models of *Caenorhabditis elegans* Infection by Bacterial and Fungal Pathogens

Jennifer R. Powell and Frederick M. Ausubel

Summary

The nematode *Caenorhabditis elegans* is a simple model host for studying the relationship between the animal innate immune system and a variety of bacterial and fungal pathogens. Extensive genetic and molecular tools are available in *C. elegans*, facilitating an in-depth analysis of host defense factors and pathogen virulence factors. Many of these factors are conserved in insects and mammals, indicating the relevance of the nematode model to the vertebrate innate immune response. Here, we describe pathogen assays for a selection of the most commonly studied bacterial and fungal pathogens using the *C. elegans* model system.

Key Words: *Caenorhabditis elegans*; innate immunity; pathogenesis; *Cryptococcus neoformans*; *Drechmeria coniospora*; *Enterococcus faecalis*; *Microbacterium nematophilum*; *Pseudomonas aeruginosa*; *Salmonella enterica*; *Serratia marcescens*; *Staphylococcus aureus*; *Yersinia pestis*; *Yersinia pseudotuberculosis*.

1. Introduction

The self-fertilizing nematode *Caenorhabditis elegans* lives in the soil, where it feeds on microorganisms. Its small size, short generation time, powerful genetics, ease of handling, invariant developmental lineage, sequenced genome, and reverse genetic tools have formed the basis of a worldwide community of *C. elegans* researchers since its initial popularization in 1974 by Sydney Brenner as a model genetic organism (*1*). It is now one of the best-understood metazoans. Only recently, however, has *C. elegans* been used as a model to study bacterial and fungal pathogenesis and innate immunity.

A remarkably large number of human microbial pathogens have been shown to infect *C. elegans*, including the Gram-negative bacterial pathogens *Pseudomonas aeruginosa (2)*, *Salmonella enterica (3,4)*, *Serratia marcescens (5)*, and *Yersinia pestis (6)*; the Gram-positive bacterial pathogens *Staphylococcus aureus (7)* and *Enterococcus faecalis (8)*; and the pathogenic yeast *Cryptococcus neoformans (9)*. Additionally, several putative natural pathogens of *C. elegans* have been identified, including the Gram-positive bacterium *Microbacterium nematophilum (10)* and the fungus *Drechmeria coniospora (11,12)*. When exposed to these and other pathogens *(13)*, worms show visible signs of infection (e.g., bacterial biofilm formation, tail swelling, and massive internal accumulation of pathogens) and/or die much more quickly (mean survival = 1.5–7 days) than they do when they are fed the relatively nonpathogenic *Escherichia coli* strain OP50 (mean survival = 15–17 days).

Several lines of evidence suggest the *C. elegans* host–pathogen system is a legitimate model of the infectious process and innate immune response in vertebrates and other species. First, the longevity of worms grown on heat- or antibiotic-killed pathogenic bacteria or yeast is approximately equivalent to the longevity of worms grown on the standard laboratory food source *E. coli* OP50 *(13)*. This suggests that the pathogens are not simply reducing the lifespan of the worms as a side effect of poor nutrition but rather are killing them through an infectious process that requires live microbes. Second, many mutations in pathogenic bacteria and yeast that adversely affect pathogenesis in mammalian hosts also result in diminished killing of *C. elegans*. Conversely, when *C. elegans* was used to identify less virulent mutants in transposon mutation libraries of several different pathogens, including *P. aeruginosa*, *E. faecalis*, *S. enterica*, *S. aureus*, *S. marcescens*, and *C. neoformans*, many of these mutants exhibited reduced virulence in mammalian models of infection *(13)*. Finally, components of *C. elegans* innate immune signaling pathways are conserved in insects and mammals, suggesting the existence of an ancient immune response in a common ancestor of these diverse species *(14)*. Thus, the observation that many pathogens cause disease in both *C. elegans* and mammals through a shared set of pathogen virulence factors and host defense genes validates the *C. elegans* pathogenicity model as a simple host system that can be used empirically to identify novel pathogen virulence factors and components of the innate immune system conserved in other species. For recent reviews, see **refs. *13*, *15*,** and ***16***.

At least two distinct general mechanisms exist by which pathogens kill *C. elegans*: infection- and toxin-associated killing. Strains of *P. aeruginosa* have been shown to kill worms by both mechanisms *(17–19)*. As mentioned above and described in greater detail below in **Subheadings 3.2.1** and **3.3.1**, when *P. aeruginosa* strain PA14 is grown on a low osmolarity minimal medium, killing of *C. elegans* occurs relatively slowly over the course of several days in a manner called "slow killing" (SK) *(2)*. In addition to this infection-like mode of killing, when PA14 is grown on high osmolarity rich media, it causes death of the nematodes within a matter of hours. This "fast killing," which is mechanistically distinct from SK and is not associated with the accumulation of bacteria in the worm intestine, is due to the production of one or more diffusible, low-molecular-weight toxins of the pyocyanin-phenazine class *(19)*. A different *P. aeruginosa* strain, PAO1, when grown on rich media, kills nematodes by producing a large amount of hydrogen cyanide *(17,18)*. Many other species can kill worms using a variety of toxins, including *Enterococcus faecium* through hydrogen peroxide *(20)* and *Bacillus thuringiensis* through a set of pore-forming Crystal toxins *(21,22)*. Although toxin-mediated killing is an interesting phenomenon that merits further study, this chapter will focus on the infection (i.e., SK) of *C. elegans* by various bacterial and fungal pathogens.

2. Materials

2.1. Equipment

1. Worm pick: Fashion a wire worm pick by inserting 2–3 cm of 32-gauge platinum wire into a holder. The holder can be a disposable glass Pasteur pipette, which is attached to one end of the wire by holding it over a flame to melt the glass around the wire. Alternatively, the wire can be screwed into the handle of a bacterial inoculating loop. If desired, flatten the end of the pick with pliers or a hammer.
2. Microspreader: A glass microspreader may be made by holding a disposable glass Pasteur pipette over a flame about 1–2 cm from the tip until it bends into an L shape. Seal the very end by holding the tip of the pipette just at the edge of the flame. Sterilize before each use by dipping the microspreader in >75% ethanol and quickly passing it through a flame to burn off the excess ethanol.

2.2. Antibiotics and Buffers

1. Nalidixic acid (Nal): Dissolve in 0.1 M NaOH to a stock concentration of 10 mg/ml, filter sterilize, and store in the dark at –20°C.
2. Kanamycin (Kan): Dissolve in dH_2O to a stock concentration of 80 mg/ml, filter sterilize, and store at –20°C.
3. Ampicillin (Amp): Dissolve in dH_2O to a stock concentration of 100 mg/ml, filter sterilize, and store at –20°C.
4. Streptomycin (Strep): Dissolve in dH_2O to a stock concentration of 50 mg/ml, filter sterilize, and store at –20°C.

5. Gentamycin: Dissolve in dH$_2$O to a stock concentration of 10 mg/ml, filter sterilize, and store at –20°C.
6. M9W buffer: Mix in MilliQ ddH$_2$O—0.3% (w/v) KH$_2$PO$_4$, 0.6% (w/v) Na$_2$HPO$_4$, and 0.5% (w/v) NaCl. Autoclave at 121°C for 30 min, cool to 60°C, and add filter sterilized 1 mM MgSO$_4$.

2.3. Media

1. Luria-Bertani (LB) broth: Dissolve in MilliQ ddH$_2$O—1% (w/v) Bacto-tryptone, 0.5% (w/v) Bacto-yeast, and 0.5% (w/v) NaCl. Adjust the pH to 7.0 using 1M NaOH. Autoclave at 121°C for 40 min. Store at room temperature for up to 6 months; add appropriate antibiotic just before use.
2. LB agar plates: Mix in MilliQ ddH$_2$O—1% (w/v) Bacto-tryptone, 0.5% (w/v) Bacto-yeast, 0.5% (w/v) NaCl, 1.5% agar. Adjust the pH to 7.5 and autoclave at 121°C for 40 min. Cool agar to 55°C, add appropriate antibiotic, and pour 20 ml in each 10-cm Petri dish. Store at 4°C for up to 3 months.
3. Nematode growth medium (NGM) (general *C. elegans* growth; *S. marcescens*- and *D. coniospora*-killing assays; *Yersinia*, *M. nematophilum* pathogenesis assays): Mix in MilliQ ddH$_2$O—0.25% (w/v) Bacto-Peptone (BD, Franklin Lakes, NJ, USA), 0.3% (w/v) NaCl, and 2% (w/v) Bacto-Agar (BD). Autoclave at 121°C for 40 min, and then add filter sterilized 5 μg/ml cholesterol (make 5 mg/ml stock solution in ethanol), 1 mM MgSO$_4$, 25 mM KH$_2$PO$_4$ (pH 6). If desired, also add 0.1% nystatin to reduce fungal contamination of the plates. Cool to 55°C, and then add filter sterilized 1 mM CaCl$_2$ and 0.01875% Strep. Pour 10 ml of NGM agar in each 6-cm Petri dish or 20 ml of NGM agar in each 10-cm Petri dish.
4. SK plates (*P. aeruginosa*- and *S. enterica*-killing assays): Mix in MilliQ ddH$_2$O—0.35% (w/v) Bacto-Peptone (BD), 0.3% (w/v) NaCl, 5 μg/ml cholesterol (make 5 mg/ml stock solution in ethanol), and 2% (w/v) Bacto-Agar (BD). Autoclave for 25 min, cool to 55°C, and then add 1 mM CaCl$_2$ (make 1 M stock in dH$_2$O, filter sterilize), 1 mM MgSO$_4$ (make 1 M stock in dH$_2$O, filter sterilize), and 25 mM KH$_2$PO$_4$ (make 1 M stock, pH 6, in dH$_2$O, filter sterilize). Pour 4 ml of SK agar in each 3.5-cm Petri dish (Falcon, San Jose, CA, USA). Store plates at 4°C for up to 1 month.
5. Tryptic soy broth (TSB): Dissolve TSB mix (Difco, Franklin Lakes, NJ, USA) in MilliQ ddH$_2$O and autoclave at 121°C for 20 min. Store at room temperature for up to 6 months. Add appropriate antibiotic just before use.
6. Tryptic soy agar (TSA) plates (*S. aureus*-killing assay): Dissolve TSA mix (Difco) in MilliQ ddH$_2$O and autoclave at 121°C for 20 min. Cool to 55°C and add sterile filtered NaI to a final concentration of 5–10 μg/ml. Pour 4 ml of agar in each 3.5-cm Petri dish or 20 ml in each 10-cm Petri dish. Store plates at 4°C in the dark for up to several months (*see* **Note 1**).
7. Brain–heart infusion (BHI) broth: Dissolve BHI mix (Difco) in MilliQ ddH$_2$O and autoclave at 121°C for 30 min. Store at room temperature for up to 6 months. Add appropriate antibiotic just before use.

8. BHI agar plates (*E. faecalis*- and *C. neoformans*-killing assays): Dissolve BHI agar (Difco) mix in MilliQ ddH$_2$O and autoclave at 121°C for 30 min. Cool to 55°C and add sterile filtered Kan to a final concentration of 80 μg/ml. Pour 4 ml of agar in each 3.5-cm Petri dish or 20 ml in each 10 cm Petri dish. Store plates at 4°C for up to several months.
9. YPD broth: Dissolve in Milliq ddH2O–1% (w/v) Bacto-yeast, 2% (w/v) Bacto-peptone, and 2% (w/v) dextrose. Autoclave at 121oC for 40 min. Store at room temperature for up to 6 months; add appropriate antibiotic just before use.

Table 1
Strains Most Commonly Used in Pathogenicity Assays

Species	Strain	Mean survival of *C. elegans* (13)	Strain reference
Caenorhabditis elegans	N2 Bristol (wild-type)	N/A	(1)
	fer-15(b26)		(28)
	fer-15(b26); fem-1(hc17)		(29)
Cryptococcus neoformans	H99	5–7 days	(30)
Drechmeria coniospora	NY [5]	2–4 days	(12)
Enterococcus faecalis	MMH594	3–7 days	(31)
	OG1RF		(32)
Escherichia coli	OP50	15–17 days	(1)
Microbacterium nematophilum	CBX102	N/A	(10)
Pseudomonas aeruginosa	PA14	1.5–3 days	(33)
Salmonella enterica	SL1344	4 days	(34)
Serratia marcescens	Db11	2–6 days	(35)
Staphylococcus aureus	NCTC8325	1.5–3 days	(36)
Yersinia pestis	KIM6+ (see **Note 2**)	N/A	(37)
Yersinia pseudo-tuberculosis	YPIII	N/A	(38)

2.4. Strains

Listed in **Table 1** are the strains most commonly used in pathogenicity assays. In most cases, multiple different strains of the same pathogen were tested and one was selected as the preferred strain. Different strains exhibit different phenotypes in the *C. elegans* assay, and the recommended strain was frequently chosen because it was found to be especially virulent for this assay. For example, *P. aeruginosa* strain PA14 but not PA01 is highly virulent under the conditions described below in **Subheading 3.2.1**. Similarly, biofilm formation by *Y. pestis* and by *Yersinia pseudotuberculosis* is highly strain-dependent *(23)* (C. Darby, personal communication). The growth and storage conditions of many pathogens affect their virulence. In general, the pathogens should be streaked from a frozen stock and the plates stored at 4°C for no more than a week to ensure potency.

3. Methods
3.1. C. elegans Growth
3.1.1. Worm Growth and Maintenance

1. Seed 6-cm NGM plates with 50–200 µl of a saturated culture of *E. coli* OP50 grown in LB broth at 37°C overnight. If desired, spread the culture slightly, but do not allow it to reach the sides of the plate.
2. Allow the OP50 lawn to dry and grow for 1–3 days at room temperature.
3. Maintain populations of worms by transferring wild-type (N2) hermaphrodites to a fresh OP50 plate each day (*see* **Note 3**).
4. Transfer worms to new plates by picking up two to five worms with a sterilized worm pick or by using a sterilized spatula to cut a chunk of agar from a plate on which the worms have recently starved and placing them on a fresh plate (*see* **Note 4**). *C. elegans* are typically grown at 20°C but can be grown from 15 to 25°C. The generation time of *C. elegans* at 25°C is less than at 15°C.

3.1.2. Worm Synchronization

1. Wash gravid hermaphrodites, which contain many fertilized eggs visible in the uterus, off one or more plates with M9W buffer into a 15-ml Falcon tube.
2. Spin the worms in a clinical centrifuge at $1500 \times g$ for 30 s to pellet; aspirate the supernatant until approximately 500 µl of worms and liquid remain.
3. Add 400 µl of household bleach and 100 µl of 5N NaOH. Shake vigorously.
4. Monitor the lysis of the worm cuticle by laying the Falcon tube on a dissecting microscope. The worms will first break in half, releasing their eggs. Adult and larval cuticles will continue to dissolve, but the unhatched progeny are partially protected from the bleach by their eggshells. Shake periodically. After about 3–4 min, almost but not all the adult worms should have dissolved. Do not allow the reaction to proceed longer than 4 min or the eggs will begin to die.

5. Quickly add 14 ml of M9W buffer to dilute the bleach. Pellet the eggs in a clinical centrifuge at 1500 × g for 30 s. Pour off or aspirate the supernatant.
6. Wash the eggs twice by adding 14 ml of M9W buffer, shaking to resuspend the eggs, and spinning to pellet the eggs each time. After two washes, resuspend the eggs in 5 ml of M9W buffer.
7. Rotate the 15-ml tubes on a rotisserie at room temperature for at least 12 h. The eggs will hatch and larvae will arrest development in the absence of food. Most strains can be kept as arrested L1 larvae on the rotator at room temperature for up to 5 days. Wild-type arrested L1 larvae will resume synchronous development when dropped onto NGM plates seeded with OP50 and will reach the L4 larval stage in 48 h at 20°C.

3.2. Preparation of Pathogenicity Assay Plates

3.2.1. Preparation of P. aeruginosa-Killing Plates

1. Streak *P. aeruginosa* PA14 bacteria from frozen stock onto a 10-cm LB agar plate. Grow at 37°C for 12–18 h and store the streaked plate at 4°C for no longer than a week.
2. Inoculate 2–5 ml of LB broth with a single colony of PA14. Grow with aeration at 37°C for 12–16 h (*see* **Note 5**).
3. Drop 10 μl of saturated culture onto each 3.5-cm SK plate and spread the culture using a glass microspreader or bent pipetman tip. Allow the plates to dry at room temperature for a few minutes if necessary (*see* **Notes 6** and **7**).
4. Incubate the seeded plates at 37°C for 24 h.
5. Transfer the plates to 25°C or room temperature and incubate for 24 h (*see* **Note 8**). The PA14 bacteria should form a lawn with a large halo surrounding the edges where the motile bacteria have traveled. The agar should also have a noticeable green tint from the phenazines, low-molecular-weight tricyclic compounds, pumped into the media by the bacteria.

3.2.2. Preparation of S. aureus-Killing Plates

1. Streak *S. aureus* NCTC8325 bacteria from frozen stock onto a 10-cm Petri dish containing TSA + 10 μg/ml Nal (*see* **Note 9**). Grow at 37°C overnight and store the streaked plate at 4°C for no longer than a week.
2. Inoculate 2–5 ml of TSB + 10 μg/ml Nal with a single colony of *S. aureus*. Grow with aeration at 37°C overnight.
3. Dilute the saturated *S. aureus* culture 1:10 in TSB + 10 μg/ml Nal, and drop 10 μl of this dilute culture onto each 3.5-cm TSA + Nal plate. Spread the culture using a glass microspreader or bent pipetman tip; avoid spreading the culture all the way to the edge of the plate. Allow the plates to dry at room temperature for a few minutes (*see* **Note 7**).
4. Incubate the plates at 37°C for 3–6 h.

5. Dry the plates in a sterile flow hood with their lids open until the plates equilibrate to room temperature and no condensation remains on the lid (*see* **Note 10**).

3.2.3. Preparation of E. faecalis-Killing Plates

1. Streak *E. faecalis* MMH594 bacteria from frozen stock onto a 10 cm BHI + 80 μg/ml Kan plate. Grow at 37°C overnight and store the streaked plate at 4°C for no longer than a week.
2. Inoculate 2–5 ml of BHI + 80 μg/ml Kan with a single colony of *E. faecalis*. Grow with aeration for 4–5 h at 37°C.
3. Spread 10 μl of log phase culture onto each 3.5-cm BHI + 80 μg/ml Kan plate using a glass microspreader (*see* **Note 7**).
4. Incubate the plates at 37°C overnight. Before use, equilibrate the plates to room temperature. If necessary, dry the plates in a sterile flow hood for 15–60 min with their lids open until no condensation remains on the lid (*see* **Note 11**).

3.2.4. Preparation of S. enterica-Killing Plates

1. Streak *S. enterica* SL1344 bacteria from frozen stock onto a 10-cm LB agar plate. Grow at 37°C overnight.
2. Inoculate 2–5 ml of LB broth with a single colony of *S. enterica*. Grow with aeration overnight at 37°C.
3. Spread 10 μl of saturated culture onto each 3.5-cm SK plate.
4. Incubate the plates at 37°C for 12 h. Before use, equilibrate at 25°C for 2 h.

3.2.5. Preparation of S. marcescens-Killing Plates

1. Streak *S. marcescens* Db11 bacteria from frozen stock onto a 10-cm LB agar plate. Grow at 37°C overnight and store the streaked plate at 4°C for no more than a week.
2. Inoculate 5 ml of LB broth with a single colony of *S. marcescens*. Grow the culture with shaking at 37°C for 8–10 h.
3. Drop 50 μl of the culture onto 6-cm NGM plates, taking care that the culture does not touch the edges of plate. Allow the plates to dry at room temperature for a few minutes (*see* **Note 12**).
4. Incubate the plates at 37°C overnight.

3.2.6. Preparation of Yersinia pseudotuberculosis and Y. pestis Assay Plates

1. Streak *Y. pseudotuberculosis* YPIII or *Y. pestis* KIM6+ from frozen stock onto a 10-cm LB agar plate.
2. Inoculate LB broth with a single colony of *Yersinia*. Grow with shaking overnight at 26°C (*see* **Note 13**). Grow *E. coli* identically as a control.
3. Pipette 100–120 μl of saturated culture onto each 6-cm NGM plate. Allow the plates to dry undisturbed so that the culture does not touch the edges of the plates.

4. Incubate the seeded plates for 16–24 h at 20–26°C to grow bacterial lawns. Fresh assay plates should always be used, because biofilm capability diminishes as bacterial lawns age.

3.2.7. Preparation of M. nematophilum Assay Plates

1. Streak *M. nematophilum* CBX102 bacteria from frozen stock onto a 10-cm LB agar plate. Also streak *E. coli* OP50 bacteria from frozen stock onto a 10-cm LB agar plate. Grow at 37°C.
2. Inoculate 2–5 ml of LB broth with a single colony of *M. nematophilum*. Grow at 37°C for about 48 h, until the culture reaches stationary phase. The culture may be stored at 4°C and will retain viability and virulence for up to several months.
3. Inoculate 2–5 ml of LB broth with a single colony of *E. coli* OP50. Grow with aeration at 37°C overnight.
4. Mix the stationary phase *M. nematophilum* culture with the stationary phase OP50 culture in a volumetric ratio of 1:9 (*see* **Note 14**).
5. Spread 10–20 μl of the bacterial mixture on a 6-cm NGM agar plate. Incubate at 37°C overnight.

3.2.8. Preparation of C. neoformans-Killing Plates

1. Inoculate 2 ml of YPD + 45 μg/ml Kan + 100 μg/ml Amp + 100 μg/ml Strep with *C. neoformans* H99 directly from frozen stock. Grow at 30°C for 24 h.
2. Spread 10 μl of each culture onto each 3.5 cm BHI + antibiotic (45 μg/ml Kan, 100 μg/ml Amp, or 100 μg/ml Strep) plate (*see* **Notes 7** and **15**).
3. Incubate the plates at 30°C for 24 h.
4. Transfer the plates to 25°C and incubate for another 24 h (*see* **Note 16**).

3.2.9. D. coniospora Growth

1. To harvest *D. coniospora* spores, add sterile M9W buffer to a plate of infected worms.
2. Use a sterile glass microspreader to scrape the agar surface gently. Try to remove most of the spores, but not the dead old worms, from the plate. The solution of M9W should become white and turbid.
3. Grow 1000–2000 synchronized worms to the L4 or young adult stage on a 10-cm NGM agar plate that is seeded with a lawn of OP50.
4. Add about 300 μl of the freshly harvested *D. coniospora* spore solution to the plate. Be sure that there is enough spore solution to wet the entire surface area of the agar.
5. Incubate the infection plate at 25°C for 2 days. After 1 day, check to make sure the worms still have enough OP50 food on the plate; add more if necessary to prevent the worms from starving.
6. After the 2-day incubation period, wash the worms off the plate with M9W buffer into a 15-ml Falcon tube.

7. Allow the adult worms to settle to the bottom of the 15-ml tube for about 5 min at room temperature. Decant the supernatant, which will contain uninfected L1 larvae.
8. Transfer the remaining infected adult worms to a fresh unseeded Petri dish containing NGM agar + 100 μg/ml Amp + 15 μg/ml gentamycin (see **Note 17**). Spread the worms evenly across the surface of the agar. The infected worms may crawl around a bit but will soon die.
9. Incubate the plate of infected worms at 25°C for 1 week. The *D. coniospora* will grow and each worm will become a dense spot of cotton-like hyphae.
10. Store the plates of *Drechmeria*-infected worms at 20°C for no more than 2 months.

3.3. Setting Up and Scoring Pathogenicity Assays

3.3.1. Setting Up and Scoring Pathogenicity Assays: P. aeruginosa, S. aureus, and E. faecalis Killing

1. Pick about 30 age-matched L4 or young adult hermaphrodites and place them on the prepared killing plates (*see* **Note 18**). Set the worms down on the bare agar, just outside the lawn of bacteria. Try to transfer as little OP50 as possible from the maintenance plates to the killing plates (*see* **Note 19**).
2. Incubate the killing plates at 25°C (*see* **Note 20**).
3. Score the worms as dead or alive at various points along the time course of death for the population, using a touch movement assay for death. Visually inspect the worm for movement. If it is not moving, then use a worm pick to touch the worm gently on the nose, stroke the worm along its side, and touch the tail. Watch for any movement, especially pumping in the pharynx, foraging behavior in the head, or curling of the tail. Worms that are not moving are scored as dead, counted, and picked off the killing plate. Worms that are still moving are scored as alive and are also counted (*see* **Notes 21–23**).

3.3.2. Setting Up and Scoring Pathogenicity Assays: S. enterica Killing

1. Pick about 20–30 age-matched gravid hermaphrodite worms that are 1 day past the L4/adult molt and place them on the prepared killing plates (*see* **Note 24**). Set the worms down on the bare agar, just outside the lawn of bacteria. Try to transfer as little OP50 as possible from the maintenance plates to the killing plates.
2. Incubate the killing plates at 20 or 25°C (*see* **Note 25**).
3. Transfer the infected animals to fresh *S. enterica*-killing plates daily during the first 5 days of the assay or until the end of the reproductive period (*see* **Note 26**).
4. Score the worms as dead or alive at various points along the time course of death for the population, using a touch movement assay for death (*see* **Subheading 3.3.1., step 3**). Remove the dead worms at each time point.

3.3.3. Setting Up and Scoring Pathogenicity Assays: S. marcescens Killing

1. For *S. marcescens* assays, wild-type (N2) *C. elegans* are typically not used. Infected wild-type hermaphrodites will lay eggs during the first days of the infection. In addition, some infected animals will retain eggs, which will hatch internally and kill the adult animal (bag-of-worms phenotype). Therefore, to avoid any confusion between the generations and an effect of the bag-of-worms phenotype on the survival data, it is best to use conditionally sterile mutant worms such as *fer-15* mutants. These mutants are not able to fertilize their oocytes at 25°C; thus, the strain must be maintained at 15°C.
2. Synchronize *fer-15*-mutant worms by harvesting eggs as described above in **steps 1–6** of **Subheading 3.1.2**.
3. Deposit the harvested *fer-15*-mutant eggs on plates containing OP50. Incubate the plates at 25°C. The worms will develop into sterile adults at this temperature.
4. When the *fer-15* worms have reached the L4 stage, remove the prepared killing plates from the 37°C incubator and let them cool down at room temperature for at least 30 min.
5. Pick about 20 L4 *fer-15* hermaphrodites from the 25°C maintenance plates and place them on the prepared killing plates. To obtain statistically significant results, set up five to six plates with 20 worms each for each condition tested (*see* **Notes 27** and **28**).
6. Incubate the killing plates at 25°C (*see* **Note 29**).
7. Score the worms as dead or alive at various points along the time course of death for the population, using a touch movement assay for death (*see* **Subheading 3.3.1., step 3**). Remove the dead worms at each time point (*see* **Notes 21** and **30**).

3.3.4. Setting Up and Scoring Pathogenicity Assays: Y. pseudotuberculosis and Y. pestis Growth Inhibition (see **Note 31**)

1. Pick 10 gravid adult hermaphrodite worms and place them on prepared *Yersinia* assay plates.
2. Allow the gravid hermaphrodites to lay eggs for 2 h at room temperature, and then remove the adults from the plates. The adults will have laid roughly 100–200 eggs; if larger numbers are desired, multiple plates should be used, because too many animals will titrate the bacterial polysaccharide (*see* **Note 32**).
3. Incubate the assay plates at 20°C for 48 h.
4. Score the developmental stage of the worms on the assay plates relative to the stage of worms grown in parallel on plates with control *E. coli*. It is necessary only to tally L4 and non-L4; the exact stage of the growth-inhibited animals need not be determined. Data from this assay are reported as the percentage of worms that reach the L4 developmental stage 48 h after eggs are laid on *Yersinia* lawns (*see* **Note 33**).

3.3.5. Setting Up and Scoring Pathogenicity Assays: Y. pseudotuberculosis and Y. pestis Biofilm Formation (see **Note 34**)

1. Grow the desired *C. elegans* strains on standard *E. coli* food until the L4 or young adult stage (*see* **Note 35**).
2. Pick up to 50 *C. elegans* and place them on prepared *Yersinia* assay plates (*see* **Note 36**).
3. Incubate the assay plates at room temperature for 4 h.
4. Score each animal as biofilm positive or negative. Biofilms are readily apparent as a mass of foreign material adhering to the head and moving with the animal (*see* **Note 35**).
5. To confirm the genotype of any individual hermaphrodite, place it on a *Yersinia* lawn and allow it to lay eggs. Score these progeny for the appearance of biofilms as the worms develop. Occasionally, a worm will not have a biofilm but will not be a true biofilm-absent-on-head (Bah) mutant; the appearance of biofilms on the progeny identifies the parent worm as non-Bah for Bah mutations that are recessive.

3.3.6. Setting Up and Scoring Pathogenicity Assays: M. nematophilum Infection

1. Pick 20–30 L3 or L4 hermaphrodite larvae and place them on prepared *M. nematophilum* assay plates (*see* **Note 37**).
2. Incubate the assay plates at 25°C (*see* **Note 38**).
3. Score the worms for infection with *M. nematophilum*. *M. nematophilum* bacterial cells are able to colonize the rectum of susceptible nematodes and adhere tightly to the rectal and post-anal cuticle of infected worms. As a result of the infection, worms develop greatly swollen and deformed tails, with associated constipation, slowed growth, and reduced fertility. The misshapen tail phenotype that characterizes infection by *M. nematophilum* is called the deformed anal region (Dar) phenotype. Infection of L3 or L4 larvae is rapid; a Dar phenotype is visible within 6 h of feeding on mixed *M. nematophilum* + OP50 lawns at 25°C.

3.3.7. Setting Up and Scoring Pathogenicity Assays: C. neoformans Killing

1. Pick about 25 L4 hermaphrodite worms and place them on prepared *C. neoformans*-killing plates. Set the worms down on the bare agar, just outside the lawn of yeast. Try to transfer as little OP50 as possible from the maintenance plates to the killing plates.
2. Incubate the killing plates at 25°C.
3. Score the worms as dead or alive every 24 h along the time course of death for the population, using a touch movement assay for death (*see* **Subheading 3.3.1., step 3**). Remove the dead worms at each time point (*see* **Notes 21 and 22**).

3.3.8. Setting Up and Scoring Pathogenicity Assays: D. coniospora

1. For *D. coniospora* assays, wild-type worms or worms containing an integrated transgene of *nlp-29::gfp* can be used *(24)*.
2. Synchronize L1 larvae by bleaching gravid adult hermaphrodites and allowing the resulting eggs to hatch in M9W buffer, as described above in **Subheading 3.1.2**.
3. Drop 100–200 synchronized L1 worms onto 6-cm Petri dishes containing NGM agar and seeded with OP50.
4. Incubate the worms at 25°C until they reach the L4 larval stage.
5. Add about 150 µl of freshly harvested *D. coniospora* spore solution to the plate. Be sure there is enough spore solution to wet the entire surface of the agar. The virulence of the spores depends on the age of the plate from which they were harvested, so the concentration of spores used in a pathogenicity assay should be adjusted accordingly.
6. Incubate the infection plate at 25°C.
7. If the *nlp-29::gfp* strain is used, score the infection of worms by *D. coniospora* by examining the induction of green fluorescence. Green fluorescent protein (GFP) is not produced in uninfected worms, but its expression is induced approximately 4 h after *D. coniospora* exposure (*see* **Fig. 1A and B**). If desired, the induction can be quantified using the COPAS fluorescence worm sorter manufactured by Union Biometrica Holliston, MA, USA *(24)*. Calculate the relative fluorescence of GFP (green) under control of the inducible *nlp-29* promoter with respect to a non-inducible dsRed (red) baseline control. Plot the ratio for each worm with respect to that worm's time-of-flight on a semi-log scale (*see* **Fig. 1C**).
8. Score the worms as dead or alive every 12 h along the time course of death for the population. Transfer the living worms to fresh assay plates each day for the first 3 days of the assay to prevent the progeny of infected adults obscuring the results of the assay.

3.4. Analysis of Killing Assays

3.4.1. Statistical Analysis: Graphing Survival Curves for Killing Assays

1. Score the infection by the pathogens *P. aeruginosa*, *S. aureus*, *E. faecalis*, *S. enterica*, *S. marcescens*, and *C. neoformans* by counting the number of worms that are alive and dead on a plate along a time course. Be sure to take at least one early time point when the worms have not yet begun to die, at least three time points during the exponential phase of death, and at least one late time point. Score the assay until all the worms are dead. The required frequency of time points thus varies depending on the rate at which worms die on a particular pathogen. For example, pathogens such as *P. aeruginosa* and *S. aureus* that kill worms very quickly require that time points be taken about 3 times a day during the period of rapid death. Pathogens such as *C. neoformans*, *S. marcescens*, *E. faecalis*, and

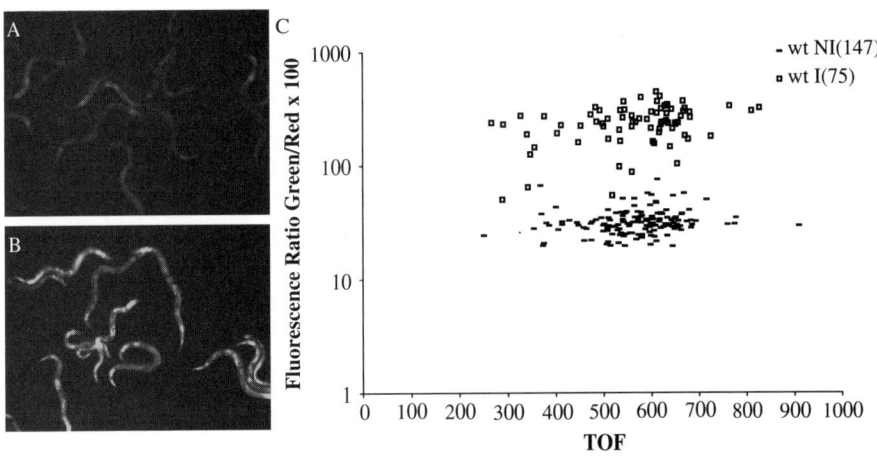

Fig. 1. Induction of *nlp-29::gfp* upon infection of *Caenorhabditis elegans* with *Drechmeria coniospora*. Worms expressing the reporter *frIs7[Pnlp-29::gfp]* show low levels of GFP fluorescence under non-infective conditions (**A**). Within 24 h after infection with *D. coniospora*, expression of the peptide *nlp-29* is induced, and thus these transgenic worms fluoresce at high levels (**B**). The level of fluorescence of infected and non-infected *frIs7[nlp-29::gfp]* worms was quantitated in a COPAS fluorescence worm sorter (Union Biometrica). Infected worms, represented by boxes, have a significantly higher fluorescence ratio than non-infected worms, represented by dashes (**C**). Figure kindly provided by N. Pujol.

S. enterica kill the worms more slowly and thus time points can be taken every 24 hours over a greater number of days.

2. Plot the Kaplan–Meier estimate of the population survival curve using statistical analysis software such as Prism (Graphpad Software San Diego, CA, USA). This software uses the Log-Rank method to compare survival curves. Alternately, the calculations may be performed manually with the help of a spreadsheet program such as Microsoft Excel. For each plate at each time point, calculate the percent alive:

Total worms = the sum of the number of worms dead at all time points

Number dead by a given time = the sum of number of worms dead at his time point and all previous time points

Number alive at a given time = total worms − number dead by this time

$$\text{Percent alive at a given time} = \frac{\text{number alive at this time}}{\text{total worms}} \times 100$$

3. Calculate the average percent alive for each condition at each time point. To do this, average the percent alive measurements for each of the replicate plates (*see* **Note 39**).

Models of Infection in C. elegans 417

4. Plot the average percent alive for each condition as a function of time. This survival curve typically has a sigmoidal decay shape and is an estimate of the survival function, the probability (on the y-axis) of an individual worm surviving a particular length of time (along the x-axis). The survival curves for different conditions assayed in parallel should be plotted on the same set of axes; however, the survival curves from assays not conducted in parallel (i.e., set up on different days) may not be compared directly on the same axes due to day-to-day variability.
5. Calculate the error for each time point. The error bars for a given time point can be presented as the standard deviation of the individual percent alive measurements across the replicates or as the standard error of the mean.

3.4.2. Statistical Analysis: Comparing Population Curves for Killing Assays

1. Calculate the mean survival for the population of worms for each of the replicates of a given condition:

   Worm survival at a given time point = (number of worms scored as dead at that time point) × (length of time elapsed since the beginning of the assay)

   $$\text{Mean survival} = \frac{\text{sum of worm survival at all time points}}{\text{total worms}}$$

2. Calculate the mean survival for a condition by averaging the mean survival values for each of the replicates.
3. Calculate the standard deviation or standard error of the means.
4. To compare two conditions, perform a student's T-test to determine whether the mean survival values for the replicates of one condition are significantly different from the mean survival values for the replicates of another condition.

3.4.3. Interpretation and Controls

1. The quantitative rate of killing of worms by a given pathogen is highly dependent on environmental conditions, including age of killing plates, age of bacterial culture, autoclave time of media, and temperature. It is impossible to repeat an experiment with exactly identical conditions; thus, direct comparisons may be made only within a given experiment. All controls and all experimental conditions must be assayed in parallel.
2. The necessary controls vary by experimental design. Most frequently, we test the phenotypes of different worm genotypes upon exposure to a given pathogen. In this case, it is mandatory to include wild-type worms as a control. Additionally, it may also be useful to include a reference mutant with a known phenotype as a measure of the dynamic range of that particular experiment. For example, if one is testing the phenotype of several mutations hypothesized to confer enhanced sensitivity to pathogens (Esp phenotype), assaying the phenotype of a known Esp mutant in parallel allows for a comparison of the relative sensitivity of the unknowns. Another

common experimental design is to test the effects of several different strains or mutants of a given pathogen species on one reference *C. elegans* genotype. In this case, it is useful to include both wild-type and strongly attenuated pathogen controls.

3. Even if the proper controls are included in parallel within the same pathogenicity assay, it is important to be careful when interpreting Esp phenotypes. Because death is a complex process affected by many variables, there are myriad processes that contribute to the rate of death. Worms that have defects in processes unrelated to the innate immune response may be generally sick and thus more likely to die whether in the presence or absence of pathogens. To eliminate general sickness as a cause for perceived sensitivity to pathogens, it is critical to determine the lifespan of each worm strain used when in the presence of benign bacteria. Typically, this is done by picking L4 worms and placing them on NGM plates seeded with the standard laboratory food source *E. coli* OP50, and scoring the survival of the worms over time in the same manner as is done for a killing assay on pathogens. Because the progeny of these worms will grow rapidly and make the assay impossible to score, the adult worms may be transferred to fresh NGM plates with OP50 during their reproductive period, generally the first 4–5 days of adulthood. Alternatively, 100 µg/ml 5-fluoro-2′-deoxyuridine (FUDR), which prevents progeny production, can be added to the assay plates after the OP50 lawn has grown but just before the worms are picked to the plates. Worms that have a wild-type lifespan on OP50 but die more quickly than wild-type worms when fed pathogens are considered bona fide immunocompromised mutants. Although the lifespan of worms on OP50 is the most common control done to measure non-specific sickness and is generally accepted by the field, a caveat is that even OP50 grown under these conditions may be slightly pathogenic *(25)*. Also, it is difficult to control for the quality of nutrition received by the worms when they are eating different species of bacteria grown on different media. Thus, a more rigorous, although less common, control for sensitivity is to compare the worms fed pathogens under the standard assay conditions described above with worms fed the same strain of bacteria grown under the same conditions but then killed just before the assay. Preliminary tests with *P. aeruginosa* and *S. aureus* suggest that this is not trivial, as the severe conditions needed to kill 100% of these bacteria also affect the integrity of the media on which they are growing (E. Troemel, and J. Powell, unpublished results). The best alternative at this time is to grow liquid cultures of pathogen, concentrate them 10-fold, heat-kill the bacteria by incubating at 100°C for 1 h, and then place the dead bacteria on standard assay plates. Alternatively, heat-kill the bacteria first, and then concentrate the culture (J. Irazoqui, C. L. Kurz, personal communication).

4. Just as other factors can contribute to a perceived Esp phenotype, processes outside the immune response can also influence the phenotype of worms that are more resistant to pathogens (enhanced resistance to pathogens, Erp phenotype). A major cause of resistance to *P. aeruginosa* and *S. aureus* is reduced fertility. This is due at least in part to the fact that worms grown on these pathogens retain eggs, which

hatch internally and kill the mother (bag-of-worms phenotype). Worms that are sterile die more slowly than worms that are fertile, regardless of whether the sterility is caused by a genetic mutation in the worm or the presence in the media of FUDR, which inhibits DNA synthesis and thus chemically sterilizes worms. To control for the effects of fertility on the rate of death of worms fed pathogens, one can use conditional sterile mutations as a genetic background. One common genotype is the double-mutant *fer-15; fem-1*. These worms are 100% sterile when grown at 25°C. Temperature-sensitive sterile strains such as *fer-15; fem-1* should be maintained at 15°C. Several days before the worms are needed for a killing assay, synchronize them as described above in **Subheading 3.1.2**. Place the synchronized L1 larvae on NGM plates with OP50 and grow to the L4 stage at 25°C. Use these sterile worms to set up a killing assay. Alternatively, FUDR may be used in *P. aeruginosa* assays. Prepare the killing plates as described. Just before placing worms on the plates, add FUDR (dissolved in water) to 100 μg/ml around the edge of the bacterial lawn. Let the FUDR soak in for a few minutes, and then pick the worms to the plate and proceed with the assay as described. A third alternative to control for the effects of sterility and prevent a bag-of-worms phenotype is to use males in the killing assay.

5. As noted above, death is a complicated phenotype and can be influenced by numerous factors. It is also a terminal readout of a complex process and is thus not as sensitive to small perturbations in some subset of the many steps of the process. Ideally, an assay for the pathogenicity of microbes and the immune response of *C. elegans* would be able to detect small changes in the response of worms in a quantitative manner and would be able to differentiate among different components of the response. Currently, the best candidates for such an assay involve examining the transcriptional regulation of putative immune effectors through reporter transgenes or quantitative RT-PCR. A nice example of a molecular reporter of infection is the induction of *nlp-29::gfp* within 12 h of infection of worms by *D. coniospora* (*see* **Fig. 1**). Multiple laboratories in the field are currently working to generate additional similar reagents.

4. Notes

4.1. Materials

1. Increasing the autoclave time of the media decreases the rate of killing of worms by NCTC 8325. As TSA plates age, the rate of killing decreases.
2. KIM6+ is avirulent for mammalian infection.

4.2. C. elegans *Growth*

3. Basic information about culturing *C. elegans* can be found at http://www.wormbook.org. A guide to *C. elegans* anatomy is located at http://www.wormatlas.org.
4. Picking two to five L4 or adult worms from a growing plate and placing them on a fresh lawn of OP50 will allow them to lay eggs, which can be used to maintain the

worm strain further or to set up pathogen assays after they reach the L4 or young adult stage. Because the eggs are laid over the course of several days, these progeny will not be synchronized. Worms can also survive several months of starvation by arresting as L1 larvae or entering an alternate stress-resistant larval stage called the dauer stage. It is not as easy to pick worms from a starved plate, so typically a chunk of agar is transferred to a fresh lawn of OP50 to resume the development of the arrested starved worms. This method can be slightly faster than picking worms for the purpose of strain maintenance; however, for pathogenicity assays, it is generally preferable to use worms that have been well fed over their entire life rather than starved and recovered.

4.3. Preparation of Pathogenicity Assay Plates

5. PA14 has a tendency to form phenotypic variants under certain growth conditions *(26)*. These variants are not well characterized but have been shown to be less virulent in standard worm killing assays and should be avoided by growing the bacteria in the absence of antibiotics and with sufficient aeration. Test tubes of liquid culture may be grown either in a rotor or in a shaker, but not in static culture. It is also important not to allow the culture to grow at 37°C for more than 16 h, or the bacteria will begin to lyse. If necessary, the culture may be removed from the 37°C incubator and stored static at room temperature for a few hours before seeding the plates for a killing assay.
6. When spreading the culture, try to avoid spreading it all the way to the edge of the plate because this will increase the difficulty of scoring the assay and will increase the chance that worms will crawl up the sides of the plate.
7. To obtain statistically significant results, typically set up assays in triplicate. Be sure to prepare enough plates so that at least three plates can be used for each condition tested. If many conditions need to be screened, performing killing assays in triplicate can be overwhelming. In this case, assays can be carried out in duplicate and those conditions with potentially interesting results can be re-tested in triplicate to ascertain statistical significance.
8. This incubation allows the bacteria to produce factors required for their full virulence. The incubation time can be reduced to as little as 6 h, but the worms will die at a slower rate. Increasing the incubation time from 24 to 48 h slightly increases the rate of worm death.
9. Concentrations of Nal higher than 10 µg/ml have a negative effect on worm health and viability. Some strains of *S. aureus* also cannot tolerate 10 µg/ml Nal. If desired, the concentration may be decreased to 7.5 µg/ml. Decreasing the concentration of Nal further increases the chance of contamination of the killing assay by OP50.
10. Drying may take 15–60 min, depending on the plate age and storage conditions. Too little drying facilitates the escape of the worms onto the sides of the plate, whereas too much drying will cause the agar to crack.

11. An alternate method is to spread 10 μl of a saturated overnight culture diluted 1:10 in BHI + Kan. Incubate these plates at 37°C for 4 h to overnight.
12. To obtain statistical significance of the results of *S. marcescens* assays, prepare enough plates to do about five to six replicates of each condition.
13. It is critical that the growth and incubation temperatures are <30°C, because biofilm formation is inhibited at higher temperatures.
14. Pure lawns of *M. nematophilum* can be grown on NGM agar plates and used for infection of worms. Pure lawns, however, may take some time to grow up and eventually become too sticky and unpalatable for the worms, so mixed bacterial lawns are normally used.
15. Any one of the three antibiotics is sufficient to reduce *E. coli* growth.
16. This incubation time allows the lawn of yeast to grow to the appropriate thickness.
17. The antibiotics will kill the remaining OP50.

4.4. Setting Up and Scoring Pathogenicity Assays

18. L4 hermaphrodites die more slowly than young adult hermaphrodites. Males also may be used in killing assays, but they die more slowly on *P. aeruginosa* or *S. aureus* than hermaphrodites of the same age. Thus, it is critical that all worms in a given experiment be of the same sex and age. Worms can be synchronized as described above in **Subheading 3.1.2** or staged by eye and picked selectively off a mixed stage plate.
19. Contamination of the killing plates with OP50 can be a serious problem for some pathogen assays because it attenuates or abrogates killing of the worms by the pathogen. In the *S. aureus* and *E. faecalis* assays, the Nal and Kan, respectively, included in the killing plates reduce but do not eliminate the chance of contamination by OP50. When setting up *S. aureus*-killing assays, be particularly careful to transfer as little OP50 as possible. Use a worm pick to wipe or dab up any OP50 that is transferred after the worms crawl away from the spot where they were deposited on the agar. Note the location of their placement on the plate; if OP50 growth is visible during the course of the assay or if worms preferentially clump in this location, then the plate is contaminated and the data cannot be interpreted. OP50 contamination is rarely a problem for PA14 or *S. marcescens* assays as these pathogens kill OP50.
20. Changing the temperature of the killing assay slightly can change the rate of worm death. For example, incubating the killing plates at 23°C instead of 25°C can help resolve differences between populations with very similar phenotypes by extending the range of the assay.
21. When scoring killing assays, be careful not to miss ghosts, worm corpses that have begun to decompose. Ghosts are frequently almost completely clear, with virtually no discernable internal structure, and can be difficult to see on the lawn of bacteria or yeast. Ghosts are often easier to see if the angle of the transmitted light on the microscope is adjusted.

22. Virtually all worms on *S. aureus* and many of the worms on other pathogens such as *P. aeruginosa* PA14 and *C. neoformans* will retain their eggs, which will then hatch internally and lead to a bag-of-worms phenotype. When using a touch assay for death, be careful that you are looking at the movement of the mother worm and not movement due to the internally hatched progeny crawling around inside the mother. Observation of internal larvae frequently corresponds to rapid death of the mother, but it is not uncommon to see a hermaphrodite that can still pump or forage independently and thus should be scored alive but that also has an early-stage bag-of-worms phenotype with a few hatched larvae in her uterus. Under some circumstances, it may be desirable to assay sterile worms to prevent the bag-of-worms phenotype. Be cautious, however, because sterility makes the worms more resistant to pathogens through uncharacterized mechanisms. All conditions must therefore be assayed in the same strain background. See **Subheading 3.4.3., point 4**, for more details.
23. Worms infected with *P. aeruginosa* are capable of producing progeny, which can confound the scoring of the killing assay if the infected parents do not die before the progeny develop. A major advantage of using PA14 compared with other strains of *P. aeruginosa* is that PA14' kills the worms very quickly before many eggs are produced so progeny growth is not typically a problem. This phenotypic byproduct of PA14's strong virulence is particularly fortuitous because it allows for the relatively easy screening for attenuated mutants of PA14 that do permit progeny production and development. *S. aureus* and *E. faecalis* inhibit the development of the progeny, so the arrested larvae rarely confound the scoring of killing assays with these pathogens.
24. Gravid adult hermaphrodites are more susceptible to *S. enterica* than larvae or young adults and thus are used to increase the rate of death in this assay. It is critical that all worms in an experiment be of the same sex and age. Worms can be synchronized as described above in **Subheading 3.1.2** or staged by eye and picked selectively off a mixed stage plate.
25. Incubating the killing plates at 25°C will increase the rate of killing relative to incubating them at 20°C.
26. *S. enterica* does not kill or cause the arrest of the progeny of the infected worms. If the worms are not transferred to fresh plates during the reproductive period, the progeny of the infected worms will grow to adulthood and confuse the scoring of the original infected worms. Worms do not need to be transferred after the reproductive period; however, the absence of transfers decreases the rate of killing by *S. enterica*. The lifespan of worms on OP50 lawns is also reduced by transferring, suggesting that both fresh bacteria and the transferring process may have a negative impact on the lifespan of worms.
27. It is critical to use synchronized worms to set up the killing assays, because the age of a worm affects its susceptibility to infection *(27)*.

28. Sometimes, the worms will burrow into the agar medium. This is particularly a problem if the surface tension of the agar is broken by a scratch or nick in the agar. Take care when depositing worms on the plates not to damage the agar surface. When the worms burrow into the agar, they cannot be scored easily using the standard touch movement assay. If the worms do burrow, transfer the remaining worms to an extra assay plate that was prepared in parallel.
29. Temperatures below 25 °C can be used in order to accentuate a difference seen at 25°C. Reducing the temperature of the assay slows down the bacterial and the nematode metabolisms, thus increasing the resolution of the assay. It is important that the worms to be assayed are raised at the same temperature at which the assay will be conduced to prevent the effects of thermotaxis. Thus, if the killing assay will be performed at a lower temperature such as 20°C, grow the worms at 20°C also.
30. Worms infected with *S. marcescens* are particularly prone to leaving the bacterial lawn. This phenomenon is more marked if the bacteria are cultured at 25°C. Some animals crawl onto the sides of the Petri dish, desiccate, and die. These worms are discarded from the analysis.
31. In the growth-inhibition assay, a biofilm covers the nematode mouth and blocks feeding partially or completely, causing developmental arrest (but not death). Direct measurement of biofilm mass on worms would be difficult; the nematode growth assay, although indirect, quantifies biofilm with enough precision that partially defective *Yersinia* mutants can be distinguished from wild-type bacteria.
32. It is best to count the parents as they are put down on the assay plates, and again when they are removed, because worms that remain and continue laying eggs will ruin the assay. Biofilm accumulation on the parental worms and aberrant locomotion by these adults should be visible at the end of the egg-laying period.
33. L4 animals are easily distinguished on *E. coli* because the developing vulva is visible at mid-body as a clear patch, about half the diameter of the animal, that contrasts with the darker intestine. On this standard food, inbred laboratory *C. elegans* grow with high synchrony, and approximately 100% of animals are L4. On *Yersinia* lawns, most animals will be one of the three earlier stages. Scoring is more difficult on *Yersinia* because starvation makes the intestine pale, and the developing vulva does not stand out strongly as it does on *E. coli* food. Some worms are so small that it is obvious they are not L4; plates should be scrutinized carefully to make sure none of these severely affected animals are overlooked in the dense bacterial lawn. For the larger animals, the major difficulty is discriminating L3s and L4s; with practice on synchronous *E. coli*-grown animals, one can learn to do this with high confidence.
34. The nematode biofilm formation assay is used in *C. elegans* genetic analysis. Approximately 100% of wild-type *C. elegans* acquire biofilms within 4 h of transfer from *E. coli* to *Y. pseudotuberculosis*; for most mutants with the Bah phenotype, 0%

will have biofilms. Because rare wild-type animals are Bah, definitive genotyping of an animal requires examination of its brood on *Yersinia*.
35. In most cases, the Bah phenotype is not stage-specific, and assays are most convenient with young adults or L4s. The most difficult stage to score is the L1 larval stage, because the animals are so small that observing biofilms with a stereomicroscope is difficult. A few mutants have different biofilm phenotypes at different stages, so initial characterization of a new mutant should include testing each stage for Bah.
36. It is not advisable to collect worms in liquid and pipette them to the plates, because standing liquid interferes with biofilm attachment. No more than about 50 animals should be placed on a plate to ensure that biofilm exopolysaccharide is not limiting.
37. L1 and L2 worms can become infected by *M. nematophilum* but exhibit a weaker phenotype than later larval stages. Worms that are exposed to the bacteria as adults show little or no infection phenotype, although adults in which an infection was established at the L4 stage or earlier have a strong phenotype. Males may also be used in the assay; males typically exhibit a more severe phenotype than hermaphrodites upon infection with *M. nematophilum*.
38. Infection is strongest and most easily assayed at 25°C but also occurs at lower growth temperatures such as 15 and 20°C. However, the Dar response (*see* **step 3**) to *M. nematophilum* is somewhat variable and attenuated at 15°C.

4.5. Analysis of Killing Assays

39. If there is a large difference in the number of worms on each of the replicate plates, averaging the percents will give greater weight to the survival of worms on plates with fewer individuals. This can be avoided by pooling the data for the replicate plates as if the worms were all on one plate and then calculating a single percent alive for the pooled data. Based on our experience, however, we suspect that the plate effect, or the slightly different conditions experienced by worms on different plates, is greater than the potential skew resulting from weighting individual worms unequally. Thus, we favor calculating the percent alive for each replicate and then averaging those values.

Acknowledgments

The authors thank the following people for generously contributing protocols and for their helpful comments on the manuscript: Alejandro Aballay, Creg Darby, Rhonda Feinbaum, Jonathan Hodgkin, Javier Irazoqui, C. Leopold Kurz, Sachiko Miyata, Terrence Moy, Eleftherios Mylonakis, Nathalie Pujol, and Emily Troemel.

References

1. Brenner, S. (1974) The genetics of Caenorhabditis elegans. *Genetics*, **77**, 71–94.

2. Tan, M.W., Mahajan-Miklos, S. and Ausubel, F.M. (1999) Killing of Caenorhabditis elegans by Pseudomonas aeruginosa used to model mammalian bacterial pathogenesis. *Proc Natl Acad Sci USA*, **96**, 715–720.
3. Aballay, A., Yorgey, P. and Ausubel, F.M. (2000) Salmonella typhimurium proliferates and establishes a persistent infection in the intestine of Caenorhabditis elegans. *Curr Biol*, **10**, 1539–1542.
4. Labrousse, A., Chauvet, S., Couillault, C., Kurz, C.L. and Ewbank, J.J. (2000) Caenorhabditis elegans is a model host for Salmonella typhimurium. *Curr Biol*, **10**, 1543–1545.
5. Kurz, C.L. and Ewbank, J.J. (2000) Caenorhabditis elegans for the study of host-pathogen interactions. *Trends Microbiol*, **8**, 142–144.
6. Darby, C., Hsu, J.W., Ghori, N. and Falkow, S. (2002) Caenorhabditis elegans: plague bacteria biofilm blocks food intake. *Nature*, **417**, 243–244.
7. Sifri, C.D., Begun, J., Ausubel, F.M. and Calderwood, S.B. (2003) Caenorhabditis elegans as a model host for Staphylococcus aureus pathogenesis. *Infect Immun*, **71**, 2208–2217.
8. Garsin, D.A., Sifri, C.D., Mylonakis, E., Qin, X., Singh, K.V., Murray, B.E., Calderwood, S.B. and Ausubel, F.M. (2001) A simple model host for identifying Gram-positive virulence factors. *Proc Natl Acad Sci USA*, **98**, 10892–10897.
9. Mylonakis, E., Ausubel, F.M., Perfect, J.R., Heitman, J. and Calderwood, S.B. (2002) Killing of Caenorhabditis elegans by Cryptococcus neoformans as a model of yeast pathogenesis. *Proc Natl Acad Sci USA*, **99**, 15675–15680.
10. Hodgkin, J., Kuwabara, P.E. and Corneliussen, B. (2000) A novel bacterial pathogen, Microbacterium nematophilum, induces morphological change in the nematode C. elegans. *Curr Biol*, **10**, 1615–1618.
11. Jansson, H.B., Jeyaprakash, A. and Zuckerman, B.M. (1985) Differential adhesion and infection of nematodes by the endoparasitic fungus *Meria coniospora* (Deuteromycetes). *Appl Environ Microbiol*, **49**, 552–555.
12. Jansson, H.B. (1994) Adhesion of conidia of *Drechmeria coniospora* to *Caenorhabditis elegans* wild type and mutants. *J Nematology*, **26**, 430–435.
13. Sifri, C.D., Begun, J. and Ausubel, F.M. (2005) The worm has turned–microbial virulence modeled in Caenorhabditis elegans. *Trends Microbiol*, **13**, 119–127.
14. Kim, D.H. and Ausubel, F.M. (2005) Evolutionary perspectives on innate immunity from the study of Caenorhabditis elegans. *Curr Opin Immunol*, **17**, 4–10.
15. Schulenburg, H., Kurz, C.L. and Ewbank, J.J. (2004) Evolution of the innate immune system: the worm perspective. *Immunol Rev*, **198**, 36–58.
16. Nicholas, H.R. and Hodgkin, J. (2004) Responses to infection and possible recognition strategies in the innate immune system of Cacnorhabditis elegans. *Mol Immunol*, **41**, 479–493.
17. Gallagher, L.A. and Manoil, C. (2001) Pseudomonas aeruginosa PAO1 kills Caenorhabditis elegans by cyanide poisoning. *J Bacteriol*, **183**, 6207–6214.
18. Darby, C., Cosma, C.L., Thomas, J.H. and Manoil, C. (1999) Lethal paralysis of

Caenorhabditis elegans by Pseudomonas aeruginosa. *Proc Natl Acad Sci USA*, **96**, 15202–15207.
19. Mahajan-Miklos, S., Tan, M.W., Rahme, L.G. and Ausubel, F.M. (1999) Molecular mechanisms of bacterial virulence elucidated using a Pseudomonas aeruginosa-Caenorhabditis elegans pathogenesis model. *Cell*, **96**, 47–56.
20. Moy, T.I., Mylonakis, E., Calderwood, S.B. and Ausubel, F.M. (2004) Cytotoxicity of hydrogen peroxide produced by Enterococcus faecium. *Infect Immun*, **72**, 4512–4520.
21. Marroquin, L.D., Elyassnia, D., Griffitts, J.S., Feitelson, J.S. and Aroian, R.V. (2000) Bacillus thuringiensis (Bt) toxin susceptibility and isolation of resistance mutants in the nematode Caenorhabditis elegans. *Genetics*, **155**, 1693–1699.
22. Huffman, D.L., Bischof, L.J., Griffitts, J.S. and Aroian, R.V. (2004) Pore worms: using Caenorhabditis elegans to study how bacterial toxins interact with their target host. *Int J Med Microbiol*, **293**, 599–607.
23. Joshua, G.W., Karlyshev, A.V., Smith, M.P., Isherwood, K.E., Titball, R.W. and Wren, B.W. (2003) A Caenorhabditis elegans model of Yersinia infection: biofilm formation on a biotic surface. *Microbiology*, **149**, 3221–3229.
24. Couillault, C., Pujol, N., Reboul, J., Sabatier, L., Guichou, J.F., Kohara, Y. and Ewbank, J.J. (2004) TLR-independent control of innate immunity in Caenorhabditis elegans by the TIR domain adaptor protein TIR-1, an ortholog of human SARM. *Nat Immunol*, **5**, 488–494.
25. Garigan, D., Hsu, A.L., Fraser, A.G., Kamath, R.S., Ahringer, J. and Kenyon, C. (2002) Genetic analysis of tissue aging in Caenorhabditis elegans: a role for heat-shock factor and bacterial proliferation. *Genetics*, **161**, 1101–1112.
26. Drenkard, E. and Ausubel, F.M. (2002) Pseudomonas biofilm formation and antibiotic resistance are linked to phenotypic variation. *Nature*, **416**, 740–743.
27. Kurz, C.L., Chauvet, S., Andres, E., Aurouze, M., Vallet, I., Michel, G.P., Uh, M., Celli, J., Filloux, A., De Bentzmann, S. et al. (2003) Virulence factors of the human opportunistic pathogen Serratia marcescens identified by in vivo screening. *EMBO J*, **22**, 1451–1460.
28. Friedman, D.B. and Johnson, T.E. (1988) A mutation in the age-1 gene in Caenorhabditis elegans lengthens life and reduces hermaphrodite fertility. *Genetics*, **118**, 75–86.
29. Nelson, G.A., Lew, K.K. and Ward, S. (1978) Intersex, a temperature-sensitive mutant of the nematode Caenorhabditis elegans. *Dev Biol*, **66**, 386–409.
30. Heitman, J., Casadevall, A., Lodge, J.K. and Perfect, J.R. (1999) The Cryptococcus neoformans genome sequencing project. *Mycopathologia*, **148**, 1–7.
31. Huycke, M.M., Spiegel, C.A. and Gilmore, M.S. (1991) Bacteremia caused by hemolytic, high-level gentamicin-resistant Enterococcus faecalis. *Antimicrob Agents Chemother*, **35**, 1626–1634.
32. Murray, B.E., Singh, K.V., Ross, R.P., Heath, J.D., Dunny, G.M. and Weinstock, G.M. (1993) Generation of restriction map of Enterococcus faecalis OG1 and investigation of growth requirements and regions encoding biosynthetic

function. *J Bacteriol*, **175**, 5216–5223.
33. Rahme, L.G., Stevens, E.J., Wolfort, S.F., Shao, J., Tompkins, R.G. and Ausubel, F.M. (1995) Common virulence factors for bacterial pathogenicity in plants and animals. *Science*, **268**, 1899–1902.
34. Hoiseth, S.K. and Stocker, B.A. (1981) Aromatic-dependent Salmonella typhimurium are non-virulent and effective as live vaccines. *Nature*, **291**, 238–239.
35. Flyg, C., Kenne, K. and Boman, H.G. (1980) Insect pathogenic properties of *Serratia marcescens*: phage-resistant mutants with a decreased resistance to *Cecropia* immunity and a decreased virulence to *Drosophila*. *J Gen Microbiol*, **120**, 173–181.
36. Iandolo, J.J. (2000) In Fischetti, V. A., Novick, R. P., Ferretti, J. J., Portnoy, D. A. and Rood, J. A. (eds.), *Gram-Positive Pathogens*. ASM Press, Washington, D. C., pp. 317–325.
37. Sikkema, D.J. and Brubaker, R.R. (1987) Resistance to pesticin, storage of iron, and invasion of HeLa cells by Yersiniae. *Infect Immun*, **55**, 572–578.
38. Gemski, P., Lazere, J.R., Casey, T. and Wohlhieter, J.A. (1980) Presence of a virulence-associated plasmid in Yersinia pseudotuberculosis. *Infect Immun*, **28**, 1044–1047.

25

Genetic Analysis of *Caenorhabditis elegans* Innate Immunity

Michael Shapira and Man-Wah Tan

Summary

Innate immunity is an ancient and conserved defense mechanism. The worm *Caenorhabditis elegans* provides a useful tool for studying the function of the innate immune system at the molecular and cellular levels within the context of a whole organism. The powerful genetics of the worm, combined with efficacy of gene knockdown by RNA interference (RNAi), offer complementary tools for analyzing the contribution of individual genes to innate immunity. It is important, however, to exclude pleiotropic effects that confound results. In this chapter, we will describe the procedures for performing both forward and reverse genetic screens and will discuss a number of techniques developed to resolve confounding effects, thus enhancing the power of this system.

Key Words: *Caenorhabditis elegans;* pathogen; innate immunity; genetic screen; RNAi; sterility; colony-forming-unit counts; survival.

1. Introduction

Innate immunity is an ancient and evolutionarily conserved defense mechanism, which provides the first line of defense against invading pathogens *(1,2)*. Innate immune responses are complex, involving various cell types that are local to the infected tissue or recruited to the site of infection, and numerous molecular factors. It is therefore useful to study innate immunity in the context of a whole organism. *Caenorhabditis elegans* provides a genetically tractable model system in which questions relating to innate immunity can be addressed in such a context *(3,4)*.

From: *Methods in Molecular Biology, vol. 415: Innate Immunity*
Edited by: J. Ewbank and E. Vivier © Humana Press Inc., Totowa, NJ

Caenorhabditis elegans can be infected by various human pathogens, simply through exposure to the pathogen as a food source. The process of infection is conserved both with respect to virulence factors deployed by the pathogen to establish infection and with respect to the molecular mechanisms employed by the host to fight off infection *(5,6)*. As one of the most genetically tractable model organisms, *C. elegans* provides powerful tools to interrogate the importance of a specific gene to protecting the animal from infection. The most fundamental of these is the genetic screen, in which random mutations are tested for their contribution to a certain phenotype. In addition, *C. elegans* offers a unique opportunity to perform RNA interference (RNAi) screens in the context of a whole organism. The phenomenon of gene silencing by inhibitory RNAi was first identified in *C. elegans (7–9)* and is easy to achieve in this system due to efficient uptake of double-stranded RNA (dsRNA) from ingested bacteria. This has been facilitated by the construction of feeding RNAi libraries, in which specific dsRNAs are expressed in *Escherichia coli* clones *(10)*, and aided by efficient amplification of the RNAi effect in the worm *(11)* leading to a systemic effect.

The power of such forward and reverse genetic screens is as great as the strength of the observed phenotype, and the clearest of all phenotypes is survival. However, survival, as a phenotype in genetic screens, has its drawbacks: Often, results of such screens are confounded by non-relevant effects on survival, or by pleiotropic effects of gene disruption. Such confounding effects include changes in eating/defecation, general stress sensitivity, or reproduction. It is possible to overcome some of these problems by using the appropriate genetic background for the screen. When one desires to use wild-type animals, however, careful post-screen verification is required. This is typically achieved with a detailed survival analysis, in which a cohort of mutant animals is fed with the pathogen of interest, and survivors are counted daily and statistically compared with wild-type animals for their survival. It is at this point that confounding effects may be resolved. In this chapter, we will discuss how to resolve several types of confounding effects, focusing on those associated with reproduction.

In adult *C. elegans* hermaphrodites it is only germline cells that continue to divide. The resulting oocytes mature as they are pushed toward the vulva, fertilized as they pass through the spermatheca, and undergo early embryogenesis before they are laid. Eggs hatch approximately 8 h after being laid. Most simply, reproduction confounds results of survival analysis by complicating counting, as progeny grow and blend with the cohort of originally infected worms. Secondly, mutations can affect egg-laying characteristics that may also affect lifespan and, accordingly, influence survival on pathogen. Thirdly, some infections impair egg-laying, resulting in the hatching of eggs inside the worm, a phenotype termed "bagging," killing the worm and complicating the interpretation of the results.

In this chapter, we will describe one of the most common methods for forward genetic screening in *C. elegans*, utilizing the potent mutagen ethyl methane sulfonate (EMS). As a reverse genetics method, we will describe one of a few variations of targeted RNAi screening. Lastly, we will describe a set of tools aimed to resolve typical confounding effects and help better understanding of the role of individual genes in pathogenesis and survival. We will consider mutants, RNAi clones, and treatments, each useful in resolving a certain problem. Throughout this chapter, we will focus on *C. elegans* infection by PA14, a clinical isolate of the human Gram-negative bacterial pathogen *Pseudomonas aeruginosa (12)*, although these methods are equally useful for other bacterial and fungal pathogens.

2. Materials

2.1. Strains

1. Worm strains discussed in this chapter are *him-5(e1467), him-8(e1489), aex-2(sa3), unc-25(e156), spe-26(it112), rrf-3(pk1426), and glp-4(bn2)* mutants, all available through the *Caenorhabditis* Genetics Center (CGC) (*see* **Note 1**). An additional strain unavailable through CGC, *esp-1(aj3)*, which contains a point mutation in the *tnt-3* gene, is described elsewhere *(6)*. All double mutants were generated by crossing original strains obtained from the CGC.
2. Bacteria expressing dsRNA (RNAi clones) were part of a *C. elegans* RNAi library expressed in *E. coli*. Two independently generated libraries are available from Geneservice, Cambridge, UK, and Open Biosystems, Inc, Huntsville, AL.
3. *Escherichia coli* bacteria on which *C. elegans* typically graze are of the OP50 strain. We use a streptomycin-resistant derivative of this strain, OP50-1, and an *E. coli* strain that expresses green fluorescent protein (GFP) (OP50-GFP) *(13)*. Both strains are available through CGC.

2.2. Media and Plates

1. Nematode growth medium (NGM) plates *(14)*: 3 g NaCl, 2.5 g Bactopeptone, and 17 g Bacto-agar are dissolved in 975 ml distilled dH_2O. After autoclaving, 1 ml cholesterol in ethanol (5 mg/ml), 1 ml 1M $CaCl_2$, 1 ml 1M $MgSO_4$, and 25 ml 1M potassium phosphate buffer (pH 6.0) are added in order. For NGM/IPTG plates, IPTG is added to a final concentration of 1 mM.
2. Modified NGM plates (slow-killing plates, SKP) *(3)*: same as NGM plates, only with 3.5 instead of 2.5 g/l Bactopeptone and 20 instead of 17 g/l Bacto-agar.
3. M9 buffer: 6 g/l Na_2HPO_4, 3 g/l KH_2PO_4, 5 g/l NaCl, and 0.25 g/l $MgSO_4 \cdot 7H_2O$.

4. EMS-M9 buffer: 5.8 g/l Na_2HPO_4, 3 g/l KH_2PO_4, 0.5 g/l NaCl, and 1 g/l NH_4Cl.
5. 2× Worm lysis solution: 40 ml Clorox, 10 ml 10 M NaOH, and 50 ml dH_2O.
6. King's broth (*Pseudomonas* growth media): 20 g/l Proteose peptone 3 (DIFCO, BD Diagnostics), 1.5 g/l KH_2PO_4, 15 ml pure glycerol, and 15 g/l Bacto-agar. After autoclaving, add 6 ml 1 M $MgSO_4$ and rifampicin to a total of 100 µg/ml.
7. LEV: M9 + 25 mM levamisole.
8. LEV/AMP/GENT: LEV solution containing 1 mg/ml Ampicillin and 100 µg/ml Gentamycin. AMP/GENT refers to the antibiotic cocktail alone dissolved in M9.
9. OP50-NGM plates: NGM agar containing 300 µg/ml streptomycin seeded with OP50-1.

3. Methods
3.1. General Techniques
3.1.1. Egg-Prep

1. Wash worms in dH_2O in a 15-ml conical tube, mix with equal volume of 2× lysis solution, and start time. The entire procedure should not take more than 6–7 min to prevent egg lysis.
2. Vortex briefly at ends of first and second minute.
3. At end of fourth minute, pound the tube against soft plastic or Styrofoam tube stand until worms appear dissolved.
4. Spin in a clinical centrifuge for 1 min at ~1000 × *g*.
5. Carefully decant the supernatant.
6. Wash once with 10 ml dH_2O.
7. Carefully decant dH_2O leaving a little to resuspend eggs in, and add to OP50-NGM plates.

3.1.2. Bacterial Culture Concentration

Saturated overnight cultures are concentrated by centrifugation at approximately 6000 × *g* for 10 min. Pellets are subsequently resuspended in the desired fraction of fresh LB (the supernatant will do, too).

3.1.3. Seeding Plates with PA14

Drop 10 µl of an overnight culture onto an SKP plate and spread over a small area using a bent 200-µl pipette tip (*see* **Note 2**). Incubate plates for 24 h at 37°C and use after cooling to room temperature.

3.1.4. UV-Killing of Bacteria (Modified from **ref. 15**)

1. Concentrate a saturated overnight *E. coli* culture × 20 in its own growth media.
2. Spread 20 µl of bacteria evenly on a 30-mm NGM plate (or 80 µl on a 60-mm plate).
3. Incubate 8 h at 37°C, or overnight at room temperature.

4. Irradiate 4 min with cover off in a UV crosslinker (Stratagene, La Jolla, CA).
5. Store at 4°C for up to a week.

3.1.5. Seeding RNAi Plates

1. Grow 3 ml culture per plate of dsRNA-expressing *E. coli* in LB with 50 µg/ml AMP.
2. Spin down each culture in 2 × 1.5 ml Eppendorf tubes on a benchtop centrifuge at maximum speed for 2 min. Discard supernatant to leave 150 µl in each tube (10× concentration of original). Resuspend the pellet and combine tubes.
3. Transfer 200 µl to the center of NGM/IPTG plate. Let dry at room temperature with top ajar at flame base for 20 min. Then, close top and let dry overnight at room temperature. Store at room temperature for up to 7 days.

3.1.6. Survival Assays

Survival assays are usually performed in triplicates, where each plate contains 30–50 worms. At a predetermined interval, dead and live worms are counted and documented (e.g., on an Excel spreadsheet). While counting, remove dead worms from plates. Worms found dehydrated on plate walls, or with bursting vulvas, are censored. Data are analyzed using Kaplan–Meyer survival analysis, and scoring for significance using the LogRank test; statistics program packages useful for such analyses are Statview (for PC's; SAS, Cary, NC), JMP (for Apple computers; SAS), or Prism (GraphPad).

3.2. EMS Screening

3.2.1. Mutagenesis

1. Egg-prep gravid worms (*see* **Subheading 3.1.1.**) to obtain several plates with a synchronized worm population. Start exposure to the mutagen when worms are predominantly at the early L4 larval stage (*see* **Note 3**).
2. Wash off plates with EMS-M9 into a 15-ml sterile conical tube. Spin down in clinical centrifuge at approximately 1000 × *g* for 30 s. Remove supernatant. Repeat several times to remove bacteria. Resuspend worm pellet in EMS-M9 to a final volume of 3 ml.
3. Place 1 ml EMS-M9 in separate 15-ml tube. Wearing gloves, in hood, add 20 µl EMS (methanesulfonic acid ethyl ester; Sigma M-0880). Shake EMS suspension until no longer cloudy (*see* **Note 4**).
4. Transfer the 3 ml of worms into the tube with the EMS/EMS-M9 solution (final concentration of EMS is 47 mM). Gently rock the tube at 20°C for 4 h.
5. Spin down worms and remove supernatant to a 15-ml plastic centrifuge tube. Wash worms twice with 3 ml of M9, transferring the supernatants to the other 15-ml tube. Using a sterile glass pipette, transfer the worms in a few drops of M9 to several OP50-NGM plates.

6. Inactivate EMS from all washes by adding a few pellets of KOH to the 15-ml tube containing EMS waste liquid. Agitate gently to mix. After a day, the EMS should all be inactivated and the liquid can be poured down the sink.
7. Following mutagenesis, allow the mutagenized worms (the P0 generation) to recover on OP50-NGM plates for 24 h. They should begin egg-laying during this period.

3.2.2. F2 Screen for Enhanced Susceptibility to Pathogen

1. Single 45 healthy looking gravid P0 worms onto OP50-NGM plates. Allow worms to lay eggs for 3 h at 25°C. Every 3 h, transfer worms to a new plate for a total of three sequential egg-lays. Carefully label which P0 worms correspond to which plate. For example, plates A1, A2, and A3 would be a set obtained from three sequential egg-lays with the same P0 gravid worm (*see* **Note 5**).
2. When F1 become gravid adults (approximately 2 days later at 25°C, or 3 days at 20°C), egg-prep to get synchronized F2 populations. Perform individual egg-preps for each F1 plate in order to keep track of which sets of F2 may be siblings.
3. Grow F2 to the L4 stage. Meanwhile, prepare plates with lawns of the pathogen of interest.
4. When F2 worms reach L4 stage, wash worms off plate with M9. Wash several times to remove excess OP50. After the last wash, remove most of the supernatant. Using a sterile Pasteur pipette, transfer the F2 in a small volume of liquid to the pathogen plate. Allow infection to proceed at 25°C; ensure that each F2 set is kept separately.
5. Pick off corpses of F2 which die early on the pathogen. Put each individual corpse on a different OP50-NGM plate. Record which F2 set (A1, C2, etc.) each corpse came from. The time-frame for checking for worms that die prematurely due to the infection is pathogen-dependent. As a general rule, however, this step should be performed before unmutagenized N2 worms begin to die. In the case of *P. aeruginosa*, we use a 24–34 h time window.
6. Keep plates with corpses at 15°C (in case the worm has any temperature-dependent lethality mutations). Because most worms are gravid by the time they are killed by the pathogen, live clonal progeny will hatch from the corpse allowing for the recovery of the mutant of interest.
7. Retest all mutants to determine if they are indeed of enhanced sensitivity to pathogen (Esp) type (*see* **Subheading 3.4.**). Remember that mutants from the same F2 sets may be siblings.

3.3. RNAi Knockdown

Three main strategies are commonly employed for RNAi knockdown by feeding: maternal exposure (affecting maternal-effect genes and early embryogenesis of progeny), exposure of eggs (affecting larval development), and exposure of adults (for evaluation of post-developmental gene functions). Each

RNAi clone found to increase susceptibility to the pathogen should be tested for changes in lifespan on non-pathogenic *E. coli* to ensure a specific contribution to immunity rather than general malaise or sensitivity. Many RNAi screens employ sterile strains such as *glp-4* or *spe-26* (see below) to simplify screens based on survival. In this section, we will describe a procedure implementing the third strategy. Minor changes are necessary for the other two strategies.

1. Egg-prep gravid worms. Lay eggs on OP50-NGM. Grow at 25°C until L4 (~48 h).
2. Wash off L4 worms with sterile worm M9 solution (autoclaved or filter-sterilized) and place approximately 100 worms on RNAi plates (*see* **Subheading 3.1.5.**).
3. Grow for 48 h. Make sure that worms do not get starved or that RNAi culture gets contaminated.
4. Transfer 30 worms to each of three pre-seeded PA14 plates (*see* **Subheading 3.1.3.**).
5. For lifespan analyses, wash worms with AMP/GENT (*see* **Note 6**) and lay on UV-killed *E. coli* (*see* **Subheading 3.1.4.**) for survival analyses.
6. For small-scale screens, testing 20–30 RNAi clones at a time, we count worms 48 and 96 h after beginning the exposure to the pathogen. Post-screen verification necessitates more frequent counting.

3.4. Resolving Confounding Effects

Analyses of existing *C. elegans* mutants have revealed that worms with certain defects in pharyngeal grinding, defecation, and reproduction are more susceptible to *P. aeruginosa*, and these are likely to be identified in an Esp screen. Some of these may have mutations in cellular functions or structures not directly related to defense responses and will have to be distinguished from bona fide innate immune mutants. Indeed, *esp-1(aj3)*, which was identified in the Esp screen, turned out to be grinder defective because it carries a lesion in the *tnt-3* gene encoding for troponin T, a structural component of the grinder *(6)*. In the following section, we describe these confounding effects and post-screening steps to resolve them.

3.4.1. Defects in Pharyngeal Grinding

Caenorhabditis elegans is a filter-feeder: taking in liquid with bacteria and then spitting out the liquid while retaining and grinding up the bacteria when they reach the terminal bulb. *C. elegans* mutants defective in grinding such as *esp-1(tnt-3)*, *eat-13*, and *phm-2* receive a higher bacteria inoculum than wild-type worms because they allow the entry of more intact pathogens into the intestine. Consequently, these mutants are more sensitive to pathogen-mediated killing.

To determine if a mutant has defects in pharyngeal grinding, expose the mutant and the parental strain at the adult stage to an OP50 strain that expresses GFP (OP50-GFP) for 6 h. Compare the intestinal lumen content of the mutant with its parental strain under a fluorescent stereo-dissecting microscope. A mutant with grinding defects, such as *esp-1*, will have the entire lumen filled with OP50-GFP compared with few to none for those with an effective grinder.

3.4.2. Defecation Defects

Caenorhabditis elegans defecates at regular intervals by means of a temperature-independent defecation motor program (DMP) *(16)*. At temperatures from 19 to 30°C, in the presence of food, defecation occurs every 45 s, with an SD of <3 s *(17,18)*. DMP consists of three sequential steps of muscle contractions resulting in the opening of the anus and expulsion of intestinal contents. Defects in any of these steps lead to failure to remove intestinal contents at regular intervals and causes a "constipated" (Con) phenotype *(16)*.

It is expected that some Con mutants would be more sensitive than wild-type worms to pathogen-mediated killing because they retain pathogens in the intestine for a longer period rather than due to immune defects. This is clearly illustrated by comparing the sensitivity of males and hermaphrodites harboring loss-of-function mutation in either the *aex-2* or the *unc-25* genes (*see* **Fig. 1**). Sensitivity to slow killing correlated directly with the Con phenotype. Adult male and hermaphrodite of *unc-25(e156);him-5(e1467)*, which are moderately to severely Con, are more sensitive than the corresponding *him-5(e1467)* control animals. Only the Con hermaphrodite *aex-2(sa3);him-5(e1467)* showed increased sensitivity. The non-Con *aex-2(sa3);him-5(e1467)* males are indistinguishable from *him-5(e1467)* males in their sensitivity to PA14.

Parameters for assessing changes in the DMP include the measurement of defecation cycle period and percent cycles that results in the expulsion of gut contents (%Exp). A cycle period is defined as the interval between the initiation of two successive contractions of the posterior body muscle (pBoc) steps, which can be easily observed as a twitch at the region around the anus. The average cycle period can be determined by observing at least 10 defecation cycles, or between 7–8 min per worm (*see* **Note 7**). Over this period, the %Exp can be determined as follows: %Exp = [100 × (number of successful Exp observed/number of pBoc)].

3.4.3. Reproduction

3.4.3.1. WORM STERILIZATION

One way to discriminate between a specific contribution to antimicrobial defense and indirect effects on survival or lifespan caused by changes in

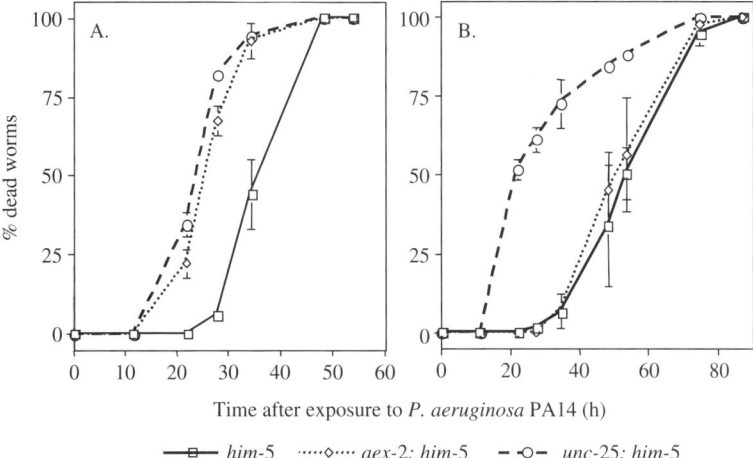

Fig. 1. Susceptibility to PA14-mediated killing correlates with constipation. The control of defecation motor program (DMP) is different between the sexes. In the hermaphrodite, both a GABA-secreting neural program (represented by the GABA-synthesizing enzyme UNC-25) and the *aex* genes are required for the appropriate muscle contraction and for defecation. In the male, on the contrary, only the GABA release apparatus is required. Consequently, both adult male and hermaphrodite *unc-25(e156)* worms (A and B, circles) that show a constipation phenotype (Con) are sensitive to PA14-mediated killing. By contrast, only adult hermaphrodites *aex-2* are Con *(25)* and are sensitive to PA14 (A, diamond), whereas *aex-2* males that are not Con are wild-type for sensitivity to PA14 (B, diamond). These strains were made in the *him-5(e1467)* background in order to obtain sufficient males for the experiment.

egg-laying is to compare the phenotype in mutant hermaphrodites versus males. Mutant strains known for having a higher incidence of males due to X chromosome nondisjunction, such as *him-5*, or *him-8*, are convenient for this purpose *(19)*. Another way that removes the need for crossing mutants of interest into this genetic background is by preventing hermaphrodites from laying eggs. Several mutants have been described that lack the germline and therefore eggs. These *glp* mutants are also long-lived, which could be of additional advantage by increasing the time resolution in survival assays. We routinely use temperature-sensitive *glp-4(bn2)* mutants *(20)*, and in many cases the double mutant *glp-4(bn2);rrf-3(pk1426)*, which has an increased gene-silencing efficiency *(21)*:

1. Grow mutants at permissive temperature (15°C) until gravid.
2. Egg-prep onto OP50-NGM.

3. Grow until L4 at non-permissive temperature (25°C). Worms will fail to develop gonads.
4. Transfer to SKP plates pre-seeded with PA14 for survival assays (*see* **Subheading 3.1.5.**).

To avoid the need to cross *glp-4* worms, or any other sterile mutant, with a mutant of interest, it is beneficial to be able to directly sterilize those mutants of interest by epigenetic means. This can be achieved using RNAis directed against several genes (e.g., *pos-1, cdc-25.1*). We will illustrate the procedure with *cdc-25.1* RNAi. The gene *cdc-25.1* encodes a CDC25 phosphatase homolog, which affects embryonic viability and is necessary for cell proliferation, most importantly, in the germline *(22,23)*. When expression of *cdc-25.1* is knocked down by a constant exposure beginning in late embryogenesis (as it happens when gravid worms are fed on bacteria expressing dsRNA directed against *cdc-25.1*), laid eggs will develop into Glp animals. When exposure begins later in embryogenesis (as it happens when eggs are laid on *cdc-25.1* RNAi plates), eggs will develop into fertile worms. The embryos they lay, however, will fail to develop and hatch (Emb phenotype). These worms are designated as Emb-laying.

1. Place 40 gravid worms on a *cdc-25.1* RNAi plate. Allow worms to lay eggs for 4 h at 25°C. Eggs laid on this plate will result mainly in Emb-laying worms.
2. Transfer worms to a new plate, 10 gravid worms per plate, and let lay eggs for an additional 4 h. These eggs will give rise to Glp animals.
3. Remove gravid animals and raise the eggs at 25°C until those develop into obviously-Glp worms (those that are not will be gravid at this point). Use these worms for subsequent assays (*see* **Note 8**).

3.4.3.2. SPERM INACTIVATION

1. *spe-26(it112)* worms are defective in spermatogenesis but are otherwise normal. The encoded protein is a homolog of *Drosophila's* Kelch protein and human securin and is expressed in spermatogonial cells and spermatocytes, where it contributes to spermatogenesis *(24)*. In the absence of males, *spe-26* hermaphrodites will lay unfertilized eggs, making these mutants useful for assessing gene function in animals in which functional gonads still exist. As with *glp-4* worms, eggs are prepped onto new NGM plates and grown at the non-permissive temperature (25°C).
2. Sperm inactivation could be similarly achieved by a moderate heat stress. This is done by growing worms at 27.5°C for 48 h starting at the late L3 or early L4 larval stage. Because spermatogenesis in hermaphrodites takes place during the last larval stage, this heat exposure will effectively inactivate all sperm.

3.4.4. Colony-Forming Unit Counts

In many cases, when gene pleiotropic effects are suspected, one would desire to test a more direct representation of immune capabilities. One such phenotype is bacterial clearance from the intestine. When worms are inoculated with an equal intestinal load of pathogen, and then shifted to OP50 plates, uncleared pathogen could be extracted at different time points and colony-forming units (CFUs) can be determined to compare clearance efficiency between different strains.

1. Grow worms on pathogen for 24 h (*see* **Note 9**).
2. Transfer to OP50-NGM plates.
3. At different time points, remove worms for CFU counts:
   a. Add 50 µl LEV to the cap of a 1.5-ml tube (*see* **Note 10**).
   b. Pick 5–20 worms and place them on the droplet on the cap, close the cap, and spin down (*see* **Note 11**).
   c. Wash twice with 200 µl LEV/AMP/GENT (*see* **Note 12**).
   d. Incubate in LEV/AMP/GENT 45 min at room temperature.
   e. Wash three times with 200 µl LEV.
   f. Add 50 ml of M9 1% Triton. Spin down.
   g. Homogenize with a pellet pestle motor (Kimble-Kontes, Vineland, NJ) for approximately 1 min (*see* **Note 13**).
   h. Prepare serial dilutions in M9, plate (on LB agar) with appropriate antibiotic selection.
   i. Incubate overnight at 37°C and count colonies.

4. Notes

1. Strains can be requested from CGC by mailing at *Caenorhabditis* Genetics Center, University of Minnesota, 250 Biological Science Center, 1445 Gortner Avenue, St. Paul, MN 55108-1095, USA, or by fax (+1 612 625-5754) or by e-mail (cgc@umn.edu). Information about the CGC and description of strains that are available can be found at http://www.cbs.umn.edu/CGC.
2. Be careful not to scratch the surface as this will allow worms to burrow into the agar.
3. Mutagenizing before L4 will theoretically give less independently mutagenized genomes because the germline will not have undergone enough rounds of proliferation. Mutagenizing too late will be ineffective.
4. Treat all tips, Pasteur pipettes, etc. that came in contact with EMS as hazardous waste.
5. The number of sets of P0 you choose to egg lay can be modified to the size of the screen you want to carry out. As per our example, 45 P0 worms × 8 eggs/h × 3h = 1080 F1 per plate or 2160 haploid genome screened.

6. AMP/GENT is used here as a general sterilizer.
7. In wild-type worms, pBoc almost always results in expulsion, but this is not the case in some defecation mutants.
8. Although attempts to simultaneously knock down the expression of more than one gene are notoriously inefficient, sequential knockdown is a fairly efficient procedure. Therefore, *cdc-25.1*-sterilized worms are useful also for subsequent tests of the effects of specific RNAi knockdown.
9. Different mutants may need different durations of exposure to the pathogen to have pathogen colonizing them to a similar level. A rough estimate of this duration could be obtained by performing an initial experiment with GFP-expressing pathogen, following the accumulation of bacteria in the worm gut by fluorescence microscopy. In cases where bacterial mutants are tested for their capability to persist in the worm gut, we usually use a worm strain that was identified in a screen for enhanced susceptibility to pathogen (*esp-1*) *(6)*, which because of its inability to grind eaten bacteria gets colonized with a large number of bacteria in a short time. This ensures that the starting inoculum for different bacterial mutants is similar.
10. Levamisole is a nematode-specific agonist of the neuromuscular nicotinic acetylcholine neurotransmitter receptor, which causes paralysis due to muscle hyperexcitation. This paralysis includes muscles of the pharynx and anus. This prevents further eating of bacteria, as well as defecating and clearing of intestinal bacteria, and, furthermore, prevents antibiotics from entering into the animal.
11. Make sure to note the number of worms as it will be later used for normalizing the counts.
12. This antibiotic cocktail kills PA14. Other antibiotic combinations may be necessary when attempting to count other bacteria. For *Salmonella,* we use gentamycin as the sole antibiotic. This step, together with the next step (3.4), will kill extracellular bacteria enabling counting only of intestinal bacteria.
13. For homogenization, we use tubes coming from the same manufacturer as the homogenizer. The pestle fits snugly to these tubes ensuring thorough homogenization.

Reference

1. Hoffmann, J. A., Kafatos, F. C., Janeway, C. A., and Ezekowitz, R. A. (1999) Phylogenetic perspectives in innate immunity. *Science* **284,** 1313–8.
2. Janeway, C. A., Jr., and Medzhitov, R. (2002) Innate immune recognition. *Annu Rev Immunol* **20,** 197–216.
3. Tan, M. W., Mahajan-Miklos, S., and Ausubel, F. M. (1999) Killing of Caenorhabditis elegans by Pseudomonas aeruginosa used to model mammalian bacterial pathogenesis. *Proc Natl Acad Sci USA* **96,** 715–20.

4. Alegado, R. A., Campbell, M. C., Chen, W. C., Slutz, S. S., and Tan, M. W. (2003) Characterization of mediators of microbial virulence and innate immunity using the *Caenorhabditis elegans* host-pathogen model. *Cell Microbiol* **5,** 435–44.
5. Tan, M. W., Rahme, L. G., Sternberg, J. A., Tompkins, R. G., and Ausubel, F. M. (1999) Pseudomonas aeruginosa killing of *Caenorhabditis elegans* used to identify *P. aeruginosa* virulence factors. *Proc Natl Acad Sci USA* **96,** 2408–13.
6. Kim, D. H., Feinbaum, R., Alloing, G., Emerson, F. E., Garsin, D. A., Inoue, H., Tanaka-Hino, M., Hisamoto, N., Matsumoto, K., Tan, M. W., and Ausubel, F. M. (2002) A conserved p38 MAP kinase pathway in *Caenorhabditis elegans* innate immunity. *Science* **297,** 623–6.
7. Timmons, L., and Fire, A. (1998) Specific interference by ingested dsRNA. *Nature* **395,** 854.
8. Guo, S., and Kemphues, K. J. (1995) *par-1*, a gene required for establishing polarity in *C. elegans* embryos, encodes a putative Ser/Thr kinase that is asymmetrically distributed. *Cell* **81,** 611–20.
9. Fire, A., Xu, S., Montgomery, M. K., Kostas, S. A., Driver, S. E., and Mello, C. C. (1998) Potent and specific genetic interference by double-stranded RNA in *Caenorhabditis elegans*. *Nature* **391,** 806–11.
10. Kamath, R. S., Fraser, A. G., Dong, Y., Poulin, G., Durbin, R., Gotta, M., Kanapin, A., Le Bot, N., Moreno, S., Sohrmann, M., Welchman, D. P., Zipperlen P., and Ahringer, J. (2003) Systematic functional analysis of the *Caenorhabditis elegans* genome using RNAi. *Nature* **421,** 231–7.
11. Grishok, A. (2005) RNAi mechanisms in *Caenorhabditis elegans*. *FEBS Lett* **579,** 5932–9.
12. Rahme, L. G., Stevens, E. J., Wolfort, S. F., Shao, J., Tompkins, R. G., and Ausubel, F. M. (1995) Common virulence factors for bacterial pathogenicity in plants and animals. *Science* **268,** 1899–902.
13. Labrousse, A., Chauvet, S., Couillault, C., Kurz, C. L., and Ewbank, J. J. (2000) *Caenorhabditis elegans* is a model host for *Salmonella typhimurium*. *Curr Biol* **10,** 1543–5.
14. Brenner, S. (1974) The genetics of *Caenorhabditis elegans*. *Genetics* **77,** 71–94.
15. Gems, D., and Riddle, D. L. (2000) Genetic, behavioral and environmental determinants of male longevity in *Caenorhabditis elegans*. *Genetics* **154,** 1597–610.
16. Thomas, J. H. (1990) Genetic analysis of defecation in *Caenorhabditis elegans*. *Genetics* **124,** 855–72.
17. Liu, D. W., and Thomas, J. H. (1994) Regulation of a periodic motor program in *C. elegans*. *J Neurosci* **14,** 1953–62.
18. Iwasaki, K., Liu, D. W., and Thomas, J. H. (1995) Genes that control a temperature-compensated ultradian clock in *Caenorhabditis elegans*. *Proc Natl Acad Sci USA* **92,** 10317–21.
19. Hodgkin, J. A., and Brenner, S. (1977) Mutations causing transformation of sexual phenotype in the nematode *Caenorhabditis elegans*. *Genetics* **86,** 275–87.

20. Beanan, M. J., and Strome, S. (1992) Characterization of a germ-line proliferation mutation in *C. elegans*. *Development* **116,** 755–66.
21. Simmer, F., Tijsterman, M., Parrish, S., Koushika, S. P., Nonet, M. L., Fire, A., Ahringer, J., and Plasterk, R. H. (2002) Loss of the putative RNA-directed RNA polymerase RRF-3 makes *C. elegans* hypersensitive to RNAi. *Curr Biol* **12,** 1317–9.
22. Ashcroft, N. R., Srayko, M., Kosinski, M. E., Mains, P. E., and Golden, A. (1999) RNA-Mediated interference of a *cdc25* homolog in *Caenorhabditis elegans* results in defects in the embryonic cortical membrane, meiosis, and mitosis. *Dev Biol* **206,** 15–32.
23. Ashcroft, N., and Golden, A. (2002) CDC-25.1 regulates germline proliferation in *Caenorhabditis elegans*. *Genesis* **33,** 1–7.
24. Varkey, J. P., Muhlrad, P. J., Minniti, A. N., Do, B., and Ward, S. (1995) The *Caenorhabditis elegans spe-26* gene is necessary to form spermatids and encodes a protein similar to the actin-associated proteins kelch and scruin. *Genes Dev* **9,** 1074–86.
25. Reiner, D. J., and Thomas, J. H. (1995) Reversal of a muscle response to GABA during *C. elegans* male development. *J Neurosci* **15,** 6094–102.

26

Measuring Cell-Wall-Based Defenses and Their Effect on Bacterial Growth in *Arabidopsis*

Min Gab Kim and David Mackey

Summary

Plants are resistant to most potentially pathogenic microbes. Frequently, resistance results from defenses activated upon recognition of "non-self." Invasion of a variety of pathogens, including Gram-negative bacteria, into plants is betrayed by the presence of pathogen-associated molecular patterns (PAMPs). Plants challenged by a non-pathogenic bacterial strain or a purified PAMP often form cell wall modifications called papillae. These cell wall thickenings, which can be observed in the electron microscope, can more easily be visualized by staining for the component molecule callose and using fluorescent microscopy. We describe a method to measure callose following infiltration of leaves of *Arabidopsis thaliana*, a model organism for basic and applied research on plant biology, with pathogenic and non-pathogenic strains of *Pseudomonas syringae* pv. *tomato* or a purified bacterial PAMP. We also detail a method to measure the growth of bacteria infiltrated into leaves of *Arabidopsis*. These methods can be used to understand the interactions between pathogen and host plant.

Key Words: *Arabidopsis thaliana; Pseudomonas syringae;* PAMP; flg22; papillae; callose; innate immunity; type III effector protein.

1. Introduction

Gram-negative bacteria including *Pseudomonas syringae* utilize a type III secretion system (TTSS) to deliver numerous effector proteins into cells of the host. These effectors are powerful weapons that contribute to bacterial virulence *(1)*. Although the biological function of individual type III effector proteins are mostly unknown, one of their virulence functions is to manipulate the host in a way that helps the pathogen avoid or overcome induced plant defense responses *(2)*. For example, overexpression of AvrPto, a type III effector from Pto DC3000 in *Arabidopsis*, abrogated cell-wall-based basal defense in response to the TTSS-defective *hrcC* mutant of Pst DC3000 *(3)*. We have shown that two other TTSS effector proteins, AvrRpm1 and AvrRpt2, suppress plant defense responses induced by Pto DC3000*hrcC* or flg22 (a synthetic peptide derived from conserved N-terminal region of flagellin *(4)*) and thus improve conditions for bacterial growth *(5)*.

One response of an attacked plant is formation of papillae, which include structural components, such as callose, and minor amounts of other substances, such as phenolics, reactive oxygen intermediates, and proteins *(6,7)*. Callose is an amorphous, high-molecular-weight $(1 \rightarrow 3)$-β-D-glucan that can be visualized by UV light-induced fluorescence of the aniline blue fluorochrome *(8)*. Callose is widely distributed in higher plants during normal plant growth and development *(9)*. In addition, plant cells respond to wounding and microbial attacks by synthesizing and depositing callose in close proximity to the invading pathogen *(10,11)*. It has been postulated that the papillae act as a physical barrier to impede microbial penetration *(12)*.

2. Materials

2.1. Treatment of Leaves with Bacteria or Purified Pathogen-Associated Molecular Pattern

1. Four-week-old *Arabidopsis* plants grown vegetatively under daily photoperiods of 8 h.
2. Bacteria grown on King's B medium plate: 20 g peptone, 15 ml glycerol, 1.5 g K_2HPO_4, 1.5 g $MgSO_4$, and 18 g Agar/1L, supplemented with appropriate antibiotics.
3. Bacterial resuspension buffer: 10 mM $MgCl_2$.
4. Purified flg22, 22 amino acid peptide derived from the highly conserved amino terminus of flagellin from *P. syringae*, resuspended at 100 µM in sterilized distilled water. The sequence of flg22 is QRLSTGSRINSAKDDAAGLQIA, and it was synthesized by EZBiolab Company in Indiana, USA.
5. One-milliliter needleless syringe.

2.2. Leaf Staining with Aniline Blue

1. Six-well plates (Fisher, Fair Lawn, USA).
2. Lactophenol solution to remove chlorophyll: Mix one part of lactic acid : glycerol : phenol : distilled water at 1:1:1:1 ratio and two parts of ethanol (*see* **Note 1**).

3. Washing buffer: 50% ethanol in distilled water.
4. Staining solution: 0.01% aniline blue (Methyl blue (M6900) from Sigma, St. Louis, USA) in 150 mM K_2HPO_4 (pH 9.5).

2.3. Fluorescence Microscopy for Callose Observation

1. Microscope slide glasses (75 × 25 mm) and cover glasses (25 × 50 mm) from Fisher.
2. Mounting medium: 50% glycerol (v/v) in distilled water.
3. Nail polish.

2.4. Bacterial Growth Assay after Syringe Infiltration

1. Four to five week-old vegetatively grown *Arabidopsis* plants.
2. Bacteria grown on King's B medium supplemented with appropriate antibiotics.
3. Bacterial resuspension buffer: 10 mM $MgCl_2$.
4. One-milliliter needleless syringe.
5. Cork borer (8 mm in diameter) to make leaf discs.
6. Electric grinder (Heidolph, Kelheim, Germany). Electric stirrer (Heidolph) with 1.7-ml stainless steel microfuge tube grinder attachment.
7. Multi-channel pipets to handle 20–200 μl.
8. U-bottom 96-well plates (Nunc (262146), Roskilde, Denmark).
9. Replicator for 96-well plate (Sigma, Z370819).
10. King's B medium plates (150 × 15 mm).

3. Methods

In **Subheadings 3.1.–3.4.**, leaves are infiltrated with bacteria or purified pathogen-associated molecular pattern (PAMP) and callose deposition, which is an indicator of cell-wall-based defense, is visualized.

3.1. Day 1—Preparation of Plants and Bacteria

1. Use *Arabidopsis* plants grown in peat moss-based potting mix (Pro-Mix Premier) with 8 h day lengths. Mark healthy leaves of no more than 2 cm in length with an indelible ink pen so they can be identified later (*see* **Note 2**).
2. Patch bacteria onto a King's B plate containing appropriate antibiotics and grown for 24 h at 30°C.

3.2. Day 2—Leaf Infiltration

1. Bacteria from 1-day-old patch are resuspended in 10 mM $MgCl_2$. Bacteria are scraped up with a sterile toothpick and resuspended in 1 ml of 10 mM $MgCl_2$. The bacterial suspension is then diluted in 10 mM $MgCl_2$.

2. The optical density (OD) of the bacterial cell suspension is measured using spectrometer set at 600 nm. Bacterial suspension is diluted to a final $OD_{600} = 0.2$. For *Pst* DC3000, an $OD_{600} = 0.2$ is approximately 1×10^8 colony-forming units (cfu)/ml.
3. Water is misted on plants 5–10 min prior to infiltration and periodically while conducting infiltrations (*see* **Note 3**).
4. The abaxial (under) side of leaves, which were marked in **step 3.1.**, is infiltrated. A 1-ml needleless syringe containing solution with bacteria or flg22 is used to pressure-infiltrate the intercellular space of the leaf. The syringe is placed on the right or left side of leaf (*see* **Note 4**). A small amount of bacterial suspension, approximately 10–15 μl/leaf, is enough to infiltrate the leaf, although loss of non-infiltrated solution leads to the use of approximately 100 μl/leaf. Infiltrated leaf area appears water-soaked. Five leaves are infiltrated for each treatment.

3.3. Day 3—Clear Leaves of Chlorophyll

1. Fourteen to sixteen hours after infiltration, leaves are collected and placed in six-well plates. Infiltration of leaves with high-level inoculum (1×10^8 cfu/ml) of pathogenic bacterial strain sometimes causes leaf collapse. However, original leaf shape will be regained in the next step. Five leaves are the maximum that can be cleared effectively in a single well of a six-well plate (*see* **Note 5**).
2. Five milliliters of fresh lactophenol solution is added to each well accommodating leaf tissue, and the plate is gently shaken for 15 min. During this step, leaves turn from green to yellowish green.
3. Lactophenol solution is removed by aspiration from one corner. Using tweezers, leaves are then unfolded and arranged to be separated from each other (*see* **Note 6**).
4. Leaves in six-well plate are then incubated at 65°C for 40 min. Before placing six-well plate in incubator set at 65°C, the lid is removed. After 40 min, leaves are shrunken, dried, and completely yellow. Original leaf shape will be restored in the next step.
5. Five milliliters of fresh lactophenol solution is added to each well accommodating leaf tissue, and six-well plate is gently shaken overnight at room temperature. It is recommended that lactophenol solution be changed with fresh solution after 2–3 h of shaking.

3.4. Day 4—Leaf Staining and Slide Preparation

1. Samples are aspirated dry from one corner.
2. Leaves are rinsed with 50% ethanol. Five milliliters of 50% ethanol is added to each well, and the plate is gently shaken for 5 min. At this step, you will observe cleared-off yellowish white leaves.
3. Leaves are rinsed again with distilled water. Five milliliters of distilled water is added to each well, and the plate is gently shaken for 10 min (*see* **Note 7**).

4. The samples are stained with aniline blue. Water is removed and 5 ml of staining solution is added in each well. Staining solution needs to be made fresh and samples are shaken gently for 1 h (*see* **Note 8**).
5. After removing staining solution, the leaves are washed by brief shaking in mounting medium (50% glycerol). The samples are then ready to be mounted.
6. Leaves are carefully transferred one by one onto the slide glass, and cover glass is placed on it (*see* **Note 9**). Nail polish is used to seal the samples. The samples can be viewed as soon as the nail polish is dry or can be stored in the dark for up to 2 weeks.
7. The slides are viewed under epifluorescent microscope with UV filters (365 nm excitation, 400 emission, 390 dichroic). Callose is visualized by UV light-induced fluorescence of the associated aniline blue fluorochrome. Scale bar is stamped in the corner of the picture to measure the area observed. The number of callose is blind

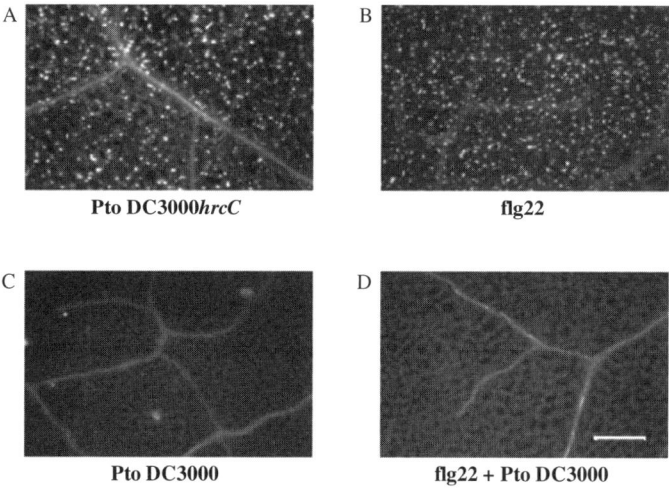

Fig. 1. Pathogen-associated molecular pattern (PAMP)-induced callose deposition in *Arabidopsis thaliana* is inhibited by type III effectors from pathogenic bacteria. Infiltration of leaves with non-pathogenic bacteria Pto DC3000*hrcC* (**A**) or synthetic peptide flg22 (**B**) induced callose deposition. However, wild-type Pto DC3000 delivers type III effectors that inhibited callose deposition (**C**). The effector proteins secreted by Pto DC3000 also suppressed callose deposition induced by co-infiltrated Pto DC3000*hrcC* (data not shown) or flg22 (**D**). Four-week-old leaves were syringe-infiltrated with 10^8 cfu/ml Pto DC3000*hrcC*, 100 μM flg22, 10^8 cfu/ml Pto DC3000, or 10^8 cfu/ml Pto DC3000 mixed with 100 μM flg22 and collected after 15 h. 10^8 cfu/ml Pto DC3000*hrcC* and 100 μM flg22 induced approximately 600–700 and 800–900 callose deposits per 1.1 mm^2 leaf area, respectively. Scale bar indicates 0.2 mm.

counted from several different leaf areas, and the average and standard deviation are calculated. QUANTITY ONE software (Bio-Rad, Hercules, USA) can be used to count the number of callose.
8. An example of the microscopic image is shown in **Fig. 1** and quantitation of the data is shown in **Fig. 2A**.

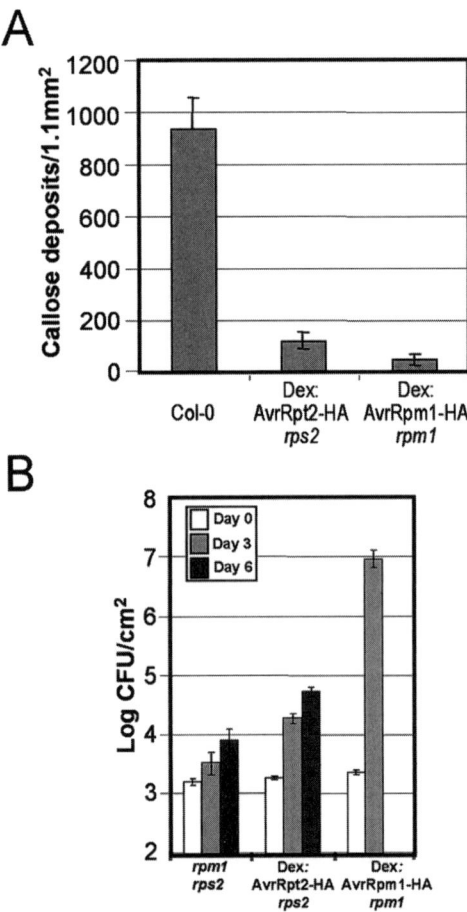

Fig. 2. The ability of type III effectors, AvrRpt2 and AvrRpm1, to inhibit pathogen-associated molecular pattern (PAMP)-induced callose deposition is correlated with enhanced growth of Pto DC3000*hrcC*. Pto DC3000*hrcC* was inoculated into transgenic *Arabidopsis* plants in which expression of either AvrRpt2 or AvrRpm1 was induced with dexamethasone 12–18 h prior to infiltration. These transgenic lines *(13,14)*

3.5. Bacterial Growth Assay

In this section, leaves are infiltrated with bacteria and bacterial growth is measured.

1. *Arabidopsis* leaves and bacteria need to be prepared 1 day before infiltration as described in **Subheading 3.1**. For growth curve, slightly larger leaves can be used (up to 4 cm in length).
2. Bacteria from fresh patch are resuspended and diluted in 10 mM $MgCl_2$ (pH 7.5). The OD of the bacterial cell suspension is measured using a spectrometer set at 600 nm. Bacterial suspension is diluted at 1×10^5 cfu/ml (*see* **Note 10**).
3. Leaves are inoculated with bacterial suspension as described in **Subheading 3.2**.
4. The excess solution (and associated bacteria) on the leaf surface is removed by dabbing with absorbent paper and the infiltrated leaves are allowed to dry.
5. Leaf discs are excised from leaves with a cork borer (8 mm in diameter) and placed in 1.5-ml microfuge tube containing 200 µl of $MgCl_2$ (pH 7.5). Three or more samples are typically needed to generate significant results. Including three leaf discs per sample helps to reduce variability between samples.
6. Samples are completely ground using electric grinder until pieces of intact leaf tissue are no longer visible (*see* **Note 11**).
7. Following the grinding of the tissue, the samples are thoroughly vortexed to ensure even distribution of bacteria within the buffer. This sample is then serially diluted in a 96-well plate (*see* **Note 12**).
8. The samples are then plated on 150 mM King's B plates supplemented with appropriate antibiotics. Plating is done using a replicator for 96-well plate, which removes a small amount of liquid from each well. The replicator is then carefully transferred onto the plate and is gently pressed-down to ensure all tips touch the

---

Fig. 2. lack the corresponding R-protein responsive to the expressed type III effector (plants expressing AvrRpt2 are *rps2* mutant and the plants expressing AvrRpm1 are *rpm1* mutant). **(A)** Graphical representation indicates that the expression of AvrRpt2 or AvrRpm1 suppresses flg22-induced callose deposition. Infiltration of 100 µM flg22 induced callose deposition in Col-0 wild type (as seen in **Fig. 1B**), but not in transgenic plants inducibly expressing AvrRpt2 or AvrRpm1. **(B)** The expression of AvrRpt2 or AvrRpm1, which suppresses PAMP-induced callose deposition (as seen in **Fig. 2A**), enhances the growth of type III secretion system (TTSS)-deficient bacteria. Leaves were infiltrated with Pto DC3000*hrcC* at 1×10^5 cfu/ml and were collected at day 0, 3, and 6. For plants expressing AvrRpm1-HA in *rpm1*, day 6 is missing because the leaves were totally collapsed. Error bars represent the standard deviation from four samples. (Reproduced from **ref. 5** with permission from Elsevier Science.)

plate. It dispenses the same volume (approximately 10 µl) of sample from each well (*see* **Note 13**).
9. The plates are placed at 30°C for 2 days and the number of colonies is counted. The cfu/leaf area is calculated based on the amount of tissue per sample, the number of colonies, and the dilution factor of each sample. From a set of serial dilutions, a single spot, which has five to 50 colonies, should be used to calculate the bacterial titer.
10. **Steps 5–9** are repeated at one or more time points after infiltration. An example of the results produced is shown in **Fig. 2B**.

4. Notes

1. Lactophenol solution needs to be mixed thoroughly before it is used because glycerol tends to sink onto the bottom and stay unmixed. It can be used for up to 6 months if it has been kept in the dark and sealed tightly.
2. To identify leaves infiltrated with different solutions, one needs different symbols to represent each treatment.
3. Humid conditions induce stomata to open and thus ease infiltration.
4. Vascular system of the leaf should be avoided for infiltration. Damage on midrib will bring about detrimental effects on the viability of leaf tissue.
5. Leaves infiltrated with different treatment should be placed in different wells because the identifying marks on leaves are removed by the lactophenol solution. Labels can be placed on the outside or bottom of the six-well plate.
6. Leaf area folded or overlapped with other leaf will remain not completely cleared off by lactophenol solution, which makes it difficult to observe even distribution of callose in whole leaf area.
7. Leaves show tendency to float on water, which will cause incomplete washing. Try to rinse them thoroughly by flipping them in the middle of shaking and by pushing them into the water.
8. Leaf-staining buffer needs to be made fresh. Fresh 0.01% aniline blue dissolved in 150 mM K_2HPO_4 (pH 9.5) appears deep blue. As time goes on, the color will be washed out.
9. It is tricky to transfer leaves from mounting solution onto slide glass because leaves are very tender after staining. Slant slide glass into mounting solution, and carefully drag leaf onto the slide glass using tweezers. Leaf can be stretched out on the slide glass. Air bubbles should be removed by holding up and gently down the leaf area trapping air bubble before placing cover slide on samples. Completely remove mounting solution from the bottom of slide glass to maintain a clean microscope stage.
10. Make bacteria at 1×10^8 cfu/ml by adjusting to $OD_{600} = 0.2$ and then dilute this suspension 1:1000 in 10 mM $MgCl_2$.
11. Cool tubes on ice prior to grinding. This reduces overheating of the samples during grinding.

12. Ninety-six-well plate is composed of 12 columns and 8 rows. Hundred microliters of leaf extract containing bacteria is mixed with 100 μl of bacterial resuspension buffer (10 mM $MgCl_2$) in the first row, and 180 μl of 10 mM $MgCl_2$ is added in each well from the second to last row. By using multi-channel pipet, 20 μl of leaf extract from the well of the first row is taken and mixed in the well of second row with 180 μl of bacterial resuspension buffer. It is necessary to mix them thoroughly by pipeting up and down several times. More reliable results are obtained with blunt-ended pipet tips. Repeating this step produces a series of 1:10 dilutions.
13. Plates need to be dried with lids off in a clean hood for 40 min before use. Drying the plates causes drops from the plating device to stay separated from one another.

Acknowledgments

This work was supported by NSF (MCB-0315673) and the Ohio Agricultural Research & Development Center (OARDC) of Ohio State University.

References

1. Gálan, J. E., and Collmer, A. (1999) Type III secretion machines: bacterial devices for protein delivery into host cells. *Science* **284,** 1322–28.
2. Chang, J. H., Goel, A. K., Grant, S. R., and Dangl, J. L. (2004) Wake of the flood: ascribing functions to the wave of type III effector proteins of phytopathogenic bacteria. *Curr Opin Microbiol* **7,** 11–8.
3. Hauck, P., Thilmony, R., and He, S. Y. (2003) A Pseudomonas syringae type III effector suppresses cell wall-based extracellular defense in susceptible Arabidopsis plants. *Proc Natl Acad Sci USA* **100,** 8577–82.
4. Felix, G., Duran, J. D., Volko, S., and Boller, T. (1999) Plants have a sensitive perception system for the most conserved domain of bacterial flagellin. *Plant J* **18,** 265–76.
5. Kim, M. G., da Cunha, L., McFall, A. J., Belkhadir, Y., DebRoy, S., Dangl, J. L., and Mackey, D. (2005) Two Pseudomonas syringae type III effectors inhibit RIN4-regulated basal defense in Arabidopsis. *Cell* **121,** 749–59.
6. Nishimura, M. T., Stein, M., Hou, B. H., Vogel, J. P., Edwards, H., and Somerville, S. C. (2003) Loss of a callose synthase results in salicylic acid-dependent disease resistance. *Science* **301,** 969–72.
7. Jacobs, A. K., Lipka, V., Burton, R. A., Panstruga, R., Strizhov, N., Schulze-Lefert, P., and Finchcr, G. B. (2003) An Arabidopsis callose synthase, GSL5, is required for wound and papillary callose formation. *Plant Cell* **15,** 2503–13.
8. Stone, B., Evans, N., Bonig, I., and Clarke, A. (1985) The application of sirofluor, a chemically defined fluorochrome from aniline blue, for the histochemical detection of callose. *Protoplasma* **122,** 191–95.

9. Stone, B., and Clarke, A. (1992) *Chemistry and Biology of (1–3)-B-D-Glucans*. La Trobe University Press, Victoria, Australia.
10. Vance, C. P., Kirk, T. K., and Sherwood, R. T. (1980) Lignification as a mechanism of disease resistance. *Annu Rev Phytopathol* **18,** 259–88.
11. Donofrio, N. M., and Delaney, T. P. (2001) Abnormal callose response phenotype and hypersusceptibility to Peronospoara parasitica in defence-compromised arabidopsis nim1-1 and salicylate hydroxylase-expressing plants. *Mol Plant Microbe Interact* **14,** 439–50.
12. Brown, I., Trethowan, J., Kerry, M., Mansfield, J., and Bolwell, G. P. (1998) Localization of components of the oxidative cross-linking of glycoproteins and of callose synthesis in papillae formed during the interaction between non-pathogenic strains of *Xanthomonas campestris* and French bean mesophyll cells. *Plant J* **15,** 333–43.
13. Mackey, D., Holt III, B. F., Wiig, A., and Dangl, J. L. (2002) RIN4 interacts with Pseudomonas syringae Type III effector molecules and is required for RPM1-mediated disease resistance in Arabidopsis. *Cell* **108,** 743–54.
14. Mackey, D., Belkhadir, Y., Alonso, J. M., Ecker, J. R., and Dangl, J. L. (2003) Arabidopsis RIN4 is a target of the type iii virulence effector AvrRpt2 and modulates RPS2-mediated resistance. *Cell* **112,** 379–89.

Index

ADCC. *See* Antibody-dependent cell cytotoxicity
Allergic reaction, 241
AMP. *See* Antimicrobial peptide
Aniline blue, 444
Anopheles, 365
  A. gambiae, 366–375
Antibody-dependent cell cytotoxicity, 128, 151, 216, 221, 232, 291, 298
Antimicrobial peptide, 379, 380, 382–383, 385–387, 391, 395
Arabidopsis thaliana
  as a model host, 443–444
Association analysis, 20–27. *See also* Mapping
  case-controlled, 21
  family-based, 21, 22–27
  population-based, 21, 38
Axenic mice, 321–324. *See also* Murine models

B-cells
  human, 204, 313
  murine, 313
BDCA2, 77, 275
Beauveria bassiana, 381–383, 385, 390
BHI. *See* Brain-heart infusion
Bioluminescence imaging, 101–118
Biomarker, 107, 109
BLI. *See* Bioluminescence imaging
BM. *See* Bone marrow
BMMC. *See* Bone marrow, derived cells
Bone marrow
  culture, 167–169, 171
  derived cells, 163, 166, 225, 242, 244, 248–252, 267, 306
  grafts, 183
  murine mast cells, 241, 245
Brain–heart infusion
  agar, 107, 327, 407
  broth, 406
BrdU
  flow cytometry, 185
  incorporation assay, 186, 189–190
  injection into mice, 189
  staining, 189–190
Bromo-deoxyuridine. *See* BrdU

Caenorhabditis elegans
  analysis of killing assays, 415–419
    confounding factors, 430, 435–436
  defecation, 436
  growth and maintenance, 408
  as a model for innate immunity, 403–405, 429–431
  pharyngeal grinding, 435
  reproduction, 436
  synchronization, 408–409
Callose, 444
Candida albicans, 381, 390
Candidate genes, 18
CD11c, 165, 173, 275, 277, 279
CD14, 276, 279, 284, 313
CD16, 198–199, 220, 275, 284–285
CD34, 66–67, 70–73, 244, 248–249, 250–252, 310, 313
CD107, 207–208, 292
CD123, 275
Cholecystokinin, 104
Chromium release assay, 201, 292
Cryptococcus neoformans, 404
  C. elegans survival on, 407
  killing assay, 414
  killing plates, 411
Cytokines, 49, 128, 165, 180, 185, 199, 205, 216, 291

DC. *See* Dendritic cell
Decanal, 103
Dendritic cell, 121, 163–178, 273–288
  bone marrow culture generated, 166–167
    lentiviral transduction, 314
  conventional, 165
  counts. *See* Dendritogram
  isolation, 166–167, 170–171
    collagenase digestion and release, 164, 166–167, 170
    depletion of non-DC lineages, 164, 167, 171

453

lineage depletion cocktail, 167, 171
maturation, 164
selection of light density cells, 164, 167, 170
spontaneous activation in culture, 164
myeloid, 256, 274, 277, 280, 282, 283
natural type 1 interferon producing cell. *See*
Plasmacytoid pre–DC
plasmacytoid, 165, 273, 277, 280, 282, 283
quantification. *See* Dendritogram
subtypes, 165–166
lymph node, 165–166
spleen, 165
thymus, 165
Dendritogram, 273–288
dual platform, 275, 279
PBMC. *See* Peripheral mononuclear cells
peripheral mononuclear cells, 276, 281, 282
rare event analysis, 277
single platform, 275, 277
Trucount. *See* Single platform
whole blood, 275, 277, 279
Diptericin, 382, 387, 389
Disease
common, 18, 21
complex, 18
multi-factorial, 18
Disease association analysis, 58–60
Dpt. See Diptericin
Drechmeria coniospora
C. elegans pathogenicity assay, 415
C. elegans survival on, 407
growth, 411–412
Drosomycin, 382, 387, 389
Drosophila, 379–392
infection, 383–385
natural bacterial infection, 384–385
natural fungal infection, 385
septic injury, 384
survival analysis, 390
cells, 395, 397, 399, 401
Drs. See Drosomycin
DsRNA, 369, 430
synthesis, 396, 399

Ecc15. See Erwinia carotovora carotovora
ENSEMBL, 13, 27
Enterococcus faecalis, 381, 390
C. elegans survival on, 407
killing assay, 412
killing plates, 410

ENU, 2, 3, 5. *See also* Mutagenesis
Eosinophil activation
cytotoxicity assays, 232–234
parasite killing, 216, 221
tumoricidal activity, 216, 221
mediator release, 231–232
eosinophil peroxidase (EPO), 216, 221
reactive oxygen species (ROS), 216, 221, 235
Eosinophil culture, 226–227
eosinophilic cell lines, 227–228
Eosinophil mediators, 216, 221
cationic proteins, 227–228.
eosinophil peroxidase (EPO), 216, 221, 227. *See also* Eosinophil activation
detection using DAB, 219, 228, 234
measurement by chemiluminescence, 231–232, 237
immunocytochemistry, 230
intracellular staining, 229, 236
reactive oxygen species (ROS). *See* Eosinophil activation
Eosinophil membrane receptor, 215–216
flow cytometry analysis, 228–229
staining, 236
Ig receptors, 222
immunofluorescence, 230–231
autofluorescence, 237
turnover, 220, 227
Eosinophil purification, 220
human, 220, 222–224
metrizamide, 217–218, 222–223, 234, 235
Hypodense/normdense, 220, 221, 223, 238
MACS, 218, 220, 224–226
percoll, 217, 223–224, 234, 235
CD16, subpopulation, 235
mouse, 225
rat, 225–226
Erwinia carotovora carotovora, 380, 384, 390

Family-Based Association Tests, 25–27, 39
genetic model, 26, 40
haplotypes, 27, 42
mendelian inconsistencies, 40
Monte Carlo permutation, 43
null hypothesis, 25
parental genotypes, 25
FBAT. *See* Family-Based Association Tests
Fc-receptor, 151–161
FcεRI, 241, 246, 251

Index

FcR. *See* Fc-receptor
Flow cytometry, 70, 140, 169, 172, 188, 190, 205, 208, 219, 228, 260, 264, 276, 292
Flt3L, 167, 171–172
    bone marrow cultures. *See* Dendritic cell, bone marrow culture generated
Fms-like tyrosine kinase, 3 ligand. *See* Flt3L

Gastrointestinal development, 323–324
GENEHUNTER, 27, 34
Genetic epidemiology, 18
Genetic marker. *See* Mapping
Genetic variation, 18, 20. *See also* Variance
Genomic DNA
    isolation, 4, 11, 306
Geobacillus stearothermophilus, 326, 329
GFAP, 107–108
    GFAP-luc, 109
Glial fibrillary acidic protein. *See* GFAP
Granulocyte, 224, 352, 357, 358, 359

Hematopoiesis, 352
Hematopoietic progenitors, 66–68, 244, 248
Heritability, 20. *See also* Variance
HLA-DR-positive DC, 275
HMC. *See* Mast cells, human
Human cord blood, 248
Human immune system mice, 65–81
    genetic background, 66
        BALB/c Rag-$2^{-/-}\gamma_c^{-/-}$, 66, 69–70, 77–78
        non-obese diabetic, 66
        severe combined immuno-deficiency, 66
    hematopoietic stem cells, 66–68
        $CD34^+CD38^-$ cell sorting, 67–68, 70, 72–73, 77, 79
        fetal liver, 66–67, 69–70, 77
        inoculation, 68, 73, 79
        MACS enrichment, 67, 70, 71–72, 78–79
        source of hematopoietic stem cells, 67, 77
        umbilical cord blood, 66–67, 69–70, 77–78
    monitoring, 68, 73–76
        blood puncture, 68, 73
        flow cytometry, 68, 70, 73, 76, 77–78
    production, 66–73
        age, 66, 69–70, 78
        intra-hepatic injection, 73
Humanized mice. *See* Human immune system mice
Human adaptative immune system Rag$2^{-/-}\gamma_c^{-/-}$ mice, 66. *See also* Human immune system mice

IgE, 251
IIPGA, 29
IL-3, 245, 251
    receptor. *See* CD123
ILT3, 275, 282
Imaging systems
    IVIS 100, 108
    IVIS 108, 200
        custom lens, 109
        two wavelength study, 108
        minimum detectable photons, 108
        quantum efficiency, 108
        two-photon laser-scanning microscopy, 119
        ultrasound, 109
Imd, 379, 380, 382–383
Immune complexes, 152, 153, 155
    binding to FcR-expressing cells, 159
    cell activation, 160
    cytokine release, 160
    staining, 159
    with $F(ab)_2$ fragments, 158
    with TNP, 156
Immunofluorescence, 204, 230, 246, 264
    fluorochrome conjugation, 169
    microscopy, 397, 400
    monoclonal antibody, 169, 171–172, 250
    staining, 169, 172, 245
Immunoglobulin receptor. *See* Fc-receptor and FcεRI
Immunophenotyping, 292–293
Intestinal flora, 321–324
Intravital imaging. *See* Imaging systems, Two-photon laser-scanning microscopy
Isolator, 324–326, 327–329

K562, 200, 293, 299
Kainic acid, 107, 109
Killer immunoglobulin-like receptor. *See* KIR
KIR, 198–199
    database, 45, 61
    genes, 18, 49–64
    gene locus, 49–64
    genotyping, 50
        gel electrophoresis, 56
        PCR, 52
        primers, 53–55
Kit, 246, 251

LD. *See* Linkage disequilibrium
LD50 9, 137

Lentiviral vectors, 301–316
  central DNA flap, 302
  pseudotyping, 303
    endogenous retroviral envelopes, 303
    vesicular stomatitis G (VSV-G) protein, 303, 315
Linkage disequilibrium, 19, 20, 21–22
Linkage. *See* Mapping
Listeria, 395, 397, 399
  persistent infection, 104
  replication, 104
LOD score. *See* Mapping
LPS, 395
Luciferase, 103
  bacterial, 103, 108
  firefly, 103, 108
Luciferin, 103
Luria-Bertani agar, 406
Luria-Bertani broth, 327, 406

Macrophage, 104, 255–256, 352, 357, 360, 395
  analysis by flow cytometry, 260, 264
  antibody staining, 257
  bone marrow-derived, 267
  immunolabeling
    alkaline phosphatase, 264
    peroxidase, 261
  peritoneal, 260–261
    harvesting, 6–7, 266–267
  preparation, 264
Magnetic cell separation, 249, 250
Malaria, 365
Mapping, 4
  genome-wide, 10
  linkage, 11, 19, 20, 22–27
  LOD score, 12
  markers, 11, 19, 20
  microsatellite, 11, 19
Mast cells, 241–252
  human (HMC), 242, 247, 248
  murine, 242
    purification, 244
  rat, 243, 246
MCMV
  deletion mutants, 132, 134
    attenuation, 136–137
  escape mutants, 128
  evasion of NK cells, 131
  immunoevasins, 135
  infection, 4, 135

  of newborn mice, 135
  infectivity, 9
  salivary gland-derived virus, 135, 138
  stock preparation and titration, 7–9, 138
  TC-derived, 135, 137
MDC. *See* Dendritic cell, myeloid
Meningitis, 107, 109
  neuronal damage, 107
MERLIN, 27, 34
Metrizamide, 242
Mice. *See* Murine models
Microbacterium nematophilum
  infection of *C. elegans*, 414
  pathogenicity assay plates, 411
Micrococcus luteus, 381
Microglial, 357
Microinjection
  equipment, 87–88
  of zebrafish, 345–346, 348–349
  of zygotes, 85–86, 88
Monocytes, 275
  classical, 283
  intermediate, 275, 283
Morpholinos, 338, 345, 346, 347, 360
Mosquito. *See also* Anopheles
  midgut, 369
  responses to infection, 366
Mouse cytomegalovirus. *See* MCMV
Mucosal immune defenses, 323–324
Murine models, 215, 221–222
  anesthesia, 88, 108, 122
    isoflurane, 108
  aseptic hysterectomy, 327, 330
  breeding, 10
  cesarian, 331
  foster nursing, 330–331
  genetic background, 66, 86
  humanized mice. *See* Human immune system mice
  monoxenic, 324
  outcrossing, 10
    dominant mutations, 10
    recessive mutations, 10
  pseudopregnant, 88–89
  transgenic, 83–85, 109
  vasectomized, 88
Mutagenesis
  EMS, 433
  ENU, 3
Mycoplasma, 314

Index

N-ethyl-N-nitrosourea. *See* ENU
Natural killer cells. *See* NK cells
Nematode growth medium (NGM), 406, 431
Nematode. *See Caenorhabditis elegans*
NK cells, 121, 197, 291
  activation assay, 185–188
    antibody coating, 187
    flow cytometry, 188
    preparation of cell suspension from spleen, 187
    staining of cell surface antigens, 187
    staining of intracellular antigens, 188
    stimulation, 187
  adoptive transfer, 139
  antiviral response, 127–143, 184–185
    cytotoxic NK assay, 139, 201
  cytolytic activity, 200, 291
  cytotoxicity assay, 139, 201, 205, 207
  depletion, 134
    with anti- asialo-GM1, 136
    with anti-NK1.1, 136
  evasion by MCMV, 131
  individual NK cell responses, 181–191
  isolation, 199, 202
  licensing, 182–184
  phenotypic analysis, 200, 204
  proliferation assay, 200, 208
  receptors, 128, 136, 294. *See also* KIR
    blocking, 136
    Ly49H, 131
    NKG2D, 131
  subset, 198, 199
NKG2D ligands
  confocal analysis, 142
  down-modulation by MCMV, 140
  flow cytometric screening, 141
  H60, 131
  MULT-1, 131, 142
  RAF-1, 131
  surface resident ligands, 142
    uptake assay, 142

Optical probes, 103

P293, 815
Packaging cell line, 307
Papillae, 444
PBMC, 248, 249, 285, 293, 298, 313
PDC. *See* Dendritic cell, plasmacytoid
Percoll gradients, 217, 220, 223, 247
Peripheral blood mononuclear cells. *See* PBMC

Peritoneal Macrophages. *See* Macrophages
Peritoneal washing, 244
Phagocyte, 397
Phagocytosis, 342, 348, 352, 379
Phenotype
  binary, 20
  quantitative, 20
Plasmodium, 365
  P. berghei, 366
  P. falciparum, 369
  P. yoelii, 366
Polymorphism information content, 19
Populations
  admixture, 21
  stratification, 21, 38
Pseudomonas aeruginosa
  C. elegans survival on, 407
  killing assay, 412
  killing plates, 409, 432
Pseudomonas entomophila, 380, 384
Pseudomonas syringae, 433–444
PupasSNP, 28
Pupasview, 28

QTDT. *See* Quantitative trait Transmission/Disequilibrium Tests
Quantitative trait Transmission/Disequilibrium Tests, 22–27
  additive effects, 24
  covariates, 24
  dominant effects, 24
  likelihood-ratio, 24
  Monte Carlo permutation, 37
  variance components, 22, 24, 36

Relish, 380, 390
Retronectin, 311
RNA
  double-stranded. *See* DsRNA
  interference. *See* RNAi
  reverse transcription, 398
RNAi, 366, 430, 431, 434
  high-throughput, 395

Salmonella enterica
  C. elegans survival on, 407
  killing assay, 412
  killing plates, 410
SCF. *See* Stem cell factor

Serratia marcescens
  C. elegans survival on, 407
  killing assay, 413
  killing plates, 410
Single nucleotide polymorphism. *See* SNP
SK agar, 406
Small GTPase, 398
SNP
  database, 28
  for mapping, 11, 19
SNPper, 28
Spätzle, 380, 390
Staphylococcus aureus
  C. elegans survival on, 407
  killing assay, 412
  killing plates, 409
Stem cell factor, 248, 252
Streptococcus pneumoniae, 108
  meningitis, 109

T cell
  human, 274, 307, 312
    activation with anti-CD3/ IL-2, 316
    expansion with CD3/CD28 beads, 312
  murine
    expansion with concanavalin, 312
TLR, 216, 241, 323–324
  ligands, 3
  polymorphisms, 18
  signaling in macrophages, 3–4, 6
TNF α, 356
Toll, 379, 380, 382–383
Toll genes, 18
Toll-like receptor. *See* TLR
Toluidine blue, 242, 245, 252
Trait. *See* Phenotype
Transgenesis, 83–85
  BACs, 85, 88
  microinjection of zygotes, 85–86, 88
Transgenic mice. *See* Murine models, Transgenesis

Transmission/Disequilibrium Test, 22–27
Tryptase, 251
Tryptic soy agar (TSA), 406
Tryptic soy broth (TSB), 406
Two-photon laser-scanning microscopy, 119–125

Ultrasound. *See* Imaging systems

Variance
  component analysis, 32
  environmental, 20
  genetic, 18, 20
    additive, 20
    dominant, 20
    heritability, 20
    phenotypic, 20
Viral "titer", 308
  by qPCR, 310

Worm. *See Caenorhabditis elegans*
Wright–Giemsa staining, 245, 252

Yersinia pestis
  biofilm formation, 414
  inhibition of *C. elegans* growth, 413
  pathogenicity assay plates, 410–411
Yersinia pseudotuberculosis
  biofilm formation
  inhibition of *C. elegans* growth, 413
  pathogenicity assay plates, 410–411
YPD broth, 407

Zebrafish, 337–364
  infection, 347
    intravenous, 349
    local, 349
  microinjection, 339, 345, 346

Printed in the United States of America.